Elements of Transitional Boundary-Layer Flow

Robert Edward Mayle

Professor Emeritus of Mechanical Engineering
Rensselaer Polytechnic Institute

2018

Bibliographic information published by the Deutsche Nationalbibliothek

The Deutsche Nationalbibliothek lists this publication in the Deutsche Nationalbibliografie; detailed bibliographic data are available on the Internet at http://dnb.d-nb.de .

Second Enhanced Edition, 2018

ISBN 978-3-8325-4598-7

Logos Verlag Berlin GmbH
Comeniushof, Gubener Str. 47,
10243 Berlin
Tel.: +49 (0)30 42 85 10 90
Fax: +49 (0)30 42 85 10 92
INTERNET: http://www.logos-verlag.de

Preface

Transition is one of the last true frontiers in fluid mechanics. A century ago, the frontier was boundary layer theory. Fifty years ago it was turbulent flow. Now, laminar and turbulent boundary layers are relatively easy to calculate, but calculating the transition from one to the other is still somewhat of an art fraught with inadequate empiricism. Indeed, my motivation for writing this book is the simple fact that we still can't *predict* the onset of transition.

The primary aim of the book is to introduce transitional flow to engineers by providing the basic aspects of transition in a logical sequence – hence the word "Elements" in the title. Since the most relevant material on transition is scattered throughout decades of journal articles, and since there are an enormous number of articles to sort through, pursuing an independent study of transitional flow becomes a daunting task for anyone. My hope with this book is to not only shorten this pursuit, but also provide a prologue to the more difficult aspects of the subject.

Because transition occurs in a large variety of flows, I have limited the scope of this book to boundary layers near a solid surface – hence the words "Transitional Boundary-Layer Flow" in the title. Furthermore, because almost everything affects transition, and something about almost all of it has been examined, I have primarily limited the subject matter to the effects of free-stream turbulence and pressure gradients on transitional flow. While the effects of surface curvature and roughness on transition are briefly considered, the effects of heat transfer, compressibility, sound, noise, and surface vibrations are not considered. Although this may be seen as a shortcoming by some, my intent is solely to provide the fundamentals of transitional flow, some methods by which it may be calculated, and the problems associated with *predicting* it. To achieve this end most straightforwardly, the book is mainly concerned with transition in a boundary layer of an isothermal, incompressible fluid on a smooth flat plate with a zero pressure gradient and elevated free-stream turbulence.

I wrote the book for those with a basic knowledge of boundary-layer theory, both laminar and turbulent, and some knowledge of modeling turbulent boundary layers. In addition, I have presumed some proficiency in applied mathematics, and some familiarity with probability and set theory. Even so, I have tried to explain enough of the mathematical details and/or thinking in the more difficult areas of analysis so that most readers can reproduce and understand the results. In this regard, the reader will find that I explain more at the beginning of the book than at its end, and wherever a new idea is introduced.

The book is divided into three main parts entitled "Before and After," "In Between," and "And Beyond." Part I is dedicated to review. In Chapter 1, the basic ideas of fluid mechanics and boundary-layer theory are summarized. In Chapter 2, several laminar boundary-layer solutions with an emphasis on forced unsteady flows are examined, while in Chapter 3, turbulence, turbulent boundary layers, and their modern method of analyses are reviewed. Part II is dedicated to pre-transitional and transitional flow. In Chapter 4, stability theory, laminar breakdown, and bypass pre-transition are considered. In Chapter 5, turbulent spots, turbulent-spot theory, and the production and propagation of turbulent spots are considered. Chapter 6 contains the generally accepted correlations for the onset of transition and spot production rate. In Chapter 7, several methods for calculating transitional boundary-layer flow are presented, compared with experimental results, and discussed. Part III, containing only Chapter 8, is dedicated to the other modes of transition. In this chapter, wake-induced, multimode, and separated-flow transition, as well as early post-transition and reverse transition are all introduced and briefly discussed. Besides these, sections containing a list of frequently used symbols, a list of milestones in boundary-layer transition, and a preview of transition are provided at the beginning of the book. Furthermore, to complement the references listed throughout the book, a bibliography of additional material is provided at its end.

Direct numerical simulation (DNS) of transitional flow is not covered. While those actively working in the area will certainly consider this shortsighted, it is beyond my scope for this book. Moreover, since transitional-flow DNS has yet become a predictive tool in spite of some impressive recent results, we are still "stuck" with using methods similar to those described in the book for calculating it.

However, in keeping with the elemental theme of the book, some new ideas, solutions, views and conclusions are presented. These include solutions for high-frequency propagating oscillations in a laminar boundary layer, an asymptotically correct model for the wall dissipation terms in the Reynolds-stress equations, a differential analysis for a shear-free turbulent boundary layer, a relevant set of Reynolds-stress equations for pre-transitional flow, a new correlation for laminar breakdown and the onset of transition, a solution for a point source of turbulent spots, a solution for a line source of three-dimensional turbulent spots, a new method for calculating the expansion of a spot in flows with a variable free-stream velocity, a set of conditionally averaged Reynolds-stress equations for transitional flow, and a variety of other new correlations, formulations, investigations, and approaches. Since none of these new items have been officially reviewed, I advise all readers to carefully check my work – which in some cases might be agonizing.

The idea for this book came to me some twenty years ago when I retired. In hind sight I should have written it then because the amount of material to read and write about was an order of magnitude less. Nonetheless, I began outlining and writing various parts of it about ten years ago, only to realize that a more comprehensive treatment of the subject matter was needed than I had initially planned. As a result, this book ever so slowly morphed into a full-time job – one that I am now happy is over. So this is it. There will be no second editions nor errata! I trust my errors will soon become well known.

Finally, for those interested in such things, the book is set in Bodoni SvtyTwo ITC TT and Helvetica. For the word processor and equations, I used Pages™ by Apple Inc., and MathType™ by Design Science, Inc. Furthermore, I used either Mathematica™ by Wolfram Research, Inc., or ProFortran™ by the Absoft Corporation for all the calculations, and either Mathematica or Pages for all the figures.

REM, 2015

I have since rectified numerous spelling, grammatical, and "cut and paste" errors, revised some notation for the sake of consistency, corrected a few obviously incorrect equations, and clarified several confusing statements in the book. Yes, I know I said I wouldn't do this. Furthermore, I have completely rewritten and clarified the discussion of the results in §3.5.1 on shear-free turbulent-boundary-layers, updated, expanded, and upgraded the two-page subsection about the effect of becalmed zones on spots to a full section entitled "Spots and Becalmed Matters," and added a brief but convenient review of tensor notation, algebra, and calculus entitled "Supplement on Cartesian Tensors," for the benefit of those readers who want to refresh their memory about tensors. Otherwise, the book remains the same.

As it turns out, I rather enjoyed working on "the" book again – first because I hated to see all my previous errors and second because I simply hadn't, and perhaps never will finish it. I hope you will enjoy my work as well.

REM, 2018

Acknowledgements

In 1987, I had the good fortune to receive an Alexander Humboldt Senior Scientist Award and took a sabbatical in Karlsruhe, Germany. It was then, when Dr. Dullenkopf showed me his latest measurements for periodic wakes passing over a turbine airfoil, that my research in transitional flow began. Within a month, using Emmons' turbulent-spot theory, we wrote "A Theory for Wake-Induced Transition." And within the following ten years, after many beers at the Vögelbräu discussing turbulence and transition among other things, Dr. Dullenkopf, Dr. Schulz and I wrote "More on the Turbulent-Strip Theory for Wake-Induced Transition," "The Effects of Incident Turbulence and Moving Wakes on Laminar Heat Transfer in Gas Turbines," "An Account of Free-Stream Turbulence Length Scale on Laminar Heat Transfer," "The Path to Predicting Bypass Transition" and "The Turbulence That Matters," which contain many of the ideas presented in this book. For those incredibly exciting and fortunate collaborative years at the Institut für Thermische Strömungsmaschinen, Universität Karlsruhe, I am indebted to Prof. Wittig's wholehearted support and hospitality. So for the many good times and memories, thanks Klaus, thanks Achmed, and thanks Sigmar.

Table of Contents

Frequently Used Symbols

Roman

a_{ij}	anisotropic stress tensor, $a_{ij} = \overline{u_i' u_j'} \big/ k - \frac{2}{3}\delta_{ij}$.
A	area [m^2]; also anisotropic stress tensor invariant $A = 1 - \frac{9}{8}(A_2 - A_3)$.
A_1, A_2, A_3	anisotropic stress tensor invariants, $A_1 = \sqrt{6}\, A_3 / A_2^{3/2}$, $A_2 = a_{ij}a_{ji}$, $A_3 = a_{ij}a_{jk}a_{ki}$.
$A_{s,xz}$	area of the turbulent-spot propagation cone in the xz-plane [m^2].
$A_{s,zt}$	area of the turbulent-spot propagation cone in the zt-plane [ms].
b	breadth of a turbulent spot [m].
c	wave celerity [m/s].
c_f	skin-friction coefficient, $c_f = \tau_0 \big/ \frac{1}{2}\rho u_\infty^2$.
$E(k)$	three-dimensional energy spectrum [m^3/s^2].
$E_1(k_1)$	one-dimensional energy spectrum of the streamwise fluctuations [m^3/s^2].
$E_2(k_1)$	one-dimensional energy spectrum of the transverse fluctuations [m^3/s^2].
f	circular frequency, [Hz].
f_i	body force per unit mass [N/kg].
I	intermittency function equal to either one or zero.
k	kinetic energy per unit mass of the fluctuations [m^2/s^2]; also, wave number [1/m].
k_s	equivalent height of sand-grain roughness [m].
K	kinetic energy per unit mass of the mean flow [m^2/s^2]; also, acceleration parameter, $K = (\nu/u_\infty{}^2) du_\infty/dx$.
l	length [m]; also, length of a turbulent spot [m].
l_d	Kolmogorov's length scale [m], $l_d = (\nu^3/\varepsilon)^{1/4}$.
L	characteristic length [m].
L_e	energy-containing eddy length scale [m], $L_e = k^{3/2}/\varepsilon$.
n_i	unit vector normal to the wall.
$\dot{n}$	number of turbulent spots produced per unit time [1/s].
$\dot{n}'$	number of turbulent spots produced per unit distance per unit time [1/ms].
$\dot{n}''$	number of turbulent spots produced per unit area per unit time [1/m^{2}s].
$\hat{n}$	turbulent spot production parameter, $\hat{n} = \dot{n}'\nu^2 \big/ u_\infty^3$.
p	pressure [N/m^2].
$\bar{p}$	time-averaged pressure component [N/m^2].
p'	fluctuating pressure component [N/m^2].
P	a general dependent variable of the flow.
Q	a general dependent variable of the flow; also, volumetric flow rate [m^3/s].
R_{pq}	correlation coefficient, $R_{pq} = \overline{Q'P'} \big/ \sqrt{\overline{Q'Q'}}\sqrt{\overline{P'P'}}$.
Re_L	Reynolds number based on a characteristic velocity and length U and L, $Re_L = UL/\nu$.
Re_T	turbulent Reynolds number, $Re_T = k^2/\nu\varepsilon$.
Re_λ	turbulent Reynolds number based on Taylor's micro-scale, $Re_\lambda = \hat{u}\lambda \big/ \nu$.
S	Strouhal number, fL/U.

t	time [s].
Tu	free-stream turbulence level, $Tu = (\hat{u}/\overline{u})_\infty$.
u_i	velocity vector [m/s].
$\overline{u}_i$	time-averaged velocity component [m/s].
u_i'	fluctuating velocity component [m/s].
u, v, w	velocity components in the x, y, z directions [m/s].
$\hat{u}$	intensity of turbulence [m/s], $\hat{u} = \sqrt{(\overline{u'u'} + \overline{v'v'} + \overline{w'w'})/3}$; also, amplitude or root-mean-square of the streamwise velocity fluctuation.
u_τ	shear velocity [m/s], $u_\tau = \sqrt{(\tau_0/\rho)}$.
u^+	dimensionless velocity in wall coordinates, $u^+ = u/u_\tau$.
U	characteristic mean-flow velocity [m/s].
$\mathrm{U}(z)$	unit step function of z.
$\mathrm{U}(z_1, z_2, ..)$	multidimensional unit step function of z_i.
W	work done per unit volume per unit time [N/m^{2}s].
x_i	position vector [m].
x, y, z	rectangular coordinates of the position vector [m].
y^+	dimensionless distance away from the wall in wall coordinates, $y^+ = u_\tau y/\nu$.

<u>Greek</u>

α	Kolmogorov's constant; also, turbulent spot expansion half-angle.
γ	intermittency factor (intermittency), $\gamma = \gamma_t$.
$\gamma_t, \gamma_b, \gamma_{nb}$	probability the flow is turbulent, laminar becalmed, and laminar non-becalmed.
Γ	circulation [m^2/s].
$\Gamma(z)$	Euler's gamma function.
$\Gamma(\alpha, z)$	incomplete gamma function.
δ	boundary-layer thickness [m].
δ_1, δ_2	displacement, momentum thickness [m].
δ_ω	unsteady diffusion thickness [m], $\delta_\omega = \sqrt{(2\nu/\omega)}$.
δ_{ij}	unit tensor (Kronecker delta) equal to one for $i=j$, otherwise zero.
$\delta(z)$	Dirac delta function of z.
ε	dissipation of energy per unit mass associated with the fluctuations [m^2/s^3]; also, a small parameter very much less than unity.
ε_{ij}	dissipation tensor associated with the fluctuations [m^2/s^3].
ε_{ijk}	alternating tensor equal to zero when any two indices are the same, +1 when the permutation of the indices is even, and −1 when it is odd.
E	dissipation of energy per unit mass associated with the mean flow [m^2/s^3].
ζ	vorticity component in the z direction [1/s].
θ	turbulent spot leading edge half-angle.
κ	von Kármán's constant.
λ	Taylor's micro-scale of turbulence [m]; also, wavelength [m].

$\lambda_{\delta 2}$	pressure gradient parameter, $\lambda_{\delta 2} = (\delta_2^2/\nu)du_\infty/dx$.
Λ	integral length scale of turbulence [m].
μ	dynamic viscosity [kg/ms].
μ'	bulk viscosity [kg/ms].
μ_T	eddy viscosity [kg/ms].
ν	kinematic viscosity [m^2/s], $\nu = \mu/\rho$.
ν_T	kinematic eddy viscosity [m^2/s], $\nu_T = \mu_T/\rho$.
Π	Coles' wake parameter.
ρ	density [kg/m^3].
σ	Emmons' spot propagation parameter.
σ_b	becalmed zone propagation parameter.
σ_{ij}	stress tensor [N/m^2].
τ_{ij}	viscous stress tensor [N/m^2].
υ_d	Kolmogorov's velocity scale [m/s], $\upsilon_d = (\nu\varepsilon)^{1/4}$.
ϕ	dimensionless stream function.
ϕ_{ij}	pressure-strain rate tensor [m^2/s^3].
Φ	dissipation of kinetic energy [N/m^{2}s].
ψ	stream function [m^2/s].
$\psi(P, P')$	spot probability density function.
ω	angular frequency [1/s].
ω_i	vorticity vector [1/s].
ω_s	spot passing frequency [1/s].
Ω	frequency parameter, $\Omega = \omega\nu/u_\infty^2$.

Subscripts

0	value at the wall; also, an initial value.
∞	value in the free stream.
b	becalmed zone.
d	dissipative, small scale eddies.
e	energy-containing, large scale eddies.
KH	Kelvin-Helmholtz.
le	leading edge.
L	laminar.
LT	length of transition.
s	turbulent spot.
t	turbulent; also, beginning of transition.
T	turbulent; also, end of transition.
TS	Tollmien-Schlichting.
te	trailing edge.

Milestones in Boundary-Layer Transition

Since 1883 when Reynolds provided our first glimpse of transition by injecting a streak of dye into water flowing through a glass tube, the path to understanding transition has neither been obvious, continuous, nor straightforward. Nevertheless, several investigations clearly stand out as being fundamental to our current understanding. The following is the author's distilled chronological account of these investigations:

1883 – Reynolds illustrates that flow in a pipe is laminar or turbulent and experimentally determines what we now call the transition Reynolds number for pipe flow.

1887 – Rayleigh considers the inviscid response of a shear layer to a "small" fluctuation in the flow and determines the necessary conditions for a laminar shear layer to become unstable.

1895 – Reynolds develops the method now called Reynolds averaging for treating unsteady fluctuating flow, introduces the "additional stress" now called Reynolds stress that arises from the unsteady motion, and uses an energy method to calculate a critical Reynolds number for unstable flow in a pipe.

1914 – Prandtl demonstrates boundary-layer transition using "trip wires" on a sphere to reduce its drag coefficient.

1916 – Rayleigh solves for the instability of a horizontal layer of fluid heated from below, calculates an intermediate laminar flow configuration for the layer, and theoretically determines what we now call the critical Rayleigh number for its appearance.

1923 – Taylor solves for the instability of the flow between two rotating cylinders, calculates the first of several intermediate laminar configurations for the flow, and theoretically determines what we now call the critical Taylor number for its appearance.

1929 – Tollmien solves for the instability of the flow in a laminar boundary layer without a pressure gradient, calculates the neutral stability curve for what we now call Tollmien-Schlichting waves, and determines a critical Reynolds number for boundary layer instability.

1940 – Görtler solves for the instability of the flow in a laminar boundary layer on a concave surface, calculates an intermediate laminar configuration for the flow, and theoretically determines what we now call the critical Görtler number for its appearance.

1943 – Schubauer and Skramstad experimentally confirm Tollmien's analysis using a wind tunnel with such a low free-stream turbulence that only the instabilities predicted by Tollmien could develop.

1951 – Emmons experimentally demonstrates that transition begins with the random formation of turbulent spots within the boundary layer, which subsequently propagate downstream and grow until all the boundary layer becomes turbulent. In addition, he develops a stochastic turbulent spot theory, introduces the probability function we now call intermittency, and describes a practical method for calculating the flow through transition.

1976 – Dyban, Epik, and Suprun conduct a series of experiments on the response of a laminar boundary layer to free-stream turbulence and demonstrate that the resulting fluctuations within the boundary layer grow more or less linearly with distance along the surface at a rate that depends on the turbulence level.

1990 – Roach and Brierley conduct a comprehensive set of experiments on laminar-to-turbulent boundary-layer transition for various free-stream pressure gradients and turbulence levels, providing a wealth of transitional flow data.

1997 – Mayle and Schulz demonstrate that the mechanism responsible for initially amplifying the fluctuations in a laminar boundary layer with free-stream turbulence is the work done by the unsteady pressure forces, show that this depends on the turbulent energy near the Kolmogorov frequency range, and develop an analysis to calculate their amplification rate.

ca. 2000 – Direct numerical simulation (DNS) experiments of transitional flow begin.

A Preview of Transition

tran·si·tion | tran'zi sh ən | *noun: a process of changing from one state to another.*

In fluid mechanics, transition is the process of changing from laminar to turbulent flow. It occurs in pipe flow and other confined *internal* flows, as well as in bounded and unbounded *external* flows. Unlike laminar or turbulent flow, however, it is not well understood even in its simplest form. This book is about *boundary-layer transition* – that is, transition occurring within a shear layer next to a wall.

Briefly, boundary-layer transition is an unsteady, three-dimensional, stochastic process. It begins in an *unsteady laminar pre-transitional* region, passes through an *intermittent laminar-turbulent transitional* region, and ends in a *fully turbulent post-transitional* region. The *mode* of transition is chiefly determined by the mechanism responsible for amplifying the fluctuations in the pre-transitional region. Broadly speaking, this mechanism depends on whether the laminar flow is stable or not, and whether disturbances external to the boundary layer exist or not. When external disturbances, such as free-stream turbulence and mechanical vibrations, are nonexistent (or *very* small) and when the flow is unstable, *self-excited* fluctuations within the boundary layer amplify exponentially, and eventually "break down" into turbulence. When external disturbances are significant, however, they both cause and excite the laminar fluctuations within the boundary layer. While transition occurring by these mechanisms would best be characterized as being *self* and *externally excited*, they are historically known as *natural*[1] and *bypass*[2] transition.

Since 1880, when Rayleigh[3] investigated *hydrodynamic stability*, and until the late 1900s, natural transition was the overwhelming concern of investigators interested in transitional flow. There are several reasons for this. The first is Tollmien's[4] success in predicting the *neutral stability curve*, and therefore, a *critical Reynolds number* and *critical frequency*, for a laminar boundary layer. The second is Schubauer and Skramstad's[5] experimental confirmation of his result. The third, and perhaps most important reason, was the need of the fast developing aircraft industry to predict transition with *low* free-stream turbulence. For a boundary layer developing along a flat or a mild convexly-curved surface, this instability consists of the well known *Tollmien-Schlichting waves* that propagate downstream within the boundary layer. For a boundary layer developing along a concave surface, the instability consists of the well known set of *Taylor-Görtler vortices* aligned parallel with the flow within the boundary layer. Despite all the attention, however, we still can't calculate the development of an unstable boundary layer downstream from the *point of instability* to the beginning of transition.[6]

More recently, attention has shifted to bypass transition. This shift was triggered in the 1970s by the gas-turbine industry's need for predicting transition with *high* free-stream turbulence.[7] Initially studied by Taylor[8] in 1936, a mechanism describing the effect of free-stream turbulence on a laminar

---

[1] Originally considered the "natural" path to transition, this term is generally reserved for transition caused by unstable *Tollmien-Schlichting* waves.

[2] This all-inclusive term was originally coined, and now generally used to denote transition by any mechanism that "bypasses" the Tollmien-Schlichting mechanism of instability.

[3] Rayleigh, J.W.S., Lord, 1880, "On the Stability, or Instability, of Certain Fluid Motion," Scientific Papers, 1, pp. 474–487, Cambridge University Press; also Proc. Lond. Math. Soc., 11, pp. 57–70.

[4] Tollmien, W., 1929, "Über die Entstehung der Turbulenz," 1. Mitt. Nachr. Ges. Wiss. Göttingen, Math. Phys. Klasse, pp. 21–44. (Engl. Transl. in NACA TM 609, 1931.)

[5] Schubauer, G.B., and Skramstad, H.K., 1943, "Laminar Boundary-Layer Oscillations and Stability of Laminar Flow," NACA War-Time Rept. W-8.

[6] Headway in this area is now rapidly being made through *direct numerical simulation* (DNS).

[7] Mayle, R.E., 1991, "The Role of Laminar-Turbulent Transition in Gas Turbine Engines," J. Turbomachinery, 113, pp. 509–537.

[8] Taylor, G.I., 1936, "Effect of Turbulence on Boundary Layer," Proc. Roy. Soc., London, A 156, pp. 307–317.

boundary layer escaped us until sixty years later when the author and his colleagues[1] proposed both a mechanism and a corresponding set of equations for calculating pre-transitional flow. Since this analysis appears to be valid up to the onset of transition, and since a large amount of data on transitional boundary layers with a turbulent free stream is presently available, it now appears that we have a better chance of solving the problem of bypass transition than natural transition.

Transition in a nutshell.

Let's start by conducting a relatively simple experiment. Place a thin flat plate in a wind tunnel parallel to the flow and measure the *surface shear stress* along the plate. Dimensional considerations suggest that these results should be independent of wind speed if we plot the *skin-friction coefficient* (time-averaged surface shear stress divided by the dynamic pressure) against the Reynolds number based on the distance from the leading edge of the plate. The data from our experiment will look something like that plotted in Fig. A.

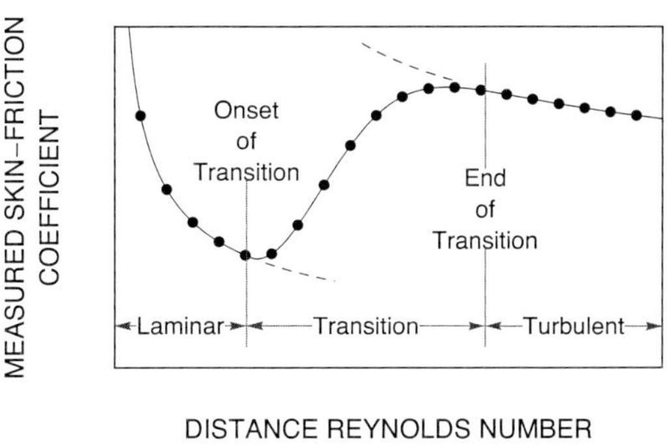

Figure A.
Streamwise variation of the skin-friction coefficient in a transitional boundary layer on a flat plate.

At *low* Reynolds numbers, corresponding to near the leading edge, the boundary layer is laminar. That is, if we place a hot-wire anemometer within the boundary layer, we will detect no velocity fluctuations there. Furthermore, we will also be able to calculate the experimental result using laminar boundary-layer theory (Chapter 2). This result is shown by the lower dashed line in the figure. At high Reynolds numbers, the boundary layer is fully turbulent. That is, we will measure relatively large velocity fluctuations within it, and be able to calculate the experimental result from a standard turbulent-flow computer program (Chapter 3). This result is shown by the upper dashed line in Fig. A.

Between these regions where the skin-friction coefficient rises with distance, the boundary layer is transitional. That is, as shown in Fig. B, we will measure an unsteady flow within the boundary layer that is laminar part of the time and turbulent the other part. For a higher free-stream turbulence level, transition will begin farther upstream than that shown in Fig. A and the length of transition will be shorter, whereas for a lower level, transition will begin farther downstream and its length will be longer. Unlike our ability to predict the laminar and turbulent flow results, however, "predicting" transitional flow is highly uncertain. This is our problem.

An intrinsic attribute of transitional flow is its unsteadiness. If we conduct our experiment with a turbulent free stream, we will also find "turbulent-looking" fluctuations within the laminar boundary layer even near the leading edge. The intensity of these fluctuations will vary across the boundary layer

[1] Mayle, R.E., and Schulz, A., 1997, "The Path to Predicting Bypass Transition," J. Turbomachinery, 119, pp. 405–411. Mayle, R.E., Dullenkopf, K., and Schulz, A., 1998, "The Turbulence That Matters," J. Turbomachinery, 120, pp. 402–409.

as shown in Fig. C with an amplitude that increases with increasing streamwise distance. Furthermore, if we conduct the experiment at different free-stream turbulence levels, we will find that increasing the turbulence level increases the growth rate of these fluctuations (Chapter 4).

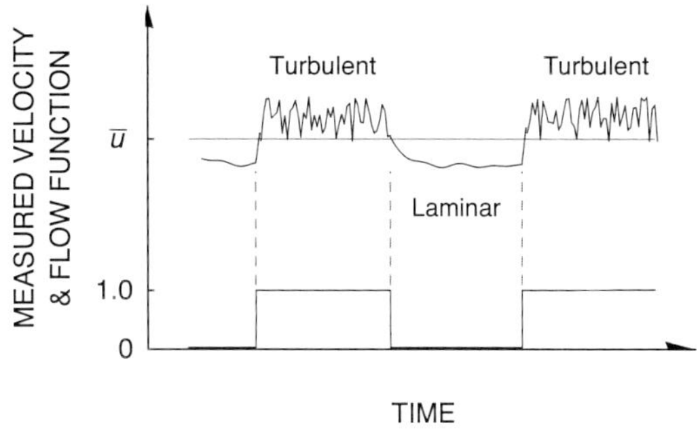

Figure B.
Velocity measurements within a transitional boundary layer and its corresponding intermittency function that equals zero when the flow is laminar and one when it is turbulent.

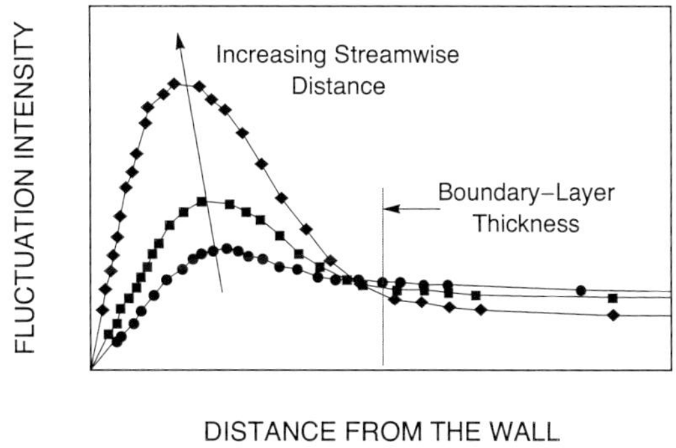

Figure C.
Intensity profiles in a pre-transitional boundary layer with free-stream turbulence.

Transition begins when these amplified fluctuations *break down* into *turbulent eddies*. This most often happens *randomly* with respect to both time and space at discrete locations scattered over the surface. It's a process yet to be fully explained. Nevertheless, the *beginning* or *onset of transition* is generally considered to be the streamwise position where the fluctuations *most often* break down. This always occurs upstream of the minimum skin-friction coefficient seen in Fig. A. The Reynolds number at this position is usually referred to as the *transition Reynolds number*. A rule of thumb dating back to boundary-layer experiments conducted some eighty years ago is "transition occurs where the Reynolds number based on the distance from the leading edge of the plate is about 350,000." Today we know that transition can begin at almost any distance downstream depending on a large number of conditions.

In the transition region, newborn turbulent eddies quickly develop into *turbulent spots* (Chapter 5). Also known as *Emmons' spots*, they encroach upon the neighboring laminar flow, expanding and coalescing across the surface as they propagate downstream until all the flow becomes turbulent. An *ensemble-averaged* shape of a typical spot is shown in Fig. D. The plan view in this figure is drawn approximately to scale, while the height in the elevation view is drawn about five times scale. Therefore, these spots are relatively flat, spread-out turbulent entities that protrude beyond the laminar boundary layer. Of course, being turbulent with eddies as large as the height of the spot itself, their interface is not as smooth as shown in the figure, but highly irregular.

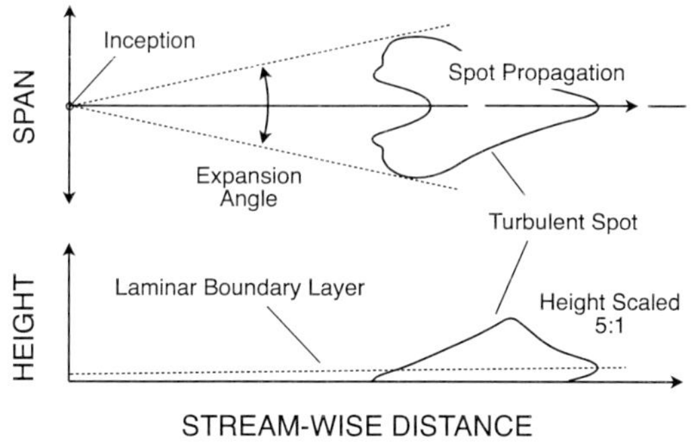

Figure D.
Plan and profile views of a turbulent spot in a laminar boundary layer.

As it turns out, the *intermittent* turbulent portions of the hot-wire signal in Fig. B correspond to the velocity within passing spots. If the sample extended over a longer period of time than shown in that figure, the randomness of the length and spacing of the individual segments with time, hence the randomness of transitional flow, would be more obvious. Nevertheless, since the spots grow with distance from their inception as shown by the dotted lines in Fig. D, we will find a mostly laminar hot-wire signal near the onset of transition and a mostly turbulent signal near its end. Assigning the number one to those parts of the signal that are turbulent and zero to those that are laminar as shown in the lower part of Fig. B, we can determine the *fraction of time the flow is turbulent* by taking the time average of this binary function over a long period of time. This fraction, most often called the *intermittency*, equals zero when the flow is *completely* laminar at a given location and one when the flow is *completely* turbulent. As a consequence, intermittency becomes the yardstick for transition.

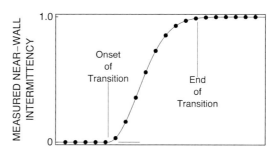

Figure E.
Streamwise distributions of intermittency in a transitional boundary layer.

If, along with measuring the surface shear stress in our experiment, we also record enough hot-wire measurements to determine the intermittency along our plate near its surface, it will vary from zero to one something like that shown in Fig. E. In the pre-transitional region of the flow, the boundary layer is laminar and the *near-wall intermittency* equals zero. In the post-transitional region, the boundary layer is turbulent and the near-wall intermittency is one. In the transitional region, where the flow is part of the time laminar and part of the time turbulent, the intermittency varies monotonically from zero to one. Presently, Emmons' *turbulent-spot theory* provides the most reliable method for predicting this near-wall behavior. Based on probability theory, it easily reproduces the behavior through transition once the *production*, *growth*, and *propagation rates* of the spots are known. Referring to Fig. D, it is easy to see that the *length of transition*, which is the distance between the beginning and end of transition, decreases as the production (number of spots per unit span) and growth rates (expansion angle) increase, and increases as the propagation rate increases. Once these three rates are known (Chapter 6), the variation of the skin-friction coefficient in the transitional portion of Fig. A can be determined (Chapter 7).

The *end of transition*, and thus the beginning of post-transitional flow, is generally considered to be where the last "remnants" of laminar flow *near the wall* most often disappear. This position is always downstream of the maximum in the skin-friction coefficient. Since a fully established turbulent flow near the wall does not necessarily imply a fully established turbulent flow in its outer *wake* region, where remnants of some spots still linger, post-transition is a region where the newly formed turbulent boundary layer matures into a fully developed one. Surprisingly, the details of this behavior have yet to be investigated.

Our "thought" experiment, that is, measuring the surface shear stress along a thin flat plate placed at zero incidence to the flow, has been actually carried out many times before. The skin-friction coefficient measurements from one of these experiments are shown in Fig. F. Unfortunately, our best guess at predicting it falls somewhere within the grey area. The conclusion of this comparison is obvious – even after a half century of research with supposedly the best correlations for the onset and length of transition, we still can't predict transitional flow even for the simplest of conditions. In this case, the maximum error in streamwise distance is roughly 20 percent of the transition length, while that in the skin-friction coefficient is an incredible 50 percent. Clearly, this is a problem. It is also why transitional flow is the last frontier in fluid mechanics.

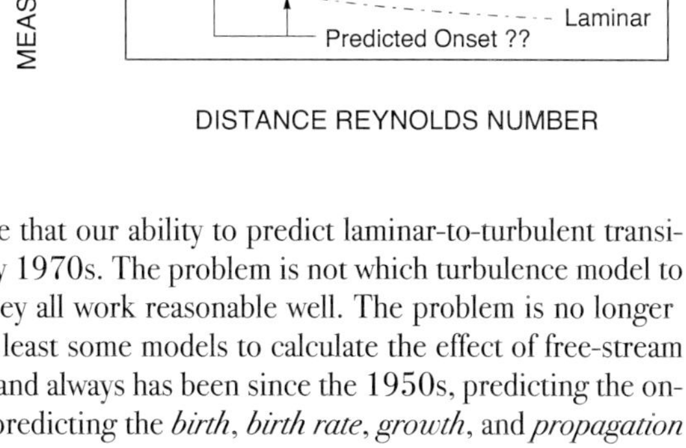

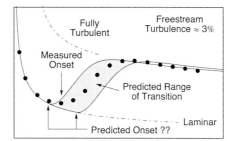

Figure F.
Measured and predicted skin-friction coefficient distribution in a transitional boundary layer.

Finally, it is the author's personal experience that our ability to predict laminar-to-turbulent transition has only slightly improved since the early 1970s. The problem is not which turbulence model to use for calculating post-transitional flow. They all work reasonable well. The problem is no longer calculating pre-transitional flow. We have at least some models to calculate the effect of free-stream turbulence on laminar flow. The problem is, and always has been since the 1950s, predicting the onset and length of transition; in other words, predicting the *birth*, *birth rate*, *growth*, and *propagation* of turbulent spots.

Part I

Before and After

1 Basic Equations

Elements of fluid mechanics. Equations for fluctuating flow. Equations for two-dimensional flow. Prandtl's boundary-layer theory. Other definitions and equations.

In this chapter, we will review the basic ideas and equations in fluid mechanics, and obtain the relevant forms of the equations for transitional boundary-layer flow. So as not to be overwhelming (and boring), the material is succinctly presented except where needed later in the book. For those interested in more detail, the books by Schlichting [1979], Rotta [1972], and Hinze [1975] should be consulted. For those familiar with the equations for turbulent flow, this chapter may be browsed.

1.1 Elements of Fluid Mechanics.

Basic quantities. Equations of motion. Kinetic energy equation.

Basic quantities. Following convention, consider a rectangular Cartesian coordinate system and define any point in space by the coordinates x_1, x_2 and x_3. Then the position vector of the point in tensor notation is x_i where $i = 1, 2, 3$. Let $u_i(x_i, t)$ and $p(x_i, t)$ denote the velocity vector and thermodynamic pressure at x_i at time t. In addition, let the tensor $\tau_{ij}(x_i, t)$ represent the viscous stress in the j direction acting on the surface element normal to the i direction.[1]

If we consider a Newtonian fluid, the viscous stress is *linearly* related to the fluid deformation rate[2] which, in its most general form, yields

$$\tau_{ij} = \mu\left(\frac{\partial u_i}{\partial x_j} + \frac{\partial u_j}{\partial x_i}\right) + \left(\mu' - \tfrac{2}{3}\mu\right)\left(\frac{\partial u_k}{\partial x_k}\right)\delta_{ij} \tag{1.1}$$

where μ and μ' are the dynamic and bulk viscosity of the fluid, and δ_{ij} is the unit tensor (Kronecker delta) which equals one when $i = j$ and zero when $i \neq j$. By exchanging the indices in the above expression, we easily see that τ_{ij} is a symmetric tensor. The diagonal elements of this tensor are the viscous *normal* stresses whereas the off-diagonal elements are the viscous *shear* stresses.

Let σ_{ij} represent the *total* stress acting on a fluid element. Since it is the sum of the thermodynamic pressure and viscous stresses, and since pressure acts normal to a surface, we obtain

$$\sigma_{ij} = -p\delta_{ij} + \mu\left(\frac{\partial u_i}{\partial x_j} + \frac{\partial u_j}{\partial x_i}\right) + \left(\mu' - \tfrac{2}{3}\mu\right)\left(\frac{\partial u_k}{\partial x_k}\right)\delta_{ij} \tag{1.2}$$

If we consider that the thermodynamic pressure equals the average of the normal stresses acting on the fluid element, that is, if $p = -\sigma_{ii}/3$, then according to the above expression we find that the bulk viscosity *must* equal zero. This is known as *Stokes' hypothesis*.[3]

To simplify matters, we will henceforth regard μ and μ' as constants. Since these quantities depend primarily on temperature, this assumption is more restrictive for liquids than for gasses. Moreover, we will also regard the density of the fluid to be a constant. This is more restrictive for gasses than for liquids. Therefore, the equations and analyses presented in this book are only valid for "nearly" isothermal, incompressible flows.[4]

Equations of motion. The fundamental equations of motion for a fluid are the conservation equation for mass and the well-known Navier-Stokes equation. For an incompressible fluid they are

---

[1] See "Supplement on Cartesian Tensors" for a brief review of tensor notation, algebra and calculus.

[2] See §1.5.

[3] Stokes [1845].

[4] This is generally true for flows with no or "very little" external heating (or cooling) and Mach numbers less than roughly one-third.

$$\frac{\partial u_i}{\partial x_i} = 0 \tag{1.3}$$

and

$$\rho\left(\frac{\partial u_i}{\partial t} + u_j \frac{\partial u_i}{\partial x_j}\right) = -\frac{\partial p}{\partial x_i} + \frac{\partial}{\partial x_j}\left[\mu\left(\frac{\partial u_i}{\partial x_j} + \frac{\partial u_j}{\partial x_i}\right)\right] + \rho f_i \tag{1.4}$$

where ρ is the density of the fluid and f_i is the body force per unit mass.[1] The first equation, better known as the *continuity equation*, asserts that the divergence of velocity for an incompressible fluid must be zero if mass is to be conserved. The second is Newton's second law of motion, or the *momentum equation*, which equates the mass times acceleration of a fluid element in a flow to the sum of the pressure, viscous and body forces acting on that element.

In addition, for an incompressible fluid, we obtain from Eqs. (1.1) and (1.3) that

$$\tau_{ij} = \mu\left(\frac{\partial u_i}{\partial x_j} + \frac{\partial u_j}{\partial x_i}\right) \tag{1.5}$$

which no longer involves the bulk viscosity. Consequently, given ρ, μ and f_i, Eqs. (1.3)–(1.5) provide a complete set of equations for the velocity, pressure, and viscous stress throughout the flow.

Kinetic energy equation. For isothermal flow, conservation of energy asserts that the increase of kinetic energy in a fluid element must equal the work done by the forces acting on the element minus the dissipation of mechanical energy within the element. We can formally express this by the equation

$$\frac{D}{Dt}\left(\tfrac{1}{2}\rho u_i u_i\right) = W - \Phi \tag{1.6}$$

where the operator $D/Dt = \partial/\partial t + u_j\partial/\partial x_j$ represents the change with respect to time *following* an infinitesimal fluid element in the flow, and W and Φ represent the work and dissipation per unit volume per unit time.

Concerned only with the surface stress σ_{ij} and body force f_i, the work done on the element is

$$W = \frac{\partial \sigma_{ij} u_j}{\partial x_i} + \rho u_i f_i = \frac{\partial}{\partial x_i}\left\{u_j\left[-p\delta_{ij} + \mu\left(\frac{\partial u_i}{\partial x_j} + \frac{\partial u_j}{\partial x_i}\right)\right]\right\} + \rho u_i f_i \tag{1.7}$$

The dissipation of energy within the element, which equals the viscous stress times the strain rate, is given by

$$\Phi = \tau_{ij}\frac{\partial u_j}{\partial x_i} = \mu\left(\frac{\partial u_i}{\partial x_j} + \frac{\partial u_j}{\partial u_i}\right)\frac{\partial u_j}{\partial x_i} \tag{1.8}$$

Although not immediately obvious, it can be shown that Φ is a sum of squared terms and therefore a positive quantity. Furthermore, by considering the second law of thermodynamics, it can be shown that the entropy of a fluid element in an isothermal flow always increases because $\Phi > 0$.

Substituting the above expressions into Eq. (1.6), the energy equation for isothermal flow becomes

$$\frac{\partial \frac{1}{2}\rho u_i u_i}{\partial t} + u_j \frac{\partial \frac{1}{2}\rho u_i u_i}{\partial x_j} = \frac{\partial}{\partial x_i}\left\{u_j\left[-p\delta_{ij} + \mu\left(\frac{\partial u_i}{\partial x_j} + \frac{\partial u_j}{\partial x_i}\right)\right]\right\} + \rho u_i f_i - \mu\left(\frac{\partial u_i}{\partial x_j} + \frac{\partial u_j}{\partial x_i}\right)\frac{\partial u_j}{\partial x_i} \tag{1.9}$$

[1] See Schlichting for a detailed derivation of these equations.

This is known as either the *kinetic energy* or *mechanical energy* equation. Since it can also be obtained by multiplying the momentum equation by u_i, this and the momentum equation are not independent.

1.2 Equations for Fluctuating Flow.

Reynolds' averaging. Continuity, momentum, and kinetic energy equations for the time-averaged and fluctuating motion. Reynolds-stress equation.

Reynolds' averaging. Let's now consider an unsteady flow that fluctuates in some manner about a *time-independent mean* flow. Following Reynolds [1895], suppose that the *instantaneous* value of any dependent variable of the flow, say Q, may be expressed as

$$Q(x_i,t)=\bar{Q}(x_i)+Q'(x_i,t) \tag{1.10}$$

where

$$\bar{Q}(x_i)=\frac{1}{T}\int_{t_0}^{t_0+T} Q(x_i,t)\,dt \tag{1.11}$$

is the time-averaged component and Q' is the fluctuating component about this mean.

Inherent in our supposition is the ability to choose a time interval T that provides a meaningful average. Clearly, when the fluctuation oscillates at one frequency, the value of T must equal a multiple of its period to yield the "correct" average. It is not so clear, however, what T should be when the spectrum of the fluctuation contains several frequencies. In this case, if we choose T about three times larger than the largest period of the fluctuation, it can be shown that the time-averaged component is accurate to within ten percent of *that* fluctuation's amplitude, and choosing T ten times larger reduces the error to about three percent. Therefore, in practice, T is usually chosen to be as large as possible compared to the period of the highest amplitude or *most relevant* fluctuation. For transitional flow where several relevant fluctuations are involved, T must always be chosen to include that with the largest period.[1]

If we assume that an appropriate T can be determined, substituting Eq. (1.10) into Eq. (1.11) yields

$$\overline{Q'}(x_i)=\frac{1}{T}\int_{t_0}^{t_0+T} Q'(x_i,t)\,dt=0 \tag{1.12}$$

From Eqs. (1.10)–(1.12) it then follows that the *variance* of the fluctuations, namely,

$$\overline{\left(Q-\bar{Q}\right)^2}=\overline{Q'^2}=\overline{Q'Q'}$$

is given by

$$\overline{Q'Q'}(x_i)=\frac{1}{T}\int_{t_0}^{t_0+T} Q'(x_i,t)Q'(x_i,t)\,dt \tag{1.13}$$

This is sometimes referred to as the "second central moment" of the fluctuation's distribution. Likewise, if P is another unsteady dependent variable of the flow, it follows that the *covariance* or *correlation* of Q and P is given by

$$\overline{Q'P'}(x_i)=\frac{1}{T}\int_{t_0}^{t_0+T} Q'(x_i,t)P'(x_i,t)\,dt \tag{1.14}$$

This is sometimes called the "first central moment of the fluctuations' joint distributions." Dividing by the square root of each variable's variance, we obtain the *correlation coefficient*, namely,

---

[1] As we will see in §7.2, the averaging process in transitional flow becomes much more complex if we want to separate the laminar and turbulent portions of the flow.

$$R_{pq}(x_i)=\overline{Q'P'}\Big/\sqrt{\overline{Q'Q'}}\sqrt{\overline{P'P'}}$$

Continuity and momentum equations. If we introduce the instantaneous velocity as the sum of its time-averaged and fluctuating components

$$u_i(x_i,t)=\bar{u}_i(x_i)+u_i'(x_i,t) \tag{1.15}$$

together with similar expressions for the pressure and body forces into Eqs. (1.3) and (1.4), the continuity and momentum equations become

$$\frac{\partial\bar{u}_i}{\partial x_i}+\frac{\partial u_i'}{\partial x_i}=0 \tag{1.16}$$

and

$$\rho\frac{\partial u_i'}{\partial t}+\rho\bar{u}_j\frac{\partial(\bar{u}_i+u_i')}{\partial x_j}+\rho u_j'\frac{\partial(\bar{u}_i+u_i')}{\partial x_j}=$$

$$+\frac{\partial}{\partial x_j}\left\{-\left(\bar{p}+p'\right)\delta_{ij}+\mu\left[\frac{\partial(\bar{u}_i+u_i')}{\partial x_j}+\frac{\partial(\bar{u}_j+u_j')}{\partial x_i}\right]\right\}+\rho\left(\bar{f}_i+f_i'\right) \tag{1.17}$$

The *time-independent* components of the continuity and momentum equations are obtained by integrating these equations with respect to time according to Eqs. (1.11)–(1.14). Subsequently, their *time-dependent* components are obtained by subtracting these time-independent equations from those above. Thus, the time-independent and time-dependent components of the continuity equation are easily found to be

$$\frac{\partial\bar{u}_i}{\partial x_i}=0 \tag{1.18}$$

and

$$\frac{\partial u_i'}{\partial x_i}=0 \tag{1.19}$$

which imply that the divergence of both the mean and fluctuating velocity components of the flow must be zero for an incompressible fluid.

In addition, the time-independent and time-dependent components of the momentum equation are found to be

$$\rho\bar{u}_j\frac{\partial\bar{u}_i}{\partial x_j}=-\overline{\rho u_j'\frac{\partial u_i'}{\partial x_j}}+\frac{\partial}{\partial x_j}\left[-\bar{p}\delta_{ij}+\mu\left(\frac{\partial\bar{u}_i}{\partial x_j}+\frac{\partial\bar{u}_j}{\partial x_i}\right)\right]+\rho\bar{f}_i \tag{1.20}$$

and

$$\rho\frac{\partial u_i'}{\partial t}+\rho\bar{u}_j\frac{\partial u_i'}{\partial x_j}=-\rho u_j'\frac{\partial\bar{u}_i}{\partial x_j}-\rho u_j'\frac{\partial u_i'}{\partial x_j}+\overline{\rho u_j'\frac{\partial u_i'}{\partial x_j}}+\frac{\partial}{\partial x_j}\left[-p'\delta_{ij}+\mu\left(\frac{\partial u_i'}{\partial x_j}+\frac{\partial u_j'}{\partial x_i}\right)\right]+\rho f_i' \tag{1.21}$$

In this case, the first term on the right-hand side of the first equation and the first three terms of the second are transposed *convective* terms. Unlike the two components of the continuity equation, the components of the momentum equation are *coupled* by the first and third term on the right-hand side of Eqs. (1.20) and (1.21) respectively. Generally, this term cannot be neglected for transitional flow.

To illustrate this, let's consider that 1) the mean velocity is the same order of magnitude as the free-stream velocity u_∞; 2) the fluctuations are the order of the root-mean-square of their amplitude $\hat{u}$; and

3) the boundary-layer thickness δ is several orders of magnitude smaller than the distance measured along the surface. Then the largest change of the mean and fluctuating velocities with respect to distance occurs *across* the boundary layer and these are of the magnitude u_∞/δ and $\hat{u}/\delta$, respectively. It also follows that the largest second derivative of the mean velocity is that across the boundary layer, and it is of magnitude u_∞/δ^2. Then, if we realize that the mean convective term on the right-hand side of Eq. (1.20) must be the same magnitude as the viscous terms somewhere within the boundary layer, the first term on the right-hand side of Eq. (1.20) *cannot* be neglected unless $Re_\delta(\hat{u}/u_\infty)^2 << 1$, where $Re_\delta = \rho u_\infty \delta/\mu$ is the Reynolds number based on the boundary layer thickness. Since $\hat{u}/u_\infty$ is typically the order of 10^{-1} or less and Re_δ is typically the order of 10^2 or more, this condition is not met except near the leading edge of the surface in the earliest stage of pre-transition. A similar analysis of Eq. (1.21) indicates that its second and third terms cannot be neglected unless $Re_\delta(\hat{u}/u_\infty) << 1$. This condition, which is even more restrictive than that for Eq. (1.20), is almost *never* met except when the external disturbances are either nonexistent or extremely small. Hence, for transitional flow, these terms should be retained.[1]

Keeping this in mind, Eq. (1.20) may be rewritten as

$$\rho \bar{u}_j \frac{\partial \bar{u}_i}{\partial x_j} = \frac{\partial}{\partial x_j}\left[-\bar{p}\delta_{ij} + \mu\left(\frac{\partial \bar{u}_i}{\partial x_j} + \frac{\partial \bar{u}_j}{\partial x_i}\right) - \rho\overline{u_i' u_j'}\right] + \rho \bar{f}_i \tag{1.22}$$

where the second-order symmetric tensor $-\rho\overline{u_i' u_j'}$ is dimensionally equivalent to a stress. Known as the *Reynolds stress*, it may be physically interpreted as the stress resulting from the unsteady momentum exchange across the surface of a fluid element. If we consider an unsteady motion in a flow with a positive mean-flow shear, it's possible to show that its off-diagonal components cause positive *shear* stresses. The diagonal components on the other hand are *normal* stresses that always act in a negative direction, that is, toward the surface of an elemental control volume. Indeed, we may interpret the average value of these stresses, which equals $-\frac{1}{3}\rho\overline{u_i' u_i'}$, as an *apparent pressure* caused by the unsteadiness. But this view is seldom taken.

Finally, by introducing the instantaneous velocity Eq. (1.15) into Eq. (1.5) and *averaging*, we obtain

$$\bar{\tau}_{ij} = \mu\left(\frac{\partial \bar{u}_i}{\partial x_j} + \frac{\partial \bar{u}_j}{\partial x_i}\right) \quad \text{and} \quad \tau_{ij}' = \mu\left(\frac{\partial u_i'}{\partial x_j} + \frac{\partial u_j'}{\partial x_i}\right) \tag{1.23}$$

which are of course the time-averaged and fluctuating components of the viscous shear stress.

Kinetic energy equations. The time-averaged component of kinetic energy per unit mass is easily obtained by squaring the expression for the instantaneous velocity Eq. (1.15) and taking one half of its average. This yields

$$\bar{\mathrm{K}} = \tfrac{1}{2}\overline{u_i u_i} = \tfrac{1}{2}\bar{u}_i \bar{u}_i + \tfrac{1}{2}\overline{u_i' u_i'}$$

which is the sum of the *kinetic energy of the mean motion* and the *average kinetic energy of the fluctuations* (per unit mass). It is customary to denote these quantities by

$$K = \tfrac{1}{2}\bar{u}_i \bar{u}_i \quad \text{and} \quad k = \tfrac{1}{2}\overline{u_i' u_i'} \tag{1.24}$$

The fluctuating component of the kinetic energy per unit mass, obtained by subtracting $\bar{\mathrm{K}}$ from one half the instantaneous velocity squared, is

[1] An interesting observation here, particularly for amplified disturbances in pre-transitional flow, is that the nonlinearities affect the fluctuations before they affect the mean flow.

$$\mathrm{K}' = \bar{u}_i u_i' + \tfrac{1}{2} u_i' u_i' - \tfrac{1}{2} \overline{u_i' u_i'}$$

which is seldom mentioned.

An equation for K, the kinetic energy of the mean motion, is obtained by multiplying the time-averaged component of the momentum equation, Eq. (1.20), by $\bar{u}_i$. After some rearrangement, we may write this equation as

$$\rho \bar{u}_j \frac{\partial K}{\partial x_j} = \rho \overline{u_i' u_j'} \frac{\partial \bar{u}_i}{\partial x_j} - \rho \frac{\partial}{\partial x_j} \left[\overline{u_j' \left(\mathrm{K}' - \tfrac{1}{2} u_k' u_k' \right)} \right]$$

$$+ \frac{\partial}{\partial x_j} \left\{ \bar{u}_i \left[-\bar{p} \delta_{ij} + \mu \left(\frac{\partial \bar{u}_i}{\partial x_j} + \frac{\partial \bar{u}_j}{\partial x_i} \right) \right] \right\} + \rho \bar{u}_i \bar{f}_i - \rho \mathrm{E} \tag{1.25}$$

where E is the dissipation of energy per unit mass associated with the *mean flow*, namely,

$$\mathrm{E} = \nu \left(\frac{\partial \bar{u}_i}{\partial x_j} + \frac{\partial \bar{u}_j}{\partial x_i} \right) \frac{\partial \bar{u}_i}{\partial x_j}$$

and $\nu = \mu/\rho$ is the kinematic viscosity of the fluid. The first two terms on the right-hand side of this equation are transposed convective terms. The significance of the next two terms follow directly from Eq. (1.7). They are the work done by the surface and body forces associated with the mean flow.

The equation for k is obtained by substituting the expression for the instantaneous velocity Eq. (1.15) and associated expressions for the pressure and body force into Eq. (1.9), taking its average (which is the equation for the mean kinetic energy $\overline{\mathrm{K}} = K + k$), and subtracting Eq. (1.25). After some effort, we obtain

$$\rho \bar{u}_j \frac{\partial k}{\partial x_j} = -\rho \overline{u_i' u_j'} \frac{\partial \bar{u}_i}{\partial x_j} - \rho \frac{\partial}{\partial x_j} \left[\overline{u_j' \left(\tfrac{1}{2} u_k' u_k' \right)} \right]$$

$$+ \frac{\partial}{\partial x_j} \left\{ \overline{u_i' \left[-p' \delta_{ij} + \mu \left(\frac{\partial u_i'}{\partial x_j} + \frac{\partial u_j'}{\partial x_i} \right) \right]} \right\} + \rho \overline{u_i' f_i'} - \rho \varepsilon \tag{1.26}$$

where ε is the dissipation of energy per unit mass associated with the *fluctuations*, namely,

$$\varepsilon = \nu \overline{\left(\frac{\partial u_i'}{\partial x_j} + \frac{\partial u_j'}{\partial x_i} \right) \frac{\partial u_i'}{\partial x_j}} \tag{1.27}$$

As above, the first two terms on the right-hand side of this equation are transposed convective terms. Also the next two terms in the equation are the work done by the surface and body forces, but in this case the work is that associated with the unsteady fluctuations. Noting that the first term on the right-hand side is the negative of the first term in Eq. (1.25), it represents energy *transferred* between the mean motion and fluctuations or *converted* from one to the other. When positive in either equation it is an energy *gain* for that motion and "appears" as a *source* term, whereas when negative it is an energy *loss* or a *sink* term. The second transposed term in Eq. (1.26) represents the transport or diffusion of the kinetic energy by the fluctuations.

For constant viscosity, Eq. (1.26) may be written as

$$\rho \bar{u}_j \frac{\partial k}{\partial x_j} = -\rho \overline{u_i' u_j'} \frac{\partial \bar{u}_i}{\partial x_j} + \frac{\partial}{\partial x_j} \left[\mu \frac{\partial k}{\partial x_j} - \rho \overline{u_j' \left(\tfrac{1}{2} u_k' u_k' \right)} - \overline{u_i' p'} \delta_{ij} \right] + \rho \overline{u_i' f_i'} - \rho \hat{\varepsilon} \tag{1.28}$$

where

$$\hat{\varepsilon}=\nu\overline{\frac{\partial u_i'}{\partial x_j}\frac{\partial u_i'}{\partial x_j}} \tag{1.29}$$

The second term now represents the diffusion of energy by viscosity and unsteady convective motion. Although this is one of the most familiar forms of the *k*-equation, the last term on the right-hand side is *not* the viscous dissipation of *k unless* the fluctuations happen to be homogeneous. This can be seen by noting that the second term in Eq. (1.27) for homogeneous fluctuations is

$$\overline{\frac{\partial u_i'}{\partial x_j}\frac{\partial u_j'}{\partial x_i}}=\frac{\partial^2\overline{u_i'u_j'}}{\partial x_j\partial x_i}-\overline{u_j'\frac{\partial^2 u_i'}{\partial x_j\partial x_i}}-\overline{u_i'\frac{\partial^2 u_j'}{\partial x_j\partial x_i}}-\overline{\frac{\partial u_i'}{\partial x_i}\frac{\partial u_j'}{\partial x_j}}=0$$

Reynolds-stress equation. We can obtain an equation for the Reynolds stresses by writing the time-dependent component of the momentum equation, Eq. (1.21), twice; once for the unsteady velocity component u_i' and once for u_j'. Each may then be multiplied by the other unsteady velocity component and added to form one equation for $u_i'u_j'$. Taking the time-average of this result yields, after some effort,

$$\rho\bar{u}_k\frac{\partial\overline{u_i'u_j'}}{\partial x_k}=-\rho\overline{u_i'u_k'}\frac{\partial\bar{u}_j}{\partial x_k}-\rho\overline{u_j'u_k'}\frac{\partial\bar{u}_i}{\partial x_k}-\rho\frac{\partial\overline{u_i'u_j'u_k'}}{\partial x_k}+\frac{\partial}{\partial x_k}\left\{\overline{u_i'\left[-p'\delta_{jk}+\mu\left(\frac{\partial u_j'}{\partial x_k}+\frac{\partial u_k'}{\partial x_j}\right)\right]}\right.$$

$$\left.+\overline{u_j'\left[-p'\delta_{ik}+\mu\left(\frac{\partial u_i'}{\partial x_k}+\frac{\partial u_k'}{\partial x_i}\right)\right]}\right\}+\overline{p'\left(\frac{\partial u_i'}{\partial x_j}+\frac{\partial u_j'}{\partial x_i}\right)}+\rho\left(\overline{u_i'f_j'}+\overline{u_j'f_i'}\right)-\rho\varepsilon_{ij} \tag{1.30}$$

where

$$\varepsilon_{ij}=\nu\left[\overline{\frac{\partial u_i'}{\partial x_k}\left(\frac{\partial u_j'}{\partial x_k}+\frac{\partial u_k'}{\partial x_j}\right)}+\overline{\frac{\partial u_j'}{\partial x_k}\left(\frac{\partial u_i'}{\partial x_k}+\frac{\partial u_k'}{\partial x_i}\right)}\right] \tag{1.31}$$

is the *Reynolds-stress dissipation tensor*. Contracting the above equations by replacing index *j* with *i* yields Eqs. (1.26) and (1.27) with $\varepsilon=\varepsilon_{ii}/2$. Therefore Eqs. (1.26) and (1.30) are not independent. If we consider constant viscosity, the above equations may be written as

$$\rho\bar{u}_k\frac{\partial\overline{u_i'u_j'}}{\partial x_k}=-\rho\overline{u_i'u_k'}\frac{\partial\bar{u}_j}{\partial x_k}-\rho\overline{u_j'u_k'}\frac{\partial\bar{u}_i}{\partial x_k}$$

$$+\frac{\partial}{\partial x_k}\left(\mu\frac{\partial\overline{u_i'u_j'}}{\partial x_k}-\rho\overline{u_k'u_i'u_j'}\right)+\left[\overline{u_i'\left(\rho f_j'-\frac{\partial p'}{\partial x_j}\right)}+\overline{u_j'\left(\rho f_i'-\frac{\partial p'}{\partial x_i}\right)}\right]-\rho\hat{\varepsilon}_{ij} \tag{1.32}$$

where

$$\hat{\varepsilon}_{ij}=2\nu\overline{\frac{\partial u_i'}{\partial x_k}\frac{\partial u_j'}{\partial x_k}} \tag{1.33}$$

In this equation, the first two terms on the right-hand side may be interpreted as the production of Reynolds stresses by the mean-flow strain rate, the next as the viscous and convective diffusion of these stresses, and the fourth as the production by fluctuating body and pressure forces. By contracting the above equations and comparing them with Eqs. (1.28) and (1.29), we obtain $\hat{\varepsilon}=\hat{\varepsilon}_{ii}/2$.

1.3 Equations for Two-Dimensional Flow.

Continuity, momentum, kinetic energy, and Reynolds-stress equations.

The equations in the preceding section simplify considerably for a two-dimensional mean flow. Using conventional notation, we denote the components of the position and velocity vectors by x, y and z, and u, v and w. Furthermore, if we arrange the coordinate system such that $\overline{w} = 0$, then all time-averaged quantities involving an odd power of the fluctuation w' will equal zero, all derivatives of any time-averaged quantity with respect to z will equal zero, and the time-independent components of the continuity and momentum equations, Eqs. (1.18) and (1.20), become

$$\frac{\partial \overline{u}}{\partial x}+\frac{\partial \overline{v}}{\partial y}=0 \tag{1.34}$$

$$\overline{u}\frac{\partial \overline{u}}{\partial x}+\overline{v}\frac{\partial \overline{u}}{\partial y}=-\frac{1}{\rho}\frac{\partial \overline{p}}{\partial x}+\frac{\partial}{\partial x}\left(\nu\frac{\partial \overline{u}}{\partial x}\right)+\frac{\partial}{\partial y}\left(\nu\frac{\partial \overline{u}}{\partial y}\right)-\overline{u'\frac{\partial u'}{\partial x}}-\overline{v'\frac{\partial u'}{\partial y}} \tag{1.35}$$

$$\overline{u}\frac{\partial \overline{v}}{\partial x}+\overline{v}\frac{\partial \overline{v}}{\partial y}=-\frac{1}{\rho}\frac{\partial \overline{p}}{\partial y}+\frac{\partial}{\partial x}\left(\nu\frac{\partial \overline{v}}{\partial x}\right)+\frac{\partial}{\partial y}\left(\nu\frac{\partial \overline{v}}{\partial y}\right)-\overline{u'\frac{\partial v'}{\partial x}}-\overline{v'\frac{\partial v'}{\partial y}} \tag{1.36}$$

where, since they are easily included if needed, we have considered $f_i = 0$. Since the unsteady motion may still be three-dimensional, the kinetic energy equation for the fluctuations, Eq. (1.28), becomes

$$\begin{aligned}\overline{u}\frac{\partial k}{\partial x}+\overline{v}\frac{\partial k}{\partial y}&=-\overline{u'v'}\left(\frac{\partial \overline{u}}{\partial y}+\frac{\partial \overline{v}}{\partial x}\right)-\left(\overline{u'u'}-\overline{v'v'}\right)\frac{\partial \overline{u}}{\partial x}-\frac{1}{\rho}\left(\overline{u'\frac{\partial p'}{\partial x}}+\overline{v'\frac{\partial p'}{\partial y}}\right)\\&+\frac{\partial}{\partial x}\left[\nu\frac{\partial k}{\partial x}-\tfrac{1}{2}\overline{u'\left(u'u'+v'v'+w'w'\right)}\right]+\frac{\partial}{\partial y}\left[\nu\frac{\partial k}{\partial y}-\tfrac{1}{2}\overline{v'\left(u'u'+v'v'+w'w'\right)}\right]-\hat{\varepsilon}\end{aligned} \tag{1.37}$$

where

$$k=\frac{(\overline{u'u'}+\overline{v'v'}+\overline{w'w'})}{2} \quad \text{and} \quad \hat{\varepsilon}=\nu\sum_{u_n'=u',v',w'}\left[\overline{\left(\frac{\partial u_n'}{\partial x}\right)^2}+\overline{\left(\frac{\partial u_n'}{\partial y}\right)^2}+\overline{\left(\frac{\partial u_n'}{\partial z}\right)^2}\right] \tag{1.38}$$

Finally, the equations for the nonzero Reynolds stress components become

$$\begin{aligned}\overline{u}\frac{\partial \overline{u'u'}}{\partial x}+\overline{v}\frac{\partial \overline{u'u'}}{\partial y}&=-2\overline{u'u'}\frac{\partial \overline{u}}{\partial x}-2\overline{u'v'}\frac{\partial \overline{u}}{\partial y}-2\overline{u'\frac{\partial p'/\rho}{\partial x}}\\&+\frac{\partial}{\partial x}\left(\nu\frac{\partial \overline{u'u'}}{\partial x}-\overline{u'u'u'}\right)+\frac{\partial}{\partial y}\left(\nu\frac{\partial \overline{u'u'}}{\partial y}-\overline{v'u'u'}\right)-\hat{\varepsilon}_{11}\end{aligned} \tag{1.39}$$

$$\begin{aligned}\overline{u}\frac{\partial \overline{v'v'}}{\partial x}+\overline{v}\frac{\partial \overline{v'v'}}{\partial y}&=-2\overline{u'v'}\frac{\partial \overline{v}}{\partial x}-2\overline{v'v'}\frac{\partial \overline{v}}{\partial y}-2\overline{v'\frac{\partial p'/\rho}{\partial y}}\\&+\frac{\partial}{\partial x}\left(\nu\frac{\partial \overline{v'v'}}{\partial x}-\overline{u'v'v'}\right)+\frac{\partial}{\partial y}\left(\nu\frac{\partial \overline{v'v'}}{\partial y}-\overline{v'v'v'}\right)-\hat{\varepsilon}_{22}\end{aligned} \tag{1.40}$$

$$\overline{u}\frac{\partial \overline{w'w'}}{\partial x}+\overline{v}\frac{\partial \overline{w'w'}}{\partial y}=-2\overline{w'\frac{\partial p'/\rho}{\partial z}}$$

$$+\frac{\partial}{\partial x}\left(\nu\frac{\partial\overline{w'w'}}{\partial x}-\overline{u'w'w'}\right)+\frac{\partial}{\partial y}\left(\nu\frac{\partial\overline{w'w'}}{\partial y}-\overline{v'w'w'}\right)-\hat{\varepsilon}_{33} \tag{1.41}$$

$$\bar{u}\frac{\partial\overline{u'v'}}{\partial x}+\bar{v}\frac{\partial\overline{u'v'}}{\partial y}=-\overline{u'u'}\frac{\partial\bar{v}}{\partial x}-\overline{v'v'}\frac{\partial\bar{u}}{\partial y}-\overline{v'\frac{\partial p'/\rho}{\partial x}}-\overline{u'\frac{\partial p'/\rho}{\partial y}}$$

$$+\frac{\partial}{\partial x}\left(\nu\frac{\partial\overline{u'v'}}{\partial x}-\overline{u'u'v'}\right)+\frac{\partial}{\partial y}\left(\nu\frac{\partial\overline{u'v'}}{\partial y}-\overline{v'u'v'}\right)-\hat{\varepsilon}_{12} \tag{1.42}$$

where

$$\hat{\varepsilon}_{11}=2\nu\left[\overline{\left(\frac{\partial u'}{\partial x}\right)^2}+\overline{\left(\frac{\partial u'}{\partial y}\right)^2}+\overline{\left(\frac{\partial u'}{\partial z}\right)^2}\right],\quad \hat{\varepsilon}_{12}=2\nu\left[\overline{\frac{\partial u'}{\partial x}\frac{\partial v'}{\partial x}}+\overline{\frac{\partial u'}{\partial y}\frac{\partial v'}{\partial y}}+\overline{\frac{\partial u'}{\partial z}\frac{\partial v'}{\partial z}}\right],\ \text{etc.} \tag{1.43}$$

1.4 Prandtl's Boundary-Layer Theory.

Equations for mean flow. Equations for fluctuating flow. Near-wall behavior.

In 1904, Prandtl observed that all the viscous effects for high Reynolds-number flow about streamlined objects are confined to an extremely thin layer next to the surface. Consequently, he surmised that the solution for the "outer" free-stream flow can be obtained by neglecting this *boundary layer* and solving the inviscid equations of motion (i.e., those obtained by setting μ or ν equal to zero), while the solution for the "inner" viscous flow can be obtained by neglecting the pressure gradient normal to the surface and solving the viscous equations of motion using the inviscid results at the surface for its free-stream boundary condition. As we will momentarily see, this changes the momentum equation for the viscous boundary-layer flow from an elliptic to a parabolic differential equation which is generally easier to solve.

Equations for the mean flow. In this section, we will develop the equations for a two-dimensional boundary layer on a solid surface with negligible curvature.[1] Following the usual convention, we will consider a curvilinear coordinate system attached to the surface with x measured from its leading edge in the direction of the flow, y measured normal to the surface into the flow, and z measured along the surface transverse to the flow. Then, as in the previous section, the corresponding mean-flow velocity components are $\bar{u}(x,y)$ and $\bar{v}(x,y)$ with $\bar{w} = 0$. In addition, the free-stream velocity and the boundary-layer thickness, denoted by u_∞ and δ respectively, will be functions of x only.[2]

Since a reasonable estimate for the angle of the streamline at the edge of the boundary layer relative to the surface is $d\delta/dx$, the quantity $d\delta/dx$ for a "thin" boundary layer must be very much less than unity *everywhere* along the surface. Furthermore, given that this angle is small, the *normal* component of the mean velocity within the boundary layer must be small compared to the *streamwise* component. In fact, we can determine geometrically that the normal component is at most of magnitude $u_\infty d\delta/dx$.

By comparing Eqs. (1.35) and (1.36), we then conclude that the convective and viscous terms in the y-momentum equation are an order of magnitude $d\delta/dx$ smaller than their counterparts in the x-momentum equation. Neglecting these terms reduces the y-momentum equation to

---

[1] If the boundary layer is extremely thin compared to the radius of curvature of the surface, the boundary layer can be treated as though it was developing on a flat surface. See Schlichting [1979] regarding the use of a curvilinear coordinate system.

[2] In this instance, and others where its use cannot be misinterpreted, the unbarred quantity u_∞ refers to the time-averaged free-stream velocity.

$$0=-\frac{1}{\rho}\frac{\partial\bar{p}}{\partial y}-\overline{u'\frac{\partial v'}{\partial x}}-\overline{v'\frac{\partial v'}{\partial y}}$$

Since the magnitude of the unsteady velocity components will usually be all the same order, and since the characteristic streamwise distance associated with a boundary layer of thickness δ, which grows at the rate $d\delta/dx$, is of the magnitude $\delta(d\delta/dx)^{-1}$, we determine that the next-to-the-last term in the above equation is an order of magnitude $d\delta/dx$ smaller than the last and therefore can be neglected. Hence the y momentum equation for a boundary layer reduces to

$$\frac{1}{\rho}\frac{\partial\bar{p}}{\partial y}=\frac{\partial(-\overline{v'v'})}{\partial y}$$

Integrating this expression and evaluating the integration constant at the edge of the boundary layer, where $\bar{p}=\bar{p}_\infty(x)$ and $\overline{v'v'}=\overline{v_\infty'^2}$, we obtain

$$\bar{p}(x,y)=\bar{p}_\infty-\rho\left(\overline{v'v'}-\overline{v_\infty'^2}\right) \tag{1.44}$$

Consequently, since $\overline{v'v'}=0$ at the surface, the mean pressure difference across a boundary layer is $\rho\overline{v_\infty'^2}$, which is small within the boundary layer approximation.

Using Eq. (1.44) to eliminate the pressure in Eq. (1.35), the x-momentum equation becomes

$$\bar{u}\frac{\partial\bar{u}}{\partial x}+\bar{v}\frac{\partial\bar{u}}{\partial y}=-\frac{1}{\rho}\frac{d\bar{p}_\infty}{dx}+\frac{\partial}{\partial x}\left(\nu\frac{\partial\bar{u}}{\partial x}\right)+\frac{\partial}{\partial y}\left(\nu\frac{\partial\bar{u}}{\partial y}\right)-\frac{\partial(\overline{u'u'}-\overline{v'v'}+\overline{v_\infty'^2})}{\partial x}+\frac{\partial(-\overline{u'v'})}{\partial y}$$

Since the first of the viscous terms in this equation is an order of magnitude $(d\delta/dx)^2$ smaller than the second, we can neglect it when $d\delta/dx \ll 1$. Hence, the x-momentum equation for flow within a boundary layer reduces to

$$\bar{u}\frac{\partial\bar{u}}{\partial x}+\bar{v}\frac{\partial\bar{u}}{\partial y}=-\frac{1}{\rho}\frac{d\bar{p}_\infty}{dx}+\frac{\partial}{\partial y}\left(\nu\frac{\partial\bar{u}}{\partial y}\right)-\frac{\partial(\overline{u'u'}-\overline{v'v'}+\overline{v_\infty'^2})}{\partial x}+\frac{\partial(-\overline{u'v'})}{\partial y}$$

Realizing that the convective terms in the above equation must be the same order of magnitude as the viscous term, at least somewhere within the boundary layer, we obtain $Re_\delta(d\delta/dx)$ must be of order unity. Therefore, the condition $d\delta/dx \ll 1$ is equivalent to Prandtl's condition that $Re_\delta \gg 1$.

In summary then, the appropriate equations for the mean flow within a boundary layer are

$$\frac{\partial\bar{u}}{\partial x}+\frac{\partial\bar{v}}{\partial y}=0 \tag{1.45}$$

$$\bar{u}\frac{\partial\bar{u}}{\partial x}+\bar{v}\frac{\partial\bar{u}}{\partial y}=-\frac{1}{\rho}\frac{d\bar{p}_\infty}{dx}+\frac{\partial}{\partial y}\left(\nu\frac{\partial\bar{u}}{\partial y}-\overline{u'v'}\right)-\left\langle\frac{\partial}{\partial x}\left(\overline{u'u'}-\overline{v'v'}+\overline{v_\infty'^2}\right)\right\rangle \tag{1.46}$$

Here, and in the equations that follow, we can neglect the terms enclosed by the angled brackets when $\hat{u}/u_\infty \ll 1$, where $\hat{u}$ is the intensity of the fluctuations within the boundary layer. The Reynolds shear-stress term, on the other hand, cannot be neglected as shown in §1.2 unless $Re_\delta(\hat{u}/u_\infty)^2 \ll 1$.

Expanding the streamwise velocity using a Taylor's series in y, the functional dependence of $\bar{u}$ near the wall is found to be

$$\bar{u}(x,y)=\frac{\bar{\tau}_0}{\mu}y+\frac{1}{2\mu}\frac{d\bar{p}_\infty}{dx}y^2+O(y^4) \tag{1.47}$$

where $\bar{\tau}_0(x)$ is the mean shear stress $\bar{\tau}_{xy}$ evaluated at the wall. The values at the wall for the higher derivatives of $\overline{u}$ in the series can be determined by taking successive derivatives of the momentum equation and evaluating it at the wall. As it turns out, the y^4 term in the above expression is the first term in which the fluctuations appear.

From continuity, we obtain

$$\overline{v}(x,y)=-\frac{1}{2\mu}\frac{d\overline{\tau}_0}{dx}y^2-\frac{1}{6\mu}\frac{d^2\overline{p}_\infty}{dx^2}y^3-O(y^5) \tag{1.48}$$

Hence, the streamwise and normal velocity components near the wall are linear and quadratic functions of y respectively. Where the wall shear stress equals zero, however, $\overline{u}$ becomes quadratic in y, and the flow is either *separating* (leaving the wall) or *reattaching*. When $d\overline{\tau}_0/dx$ is negative, $\overline{v}$ is positive near the wall and the flow is separating, otherwise $\overline{v}$ is negative and the flow is reattaching.

From Eq. (1.47) we also see that the curvature of the velocity profile at the wall is positive for *adverse* pressure gradients (decelerating free streams) and negative for *favorable* gradients (accelerating free streams). With the streamwise velocity asymptotically approaching its free-stream value from below, we can expect the curvature of the profile to be negative near the free stream. Hence we conclude that velocity profiles in an adverse pressure gradient will exhibit an *inflection point*, while those in a favorable pressure gradient will not. According to Eq. (1.47), the inflection point resides at the wall when $d\overline{p}_\infty/dx = 0$, that is, when the free-stream velocity is constant. Much of this can be deduced from a physical argument when we realize that the slower moving fluid near the wall is most affected by a free-stream pressure gradient because its inertia is less than that farther away.

From Eq. (1.46), we can show that the total shear stress near the wall is given by

$$\tau=\mu\frac{\partial\overline{u}}{\partial y}-\overline{u'v'}=\overline{\tau}_0+\frac{d\overline{p}_\infty}{dx}y+\frac{1}{6}\frac{\rho\overline{\tau}_0}{\mu^2}\frac{d\overline{\tau}_0}{dx}y^3+\ldots \tag{1.49}$$

Clearly, the shear stress may either increase or decrease away from the wall depending on the streamwise gradients in pressure and wall shear stress. For flow in a constant width channel, the first of these is a negative constant whereas the second equals zero. In this case the shear stress decreases linearly from the wall. For a boundary layer with a zero pressure gradient, the wall shear stress decreases in the streamwise direction as the boundary layer grows. Consequently, the shear stress decreases away from the wall according to the third power of y.

At the edge of the boundary layer, where the various quantities assume their free stream values and all derivatives with respect to y equal zero, the momentum equation, Eq. (1.46), reduces to

$$\overline{u}_\infty\frac{d\overline{u}_\infty}{dx}=-\frac{1}{\rho}\frac{d\overline{p}_\infty}{dx}-\left\langle\frac{d}{dx}\left(\overline{u_\infty'^2}\right)\right\rangle \tag{1.50}$$

which may then be used to eliminate the free-stream pressure gradient in Eq. (1.46).

Equations for the fluctuating flow. Applying the boundary layer approximation to the kinetic-energy equation and the Reynolds-stress equations, Eqs. (1.37) and (1.39)–(1.42), the kinetic-energy equation for the flow in a boundary layer becomes

$$\overline{u}\frac{\partial k}{\partial x}+\overline{v}\frac{\partial k}{\partial y}=-\overline{u'v'}\frac{\partial\overline{u}}{\partial y}-\frac{1}{\rho}\left(\overline{u'\frac{\partial p'}{\partial x}}+\overline{v'\frac{\partial p'}{\partial y}}\right)+\frac{\partial}{\partial y}\left[\nu\frac{\partial k}{\partial y}-\overline{v'k}\right]-\hat{\varepsilon}-\left\langle\left(\overline{u'u'}-\overline{v'v'}\right)\frac{\partial\overline{u}}{\partial x}\right\rangle \tag{1.51}$$

and the Reynolds-stress equations become

$$\overline{u}\frac{\partial\overline{u'u'}}{\partial x}+\overline{v}\frac{\partial\overline{u'u'}}{\partial y}=-2\overline{u'v'}\frac{\partial\overline{u}}{\partial y}-2\overline{u'\frac{\partial p'/\rho}{\partial x}}$$

$$+\frac{\partial}{\partial y}\left(\nu\frac{\partial\overline{u'u'}}{\partial y}-\overline{v'u'u'}\right)-\hat{\varepsilon}_{11}-\left\langle 2\overline{u'u'}\frac{\partial\bar{u}}{\partial x}+\frac{\partial}{\partial x}\left(\overline{u'u'u'}\right)\right\rangle \tag{1.52}$$

$$\bar{u}\frac{\partial\overline{v'v'}}{\partial x}+\bar{v}\frac{\partial\overline{v'v'}}{\partial y}=-2\overline{v'\frac{\partial p'/\rho}{\partial y}}+\frac{\partial}{\partial y}\left(\nu\frac{\partial\overline{v'v'}}{\partial y}-\overline{v'v'v'}\right)-\hat{\varepsilon}_{22}-\left\langle 2\overline{v'v'}\frac{\partial\bar{v}}{\partial y}+\frac{\partial}{\partial x}\left(\overline{v'u'v'}\right)\right\rangle \tag{1.53}$$

$$\bar{u}\frac{\partial\overline{w'w'}}{\partial x}+\bar{v}\frac{\partial\overline{w'w'}}{\partial y}=-2\overline{w'\frac{\partial p'/\rho}{\partial z}}+\frac{\partial}{\partial y}\left(\nu\frac{\partial\overline{w'w'}}{\partial y}-\overline{v'w'w'}\right)-\hat{\varepsilon}_{33}-\left\langle\frac{\partial}{\partial x}\left(\overline{v'w'w'}\right)\right\rangle \tag{1.54}$$

$$\bar{u}\frac{\partial\overline{u'v'}}{\partial x}+\bar{v}\frac{\partial\overline{u'v'}}{\partial y}=-\overline{v'v'}\frac{\partial\bar{u}}{\partial y}-\overline{v'\frac{\partial p'/\rho}{\partial x}}-\overline{u'\frac{\partial p'/\rho}{\partial y}}$$

$$+\frac{\partial}{\partial y}\left(\nu\frac{\partial\overline{u'v'}}{\partial y}-\overline{v'u'v'}\right)-\hat{\varepsilon}_{12}+\left\langle\frac{\partial}{\partial x}\left(\overline{u'u'v'}\right)\right\rangle \tag{1.55}$$

where the various components of the dissipation reduce to

$$\hat{\varepsilon}_{11}=2\nu\overline{\left(\frac{\partial u'}{\partial y}\right)^2},\quad \hat{\varepsilon}_{22}=2\nu\overline{\left(\frac{\partial v'}{\partial y}\right)^2},\quad \hat{\varepsilon}_{33}=2\nu\overline{\left(\frac{\partial w'}{\partial y}\right)^2}\quad\text{and}\quad \hat{\varepsilon}_{12}=2\nu\overline{\frac{\partial u'}{\partial y}\frac{\partial v'}{\partial y}} \tag{1.56}$$

Finally, Eqs. (1.19) and (1.21) for the unsteady motion in a boundary layer become

$$\frac{\partial u'}{\partial x}+\frac{\partial v'}{\partial y}+\frac{\partial w'}{\partial z}=0 \tag{1.57}$$

$$\frac{\partial u'}{\partial t}+\bar{u}\frac{\partial u'}{\partial x}+\bar{v}\frac{\partial u'}{\partial y}=-\frac{1}{\rho}\frac{\partial p'}{\partial x}+\nu\left(\frac{\partial^2 u'}{\partial x^2}+\frac{\partial^2 u'}{\partial y^2}+\frac{\partial^2 u'}{\partial z^2}\right)$$

$$-u'\frac{\partial\bar{u}}{\partial x}-v'\frac{\partial\bar{u}}{\partial y}-\frac{\partial u'v'}{\partial y}+\frac{\partial\overline{u'v'}}{\partial y}-\left\langle\frac{\partial u'u'}{\partial x}-\frac{\partial\overline{u'u'}}{\partial x}+\frac{\partial u'w'}{\partial z}\right\rangle \tag{1.58}$$

$$\frac{\partial v'}{\partial t}+\bar{u}\frac{\partial v'}{\partial x}+\bar{v}\frac{\partial v'}{\partial y}=-\frac{1}{\rho}\frac{\partial p'}{\partial y}+\nu\left(\frac{\partial^2 v'}{\partial x^2}+\frac{\partial^2 v'}{\partial y^2}+\frac{\partial^2 v'}{\partial z^2}\right)$$

$$-u'\frac{\partial\bar{v}}{\partial x}-v'\frac{\partial\bar{v}}{\partial y}-\frac{\partial v'v'}{\partial y}+\frac{\partial\overline{v'v'}}{\partial y}-\left\langle\frac{\partial u'v'}{\partial x}-\frac{\partial\overline{u'v'}}{\partial x}+\frac{\partial u'w'}{\partial z}\right\rangle \tag{1.59}$$

$$\frac{\partial w'}{\partial t}+\bar{u}\frac{\partial w'}{\partial x}+\bar{v}\frac{\partial w'}{\partial y}=-\frac{1}{\rho}\frac{\partial p'}{\partial z}$$

$$+\nu\left(\frac{\partial^2 w'}{\partial x^2}+\frac{\partial^2 w'}{\partial y^2}+\frac{\partial^2 w'}{\partial z^2}\right)-\frac{\partial v'w'}{\partial y}-\left\langle\frac{\partial u'w'}{\partial x}+\frac{\partial w'w'}{\partial z}\right\rangle \tag{1.60}$$

where the fluctuations are generally three-dimensional even though the mean flow is not.

Expanding the velocity fluctuations u' and w' using a Taylor's series in y, we find their functional dependence near the wall are

$$u'(x,y,z,t)=\frac{(\tau'_{xy})_0}{\mu}y+\frac{1}{2\mu}\left(\frac{\partial p'}{\partial x}\right)_0 y^2+\frac{\rho}{12\mu^2}\left(\frac{\partial \tau'_{xy}}{\partial t}\right)_0 y^3+\cdots \tag{1.61}$$

$$w'(x,y,z,t)=\frac{(\tau'_{yz})_0}{\mu}y+\frac{1}{2\mu}\left(\frac{\partial p'}{\partial z}\right)_0 y^2+\frac{\rho}{12\mu^2}\left(\frac{\partial \tau'_{yz}}{\partial t}\right)_0 y^3+\cdots \tag{1.62}$$

where the subscript zero denotes the value of the quantity at the wall. The third term in each of these was obtained by considering $\partial p'/\partial y=0$ at $y=0$. Then from continuity

$$v'(x,y,z,t)=-\frac{1}{2\mu}\left(\frac{\partial \tau'_{xy}}{\partial x}+\frac{\partial \tau'_{yz}}{\partial z}\right)_0 y^2-\frac{1}{6\mu}\left(\frac{\partial^2 p'}{\partial x^2}+\frac{\partial^2 p'}{\partial z^2}\right)_0 y^3+\cdots \tag{1.63}$$

Hence the near-wall velocity fluctuations parallel and normal to the wall are linear and quadratic in y respectively.

It then follows, without writing the specific values for the coefficients, that the most general relation between the Reynolds stresses and y near the wall take the form

$$\begin{aligned}
\overline{u'u'}&=\overline{a_1a_1}y^2+2\overline{a_1a_2}y^3+\cdots\\
\overline{v'v'}&=\overline{b_2b_2}y^4+\cdots\\
\overline{w'w'}&=\overline{c_1c_1}y^2+2\overline{c_1c_2}y^3+\cdots\\
\overline{u'v'}&=\overline{a_1b_2}y^3+\overline{a_2b_2}y^4+\cdots\\
\overline{u'w'}&=\overline{a_1c_1}y^2+\left(\overline{a_2c_1}+\overline{a_1c_2}\right)y^3+\cdots\\
\overline{v'w'}&=\overline{b_2c_1}y^3+\overline{b_2c_2}y^4+\cdots
\end{aligned} \tag{1.64}$$

where all the quantities on the left-hand-side are functions of x, y and z, and the quantities a_n, b_n and c_n, which are the coefficients of y^n in the expansions for u', v' and w' respectively, are usually functions of x, z and t. From continuity we can show that $b_n = (a_{nx}+c_{nz})/(n+1)$, where the notation a_{nx} stands for the derivative of a_n with respect to x, etc. Although not immediately obvious, it is possible in certain situations that the Reynolds shear stresses involving v' become quartic in y. This follows from the relation between b_n and the other coefficients a_n and c_n.

By taking one half the sum of the normal stress components, we obtain

$$k=\frac{1}{2}\left(\overline{a_1a_1}+\overline{c_1c_1}\right)y^2+\left(\overline{a_1a_2}+\overline{c_1c_2}\right)y^3+\cdots \tag{1.65}$$

In addition, from Eqs. (1.27) and (1.29), we can determine that the dissipation functions near the wall are given by

$$\varepsilon/\nu=\hat{\varepsilon}/\nu+\left(4\overline{b_2b_2}+2\overline{c_1b_{2z}}+\overline{c_{1z}c_{1z}}+\overline{a_{1x}a_{1x}}+2\overline{a_1b_{2x}}+2\overline{a_{1z}c_{1x}}\right)y^2+\cdots \tag{1.66}$$

where

$$\begin{aligned}
\hat{\varepsilon}/\nu=&\left(\overline{a_1a_1}+\overline{c_1c_1}\right)+4\left(\overline{a_1a_2}+\overline{c_1c_2}\right)y\\
&+\left(4\overline{a_2a_2}+4\overline{b_2b_2}+4\overline{c_2c_2}+2\overline{a_{1x}a_{1x}}+2\overline{c_{1x}c_{1x}}+2\overline{a_{1z}a_{1z}}+2\overline{c_{1z}c_{1z}}\right)y^2+\ldots
\end{aligned}$$

Consequently, the difference between ε and $\hat{\varepsilon}$ occurs in the quadratic and higher order terms. Similarly, from either Eq. (1.31) or Eq. (1.43), we obtain

$$\begin{aligned}
\hat{\varepsilon}_{11}/2v &= \overline{a_1 a_1} + 4\overline{a_1 a_2}\,y + \left(3\overline{a_1 a_3} + 4\overline{a_2 a_2} + 3\overline{a_3 a_1} + \overline{a_{1x} a_{1x}} + \overline{a_{1z} a_{1z}}\right) y^2 + \cdots \\
\hat{\varepsilon}_{22}/2v &= 4\overline{b_2 b_2}\,y^2 + 12\overline{b_2 b_3}\,y^3 + \cdots \\
\hat{\varepsilon}_{12}/2v &= 2\overline{a_1 b_2}\,y + \left(3\overline{a_1 b_3} + 4\overline{a_2 b_2}\right) y^2 + \cdots \\
\hat{\varepsilon}_{13}/2v &= \overline{a_1 c_1} + 2\left(\overline{a_1 c_2} + \overline{a_2 c_1}\right) y + \left(3\overline{a_1 c_3} + 4\overline{a_2 c_2} + 3\overline{a_3 c_1} + \overline{a_{1x} c_{1x}} + \overline{a_{1z} c_{1z}}\right) y^2 + \cdots
\end{aligned} \tag{1.67}$$

from which the remaining components may be deduced.

1.5 Other Definitions and Equations.

Vorticity and circulation. Vorticity transport equations. Streamlines and the stream function. Transformed boundary-layer transport equations. An integral-momentum equation. Displacement thickness, momentum thickness, and the skin-friction coefficient.

Vorticity and circulation. By adding and subtracting a common term to the *strain-rate* tensor $\partial u_i/\partial x_j$, it can be written as

$$\frac{\partial u_i}{\partial x_j} = \frac{1}{2}\left(\frac{\partial u_i}{\partial x_j} + \frac{\partial u_j}{\partial x_i}\right) + \frac{1}{2}\left(\frac{\partial u_i}{\partial x_j} - \frac{\partial u_j}{\partial x_i}\right)$$

which is the sum of a *symmetric* and *anti-symmetric* tensor.[1] The symmetric component represents the rate of deformation of a fluid element in the flow, and hence is responsible for the viscous shear stress as shown in Eq. (1.1). The anti-symmetric component represents its angular velocity of rotation, which can be shown by considering the rigid-body rotation of a fluid element.

The *vorticity* of a fluid element is defined as twice the angular velocity of an infinitesimal sphere of fluid in the flow. In tensor form, we can write this as

$$\omega_i = \varepsilon_{ijk}\left(\partial u_k/\partial x_j\right) \tag{1.68}$$

where ε_{ijk} is the *alternating tensor* equal to zero when any two of the indices are the same, and either +1 or −1 depending on whether the indices (i, j, k) are respectively either an even or odd permutation of (1, 2, 3). Flow with vorticity is called "rotational flow," while that with zero vorticity is called "irrotational." Regardless, we can easily show that

$$\frac{\partial \omega_i}{\partial x_i} = 0$$

A line drawn from any point through the flow that is everywhere tangent to the *instantaneous* axis of fluid rotation is called a "vortex line." The differential equations of this line are given by

$$\frac{dx_1}{\omega_1} = \frac{dx_2}{\omega_2} = \frac{dx_3}{\omega_3}$$

In addition, the cylindrical surface formed by all vortex lines passing through a closed curve within the flow is called a "vortex tube." When all the vortex lines in the flow thread though the tube, that is, the flow outside the tube is irrotational, the aggregate is simply called a "vortex."

---

[1] See "Supplement on Cartesian Tensors."

A related quantity is the *circulation*, which is defined as the line integral of the velocity around any closed curve C within the flow. Usually denoted by Γ, it can be determined from

$$\Gamma = \int_C u_i \, dx_i \tag{1.69}$$

Using *Stokes' theorem*,[1] the circulation can also be expressed as $\Gamma = \int_A \omega_i \, dA_i$. Thus, circulation equals the *strength* of all the vortex lines threading through a vortex tube, or simply, the strength of the vortex if it is isolated in the flow.

Vorticity transport equations. An equation for the vorticity can be obtained by applying the operator $\varepsilon_{ijk}(\partial/\partial x_j)$ to the momentum equation, Eq. (1.4), after writing the equation for the velocity u_k. For constant viscosity, this yields

$$\frac{\partial \omega_i}{\partial t} + u_j \frac{\partial \omega_i}{\partial x_j} = \omega_j \frac{\partial u_i}{\partial x_j} + \nu \frac{\partial^2 \omega_i}{\partial x_j^2}$$

The first term on the right-hand side, known as the *vortex-stretching term*, represents the increase or decrease of vorticity caused by the local straining of the flow, while the second term represents the viscous diffusion of vorticity. By setting ν equal to zero, *Helmholtz' equation* for inviscid flow is obtained, which leads to several well-known laws about vorticity.[2] One of them, known as the *Kelvin-Helmholtz law*, states that when the initial vorticity is zero throughout the flow, it will remain zero. Another is that vortex lines cannot begin nor end within the fluid. That is, they must either form closed curves or begin and end at the fluid's boundaries.

Introducing the instantaneous velocity and vorticity as the sum of their time-averaged and fluctuating components, we can obtain an instantaneous vorticity transport equation. By taking its average, we obtain the time-averaged component of the transport equation, namely,

$$\bar{u}_j \frac{\partial \bar{\omega}_i}{\partial x_j} = \bar{\omega}_j \frac{\partial \bar{u}_i}{\partial x_j} + \overline{\omega'_j \frac{\partial u'_i}{\partial x_j}} + \frac{\partial}{\partial x_j}\left(\nu \frac{\partial \bar{\omega}_i}{\partial x_j} - \overline{u'_j \omega'_i} \right)$$

and by subtraction, we obtain its time-dependent companion

$$\begin{aligned} \frac{\partial \omega'_i}{\partial t} + \bar{u}_j \frac{\partial \omega'_i}{\partial x_j} &= \bar{\omega}_j \frac{\partial u'_i}{\partial x_j} + \omega'_j \frac{\partial \bar{u}_i}{\partial x_j} \\ &\quad + \omega'_j \frac{\partial u'_i}{\partial x_j} - \overline{\omega'_j \frac{\partial u'_i}{\partial x_j}} + \nu \frac{\partial^2 \omega'_i}{\partial x_j^2} - u'_j \frac{\partial \bar{\omega}_i}{\partial x_j} - u'_j \frac{\partial \omega'_i}{\partial x_j} + \overline{u'_j \frac{\partial \omega'_i}{\partial x_j}} \end{aligned} \tag{1.70}$$

This last equation, or rather a linearized two-dimensional version of it, is the starting point for most laminar-flow stability analyses.

Streamlines and the stream function. A line drawn from any point through the flow that is everywhere tangent to the *instantaneous* velocity of the fluid is called a "streamline." The differential equations of this line is given by

$$\frac{dx_1}{u_1} = \frac{dx_2}{u_2} = \frac{dx_3}{u_3}$$

Analogous to vortex tubes, the cylindrical surface formed by all streamlines passing through a closed curve within the flow is called a "stream tube." In this case, if A is the area bounded by a closed curve,

---

[1] Ibid.

[2] See Lamb [1945].

the volumetric flow rate through the stream tube is $Q=\int_A u_i\, dA_i$. Since no flow crosses the surface of a stream tube, the flow rate through the tube at any instant equals Q everywhere along its length.

For a two-dimensional mean flow, the continuity equation, Eq. (1.34), is identically satisfied by introducing Lagrange's[1] *stream function* $\psi(x, y, t)$ defined by

$$\frac{\partial\psi}{\partial x}=-\bar{v} \quad \text{and} \quad \frac{\partial\psi}{\partial y}=\bar{u} \tag{1.71}$$

Since $d\psi=-\bar{v}dx+\bar{u}dy$, curves of constant ψ are streamlines. In addition, since the flow rate between any two points in the flow is $Q=\int_1^2 u_i n_i\, ds$ per unit distance in the z direction, where s is the distance along any curve connecting points 1 and 2, and n_i is the outward normal to the curve, we obtain $Q=\int_1^2 d\psi=\psi_2-\psi_1$. Of course, since no flow crosses the streamlines ψ_1 and ψ_2, the flow rate between them equals Q everywhere along their length.

Also for two-dimensional flow, the only component of vorticity is in the z-direction. From Eq. (1.68), it is given by

$$\bar{\zeta}=\frac{\partial\bar{v}}{\partial x}-\frac{\partial\bar{u}}{\partial y} \tag{1.72}$$

In terms of the stream function, we can easily show that this component equals $-\nabla^2\psi$, where ∇^2 is the Laplace operator.[2]

Transformed boundary-layer transport equations. It just so happens that the mean velocity component $\bar{v}$ can be eliminated from the boundary-layer equations by using the coordinate system $\xi=x$ and $\eta=\psi$ instead of x and y. Known as *von Mises' [1927] transformation*, we then have

$$\frac{\partial\bar{u}}{\partial x}=\frac{\partial\bar{u}}{\partial\xi}-\bar{v}\frac{\partial\bar{u}}{\partial\psi} \quad \text{and} \quad \frac{\partial\bar{u}}{\partial y}=\bar{u}\frac{\partial\bar{u}}{\partial\psi}$$

Consequently, from Eq. (1.46) we obtain

$$\bar{u}\frac{\partial\bar{u}}{\partial\xi}=-\frac{1}{\rho}\frac{dp_\infty}{d\xi}+\bar{u}\frac{\partial}{\partial\psi}\left(\nu\bar{u}\frac{\partial\bar{u}}{\partial\psi}-\overline{u'v'}\right)$$

Once the mean velocity is obtained, the physical distance y can then be determined from

$$y=\int\frac{d\psi}{\bar{u}}$$

Hence, von Mises' transformation changes the x-momentum equation into a diffusion equation with a variable diffusivity and source term. This becomes clearer after rearranging it into the form

$$\frac{\partial\bar{u}}{\partial\xi}=\frac{\partial}{\partial\psi}\left(\nu\bar{u}\frac{\partial\bar{u}}{\partial\psi}\right)+\left(-\frac{1}{\rho\bar{u}}\frac{dp_\infty}{d\xi}-\frac{\partial\overline{u'v'}}{\partial\psi}\right)$$

Since all of the previous boundary-layer transport equations can be cast into this form, the transport equation for any quantity $\phi(x, y)$ may be transformed into an equation for $\phi(\xi, \psi)$ having the form

$$\frac{\partial\phi}{\partial\xi}=\frac{\partial}{\partial\psi}\left(\nu\bar{u}\frac{\partial\phi}{\partial\psi}\right)+S_\phi \tag{1.73}$$

---

[1] Lagrange [1781].

[2] See "Supplement on Cartesian Tensors."

where S_ϕ is the appropriate source term. Most computational *boundary-layer programs* solve some version of this equation.

An integral momentum equation. Following Kármán [1921], the boundary-layer equations for two-dimensional flow can be integrated with respect to the distance away from the wall. Integrating Eq. (1.45) provides,

$$\int_0^y \frac{\partial \overline{u}}{\partial x} dy + \int_0^y \frac{\partial \overline{v}}{\partial y} dy = 0$$

which for an impervious wall yields

$$\overline{v} = -\int_0^y \frac{\partial \overline{u}}{\partial x} dy \tag{1.74}$$

After substituting this and Eq. (1.50) into the momentum equation and integrating, we obtain

$$\int_0^\infty \left(\overline{u}\frac{\partial \overline{u}}{\partial x} - \frac{\partial \overline{u}}{\partial y}\int_0^y \frac{\partial \overline{u}}{\partial x} dy - \overline{u}_\infty \frac{d\overline{u}_\infty}{dx} \right) dy = \int_0^\infty \frac{\partial}{\partial y}\left(\frac{\mu}{\rho}\frac{\partial \overline{u}}{\partial y} - \overline{u'v'} \right) dy$$

Integrating the second term of the integrand by parts and assuming the Reynolds shear stress is zero in the free stream, yields

$$\int_0^\infty \left(2\overline{u}\frac{\partial \overline{u}}{\partial x} - \overline{u}_\infty \frac{d\overline{u}}{dx} - \overline{u}_\infty \frac{d\overline{u}_\infty}{dx} \right) dy = -\frac{\overline{\tau}_0}{\rho}$$

which can be rearranged into the form

$$\frac{d}{dx}\int_0^\infty \overline{u}\left(\overline{u}_\infty - \overline{u}\right) dy + \frac{d\overline{u}_\infty}{dx}\int_0^\infty \left(\overline{u}_\infty - \overline{u}\right) dy = \frac{\overline{\tau}_0}{\rho}$$

This equation can be simplified by introducing the *displacement* and *momentum thicknesses*. Defined respectively by

$$\delta_1 = \int_0^\infty \left(1 - \frac{\overline{u}}{\overline{u}_\infty} \right) dy \quad \text{and} \quad \delta_2 = \int_0^\infty \left(\frac{\overline{u}}{\overline{u}_\infty} \right)\left(1 - \frac{\overline{u}}{\overline{u}_\infty} \right) dy \tag{1.75}$$

they represent the deficits of mass-flow and momentum (per unit mass) caused by the boundary layer relative to that for a fluid moving at the free-stream velocity. Unlike the boundary-layer thickness δ, which is a subjectively defined quantity, the displacement and momentum thicknesses are two well-defined quantities of the boundary layer. The ratio of these thicknesses, taken as $H = \delta_1/\delta_2$, is known as the *shape factor*. Here it can be deduced from the definitions of δ_1 and δ_2 that a "fuller" velocity profile corresponds to a value of H closer to unity. Therefore, since turbulence is more effective in transferring momentum toward the wall than viscosity, we expect H to be smaller for a turbulent than a laminar profile. The shape factors for laminar flow vary from about 2.3 at a stagnation point to 3.5 at separation, while those for turbulent flow vary from about 1.3 to 2.5.

Substituting the definitions given in Eq. (1.75) into the previous integral equation, we obtain

$$\frac{d}{dx}\left(\overline{u}_\infty^2 \delta_2 \right) + \overline{u}_\infty \delta_1 \frac{d\overline{u}_\infty}{dx} = \frac{\overline{\tau}_0}{\rho} \tag{1.76}$$

This is the *integral-momentum equation* for a two-dimensional boundary layer as first presented by Gruschwitz [1931]. Developed originally by Kármán for flow with a zero pressure gradient, and thus known as *von Kármán's integral equation*, it was used extensively in the mid 1900s to calculate both laminar and turbulent boundary-layer flow.[1]

For a constant free-stream velocity, the above equation reduces to

$$\overline{u}_\infty^2 \frac{d\delta_2}{dx} = \frac{\overline{\tau}_0}{\rho}$$

If we introduce the skin-friction coefficient, defined by

$$\overline{c}_f = \overline{\tau}_0 \Big/ \tfrac{1}{2}\rho \overline{u}_\infty^2 \tag{1.77}$$

this equation takes the memorable form

$$\frac{\overline{c}_f}{2} = \frac{d\delta_2}{dx} \tag{1.78}$$

which directly relates the skin-friction coefficient to the rate of change of the momentum thickness. In addition, when the free-stream velocity is constant, Eq. (1.74) yields

$$\frac{\overline{v}_\infty}{\overline{u}_\infty} = \frac{d\delta_1}{dx} \tag{1.79}$$

Since the left-hand side is the angle of the streamlines in the free stream according to the boundary-layer approximation, we see that an effect of the boundary layer is to "displace" the free-stream flow away from the wall in direct proportion to the increase in the displacement thickness.

References

Gruschwitz, H., 1931, "Die turbulente Reibungsschicht in ebener Strömung bei Druckfall und Druckanstieg," Ing.-Arch., 2, pp. 321–346.

Hinze, J.O., 1975, Turbulence, McGraw-Hill, New York.

Kármán, Th. von, 1921, "Über laminare und turbulente Reibung," ZAMM 1, pp. 233–253; also NACA TM 1092 (in English).

Lagrange, J.L., 1781, "Mémoire sur la Théorie du Mouvement des Fluides," Nouv. mém. de l'Acad. de Berlin, Oeuvres, iv, p. 720.

Lamb, H., 1945, Hydrodynamics, Dover Publications, New York.

Mises, R. von, 1927, "Bemerkungen zur Hydrodynamik," ZAMM, 7, pp. 425–431.

Prandtl, L., 1904, "Über Flüssigkeitsbewegung bei sehr kleiner Reibung," Proc. 3rd Intern. Math. Congress, Heidelberg; also NACA TM 452 (in English).

Reynolds, O., 1895, "On the Dynamical Theory of Incompressible Viscous Fluids and the Determination of the Criterion," Phil. Trans. Roy. Soc. A, 186, pp. 123–164.

Rotta, J.C., 1972, Turbulente Strömungen, B.G. Teubner, Stuttgart (in German).

Schlichting, H., 1979, Boundary-Layer Theory, McGraw-Hill, New York.

Stokes, G.G., 1845, "On the Theories of Internal Friction of Fluids in Motion," Trans. Cambridge Phil. Soc., 8, pp. 287–305.

[1] See for example Schlichting [1979].

2 Laminar Boundary Layers

A steady laminar boundary layer. Unsteady laminar boundary layers without a mean flow. Unsteady laminar boundary layers with a mean flow.

In order to understand pre-transitional flow, one must understand unsteady laminar flow, particularly the effects of unsteady traveling waves. Over the last century, since Prandtl first observed a "boundary layer" and developed the appropriate equations to describe it, interest in laminar boundary layers has waxed and waned. Beginning with Blasius' [1908] solution for a two-dimensional boundary layer on a flat plate with a zero pressure gradient, analytical and quasi-analytical solutions have been obtained for evermore complex laminar flows. This trend continued until the 1960s when computational boundary-layer programs were developed.

Today, because they can be calculated routinely using numerical flow solvers, laminar boundary layers have almost been forgotten. In this chapter we will review the fundamental solutions of steady and unsteady laminar boundary-layer theory, and obtain new solutions for laminar flow with free-stream traveling waves. Since the presentation is neither extensive nor detailed, readers interested in laminar boundary-layer theory are referred to the books by Rosenhead [1963] and Schlichting [1979] for an in-depth treatment of the subject.

Some of the important results in this chapter are:

1. The diffusion thickness for a laminar *steady* flow is $\surd(\nu x/u_\infty)$.
2. The two diffusion thicknesses for a laminar *unsteady* flow are $\surd(\nu x/u_\infty)$ and $\surd(\nu/\omega)$, where ω is the frequency of the fluctuation.
3. The effect of free-stream *standing* and *traveling* waves on a boundary layer are significantly different.
4. High frequency waves traveling *faster* than the free-stream velocity have a phase-locked component in the outer part of the boundary layer, while those traveling *slower* have a large amplified component near the wall.

Beside the references at the end of this chapter, other relevant articles may be found in the Bibliography.

2.1 A Steady Boundary Layer – Blasius Solution.

A similarity solution. Blasius' equation. Near-wall behavior. The skin-friction coefficient. Boundary-layer thicknesses.

For many reasons, Blasius' solution for a laminar boundary layer on a flat plate is *the* fundamental solution in laminar boundary-layer theory. In this section, we will examine his solution to reacquaint ourselves with his results and introduce the various boundary-layer parameters.

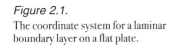

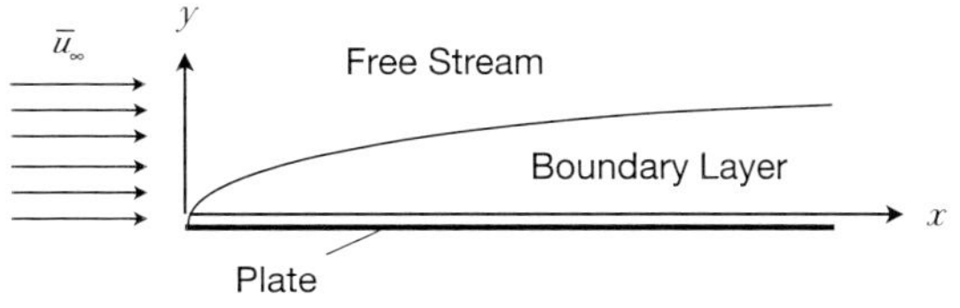

Figure 2.1.
The coordinate system for a laminar boundary layer on a flat plate.

Consider a two-dimensional laminar flow over a flat plate aligned parallel to a steady stream. With x

measured along the plate from its leading edge and y measured normal to the plate as shown in Fig. 2.1, the corresponding mean-flow velocity components are $\bar{u}$ and $\bar{v}$. For a constant free-stream velocity $\bar{u}_\infty$, the free-stream pressure gradient according to Eq. (1.50) equals zero. Then the continuity and momentum equations, Eqs. (1.45) and (1.46), reduce to

$$\frac{\partial \bar{u}}{\partial x}+\frac{\partial \bar{v}}{\partial y}=0 \tag{2.1}$$

and

$$\bar{u}\frac{\partial \bar{u}}{\partial x}+\bar{v}\frac{\partial \bar{u}}{\partial y}=\nu\frac{\partial^2 \bar{u}}{\partial y^2} \tag{2.2}$$

with the boundary conditions $\bar{u}=\bar{v}=0$ at $y=0$, and $\bar{u}=\bar{u}_\infty$ at $y=\infty$ and $x=0$.

Since the geometry is completely independent of the scale at which it is viewed and the free-stream velocity is constant, the velocity profiles must be similar at every streamwise position; that is, they may be reduced to one profile when the distance normal to the plate is expressed in terms of an appropriate "similarity" variable. The scale for the streamwise velocity is clearly $\bar{u}_\infty$. Multiplying and dividing Eq. (2.2) through by this velocity squared and forming dimensionless velocities using $\bar{u}_\infty$ renders the coefficient on the right-hand side of the equation equal to $\nu/\bar{u}_\infty$, which has the dimension of *length*. Hence each term in the equation now has the dimension 1/*length*, with the ratio of the first-to-third taking the form $\bar{u}_\infty(\Delta y)^2/\nu\Delta x$. It then follows that the similarity variable must have the form

$$\eta=y\Big/\sqrt{\nu x/\bar{u}_\infty} \tag{2.3}$$

The continuity equation is immediately satisfied by introducing a stream function as defined in §1.5, namely,

$$\bar{u}=\partial\psi/\partial y \quad \text{and} \quad \bar{v}=-\partial\psi/\partial x$$

For a similar velocity profile, that is, where $\bar{u}/\bar{u}_\infty$ is a function of the similarity variable η only, the stream function, which has the dimensions *length*^{2}/*time*, must have the form

$$\psi(x,y)=f(\eta)\sqrt{\nu\bar{u}_\infty x}$$

where $f(\eta)$ is a dimensionless stream function. Then

$$\bar{u}=\partial\psi/\partial y=\bar{u}_\infty f' \quad \text{and} \quad \bar{v}=-\partial\psi/\partial x=\tfrac{1}{2}\bar{u}_\infty Re_x^{-1/2}\left(\eta f'-f\right) \tag{2.4}$$

where the primes denote differentiation with respect to η, and $Re_x=\bar{u}_\infty x/\nu$ is the distance Reynolds number. Substituting these expressions into Eq. (2.2) we obtain

$$ff''+2f'''=0 \tag{2.5}$$

which is known as *Blasius' equation*.

The boundary conditions on $f(\eta)$ follow directly from those on $\bar{u}$ and $\bar{v}$. Noting that $\eta\rightarrow\infty$ as either $y\rightarrow\infty$ or $x\rightarrow 0$, they are $f(0)=0, f'(0)=0$ and $f'(\infty)=1$.

Blasius obtained his solution by matching a series solution in terms of η near the wall (small η's) to an *asymptotic* series solution in terms of $1/\eta$ far away from the wall (large η's). Today, the solution to Eq. (2.5) can "instantly" be obtained using standard numerical algorithms. Some of the results near the wall and at the "edge" of the boundary layer are presented in Table 2.1.[1] From this table it is easily deduced that the solution near the wall takes the form

[1] See Appendix A for a complete table.

$$f(\eta)=\frac{1}{2}(0.33206)\eta^2-\ldots \tag{2.6}$$

whereas that near the free stream is given by

$$f(\eta)=\eta-1.72076 \tag{2.7}$$

Since the stream function for a uniform flow over the wall is $f=\eta$, the effect of the boundary layer on the free stream is to reduce the stream function there by the constant amount of about 1.72. This effect may also be viewed as a *displacement* of the free stream *away* from the wall by the boundary layer.

Table 2.1

Results of Blasius' solution for small and large η.

η	f	f'	f''	η	f	f'	f''
0	0	0	0.33206	...	...	...	...
0.2	0.00664	0.06641	0.33199	7.2	5.47925	0.99996	0.00013
0.4	0.02656	0.13276	0.33147	7.4	5.67924	0.99998	0.00007
0.6	0.05974	0.19894	0.33008	7.6	5.87924	0.99999	0.00004
0.8	0.10611	0.26471	0.32739	7.8	6.07924	1	0.00002
1.0	0.14557	0.32978	0.32301	8.0	6.27924	1	0.00001
...	...	...	...	8.8	7.07923	1	0

The velocity components $\bar{u}$ and $\bar{v}$ can be obtained directly from Eq. (2.4) once the solution for $f(\eta)$ has been obtained. Their profiles are plotted in Fig. 2.2. Near the wall, the streamwise and normal velocity components vary linearly and quadratically in η as they should according to Eqs. (1.47) and (1.48). From Eqs. (2.4) and (2.6), they are given in terms of x and y by

$$\bar{u}(x,y)=0.332\frac{\bar{u}_\infty^2}{\nu}Re_x^{-1/2}y-\ldots \quad \text{and} \quad \bar{v}(x,y)=0.083\frac{\bar{u}_\infty^3}{\nu^2}Re_x^{-3/2}y^2+\ldots \tag{2.8}$$

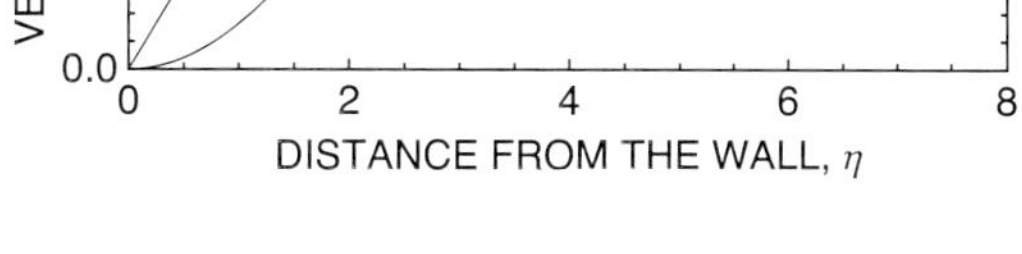

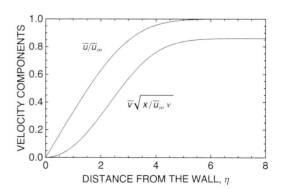

Figure 2.2.
Streamwise and normal velocity profiles for Blasius' laminar flow on a flat plate with $dp/dx=0$.

The surface shear stress follows directly from

$$\overline{\tau}_0(x) = \mu\left(\partial\overline{u}/\partial y\right)_{y=0} = 0.332\rho\overline{u}_\infty^2 Re_x^{-1/2} \tag{2.9}$$

Then by definition, the local skin-friction coefficient is given by

$$\overline{c}_f = \overline{\tau}_0 \big/ \tfrac{1}{2}\rho\overline{u}_\infty^2 = 0.664 Re_x^{-1/2} \tag{2.10}$$

Near the free stream, $\overline{u}$ approaches its free-stream value $\overline{u}_\infty$, while from Eq. (2.4) and the numbers given for large η in Table 2.1, $\overline{v}$ approaches the value

$$\overline{v}_\infty = 0.86\overline{u}_\infty Re_x^{-1/2} \tag{2.11}$$

According to Fig. 2.2, the "edge" of the boundary-layer appears to be about where η equals five.[1] This may be expressed as

$$\delta/x \approx 5 Re_x^{-1/2} \quad \text{or} \quad Re_\delta \approx 5\sqrt{Re_x} \tag{2.12}$$

Clearly, the boundary-layer thickness increases with an increase in viscosity and distance, and decreases with an increase in the free-stream velocity as physically expected. Since $\delta/x \approx 5/\sqrt{Re_x}$, it is also clear that this ratio is much less than unity when the Reynolds number is very large. To estimate the lowest value for which it is valid, recall from §1.4 that the ratio of the viscous normal force to the viscous shear force in the momentum equation is proportional to the quantity $(\delta/x)^2$, which approximately equals $25/Re_x$. Therefore, Prandtl's boundary-layer equation, and subsequently Blasius' solution, is valid when Re_x is much greater than about 25, or let's say when $Re_x \geq 25^2 = 625$. Then the viscous normal-to-shear force ratio at this limit will be four percent, the thickness of the boundary layer will be about one-fifth of the distance downstream from the leading edge, and the angle of the streamline near the edge of the boundary layer, obtained by using Eq. (2.11), will be about two degrees. Also, the boundary-layer-thickness Reynolds number will be $Re_\delta \approx 125$.

Rather than using the boundary-layer thickness δ to scale the distance y from the wall, except for making estimates, it is best to use the displacement or momentum thicknesses as defined in Eq. (1.75). In this case, they are most easily obtained by integrating Eqs. (1.78) and (1.79) with respect to x after substituting Eqs. (2.10) and (2.11). This yields

$$Re_{\delta 1} = 1.721\sqrt{Re_x} \quad \text{and} \quad Re_{\delta 2} = 0.664\sqrt{Re_x} \tag{2.13}$$

where it has been presumed that both thicknesses equal zero at the leading edge of the plate. Note that the first expression could also have been obtained from Eq. (2.7) by setting $f = 0$ and $y = \delta_1$. The shape factor for this flow is given by $H = \delta_1/\delta_2 = 2.59$. At $Re_x = 625$, our presumed lower limit for the boundary-layer approximation, the values of the displacement- and momentum-thickness Reynolds numbers are $Re_{\delta 1} = 43$ and $Re_{\delta 2} = 17$.

Finally, note that the velocity can also be scaled using

$$u_\tau = \sqrt{\overline{\tau}_0/\rho} = \overline{u}_\infty\sqrt{\overline{c}_f/2} \tag{2.14}$$

Known as the *friction velocity*, it is usually associated with turbulent flow. Its corresponding length scale, known as the *friction* or *shear thickness*, is

$$\delta_\tau = \nu/u_\tau = \left(\nu/\overline{u}_\infty\right)\sqrt{2/\overline{c}_f}$$

Note by definition that the Reynolds number based on the friction thickness and velocity equals one, while that based on the friction thickness and the free-stream velocity is given by

---

[1] While this is the most widely accepted and easily remembered estimate for Blasius' flow, the boundary-layer thickness is usually considered to be where the velocity equals ninety-nine percent of its free-stream value.

$$Re_{\delta\tau} = 1.32\, Re_x^{1/4}$$

In contrast to all the other boundary-layer-thickness Reynolds numbers, this one varies according to the one-quarter rather than the one-half power of the distance Reynolds number.

2.2 Unsteady Boundary Layers without a Mean Flow.

Shear- and pressure-driven oscillations.

Forced oscillations in a laminar boundary layer are *driven* by either unsteady shear or pressure forces. Unlike boundary layers in steady flow, where the vorticity generated at the wall convects downstream as it diffuses away from the wall, the vorticity generated at the wall by these unsteady forces diffuses outward as *transverse* or *shear waves* to form an unsteady boundary layer. In this section we will examine the motion of a fluid near a wall responding to oscillating shear and pressure forces with $\overline{u} = \overline{v} = 0$.

Shear-driven oscillations. One of the most basic problems in unsteady laminar flow is the two-dimensional motion caused by a plate oscillating parallel to itself in a quiescent fluid. Known as *Stokes' [1851] second problem*, consider a plate of infinite extent at $y = 0$ oscillating harmonically in the x-direction with a velocity given by $u'(x, 0, t) = u_0'(t) = \hat{u} \cos \omega t$, which is the real part of

$$u_0'(t) = \hat{u}\, \exp(i\omega t)$$

where ω is its angular frequency. Expecting the fluid to oscillate in lamina parallel to the plate, v' equals zero and only the time-dependent momentum equation in the x direction, Eq. (1.58), needs to be solved. In this case, it reduces to

$$\frac{\partial u'}{\partial t} = \nu \frac{\partial^2 u'}{\partial y^2} \tag{2.15}$$

The solution to this equation satisfying the boundary conditions $u'(0, t) = u_0'(t)$ and $u'(\infty, t) = 0$, is quickly determined to be

$$u'(y,t) = \hat{u}\, \exp\left[i\omega t - (1+i)\, y/\delta_\omega \right]$$

where the quantity

$$\delta_\omega = \sqrt{2\nu/\omega} \tag{2.16}$$

is best interpreted as an *unsteady-diffusion thickness*. Taking the real part of the solution, we obtain

$$u'(y,t) = \hat{u}\, e^{-y/\delta_\omega} \cos\left(\omega t - y/\delta_\omega\right) \tag{2.17}$$

which describes exponentially decaying shear waves with a wavelength $2\pi\delta_\omega$ propagating outwards from the plate with a celerity $\omega\delta_\omega = \sqrt{(2\omega\nu)}$. Clearly, these waves are the *direct* response to the oscillating *shear force* at the plate.

The solution given by Eq. (2.17) is plotted in Fig. 2.3. Shown for various instants in the *accelerating* portion of the plate's cycle, they also correspond to various negative positions in the plate's cycle relative to its mean position. Profiles for the decelerating portion of the cycle at $\omega t = \pi/4$, $\pi/2$ and $3\pi/4$ can be obtained by respectively flipping the $-3\pi/4$, $-\pi/2$ and $-\pi/4$ curves about the horizontal axis. Regardless, almost all the motion associated with this flow is clearly confined to a layer about $2\delta_\omega$ thick. In fact, the amplitude of the oscillation one wave length away, which is slightly larger than the extent of the abscissa of the figure, is about $1/500$ that at the plate.

The variation of shear stress at the plate, determined from Eq. (2.17), is

$$\tau_0'(t) = \mu\left(\partial u'/\partial y\right)_{y=0} = \sqrt{2}\left(\mu\hat{u}/\delta_\omega\right)\sin\left(\omega t - \pi/4\right)$$

which *lags* the position of the plate by forty-five degrees. Recalling the definition of δ_ω, the amplitude of this stress is found to increase with an increase in viscosity, frequency, and amplitude of the plate's velocity.

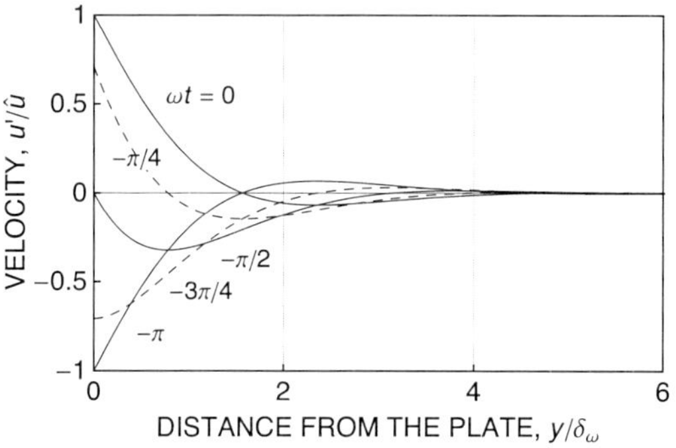

Figure 2.3.
Velocity profiles in a fluid bounded by a plate oscillating parallel to itself. Shown at various instances through one-half of the cycle, profiles for $\omega t = \pi/4$, $\pi/2$ and $3\pi/4$ can be obtained by respectively flipping the $-3\pi/4$, $-\pi/2$ and $-\pi/4$ curves about the horizontal axis.

Pressure-driven oscillations. Now consider the two-dimensional motion of a fluid bounded by a stationary wall at $y = 0$ caused by the fluid far away from the wall oscillating as a *standing wave* with the velocity in the x direction given by the real part of

$$u_\infty'(t) = \hat{u}\ \exp(i\omega t)$$

Again, expecting the fluid to oscillate in lamina parallel to the wall, v' equals zero and u' can be determined from Eq. (1.58) as in the previous case, but now the pressure gradient varies with t since the free stream oscillates. Evaluating Eq. (1.58) in the free stream, we obtain

$$-\frac{1}{\rho}\frac{\partial p'}{\partial x} = -\frac{1}{\rho}\frac{\partial p_\infty'}{\partial x} = \frac{\partial u_\infty'}{\partial t} \tag{2.18}$$

Then the equation for $u'(y, t)$ becomes

$$\frac{\partial u'}{\partial t} = \frac{\partial u_\infty'}{\partial t} + \nu\frac{\partial^2 u'}{\partial y^2} \tag{2.19}$$

The solution to this equation satisfying the boundary conditions $u'(0, t) = 0$ and $u'(\infty, t) = u_\infty'(t)$, is easily obtained by introducing the dependent variable $u_\infty' - u'$ and noting that it satisfies Eq. (2.15) together with its boundary conditions. Thus the solution to Eq. (2.19) is

$$u'(y,t) = \hat{u}\exp(i\omega t)\left\{1 - \exp\left[-\left(1+i\right)y/\delta_\omega\right]\right\}$$

which upon taking the real part yields

$$u'(y,t) = \hat{u}\left[\cos(\omega t) - e^{-y/\delta_\omega}\cos\left(\omega t - y/\delta_\omega\right)\right] \tag{2.20}$$

As we might expect, this result is the sum of pressure waves (the first term in brackets) and outward propagating shear waves (the second term). The shear waves have the same characteristics as before and are again caused by an oscillating shear force at the wall, except that they now happen to be an *indirect* response to the oscillating *pressure* force acting throughout the fluid.

The expression given in Eq. (2.20) is plotted in Fig. 2.4 for various instants in the accelerating portion of the free-stream velocity. They also correspond to that part of the cycle where the pressure force acts in the positive x-direction. In this case, the flow near the wall *leads* that farther away because its inertia is smaller and hence it accelerates faster than the fluid in the free stream. As a result, the wall shear "anticipates" or *leads* the oscillation farther away from the wall. Indeed, the variation of the wall shear stress is given by

$$\tau_0'(t) = \sqrt{2}\left(\mu\hat{u}/\delta_\omega\right)\cos\left(\omega t + \pi/4\right)$$

which clearly shows a near-wall phase lead of forty-five degrees. While leading the oscillation in velocity, it can be shown that the wall shear stress *lags* the pressure force by forty-five degrees.

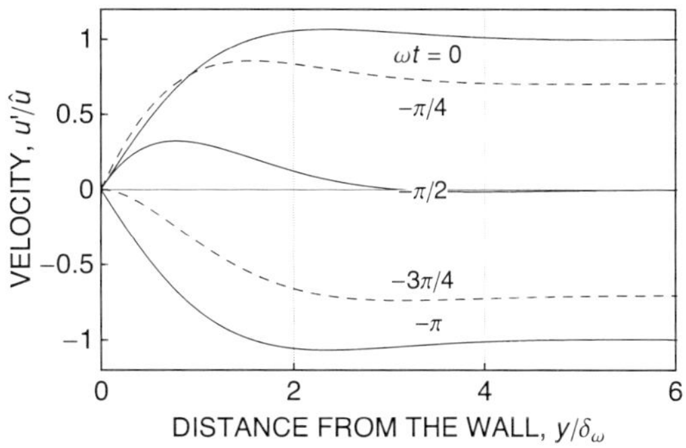

Figure 2.4.
Velocity profiles in a fluid bounded by a stationary wall with a free stream oscillating parallel to the wall. Shown at various instances through one-half of the cycle, profiles for $\omega t = \pi/4, \pi/2$ and $3\pi/4$ can be obtained by respectively flipping the $-3\pi/4, -\pi/2$ and $-\pi/4$ curves about the horizontal axis.

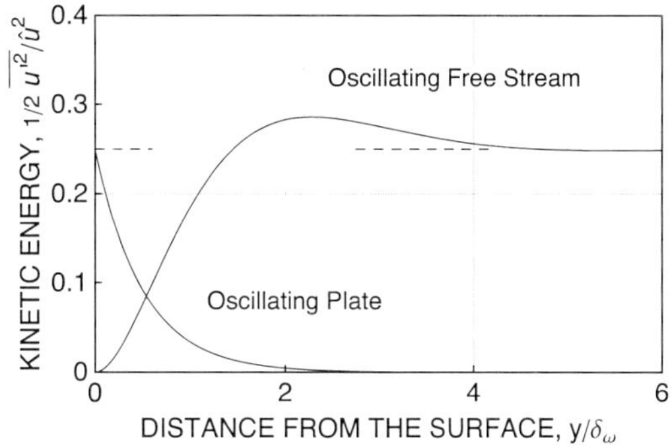

Figure 2.5.
Dimensionless kinetic energy distributions for both a stationary wall with an oscillating free stream and an oscillating plate.

The kinetic energy of this motion is given by

$$k = \frac{1}{2}\overline{u'u'} = \frac{1}{4}\hat{u}^2\left[1 + e^{-2y/\delta_\omega} - 2e^{-y/\delta_\omega}\cos\left(y/\delta_\omega\right)\right]$$

which is plotted in Fig. 2.5 together with the kinetic-energy distribution for an oscillating plate in a quiescent surrounding. In contrast to the monotonic behavior for the shear-driven flow, the pressure-driven flow exhibits an energy excess that can be attributed to the production of energy by the unsteady pressure force.

Since u' is the only nonzero velocity component, and since derivatives of a time-averaged quantity with respect to y are the only nonzero derivatives for this flow, the kinetic-energy equation, Eq. (1.37), reduces to

$$0=-\frac{1}{\rho}\overline{u'\frac{\partial p'}{\partial x}}+\nu\frac{\partial^2 k}{\partial y^2}-\hat{\varepsilon}$$

which simply expresses a balance between the production by the unsteady pressure force, and the viscous diffusion and dissipation of kinetic energy. According to Eq. (2.18), the production term is given by

$$-\frac{1}{\rho}\overline{u'\frac{\partial p'}{\partial x}}=\overline{u'\frac{\partial u'_\infty}{\partial t}} \tag{2.21}$$

from which we obtain

$$-\frac{1}{\rho}\overline{u'\frac{\partial p'}{\partial x}}=\frac{1}{2}\omega\hat{u}^2 e^{-y/\delta_\omega}\sin\left(y/\delta_\omega\right) \tag{2.22}$$

The viscous diffusion term is easily determined to be

$$\nu\frac{\partial^2 k}{\partial y^2}=\frac{1}{2}\omega\hat{u}^2\left[e^{-2y/\delta_\omega}-e^{-y/\delta_\omega}\sin\left(y/\delta_\omega\right)\right]$$

while the dissipation, which follows from Eq. (1.38), is given by

$$\hat{\varepsilon}=\nu\overline{\frac{\partial u'}{\partial y}\frac{\partial u'}{\partial y}}=\frac{1}{2\pi}\int_0^{2\pi}\frac{\partial u'}{\partial y}\frac{\partial u'}{\partial y}d\left(\omega t\right)=\nu\frac{\hat{u}^2}{\delta_\omega^2}e^{-2y/\delta_\omega}=\frac{1}{2}\omega\hat{u}^2 e^{-2y/\delta_\omega}$$

These terms, divided by $\omega\hat{u}^2/2$, are plotted as either a *gain* or *loss* of kinetic energy in Fig. 2.6. According to the kinetic-energy equation, the sum of them must equal zero. For unsteady boundary layers, this balance is not unique. At the wall, the diffusion is balanced by its dissipation, whereas away from the wall the diffusion of kinetic energy is balanced by its production, be it caused by unsteady shear and/or pressure forces. In this case, the production of energy is caused by the unsteady pressure forces, which in turn leads to the excess energy away from the wall as shown in the previous figure.

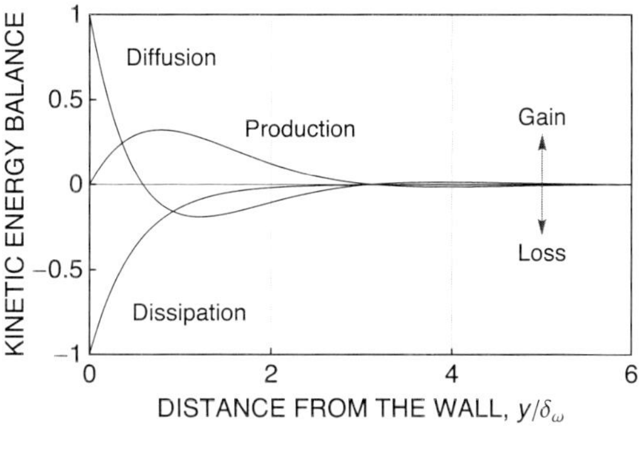

Figure 2.6.
Energy balance in the unsteady boundary layer near a stationary wall with an oscillating free stream.

2.3 Unsteady Boundary Layers with a Mean Flow.

Basic considerations. High-frequency oscillations. High-frequency traveling waves. Low frequency oscillations and traveling waves.

In this section we will investigate the effect of free-stream oscillations and traveling waves on a laminar boundary layer. The usual approach[1] to solving these problems is to assume that the amplitude of the oscillation $\hat{u}$ is small compared to the magnitude of the free-stream velocity u_∞, and expand the solution about the mean flow as a series of terms with increasing powers of the small amplitude parameter $\hat{u}/u_\infty$. Contrary to this approach, however, we will consider the high-frequency approach first proposed by Lin [1957].

We begin this section by discussing what constitutes low- and high- frequency oscillations in a boundary layer. We will then develop the appropriate equations for the steady and unsteady components of motion, solve them for high-frequency free-stream *oscillations*, and then for high-frequency *traveling waves*. For completeness, we will conclude with a very brief look at the quasi-steady nature of low-frequency oscillations and waves.

2.3.1 Basic considerations.

Effect of free-stream oscillations. A high-frequency approach to unsteady flow. Equations for the mean and unsteady flow.

Consider the response of a laminar boundary layer to oscillations in its free stream. When the frequency of these oscillations is very low compared to the viscous *rate of diffusion* through the boundary layer, the vorticity generated at the wall by the unsteady pressure force during any portion of the cycle has enough time to viscously diffuse *across* the entire thickness of the boundary layer. In this case, the flow at any instant will be nearly identical to that for steady flow, but with a free-stream velocity equal to its *instantaneous* value. On the other hand, when the frequency of the oscillations is very high compared to the viscous rate of diffusion, any change in the "wall-generated" vorticity will occur much too fast to diffuse across the boundary layer through any given cycle. As a result, the motion near the wall will always be out of phase with that in the free-stream such that viscous shear waves will always be present within the layer.

Since the viscous rate of diffusion through a steady boundary layer is proportional to ν/δ^2, it follows that the criterion for a low-frequency response is $\omega << \nu/\delta^2$, or

$$\omega\delta^2/\nu << 1 \quad \text{(low-frequency response)}$$

and that for a high-frequency response is (2.23)

$$\omega\delta^2/\nu >> 1 \quad \text{(high-frequency response)}$$

Introducing the unsteady-diffusion thickness, Eq. (2.16), these criteria can be expressed as $(\delta/\delta_\omega)^2 << 1$ and $(\delta/\delta_\omega)^2 >> 1$ respectively. Noting that the quantity $\omega\delta^2$ can be interpreted as the diffusivity of the unsteady motion through the boundary layer, the quantity $\omega\delta^2/\nu$ also represents a ratio of the unsteady-to-steady diffusivities.

Introducing a dimensionless *frequency parameter* defined by

$$\Omega = \omega\nu/\bar{u}_\infty^2 \tag{2.24}$$

the above criteria may be expressed as

$$\Omega Re_\delta^2 << 1 \quad \text{(low-frequency response)}$$

and (2.25)

$$\Omega Re_\delta^2 >> 1 \quad \text{(high-frequency response)}$$

That is, the response to an oscillation of a given frequency depends on the boundary-layer Reynolds number. Consequently, for any given frequency we can associate a low-frequency response with small

---

[1] See Lighthill [1954], Rott [1956], Rott and Rosenzweig [1960], Rott [1964], and Ackerberg and Phillips [1972].

Reynolds numbers and a high-frequency response with large Reynolds numbers.

Since Re_δ^2 is proportional to Re_x, the quantity ΩRe_δ^2 is proportional to $\omega x/u_\infty$ which is a common expansion parameter used in series solutions for low-frequency unsteady flow. A related parameter is $\omega\delta/u_\infty$ which is clearly an order Re_δ^{-1} less than the parameters above. So when the condition $\omega\delta/u_\infty << 1$ is used as the condition for low-frequency oscillations, there remains the region $Re_\delta^{-1} \approx \omega\delta/u_\infty << 1$ that still yields a *high-frequency viscous* response. This distinction will show up when we examine traveling waves.

To get an idea of what the previous parameters imply for transitional flow, consider a free-stream oscillation with a frequency *equal to or greater than* the frequency at which the Blasius boundary layer becomes *unstable*. We will investigate this instability and the oscillations known as *Tollmien-Schlichting waves*[1] in §4.2, but for now we only need to know that the critical frequency associated with this instability is approximately $\Omega_{TS} = 2.4(10)^{-4}$. Consequently, according to Eq. (2.25), a low-frequency response at this frequency occurs whenever $Re_\delta << 70$, which is less than the Reynolds number of 125 for which Prandtl's boundary-layer approximations are valid (see §2.1), and much less than the minimum Reynolds number of $Re_{\delta,TS} = 1{,}500$ at which this instability can occur. As a result, all but the boundary layer very near the leading edge of the plate responds as though oscillations with frequencies Ω_{TS} or greater are high frequency oscillations.

Lin's approach. Consider a two-dimensional boundary-layer. Then the equations for the mean flow, Eqs. (1.45) and (1.46), are

$$\frac{\partial \bar{u}}{\partial x}+\frac{\partial \bar{v}}{\partial y}=0 \tag{2.26}$$

$$\bar{u}\frac{\partial \bar{u}}{\partial x}+\bar{v}\frac{\partial \bar{u}}{\partial y}=\bar{u}_\infty\frac{d\bar{u}_\infty}{dx}+\nu\frac{\partial^2 \bar{u}}{\partial y^2}-\frac{\partial \overline{u'v'}}{\partial y}+\frac{\partial}{\partial x}\left(\overline{u_\infty'^2}-\overline{v_\infty'^2}-\overline{u'u'}+\overline{v'v'}\right) \tag{2.27}$$

where the pressure gradient has been eliminated using Eq. (1.50), the mean free-stream velocity is considered a function of x only, and the unsteady free-stream velocity is presumed to be a function of both x and t. Following Lin [1957], we will also consider two-dimensional oscillations with $w' = 0$. Then the equations for the unsteady motion, Eqs. (1.57) and (1.58), after a slight rearrangement, become

$$\frac{\partial u'}{\partial x}+\frac{\partial v'}{\partial y}=0 \tag{2.28}$$

$$\frac{\partial u'}{\partial t}+\bar{u}\frac{\partial u'}{\partial x}+\bar{v}\frac{\partial u'}{\partial y}=-\frac{1}{\rho}\frac{\partial p'}{\partial x}+\nu\frac{\partial^2 u'}{\partial x^2}+\nu\frac{\partial^2 u'}{\partial y^2}-u'\frac{\partial \bar{u}}{\partial x}-v'\frac{\partial \bar{u}}{\partial y}-u'\frac{\partial u'}{\partial x}-v'\frac{\partial u'}{\partial y}+\overline{u'\frac{\partial u'}{\partial x}}+\overline{v'\frac{\partial u'}{\partial y}} \tag{2.29}$$

with a similar equation for v'.

Since the scales associated with the mean and unsteady motion can be different, let's consider that changes in the unsteady components occur in the x direction over the distance l, and in the y direction over a distance δ'. Specifically, we will assume $\delta << l \le L$, where L is the length scale associated with changes of the mean flow in the x direction, and $\delta_\omega \le \delta' \le \delta$ depending on the frequency of the oscillation. From §1.4, we have $L \sim \delta(d\delta/dx)^{-1}$ which from §2.1 is of the order δRe_δ.

Presuming u' is the order of the oscillation's amplitude in the free stream $\hat{u}$, v' will be at most of order

[1] In contrast to the *forced oscillations* being discussed here, these waves are oscillations that develop within the boundary layer as a result of an *inherent instability*.

$\hat{u}\delta/l$ according to Eq. (2.28). Therefore $v' << u'$, which has several consequences. First, regarding the mean flow, the Reynolds-stress component $\overline{v'v'}$ can be neglected compared to the $\overline{u'u'}$ component such that Eq. (2.27) reduces to

$$\bar{u}\frac{\partial \bar{u}}{\partial x}+\bar{v}\frac{\partial \bar{u}}{\partial y}=\bar{u}_\infty\frac{d\bar{u}_\infty}{dx}+\nu\frac{\partial^2 \bar{u}}{\partial y^2}+F(x,y) \tag{2.30}$$

where

$$F(x,y)=\underbrace{\overline{u'_\infty\frac{\partial u'_\infty}{\partial x}}-\overline{u'\frac{\partial u'}{\partial x}}-\overline{v'\frac{\partial u'}{\partial y}}}_{O[(\hat{u}/u_\infty)^2(\delta/l)Re_\delta]} \tag{2.31}$$

The order of magnitude shown beneath this function is relative to the rest of the terms in the equation.

Second, when $v' << u'$, the unsteady accelerations normal to the wall are much smaller than those parallel to the wall, which yields upon comparing Eqs. (1.58) and (1.59)

$$\frac{\partial p'}{\partial y} << \frac{\partial p'}{\partial x}$$

Hence, $p' \approx p'_\infty(x,t)$, and by evaluating Eq. (2.29) in the free stream

$$-\frac{1}{\rho}\frac{\partial p'}{\partial x}=\frac{\partial u'_\infty}{\partial t}+\bar{u}_\infty\frac{\partial u'_\infty}{\partial x}+u'_\infty\frac{\partial \bar{u}_\infty}{\partial x}+u'_\infty\frac{\partial u'_\infty}{\partial x}-\overline{u'_\infty\frac{\partial u'_\infty}{\partial x}} \tag{2.32}$$

Using this to eliminate the unsteady pressure gradient in Eq. (2.29), we obtain

<u>*High-frequency ordering*</u>:

$$\underbrace{\frac{\partial u'}{\partial t}}_{O(1)}+\underbrace{\bar{u}\frac{\partial u'}{\partial x}}_{u_\infty/\omega l}+\underbrace{\bar{v}\frac{\partial u'}{\partial y}}_{\nu/\omega\delta\delta'}+\underbrace{u'\frac{\partial \bar{u}}{\partial x}}_{\nu/\omega\delta^2}+\underbrace{v'\frac{\partial \bar{u}}{\partial y}}_{u_\infty/\omega l}=\underbrace{\frac{\partial u'_\infty}{\partial t}}_{O(1)}+\underbrace{\bar{u}_\infty\frac{\partial u'_\infty}{\partial x}}_{u_\infty/\omega l}+\underbrace{u'_\infty\frac{\partial \bar{u}_\infty}{\partial x}}_{\nu/\omega\delta^2}+\underbrace{\nu\frac{\partial^2 u'}{\partial y^2}}_{\nu/\omega\delta'^2}$$
$$+\underbrace{u'_\infty\frac{\partial u'_\infty}{\partial x}-u'\frac{\partial u'}{\partial x}-v'\frac{\partial u'}{\partial y}-\overline{u'_\infty\frac{\partial u'_\infty}{\partial x}}+\overline{u'\frac{\partial u'}{\partial x}}+\overline{v'\frac{\partial u'}{\partial y}}}_{\hat{u}/\omega l} \tag{2.33a}$$

where the order of magnitude shown beneath each term is relative to the *local* acceleration. This ordering is most appropriate for high-frequency analyses. If we instead order each term relative to the *convective* acceleration, which is more appropriate for low-frequency analyses, we obtain

<u>*Low-frequency ordering*</u>:

$$\underbrace{\frac{\partial u'}{\partial t}}_{\omega l/u_\infty}+\underbrace{\bar{u}\frac{\partial u'}{\partial x}}_{O(1)}+\underbrace{\bar{v}\frac{\partial u'}{\partial y}}_{\nu l/u_\infty\delta\delta'}+\underbrace{u'\frac{\partial \bar{u}}{\partial x}}_{\nu l/u_\infty\delta^2}+\underbrace{v'\frac{\partial \bar{u}}{\partial y}}_{O(1)}=\underbrace{\frac{\partial u'_\infty}{\partial t}}_{\omega l/u_\infty}+\underbrace{\bar{u}_\infty\frac{\partial u'_\infty}{\partial x}}_{O(1)}+\underbrace{u'_\infty\frac{\partial \bar{u}_\infty}{\partial x}}_{\nu l/u_\infty\delta^2}+\underbrace{\nu\frac{\partial^2 u'}{\partial y^2}}_{\nu l/u_\infty\delta'^2}$$
$$+\underbrace{u'_\infty\frac{\partial u'_\infty}{\partial x}-u'\frac{\partial u'}{\partial x}-v'\frac{\partial u'}{\partial y}-\overline{u'_\infty\frac{\partial u'_\infty}{\partial x}}+\overline{u'\frac{\partial u'}{\partial x}}+\overline{v'\frac{\partial u'}{\partial y}}}_{\hat{u}/u_\infty} \tag{2.33b}$$

Obviously, several scenarios are possible depending on the size of $\nu/\omega\delta^2$, $\omega l/u_\infty$, δ'/δ and $\hat{u}/u_\infty$, or any combination of them. For small-amplitude oscillations, the second line in both equations can be neglected. This immediately renders the equations linear with respect to the unsteady velocity components. For either low- or high-frequency oscillations, various other terms can be neglected depending on the values of $\omega l/u_\infty$ and δ'/δ. However, whatever the case, the viscous term must always be considered. In the following subsections, we will almost exclusively investigate the more relevant problem of high-frequency oscillations and waves.

2.3.2 High-frequency oscillations.

First consider the response of a laminar boundary layer to a simple harmonic high-frequency oscillation in amplitude (a standing wave). As in our previous examples, suppose the unsteady component of the free-stream velocity is given by the real part of

$$u'(x,\infty,t)=u'_\infty(x,t)=\hat{u}(x)\exp(i\omega t)$$

where for a high-frequency oscillation, $\omega >> \nu/\delta^2$.

Following Lin, consider that l is of the same order as L. That is, the length scale of the changes in $\hat{u}$ and u_∞ with respect to x are roughly the same. Then $l = O(\delta Re_\delta)$ and $\omega l/u_\infty = O(\omega\delta^2/\nu)$, which is much larger than unity. As a result, several of the inertia terms in Eq. (2.33a) can be neglected and, as long as $\hat{u}/u_\infty$ is an order of magnitude equal to one or less, all the nonlinear terms on the second line can be neglected. Near the wall where the viscous force is important δ' must be of order δ_ω, that is, the square root of ν/ω. The only inertia term in question (third term on the left-hand side of the equation) is then of order $\sqrt{(\nu/\omega\delta^2)}$, which can also be neglected. Consequently, for high-frequency oscillations, Eq. (2.33a) reduces to

$$\frac{\partial u'}{\partial t}=\frac{\partial u'_\infty}{\partial t}+\nu\frac{\partial^2 u'}{\partial y^2};\quad (\omega\delta^2/\nu>>1) \tag{2.34}$$

which is notably independent of the mean flow. In addition, the pressure fluctuation can be obtained from Eq. (2.32), which for high-frequency oscillations becomes

$$-\frac{1}{\rho}\frac{\partial p'}{\partial x}\approx\frac{\partial u'_\infty}{\partial t};\quad (\omega\delta^2/\nu>>1) \tag{2.35}$$

Equations (2.26), (2.28), (2.30), (2.31), (2.34), and (2.35) are Lin's boundary-layer equations for high-frequency free-stream oscillations. A remarkable feature of these equations is that they form a closed set of equations for the variables $\bar{u}$, $\bar{v}$, u', v' and p', since $\bar{u}_\infty(x)$ and $u'_\infty(x,t)$ are presumed given. Besides this, they are *not* limited to small-amplitude oscillations.

Since Eq. (2.34) is identical to the equation for a free-stream oscillation *without* a mean-flow component, it follows from Eq. (2.20) that the solution for a boundary layer with a high-frequency oscillation in amplitude is

$$u'=\hat{u}(x)\left[\cos(\omega t)-e^{-y/\delta_\omega}\cos\left(\omega t-y/\delta_\omega\right)\right] \tag{2.36}$$

Therefore, as found before, the unsteady boundary-layer thickness is about $2\delta_\omega$.

From Eq. (2.28), the corresponding solution for v' is

$$v'(x,y,t)=\frac{1}{2}\delta_\omega\frac{d\hat{u}}{dx}\left[\sqrt{2}\cos\left(\omega t-\frac{\pi}{4}\right)-\sqrt{2}e^{-\frac{y}{\delta_\omega}}\cos\left(\omega t-\frac{y}{\delta_\omega}-\frac{\pi}{4}\right)-2\frac{y}{\delta_\omega}\cos(\omega t)\right]$$

where the boundary condition $v'(x, 0, t) = 0$ has been used. Clearly, this component of velocity equals zero when $\hat{u}$ is constant.

Given u' and v' the function $F(x, y)$ in the momentum equation for the mean-flow can be determined. From Eq. (2.31), we obtain

$$F(x,y)=\frac{\hat{u}}{2}\frac{d\hat{u}}{dx}\left\{\left[\left(2+\frac{y}{\delta_\omega}\right)\cos\left(\frac{y}{\delta_\omega}\right)-\left(1-\frac{y}{\delta_\omega}\right)\sin\left(\frac{y}{\delta_\omega}\right)\right]\exp\left(-\frac{y}{\delta_\omega}\right)-\exp\left(-2\frac{y}{\delta_\omega}\right)\right\} \quad (2.37)$$

The function in braces is plotted as $F^*(y/\delta_\omega)$ in Fig. 2.7. It has a maximum at the wall and, unlike the mean pressure gradient, varies across the boundary layer. Therefore, if $\hat{u}$ depends on x, the "effective" mean pressure gradient acting on the boundary layer will be greater at the wall than in the free stream. This, of course, will affect the mean wall shear stress and quite possibly impending transition or separation. On the other hand, if $\hat{u}$ is independent of x, both v' and F equal zero, and the *mean* wall shear stress will be unaffected. For this case, the oscillations u' are simply imposed on the mean flow.

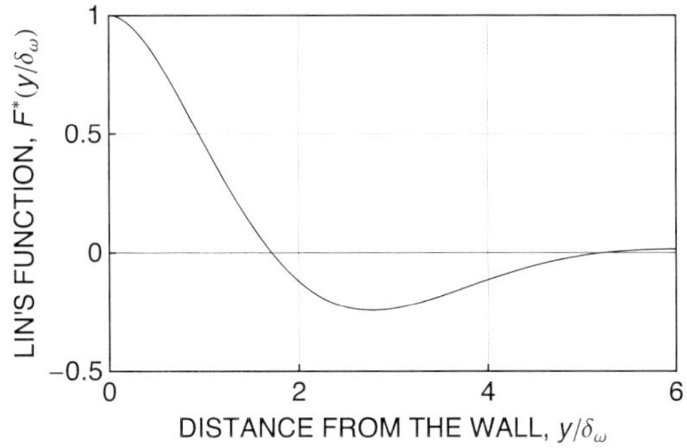

Figure 2.7.
Lin's function F^* for a harmonically oscillating free stream.

In general, the wall shear stress at any instant is given by $\tau_0=\bar{\tau}_0(x)+\tau_0'(x,t)$. The unsteady component is given by

$$\tau_0'(x,t)=\mu\left(\partial u'/\partial y\right)_{y=0}=\sqrt{2}\left(\mu/\delta_\omega\right)\hat{u}(x)\cos\left(\omega t+\pi/4\right) \quad (2.38)$$

Considering flow along a flat plate with both $\bar{u}_\infty$ and $\hat{u}$ constant, the mean wall shear stress, given by Eq. (2.9), is $\bar{\tau}_0=0.332\rho\bar{u}_\infty^2 Re_x^{-1/2}$. Hence the instantaneous wall stress for this oscillating flow is

$$\tau_0(x,t)=0.332\rho\bar{u}_\infty^2 Re_x^{-1/2}+\rho\sqrt{\omega\nu}\,\hat{u}\cos\left(\omega t+\pi/4\right)$$

Since the mean wall-stress component decreases with distance while the amplitude of the unsteady component remains constant, there is a distance at which the instantaneous stress periodically becomes zero. For unsteady flow, however, this does not necessarily imply separation.[1] Evaluating the above expression at $\omega t+\pi/4=\pi$, the condition for $\tau_0 \le 0$ is given by

$$\Omega Re_x \ge 0.332\left(\hat{u}/\bar{u}_\infty\right)^{-2}; \quad (\omega\delta^2/\nu >> 1,\ \omega l/u_\infty >> 1,\ dp_\infty/dx=0) \quad (2.39)$$

For oscillations at the Tollmien-Schlichting frequency of $\Omega_{TS}=2.4(10)^{-4}$, this occurs when the distance Reynolds number is greater than about $1400(\hat{u}/\bar{u}_\infty)^{-2}$, or when boundary-layer thickness Reynolds number is greater than $190(\hat{u}/\bar{u}_\infty)^{-1}$. With the critical Tollmien-Schlichting Reynolds number $Re_{\delta,TS}$ = 1,500, these oscillations will only affect boundary-layer stability when their amplitude is greater than about ten percent.

---

[1] See Moore [1951] for a discussion of separation in unsteady flows.

2.3.3 High-frequency traveling waves.

A practical problem of immediate interest is the response of a laminar boundary layer to a free-stream flow perturbed by high-frequency *traveling waves*. With this in mind, consider an unsteady free-stream-velocity component that varies according to the real part of

$$u'_\infty(x,t)=\hat{u}\exp\left[i\omega\left(t-x/\alpha\bar{u}_\infty\right)\right] \tag{2.40}$$

where α is a real number. This motion corresponds to a wave traveling downstream with a celerity $c=\alpha\bar{u}_\infty$ and a wave number k (not to be confused with kinetic energy) of ω/c.

The characteristic length scale of the free-stream oscillation is $l\approx 1/k=\alpha u_\infty/\omega$. If α is of order one in magnitude, then $u_\infty/\omega l=O(1)$ and several consequences immediately follow. First, the normal velocity v' will be much smaller than u' only when $\omega\delta/u_\infty << 1$. For high-frequency oscillations, where $\omega\delta^2/\nu$ must be much greater than unity, $\omega\delta/u_\infty << 1$ implies that $1 << \omega\delta^2/\nu << Re_\delta$. Second, with $u_\infty/\omega l=O(1)$, only the fourth inertia term on the left-hand side of Eq. (2.33a) and the third on the right-hand side can immediately be neglected. Third, all the nonlinear terms become of order $\hat{u}/u_\infty$ and can only be neglected when $\hat{u}/u_\infty << |1-\alpha|$. This condition arises from the fact that while the first two terms on the right-hand side of Eq. (2.33a) are of order one, their sum is of order $|1-\alpha|$.[1] Finally, of the remaining inertia terms, the third term on the left-hand side can be neglected, since it is at most of order $\sqrt{(\nu/\omega\delta^2)}$ as determined in the previous example. Consequently, assuming *small-amplitude oscillations*, Eq. (2.33a) reduces to

$$\frac{\partial u'}{\partial t}+\bar{u}\frac{\partial u'}{\partial x}+v'\frac{\partial \bar{u}}{\partial y}=\frac{\partial u'_\infty}{\partial t}+\bar{u}_\infty\frac{\partial u'_\infty}{\partial x}+\nu\frac{\partial^2 u'}{\partial y^2} \tag{2.41}$$

where $1 << \omega\delta^2/\nu << Re_\delta$, $\hat{u}/u_\infty << |1-\alpha|$ and $\alpha = O(1)$. This must be solved together with the continuity equation

$$\frac{\partial u'}{\partial x}+\frac{\partial v'}{\partial y}=0$$

The boundary conditions are $u'(x,0,t)=v'(x,0,t)=0$ and $u'(x,\infty,t)=u'_\infty(x,t)$.

Since Eq. (2.41) contains the mean velocity and its derivative with respect to y, the mean momentum equation, Eq. (2.30), which contains Lin's $F(x,y)$ function, must be solved as well. According to Eq. (2.31), however, F can be neglected compared to the other terms in the equation when $\hat{u}/u_\infty << \sqrt{(\nu/\omega\delta^2)}$, which would then render the solution for $\bar{u}$ identical to that for steady flow. In the following, we will first assume this to be true and then calculate the effect of $F(x,y)$ on the mean velocity afterwards.

With l of order u_∞/ω, L/l is of order $\omega\delta^2/\nu >> 1$, and any change in the mean flow over several wavelengths of the oscillation can be neglected. That is, as far as the oscillations are concerned, we can assume that the "local" mean flow moves parallel to the wall with $\bar{v}=0$ and $\bar{u}=fnc(y/\delta)$, where δ is a constant.[2] For Blasius' flow, the mean-flow velocity profile corresponds to that shown in Fig. 2.2, while the boundary-layer thickness is given by $\delta=\eta_\delta\sqrt{(\nu x_0/\bar{u}_\infty)}$, where $\eta_\delta\approx 5$ and x_0 is the distance from the leading edge of the plate to our point of interest.[3] The upshot of all of this is that we should

[1] An interesting consequence of this is that all the nonlinear terms must be considered when $\alpha = 1$. The reason for this becomes obvious when Eq. (2.40) with $\alpha = 1$ is substituted into Eq. (2.32) for the pressure gradient. Consequently, the solution obtained by Kestin, Maeder and Wang [1961] for $\alpha = 1$ is inappropriate for high-frequency traveling waves.

[2] Known as the *parallel flow assumption*, it will be used again when we examine boundary-layer stability in §4.2.

[3] A more precise number would result if we used one of the integral thicknesses, say, the displacement thickness. Then $\eta_{\delta 1}$ equals 1.72.

expect the unsteady velocity components to be a function of x, y/δ and t.

The solution for this problem can be obtained by first introducing an unsteady component of the stream function defined as

$$\psi'(x,y,t) = \hat{u}\delta\ \phi(\zeta)\exp\left[i\omega\left(t - x/\alpha u_\infty\right)\right] \tag{2.42}$$

where ϕ is a *complex* dimensionless stream function, $\zeta = y/\delta$, and δ is the boundary-layer thickness at x_0. The functional form of Eq. (2.42) follows directly from the free-stream behavior given in Eq. (2.40) and the definition of a stream function. Then the continuity equation is satisfied and the unsteady velocity components are given by

$$u'(x,y,t) = \frac{\partial \psi'}{\partial y} = \hat{u}\phi' \exp\left[i\omega\left(t - x/\alpha u_\infty\right)\right]$$

and (2.43)

$$v'(x,y,t) = -\frac{\partial \psi'}{\partial x} = i\left(\frac{\omega\delta}{\alpha \bar{u}_\infty}\right)\hat{u}\phi \exp\left[i\omega\left(t - x/\alpha u_\infty\right)\right]$$

where the prime on the function ϕ denotes its derivative with respect to ζ.

Defining a mean-flow stream function as $\bar{\psi} = \bar{u}_\infty \delta f_1(\zeta)$, the streamwise velocity and its derivative with respect to y is given by

$$\bar{u} = \bar{u}_\infty f_1'(\zeta) \quad \text{and} \quad d\bar{u}/dy = \bar{u}_\infty f_1''(\zeta)/\delta$$

Substituting these expressions together with Eqs. (2.40) and (2.43) into Eq. (2.41) and rearranging, an equation for $\phi(\zeta)$ is obtained, namely,

$$\varepsilon\left(\alpha\phi'''\right) - i\left(\alpha - f_1'\right)\phi' - if_1''\phi + i\left(\alpha - 1\right) = 0\ ; \quad (\varepsilon << 1) \tag{2.44}$$

where the parameter $\varepsilon = \nu/\omega\delta^2$ is much smaller than one for high-frequency waves. The boundary conditions for ϕ follow directly from the conditions imposed on the velocity components. From Eq. (2.43), they are $\phi(0) = \phi'(0) = 0$ and $\phi'(\infty) = 1$. Clearly, the solution for ϕ will depend on the parameters α and ε, and the mean-flow velocity distribution.

For Blasius' flow, $f_1(\zeta) = f(\eta)/\eta_\delta$ where $f(\eta)$ is Blasius' stream function and $\eta = \eta_\delta\zeta$. From Eq. (2.5), $f_1(\zeta)$ is found to satisfy $2f_1''' + \eta_\delta^2 f_1 f_1'' = 0$ with the boundary conditions $f_1(0) = f_1'(0) = 0$ and $f_1'(\infty) = 1$. From Eq. (2.4), the mean-velocity components are given by $\bar{u}(\zeta) = \bar{u}_\infty f_1'$ and $\bar{v}(\zeta) = \frac{1}{2}\bar{u}_\infty \eta_\delta \times Re_\delta^{-1}(\zeta f_1' - f_1)$.

Since the small parameter ε multiplies the term with the highest derivative, Eq. (2.44) is a "stiff" differential equation. Numerically this is now rather straightforward to solve as long as an appropriate algorithm is available. However, it is also possible to obtain an approximate closed-form solution using *singular perturbation theory*. While this approximate solution is relatively easy to obtain for $\alpha > 1$, that for $\alpha < 1$ is not. Therefore, we will first obtain the closed-form solution for waves traveling faster than the free-stream velocity, $\alpha > 1$, address its physical implications, and then present the numerical solution for $\alpha < 1$.

A closed-form solution for traveling waves with $\alpha > 1$. Without going into details,[1] the basic idea associated with singular perturbation theory is that as $\varepsilon \to 0$, the solution to Eq. (2.44) over much of the domain is given by neglecting the highest-derivative term except very near the boundaries where the boundary condition(s) must be satisfied. Generally, the solution near the boundary is called the "inner solution," while that farther away is referred to as the "outer solution." Within the domain

[1] See Cole [1968] and Carrier and Pearson [1968].

of the inner solution, the highest-derivative term must (for obvious reasons) be of the same order of magnitude as at least one of the other terms in the equation. As a consequence, since $\varepsilon << 1$, the function ϕ will change rapidly within this domain and change more slowly within the domain of the outer solution. Furthermore, as $\varepsilon \to 0$, the domain of the inner solution will become "vanishing small."

In the limit as ε approaches zero, presuming $\phi'''(\zeta)$ is reasonably well behaved, the first term in Eq. (2.44) can be neglected and the equation for the outer solution becomes

$$\left(\alpha - f_1'\right)h_0' - \left(\alpha - f_1'\right)' h_0 - \left(\alpha - 1\right) = 0 \tag{2.45}$$

where $h_0(\zeta)$ is the first term in an *outer expansion* for $\phi(\zeta)$; namely, $\phi(\zeta;\varepsilon) \approx h_0(\zeta) + \varepsilon h_1(\zeta) + \ldots + \varepsilon^n h_n(\zeta)$. The solution to this equation is

$$h_0(\zeta) = \left(\alpha - f_1'\right)\left[C_1 + \int_0^{\zeta} \frac{(\alpha-1)}{(\alpha - f_1')^2} d\zeta\right]$$

where C_1 is an integration constant. When $\alpha > 1$, this solution is regular (note $f_1' \le 1$), but when $\alpha < 1$ the outer solution contains a singularity where $f_1'(\zeta) = \alpha$ and changes sign across the singularity. As we will see, this has important implications regarding the amplitude and phase of the oscillations within the boundary layer.[1]

The derivative of h_0, which is proportional to the unsteady velocity component u', is given by

$$h_0'(\zeta) = \frac{\alpha - 1}{\alpha - f_1'} - f_1''\left[C_1 + \int_0^{\zeta} \frac{(\alpha-1)}{(\alpha - f_1')^2} d\zeta\right] \tag{2.46}$$

Since $f_1'(\infty) = 1$ and $f_1''(\infty) = 0$, the boundary condition $\phi'(\infty) \approx h_0'(\infty) = 1$ is satisfied regardless of C_1.

To satisfy the conditions at the wall we accentuate the first term in Eq. (2.44) near the wall, first by introducing a "stretched" independent variable $\xi = \zeta/\varepsilon^b$ and then by considering an inner solution of the form $\phi(\xi;\varepsilon) \approx \varepsilon^a g_0(\xi) + \ldots$. The exponents a and b in these expressions are determined by ensuring that the first term in Eq. (2.44) doesn't vanish as the limit $\varepsilon \to 0$ is taken. Substituting $\phi(\zeta) = h_0(\zeta) + \varepsilon^a g_0(\xi)$ into Eq. (2.44), using the fact that h_0 satisfies Eq. (2.45), and taking the limit as $\varepsilon \to 0$, we find that $a = b = 1/2$, and the equation for $g_0(\xi)$ is given by

$$g_0''' - i g_0' = 0$$

The bounded solution to this equation as $\xi \to \infty$ is

$$g_0(\xi) = C_2 + C_3 \exp\left[-\frac{1+i}{\sqrt{2}}\xi\right]$$

where C_2 and C_3 are integration constants.

Two of the three integration constants can be determined from the boundary conditions on ϕ at the wall. These conditions are $h_0(0) + \varepsilon^{1/2} g_0(0) = 0$ and $h_0'(0) + g_0'(0) = 0$. The remaining constant is determined by matching the inner and outer solutions in the limit $\varepsilon \to 0$. To a first approximation, the results of satisfying both the boundary and matching conditions are $C_1 \approx 0$, $C_2 = -C_3$, and

$$C_3 \approx \frac{\sqrt{2}}{1+i}\frac{\alpha - 1}{\alpha - f_1'}$$

[1] We will find a similar situation when we consider the stability of a laminar boundary layer in §4.2.

where $f_1'(0)$ has been replaced by $f_1'(\zeta)$ to produce a smooth match between the inner and outer solutions. Since $f_1'(\zeta) = f_1'(\varepsilon^{1/2}\xi) \to 0$ as $\varepsilon \to 0$, C_2 and C_3 are essentially constant within the approximations we have already made and should be treated so regarding differentiation.

Substituting all the above expressions into $\phi(\zeta) = h_0(\zeta) + \varepsilon^{1/2} g_0(\xi)$ yields

$$\phi(\zeta) \approx (\alpha - f_1')\int_0 \frac{(\alpha-1)}{(\alpha - f_1')^2} d\zeta + \sqrt{\varepsilon}\frac{\sqrt{2}}{1+i}\frac{\alpha-1}{\alpha - f_1'}\left[\exp\left(-\frac{1+i}{\sqrt{2}}\frac{\zeta}{\sqrt{\varepsilon}}\right)-1\right]; \quad (\alpha>1,\ \varepsilon<<1)$$

and

$$\phi'(\zeta) \approx \frac{\alpha-1}{\alpha - f_1'}\left[1-\exp\left(-\frac{1+i}{\sqrt{2}}\frac{\zeta}{\sqrt{\varepsilon}}\right)\right] - f_1''\int_0 \frac{(\alpha-1)}{(\alpha-f_1')^2} d\zeta \tag{2.47}$$

Substituting this into Eq. (2.43) and taking its real part, we obtain

$$\frac{u'(y,t^*)}{\hat{u}} \approx \frac{\alpha-1}{\alpha-f_1'}\left\{\cos(\omega t^*) - e^{-y/\delta_\omega}\cos\left(\omega t^* - y/\delta_\omega\right)\right\} - f_1''\left[\int_0^{y/\delta_\omega} \frac{(\alpha-1)}{(\alpha-f_1')^2} d\zeta\right]\cos(\omega t^*) \tag{2.48}$$

where, to simplify the above and future expressions, we have introduced the *time interval*

$$t^* = t - \left(x - x_0\right)/\alpha\bar{u}_\infty$$

This result is plotted in Fig. 2.8 for $\alpha = 3/2$, $\varepsilon = 1/200$ ($\omega\delta^2/\nu = 200$) and $Re_\delta = 2000$. Shown only for the accelerating portion of the free-stream oscillation, which also corresponds to that part of the cycle where the pressure force acts in the positive x-direction, profiles for $\omega t^* = \pi/4$, $\pi/2$ and $3\pi/4$ can be obtained by respectively flipping the $-3\pi/4$, $-\pi/2$ and $-\pi/4$ curves about the horizontal axis. The outer solution, given by Eq. (2.46) with $C_1 = 0$, is also plotted in the figure for $\omega t^* = -\pi$. By comparing the two $\omega t^* = -\pi$ profiles, the outer solution is seen to describe *most* of the solution. Furthermore, as found for the previous oscillating flows considered, the viscous effects are seen to be confined within a layer near the wall about $2\delta_\omega = \delta/5$ thick. Therefore, a high-frequency traveling wave produces two unsteady boundary layers; a *viscous* layer near the wall of thickness $2\delta_\omega$ and a larger *inertial* layer away from the wall of thickness δ. For comparison, the mean velocity profile represented by Blasius' solution is also plotted in the figure.

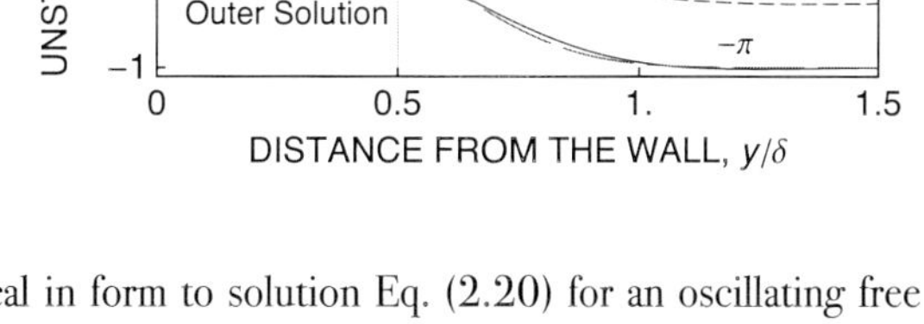

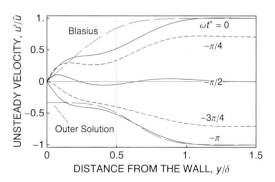

Figure 2.8.
The response of a laminar boundary layer to a free-stream flow perturbed by a high-frequency traveling wave. Plotted for $\alpha = 3/2$, $\varepsilon = 1/200$ and $Re_\delta = 2000$. Profiles for $\omega t^* = \pi/4$, $\pi/2$ and $3\pi/4$ can be obtained by flipping the $-3\pi/4$, $-\pi/2$ and $-\pi/4$ curves about the horizontal axis.

Since the first term in Eq. (2.48) is identical in form to solution Eq. (2.20) for an oscillating free stream, the oscillations near the wall are again transverse shear waves propagating away from the wall; except in this case they are *scaled* by the amplitude of the outer solution *at* the wall which equals

$\hat{u}(\alpha-1)/\alpha$, not $\hat{u}$. Consequently, the response near the wall to a traveling free-stream wave with $\alpha >$ 1 is the same as that for a free-stream oscillation *without* a mean-flow component but with a reduced "free-stream" amplitude.

The response away from the wall however is quite different. As we see, this oscillation is "locked" in phase with the free stream. Away from the wall, the inertia and pressure forces must balance. Since the inertia forces are approximately given by $Du'/Dt \sim \omega(\alpha\bar{u}_\infty - \bar{u})u'/\bar{u}_\infty$ at any instant, and since the pressure force is constant across the boundary layer, the velocity component u' through this region must vary inversely with respect to the velocity deficit $\alpha\bar{u}_\infty - \bar{u}$. Hence u' *must* decrease as the wall is approached. This process could be called "inertial repartitioning."

The real and imaginary parts of $\phi'(\zeta)$ are plotted in Fig. 2.9, where results from both the perturbation and numerical solutions are shown. Any discrepancies between the two solutions will be less for either smaller values of ε or a larger number of terms in the inner and outer expansions. Nevertheless, the agreement is seen to be quite satisfactory.

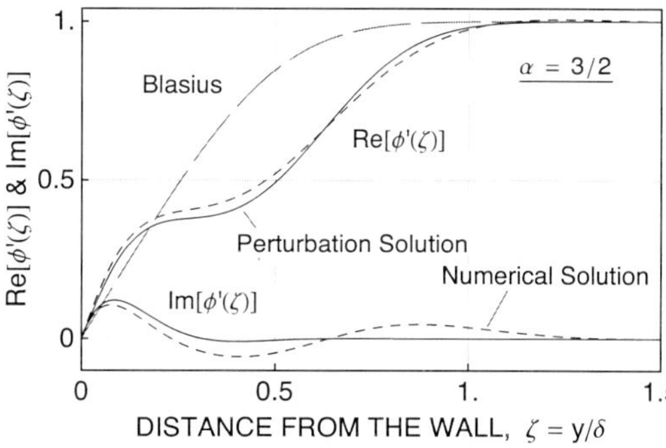

Figure 2.9.

Real and imaginary parts of $\phi'(\zeta)$ for $\alpha = 3/2$, $\varepsilon = 1/200$ and $Re_\delta = 2000$.

Traveling waves with $\alpha < 1$. In this case, although not immediately obvious, there are two regions where the highest-derivative term (the viscous term) must be considered. One of them is next to the wall as in the previous case. The other is where the inviscid solution Eq. (2.46) becomes singular; that is, where the mean velocity equals the wave speed. Thus, two "inner" solutions must be obtained with one of them "buried" within the flow. While obtained in a fashion as described above, the buried-inner solution is mathematically too involved to be considered here. Therefore, we will simply present the results from the numerical solution of Eq. (2.44) with $\alpha < 1$. In particular, except where noted, we will examine the solution for Blasius' flow with a wave traveling at one-half the free-stream velocity, that is, $\alpha = 1/2$, with $\varepsilon = 1/200$ ($\omega\delta^2/\nu = 200$) and $Re_\delta = 2000$.

The real and imaginary parts of $\phi'(\zeta)$ for this case are plotted in Fig. 2.10. These are significantly different from the results shown in the previous figure. Most notable is the phase change in the real part of $\phi'(\zeta)$ near $\zeta = 1/2$ and its amplitude near $\zeta = 1/4$. The latter position is about where the singularity in the outer perturbation solution occurs. In boundary-layer stability theory, the region near the singularity is called the "critical layer."[1] Another notable difference is the non-oscillatory nature of the imaginary part of the solution and its extent within the boundary layer. As a consequence, and in contrast to the result for $\alpha > 1$, about half of this boundary layer contains an amplified unsteadiness oscillating out of phase with the free stream.

[1] See §4.2.1, Eq. (4.15) and the ensuing discussion of the singularity in the Orr-Sommerfeld equation.

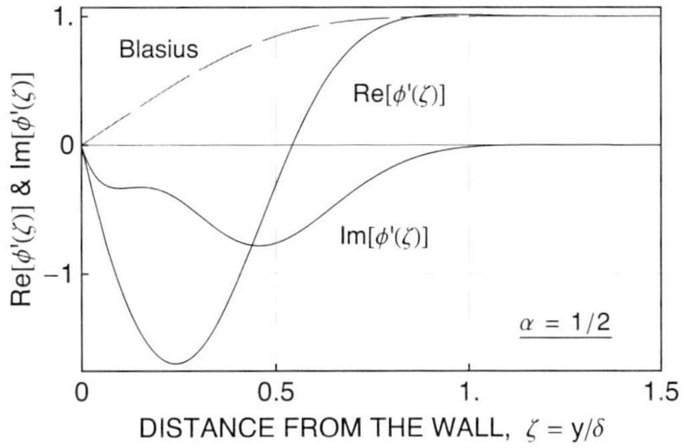

Figure 2.10.
Real and imaginary parts of $\phi'(\zeta)$ for $\alpha = 1/2$, $\varepsilon = 1/200$ and $Re_\delta = 2000$.

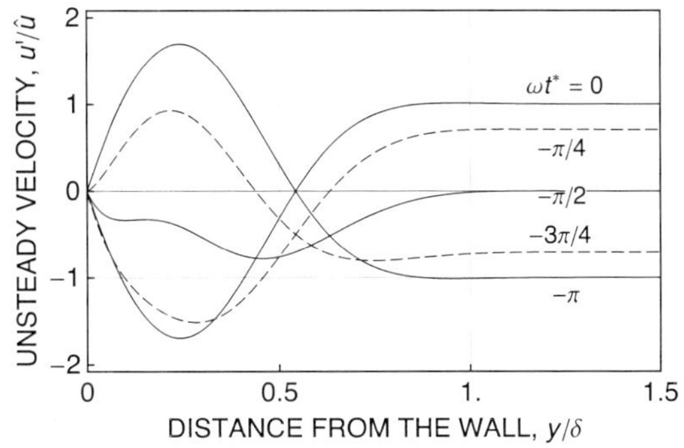

Figure 2.11.
The response of a laminar boundary layer to a free-stream flow perturbed by a high-frequency traveling wave. Plotted for $\alpha = 1/2$, $\varepsilon = 1/200$ and $Re_\delta = 2000$. Profiles for $\omega t^* = \pi/4$, $\pi/2$ and $3\pi/4$ can be obtained by flipping the $-3\pi/4$, $-\pi/2$ and $-\pi/4$ curves about the horizontal axis.

Taking the real part of Eq. (2.43), the unsteady velocity components are given by

$$\frac{u'(x,y,t)}{\hat{u}} = \phi_r'(\zeta)\cos(\omega t^*) - \phi_i'(\zeta)\sin(\omega t^*)$$

and (2.49)

$$\frac{v'(x,y,t)}{\hat{u}} = -\left(\omega\delta/\alpha\bar{u}_\infty\right)\left[\phi_r(\zeta)\sin(\omega t^*) + \phi_i(\zeta)\cos(\omega t^*)\right]$$

where ϕ_r and ϕ_i are the real and imaginary parts of $\phi(\zeta)$.

The streamwise velocity profiles are plotted in Fig. 2.11. Clearly, most of the activity occurs in the neighborhood where $f_1' = \alpha$, which for $\alpha = 1/2$ is at $\zeta \approx 0.27$. Although not particularly obvious, two unsteady waves are produced within the mean-flow boundary layer. Near the wall, transverse viscous waves propagate outward as before, while away from the wall transverse inertia waves propagate inward, first amplifying and then attenuating as they interact with the viscous waves near the singularity.

The unsteady wall shear stress is given by

$$\frac{\tau_0'(x,0,t)\,\delta}{\mu\hat{u}} = \phi_r''(0)\cos(\omega t^*) - \phi_i''(0)\sin(\omega t^*)$$

It obtains a maximum when

$$\omega t^*_{\max} = \pi - \arctan\left[\phi_i''(0)/\phi_r''(0)\right]$$

which for the present values of α and ε equals 2.50 or about 143°. The value of the wall shear stress at this point is $(\tau_0'\delta/\mu\hat{u})_{\max} \approx 16$, which was obtained from

$$\left(\frac{\tau_0'\delta}{\mu\hat{u}}\right)_{\max} = \left[\phi_r''(0)+\phi_i''(0)^2/\phi_r''(0)\right]\cos\left(\pi-\omega\, t_{\max}^*\right)$$

Several values of these quantities are given in Table 2.2 for $\alpha = 1/2$ and various values of ε. The corresponding values of δ_ω/δ are also given. For lower values of α we should expect larger values of the wall stress and visa-versa for higher values. This is deduced from the fact that the maximum amplitude of the oscillations occurs near the position where $f_1' = \alpha$ and that as the distance to this position decreases, the maximum wall-shear stress increases. In any case, the decrease in maximum wall stress as ε $(= \nu/\omega\delta^2)$ increases is caused by the increase in the unsteady diffusivity $\omega\delta^2$. A good fit for these results in terms of δ_ω/δ is

$$\left(\frac{\tau_0'\delta}{\mu\hat{u}}\right)_{\max} = 10.0+0.20\left(\delta/\delta_\omega\right)^{3/2} \;;\quad (\alpha=1/2,\, 5\le\delta/\delta_\omega\le 30)$$

Table 2.2

Results for small amplitude traveling free-stream oscillations with $\alpha = 1/2$.

ε	δ_ω/δ	ωt^*_{max}	$(\tau_0'\delta/\mu\hat{u})_{max}$	k_{max}/k_∞	y_{max}/δ
1/1600	0.035	136°	40.6	1.42	0.12
1/800	0.050	137°	28.8	1.78	0.16
1/400	0.071	139°	20.7	2.31	0.21
1/200	0.100	143°	16.0	3.05	0.25
1/100	0.141	141°	13.9	3.62	0.29

Hence, for small amplitude oscillations, the wall shear stress for flow on a flat plate at any instant is given by

$$\tau_0(x,t)=\bar{\tau}_0(x)+\tau_0'(x,t)=0.332\rho\bar{u}_\infty^2 Re_{x_0}^{-1/2}+\frac{\mu\hat{u}}{\delta}\left[\phi_r''(0)\cos(\omega t^*)-\phi_i''(0)\sin(\omega t^*)\right]$$

where Blasius' result has been substituted for the mean value. Using this it can be shown that the wall stress equals zero or less when

$$\left(\tau_0'\delta/\mu\hat{u}\right)_{\max} \ge 2\left(\hat{u}/\bar{u}_\infty\right)^{-1} \;;\quad (\text{any } \alpha,\ \text{any } \delta/\delta_\omega) \tag{2.50}$$

which is a surprisingly simple result. Using the above data fit, the wall stress will always be positive for a traveling wave with $\alpha = 1/2$ when

$$\Omega Re_x \ge 1.2\left(\hat{u}/\bar{u}_\infty\right)^{-4/3}\left\{1-\tfrac{20}{3}(\hat{u}/\bar{u}_\infty)+\ldots\right\};\quad (\alpha=1/2,\ \hat{u}/\bar{u}_\infty \ll 1,\ dp_\infty/dx=0) \tag{2.51}$$

Comparing this result with that given by Eq. (2.39) for an *oscillation* in the free-stream amplitude, this result depends on the inverse of the oscillation amplitude to the four-thirds power rather than to the inverse squared power. It is also restricted to values of $\hat{u}/\bar{u}_\infty$ less than about 0.15. Again, the reader is reminded that a wall shear stress equal to or less than zero in unsteady flow does not necessarily indicate separation.

The kinetic energy profiles for $\alpha = 1/2$ and $\alpha = 3/2$ with $\varepsilon = 1/200$ and $Re_\delta = 2000$ are plotted in

Fig. 2.12. Since $v' << u'$, the kinetic energy is approximately equal to $\frac{1}{2}\overline{u'u'}$. With $\hat{u}$ equal to the amplitude of the free-stream oscillation, the kinetic energy in the free stream equals $\hat{u}^2/4$. For α equal to $1/2$, the kinetic energy is a maximum about where $f_1'(\zeta) = \alpha$, whereas the maximum kinetic energy for α equal to $3/2$ is in the free stream. As might be expected, the position and value of the maximum when $\alpha < 1$ varies with both α and ε. For $\alpha = 1/2$, these variations are given in the last two columns of Table 2.2. Although the kinetic energy appears to increase continually with increasing ε, it decreases beyond $\varepsilon \approx 1/72$ or equivalently $\delta_\omega/\delta \approx 1/6$. Whether the solution is valid at these larger values of ε still requires investigation.

The kinetic-energy equation for this flow, Eq. (1.51), reduces to

$$\left\langle \bar{v}\frac{\partial k}{\partial y}\right\rangle = -\overline{u'v'}\frac{\partial \bar{u}}{\partial y} - \frac{1}{\rho}\overline{u'\frac{\partial p'}{\partial x}} + \nu\frac{\partial^2 k}{\partial y^2} - \hat{\varepsilon} \tag{2.52}$$

where the angled brackets indicate that advection in the direction normal to the wall is small when $\omega\delta/u_\infty << 1$. Since we assumed $\hat{u}/\bar{u}_\infty << 1$ and neglected the nonlinear unsteady terms in Eq. (2.33a), the diffusion of kinetic energy by the unsteadiness in the kinetic-energy equation is also negligible. In fact, for this flow it turns out to be identically zero.

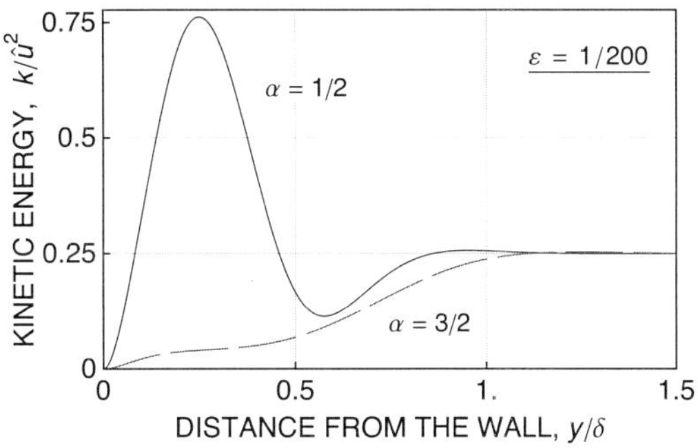

Figure 2.12.
Kinetic energy profiles for free-stream perturbations traveling with $\alpha = 1/2$ and $\alpha = 3/2$. Plotted for $\varepsilon = 1/200$ and $Re_\delta = 2000$.

Nevertheless, all the terms in Eq. (2.52) can be evaluated. For example, using Eq. (2.49), the Reynolds stress production term evaluates to

$$-\overline{u'v'}\frac{\partial \bar{u}}{\partial y} = -\frac{\hat{u}^2\bar{u}_\infty}{\delta} f_1''(\zeta)\left[\phi_r(\zeta)\phi_i'(\zeta) - \phi_i(\zeta)\phi_r'(\zeta)\right]$$

while using Eq. (2.32) with $\hat{u}/\bar{u}_\infty << 1$, the unsteady pressure production term becomes

$$-\frac{1}{\rho}\overline{u'\frac{\partial p'}{\partial x}} = \overline{u'\left(\frac{\partial u_\infty'}{\partial t} + \bar{u}_\infty\frac{\partial u'}{\partial x}\right)}$$

which evaluates to

$$-\frac{1}{\rho}\overline{u'\frac{\partial p'}{\partial x}} = \frac{1}{2}\omega\hat{u}^2\frac{\alpha-1}{\alpha}\phi_i'(\zeta)$$

The calculated variation of each term in Eq. (2.52), normalized by $\omega\hat{u}^2/2$, is plotted in Fig. 2.13 as either a *gain* or *loss* of kinetic energy. At the wall, the diffusion of kinetic energy is again balanced by its dissipation. As expected, the convection of kinetic energy by the mean flow (shown by the dashed line) is small. Contrary to most expectations, the production by the Reynolds shear stress is negative through most of the boundary layer because the shear stress is *positive*. Although not shown but easily sketched, the sum of the production terms has a maximum near $y/\delta = 1/4$, that is, where the kinetic energy of the oscillations is a maximum. Since the shear component of the production terms is relatively small in this region, the large oscillations are produced by the unsteady pressure force.

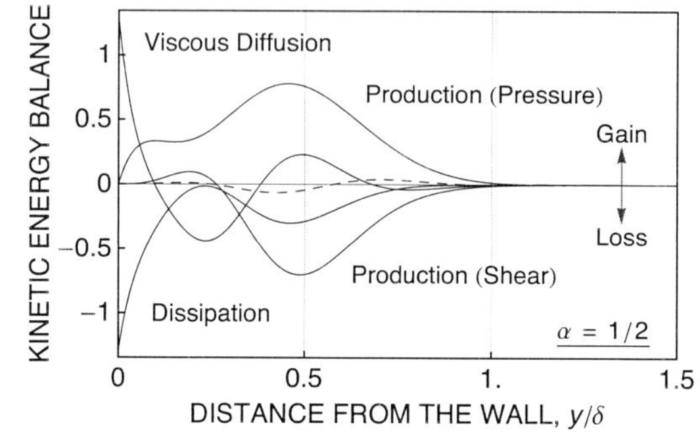

Figure 2.13.
Energy balance in a laminar boundary layer with its free-stream flow perturbed by a high-frequency traveling wave. Calculated for $\alpha = 1/2$, $\varepsilon = 1/200$ and $Re_\delta = 2000$. Normalized by $\omega\hat{u}^2/2$.

The effect of the unsteadiness on the mean flow depends on Lin's function. Defined in Eq. (2.31), its functional form for the traveling free-stream waves may be expressed as

$$F(x,y) = \overline{u'_\infty \frac{\partial u'_\infty}{\partial x}} - \overline{u' \frac{\partial u'}{\partial x}} - \overline{v' \frac{\partial u'}{\partial y}} = \frac{\omega \hat{u}^2}{\alpha \bar{u}_\infty} F^*(\zeta;\ \alpha, \varepsilon) \tag{2.53}$$

where F^* is dimensionless. A graph of F^* is presented in Fig. 2.14. Lin's result, which was plotted in Fig. 2.7, is also plotted here for comparison. The value of ε is $1/200$ for both the $\alpha = 1/2$ and $3/2$ waves. The difference between the results for the traveling waves can be explained by the large variations in the oscillations with respect to y for the $\alpha = 1/2$ case. Since $v' = 0$ at the wall, the difference between Lin's and the traveling waves at the wall can be explained by realizing that only the first two terms in the definition of F^* are relevant. In Lin's case, both u' and u_∞' and their derivatives with respect to x are in phase with respect to time, whereas for the traveling waves they are ninety degrees out of phase. Consequently, the time average of their product is finite in Lin's case and zero for the traveling free-stream waves.

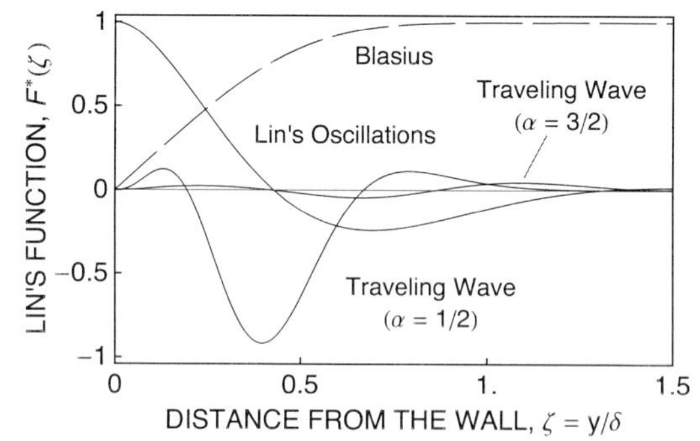

Figure 2.14.
Lin's function for free-stream perturbations traveling with $\alpha = 1/2$ and $\alpha = 3/2$, and for a free stream with an oscillating amplitude. Plotted for $\varepsilon = 1/200$ and $Re_\delta = 2000$.

For flow with a zero pressure gradient, it is relative easy to show from Eq. (2.30) that the stream function $f_1(\zeta)$ for a traveling free-stream wave satisfies a modified Blasius' equation, namely,

$$2f_1''' + \eta_\delta^2 f_1 f_1'' + 2\frac{1}{\alpha\varepsilon}\left(\frac{\hat{u}}{\bar{u}_\infty}\right)^2 F^* = 0$$

with the usual boundary conditions. This equation was numerically solved with $\alpha = 1/2$, $\varepsilon = 1/200$ and $Re_\delta = 2000$ for several values of $\hat{u}/\bar{u}_\infty$. The resulting mean-flow velocity profiles are presented in Fig. 2.15. Recalling that a negative value of F^* has the same effect as an adverse pressure gradient, the effect of the traveling wave on the mean velocity is easily understood. Here of course, in contrast to the effect of a free-stream pressure gradient, the effect of the wave varies through the boundary layer as shown in the previous figure. A notable feature in Fig. 2.15 is the nonlinear rate of increase by which the profiles deviate from one another with amplitude.

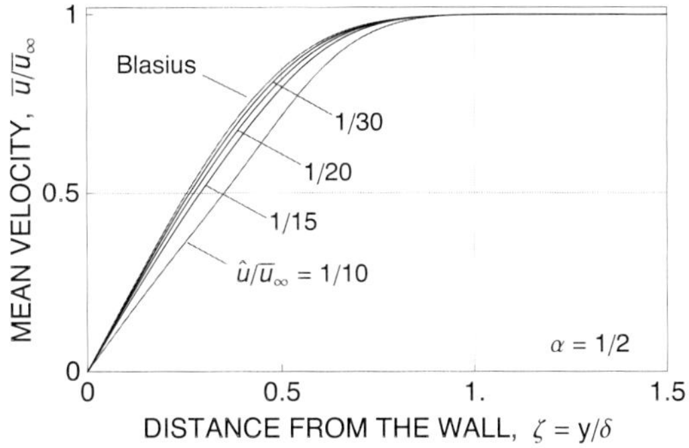

Figure 2.15.
The effect of a free-stream high-frequency traveling wave on the mean velocity profile in a laminar boundary layer. Plotted for $\alpha = 1/2$, $\varepsilon = 1/200$ and $Re_\delta = 2000$.

The effect of the wave on the *instantaneous* velocity is shown in Fig. 2.16. This was obtained by adding the fluctuating-velocity distributions shown in Fig. 2.11 for $\hat{u}/\bar{u}_\infty = 1/20$ to the corresponding mean-velocity profile presented in Fig. 2.15. Clearly, there are instances when the profile's curvature is positive, and hence instances when the profile contains a point of inflection. In this case, the inflection point lies near the singular point in the outer solution, that is, near $y/\delta = 1/4$. The instantaneous profiles for the case $\alpha = 3/2$ show no such behavior.

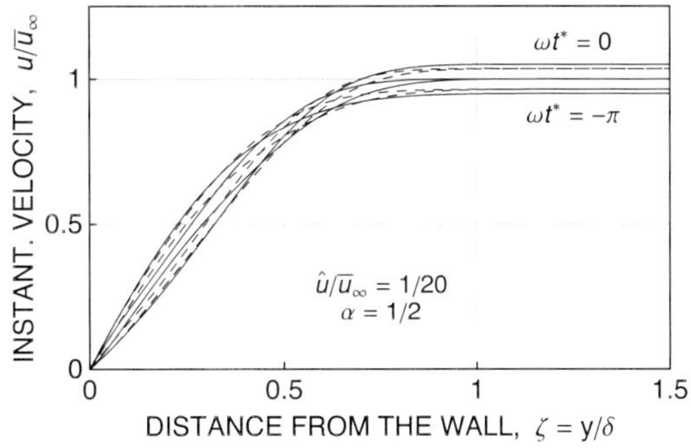

Figure 2.16.
The effect of a free-stream high-frequency traveling wave on the instantaneous velocity in a laminar boundary layer. Plotted for $\hat{u}/\bar{u}_\infty = 1/20$, $\alpha = 1/2$, $\varepsilon = 1/200$ and $Re_\delta = 2000$.

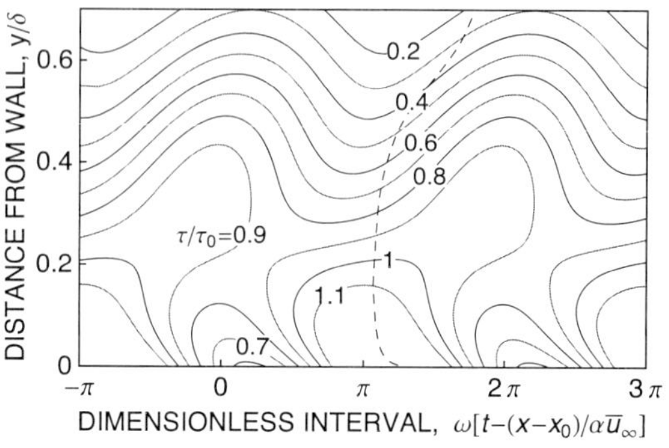

Figure 2.17.
Contours of equal shear stress at $t = 0$ for $\hat{u}/\bar{u}_\infty = 1/20$, $\alpha = 1/2$, $\varepsilon = 1/200$ and $Re_\delta = 2000$.

Although the variation in velocity profile isn't particularly notable, the variation of the instantaneous shear stress shown in Fig. 2.17 is quite remarkable. In this figure, lines of constant stress normalized by the mean wall stress are plotted for an observer moving with the wave to render them stationary. Plotted for the same conditions as in the previous figure, the phase change across the critical layer near $y/\delta = 1/4$ is quite evident in the upstream shift of the ridge of maximum stress that begins at the wall near the value π. This ridge, however, does not correspond to the ridge for the maximum instantaneous velocity shown by the dashed line in the figure, except near the wall where both are in phase.

For larger amplitude oscillations, two "peaks" in shear stress occur – one at and the other away from the wall. This is shown in Fig. 2.18 where the time-averaged stress profiles are plotted for various amplitudes of the free-stream oscillation. Here, as the amplitude increases, the shear stress develops a maximum about halfway through the boundary layer. This results from the instantaneous profiles around $\omega t^* = -\pi$ which display a region of positive curvature. For the $\hat{u}/\bar{u}_\infty = 1/8$ case, the flow according to Eq. (2.50) and Table 2.2 also reverses its direction during part of its cycle. Although the results in Fig. 2.18 for the two highest-amplitude cases in this figure are probably only approximate (since their amplitudes are not that small), the trend with increasing amplitude is obvious. Namely, traveling waves of moderate amplitude propagating *slower* than the free-stream velocity cause peaks of high shear stress about halfway through the boundary layer.

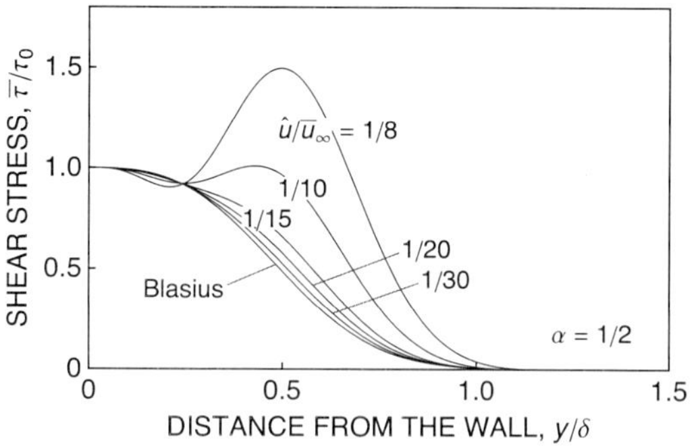

Figure 2.18.
Time-averaged shear-stress distributions for various amplitudes of a traveling wave. Evaluated for $\alpha = 1/2$, $\varepsilon = 1/200$ and $Re_\delta = 2000$.

To further appreciate the physics behind these results, the streamlines for this flow are shown in Fig. 2.19. Plotted again for an observer moving with the wave, the wall moves to the left at the speed equal to one-half the free-stream velocity since, in this case, $\alpha = 1/2$. The dashed contour, which has the value of -0.053, separates the flow moving to the right above from that moving to the left below. Clearly, upon comparing this and Fig. 2.17, the redistribution of the high and low shear stress is simply a consequence of the vortical motion within this contour. For larger amplitudes, this vortical motion extends higher within the boundary layer, eventually causing the peaks in shear stress seen in Fig. 2.18.

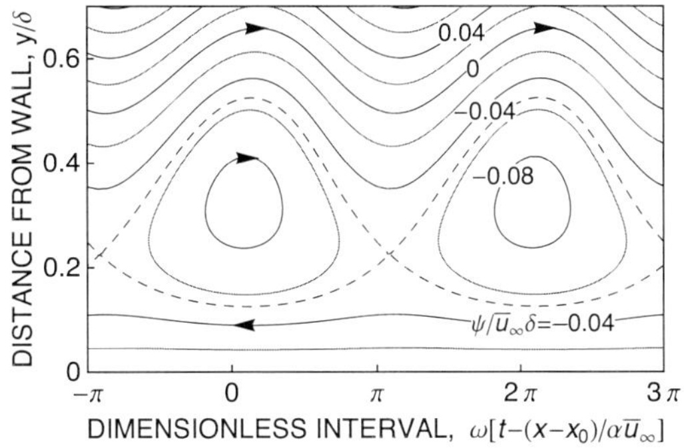

Figure 2.19.
The mean-flow streamlines at $t = 0$ for $\hat{u}/\overline{u}_\infty = 1/20$, $\alpha = 1/2$, $\varepsilon = 1/200$ and $Re_\delta = 2000$.

2.3.4 Low-frequency oscillations and waves.

Let's now consider the Reynolds-stress approach for low-frequency oscillations. Referring to Eqs. (2.28) and (2.33b), the equations for the unsteady velocity components are

$$\frac{\partial u'}{\partial x}+\frac{\partial v'}{\partial y}=0$$

and

$$\begin{aligned}\underbrace{\frac{\partial u'}{\partial t}}_{\omega l/u_\infty}+\underbrace{\overline{u}\frac{\partial u'}{\partial x}}_{O(1)}+\underbrace{\overline{v}\frac{\partial u'}{\partial y}}_{\nu l/u_\infty\delta\delta'}+\underbrace{u'\frac{\partial \overline{u}}{\partial x}}_{\nu l/u_\infty\delta^2}+\underbrace{v'\frac{\partial \overline{u}}{\partial y}}_{O(1)}=\underbrace{\frac{\partial u'_\infty}{\partial t}}_{\omega l/u_\infty}+\underbrace{\overline{u}_\infty\frac{\partial u'_\infty}{\partial x}}_{O(1)}+\underbrace{u'_\infty\frac{\partial \overline{u}_\infty}{\partial x}}_{\nu l/u_\infty\delta^2}+\underbrace{\nu\frac{\partial^2 u'}{\partial y^2}}_{\nu l/u_\infty\delta'^2}\\ \underbrace{+u'_\infty\frac{\partial u'_\infty}{\partial x}-u'\frac{\partial u'}{\partial x}-v'\frac{\partial u'}{\partial y}-\overline{u'_\infty\frac{\partial u'_\infty}{\partial x}}+\overline{u'\frac{\partial u'}{\partial x}}+\overline{v'\frac{\partial u'}{\partial y}}}_{\hat{u}/u_\infty}\end{aligned} \tag{2.54}$$

where the order of magnitude of each term is relative to the magnitude of the streamwise convective acceleration. Again several scenarios are possible depending on the parameters $\nu/\omega\delta^2$, $\omega l/u_\infty$, δ'/δ and $\hat{u}/\overline{u}_\infty$, or any combination thereof.

First consider an *oscillation* superposed on the free-stream velocity of the form given by the real part of

$$u'(x,\infty,t)=u'_\infty(x,t)=\hat{u}(x)\exp(i\omega t)$$

where now $\omega << \nu/\delta^2$, that is, for low-frequency oscillations. Furthermore suppose the changes of $\hat{u}$ and $\overline{u}_\infty$ with respect to x occur over the same distance. Then l is of order $L \approx \delta Re_\delta$ such that $O(\omega l/u_\infty) = O(\omega L/u_\infty) = O(\omega\delta^2/\nu)$ which is much less than unity. To retain the viscous term, δ' must be of order $(\nu L/u_\infty)^{1/2} \approx \delta$. Consequently, Eq. (2.54) reduces to

$$\bar{u}\frac{\partial u'}{\partial x}+\bar{v}\frac{\partial u'}{\partial y}=\bar{u}_\infty\frac{\partial u'_\infty}{\partial x}+u'_\infty\frac{\partial \bar{u}_\infty}{\partial x}-u'\frac{\partial \bar{u}}{\partial x}-v'\frac{\partial \bar{u}}{\partial y}+\nu\frac{\partial^2 u'}{\partial y^2}+O(\omega\delta^2/\nu)+O(\hat{u}/u_\infty)$$

which corresponds to the well-known equation for *quasi-steady* flow. Its general solution for the case where $\hat{u}(x)$ is proportional to the free-stream velocity $\bar{u}_\infty(x)$ and $O(\hat{u}/\bar{u}_\infty)<<1$ has been given by Lighthill [1954]. In our notation, his result is

$$u'_s=\left[\bar{u}+\tfrac{1}{2}y\left(\partial\bar{u}/\partial y\right)\right]\frac{\hat{u}}{\bar{u}_\infty}\quad\text{and}\quad v'_s=\tfrac{1}{2}\left[\bar{v}+y\left(\partial\bar{v}/\partial y\right)\right]\frac{\hat{u}}{\bar{u}_\infty}\ ;\quad(\omega\delta^2/\nu<<1,\ \hat{u}/\bar{u}_\infty<<1)$$

Higher order approximations for u' can be obtained by considering additional terms in a series solution expanded in powers of $\omega\delta^2/\nu$. Those interested in the details of this procedure should see Lighthill's paper.

Next consider a traveling wave superposed on the free-stream velocity of the form given by the real part of

$$u'(x,\infty,t)=u'_\infty(x,t)=\hat{u}\exp\left[i\omega\left(t-x/\alpha\bar{u}_\infty\right)\right]$$

In this case, as in §2.3.3, we have $l=O(\bar{u}_\infty/\omega)$. Then Eq. (2.54) reduces (which is most easily seen by multiplying the magnitude of each term by $\omega\delta^2/\nu$) to

$$\bar{v}\frac{\partial u'}{\partial y}+u'\frac{\partial \bar{u}}{\partial x}=u'_\infty\frac{\partial \bar{u}_\infty}{\partial x}+\nu\frac{\partial^2 u'}{\partial y^2}+O(\omega\delta^2/\nu)+O\left((\omega\delta^2/\nu)(\hat{u}/\bar{u}_\infty)\right)\qquad(2.55)$$

Curiously, this equation does not entail derivatives of u' with respect to either x or t, which suggests that it is valid for a wider variety of low-frequency oscillations as long as $\omega l/\bar{u}_\infty=O(1)$. Furthermore, the condition for neglecting the nonlinear terms is also less restrictive than before since $\omega\delta^2/\nu$ is already supposed to be much less than unity. Nevertheless, the solution to this equation with $\omega\delta^2/\nu<<1$ yields the quasi-steady solution as we will now see.

To keep things simple, consider uniform flow along a flat plate with a zero pressure gradient. Then

$$\bar{u}=\bar{u}_\infty f'\quad\text{and}\quad\bar{v}=\tfrac{1}{2}\bar{u}_\infty Re_x^{-1/2}\left(\eta f'-f\right)$$

Where η is the similarity variable $\eta=y\sqrt{(\bar{u}_\infty/\nu x)}$ and $f(\eta)$ is Blasius' dimensionless stream function. In addition, assume the unsteady velocity is given by

$$u'=\hat{u}\phi'(\eta)\exp\left[i\omega\left(t-x/\alpha u_\infty\right)\right]$$

where $\phi(\eta)$ is a dimensionless stream function for the unsteady component of the flow. Then Eq. (2.55) becomes

$$f\phi''+\eta\left(\phi'f''-f'\phi''\right)+2\phi'''=0$$

with the boundary conditions $\phi'(0)=0$ and $\phi'(\infty)=1$. Comparing this equation with Blasius equation, Eq. (2.5), the solution for ϕ is immediately seen to be $\phi=f$. Hence, the first order solution for u' is given by

$$u'=\hat{u}f'(\eta)\exp\left[i\omega\left(t-x/\alpha u_\infty\right)\right];\quad(\omega\delta^2/\nu<<1)$$

which is indeed the expected quasi-steady result. As with the solution for low-frequency oscillations, a higher order approximation for u' can be obtained by considering additional terms in a series solution expanded in powers of $\omega\delta^2/\nu$.

References

Ackerberg, R.C., and Phillips, J.H., 1972, "The Unsteady Laminar Boundary Layer on a Semi-Infinite Flat Plate Due to Small Fluctuations in the Magnitude of the Free Stream," J. Fluid Mech., 51, pp. 137–157.

Blasius, H., 1908, "Grenzschichten in Flüssigkeiten mit kleiner Reibung," Z. Math. Phys., 56, pp. 1–37; also NACA TM 1256.

Carrier, G.F., and Pearson, C.E., 1968, Ordinary Differential Equations, Blaisdell Publishing Company, Waltham, Massachusetts.

Cole, J.D., 1968, Perturbation Methods in Applied Mathematics, Blaisdell Publishing Company, Waltham, Massachusetts.

Kestin, J., Maeder, P.F., and Wang, W.E., 1961, "On Boundary Layers Associated With Oscillating Streams," Appl. Sci. Res. A, 10, pp. 1–22.

Lighthill, M.J., 1954, "The Response of Laminar Skin Friction and Heat Transfer to Fluctuations in the Stream Velocity," Proc. Roy. Soc. London A, 224, pp. 1–23.

Lin, C.C., 1957, "Motion in the Boundary Layer with a Rapidly Oscillating External Flow," Proc. 9th Int. Congress Appl. Mech., Brussels, 4, pp. 155–167.

Moore, F.K., 1951, "Unsteady, Laminar Boundary-Layer Flow," NACA TN 2471.

Rosenhead, L. (ed.), 1963, Laminar Boundary Layer, Clarendon Press, Oxford.

Rott, N., 1956, "Unsteady Viscous Flow in the Vicinity of a Stagnation Point," Quart. Appl. Math., 13, pp. 444–451.

Rott, N., 1964, "Theory of Time-Dependent Laminar Flows," Princeton University Series, High Speed Aerodynamics and Jet Propulsion, Princeton University Press, 4, pp. 395–438.

Rott, N., and Rosenzweig, M.L., 1960, "On the Response of the Laminar Boundary Layer to Small Fluctuations of the Free-Stream Velocity, JASS, 27, pp. 741–747, 787.

Schlichting, H., 1979, Boundary-Layer Theory, McGraw-Hill, New York.

Stokes, G.G., 1851, "On the Effect of Internal Friction of Fluids on the Motion of Pendulums," Cambridge Phil. Trans., 9, p. 8.

3 Turbulent Boundary Layers

Isotropic turbulence. A phenomenological approach. A differential approach. Free-Stream and near-wall turbulence.

Although post-transitional flow is synonymous with turbulent flow, to understand pre-transitional and transitional flow is to understand turbulence, its effect on laminar flow, and turbulent boundary layers. For more than a century, the object of turbulent flow research has been to find a practical method for calculating the Reynolds shear-stress terms in the momentum equation. Know as the *closure problem*, various methods have been proposed throughout the century to obtain these quantities. They include *mixing-length models*, *eddy-viscosity models*, *one-equation models* using various forms of the turbulent kinetic-energy equation, *two-equation models* using various forms of the turbulent kinetic-energy equation and a second equation for another characteristic quantity of turbulence such as the dissipation or length scale of turbulence, *second-moment models* using various forms of the Reynolds-stress and dissipation equations, and numerous combinations of these. All of them in their turn have significantly improved our ability to calculate turbulent boundary layers, especially during the later part of the century. Simultaneously, our basic understanding of isotropic, anisotropic, homogeneous and non-homogeneous turbulence has also improved.

In this chapter, we will first review the various scales and spectra of isotropic turbulence. We will then review some of the phenomenological approaches to obtain the mean-velocity distribution in a turbulent boundary layer. Next we will present a slightly different variation on a standard second-moment turbulence model, and finally show some calculated and experimental results. To remain brief, the material in this chapter primarily concerns turbulent flow with a zero-pressure-gradient on a flat plate. For readers interested in more details on the fundamentals of turbulence, the books by Batchelor [1953], Hinze [1959], Rotta [1972], and Tennekes and Lumley [1972] are suggested, while those interested in the phenomenological modeling of turbulent flow are referred to the book by Schlichting [1979] along with those listed above. Furthermore, those interested in the structural modeling of turbulent flow are referred to the review articles by Launder and Spalding [1974], Rodi [1978], Launder [1989], Speziale et al. [1992] and Hanjalić [1994].

Some of the important results in this chapter are:

1. Turbulent flow depends on two predominant groups of eddies – small eddies and large – and the transfer of energy between them.
2. Turbulent boundary layers are composed of an *inner* viscous region of small eddies, an *intermediate* region dominated by eddies scaled by their distance from the wall, and an *outer* intermittent region of large turbulent eddies and entrained free-stream flow.
3. The *initial* decay of free-stream turbulence depends on the transfer of energy from the large-to-small eddies, while its *final* decay depends on the viscous dissipation of the small eddies.
4. A shear-free turbulent boundary layer is composed of an *inner viscous* region that grows in proportion to the laminar diffusion thickness, and an *outer inertial* region that grows in proportion to the large free-stream eddies.

Beside the references at the end of this chapter, other relevant articles may be found in the Bibliography.

3.1 Isotropic Turbulence.

Turbulent length scales and Reynolds numbers. Turbulent energy and dissipation spectra.

In general, there are two major groups of turbulent eddies. One group contains the most energetic eddies, the other contains those most responsible for the dissipation of turbulent energy. Since dissipation depends on the viscosity of a fluid, a molecular property of the fluid, the dissipative eddies are always smaller than the energetic ones. Furthermore, they are always more isotropic than the larger energetic ones. For isotropic turbulence, the structure of turbulence is independent of direction. Therefore, the number of parameters needed to describe it is a minimum; which in turn makes isotropic turbulence the simplest form of turbulence to study.

A dimensional approach. First consider the larger, lower frequency eddies containing most of the energy, henceforth simply called the "large" eddies. Since the energy in these eddies must be transferred to the smaller eddies before it can be dissipated, the characteristic rate of energy transfer for these eddies is essentially the same as the dissipation of turbulent energy. Therefore, if we use just the turbulent kinetic energy k and its rate of dissipation $\hat{\varepsilon}$ as the two characteristic parameters for these eddies, their characteristic length, time and velocity scales from dimensional analysis are

$$L_e = k^{3/2}/\hat{\varepsilon}, \quad T_e = k/\hat{\varepsilon} \quad \text{and} \quad V_e = k^{1/2} \tag{3.1}$$

The Reynolds number associated with these eddies, now commonly referred to as the "turbulent Reynolds number," is

$$Re_T = V_e L_e/\nu = k^2/\nu\hat{\varepsilon} \tag{3.2}$$

and the *turbulent diffusion coefficient*, defined in this case by $\nu_T = V_e L_e$, is

$$\nu_T = k^2/\hat{\varepsilon} \tag{3.3}$$

As we will see, this expression takes a slightly different form for anisotropic turbulence.

Now consider the smaller, higher frequency dissipative eddies, henceforth called the "small" eddies. Using just the dissipation $\hat{\varepsilon}$ and (kinematic) viscosity ν (which is ultimately responsible for their dissipation) as the two characteristic parameters for these eddies, their characteristic length, time and velocity scales are

$$l_d = \left(\nu^3/\hat{\varepsilon}\right)^{1/4}, \quad t_d = \sqrt{\nu/\hat{\varepsilon}} \quad \text{and} \quad v_d = \left(\nu\hat{\varepsilon}\right)^{1/4} \tag{3.4}$$

The Reynolds number associated with these eddies is

$$Re_{ld} = v_d l_d/\nu = 1$$

which implies, as might be expected, that the inertia and viscous forces acting on the dissipative eddies are of the same order. The scales in Eq. (3.4) were originally defined by Kolmogorov [1941] and are known as the *Kolmogorov scales of turbulence*. The diffusion coefficient associated with these eddies is of course ν.

For isotropic turbulence, where everything is independent of direction, the above scales are sufficient to describe the turbulence and all other scales can be related to them.

Some other scales of turbulence. Frequently, the *intensity* of turbulence and various other scales based on it are found in the literature. Defined as

$$\hat{u} = \sqrt{\overline{u'u'}} = \sqrt{2k/3}$$

since the time-averaged turbulence is the same in every direction, the intensity of turbulence is clearly related to the large-scale eddies.[1] The length scale usually associated with it is defined by

$$l_e = A\hat{u}^3/\hat{\varepsilon} \tag{3.5}$$

where A, which depends on Reynolds number, is approximately equal to one. The ratio of this length

[1] The same can be said about the turbulence level which is defined by $\hat{u}/u$. See §3.4.1.

to that defined in Eq. (3.1) is $l_e/L_e = (2/3)^{3/2}A$, which roughly equals one-half. The Reynolds number corresponding to these scales is $Re_{l_e} = \hat{u}l_e/\nu$, which equals $(2/3)^2 A\ Re_T$, or roughly two-fifths Re_T.

Frequently, the turbulent Reynolds number is defined as

$$Re_\lambda = \hat{u}\lambda/\nu \tag{3.6}$$

where λ is Taylor's [1935] micro-scale.[1] For isotropic turbulence, it may be determined from

$$\lambda = \sqrt{15\nu\hat{u}^2/\hat{\varepsilon}} \tag{3.7}$$

In terms of Re_λ, it is easy to show that the four length scales L_e, l_e, l_d and λ are all related according to

$$L_e/l_d = (20/3)^{-3/4} Re_\lambda^{3/2}, \quad l_e/l_d = 15^{-3/4} A Re_\lambda^{3/2}, \quad \text{and} \quad \lambda/l_d = 15^{1/4}\sqrt{Re_\lambda} \tag{3.8}$$

Typically, Re_λ lies between 20 and 100 for grid-generated isotropic turbulence. For a Reynolds number of twenty, $L_e/l_d = 22$, $l_e/l_d = 12$, and $\lambda/l_d = 9$, whereas for $Re_\lambda = 100$, $L_e/l_d = 240$, $l_e/l_d = 130$ and $\lambda/l_d = 20$. The "large" Reynolds numbers generally referred to in the early literature on isotropic turbulence supposedly exceeded 2000. In this case, $L_e/l_d \geq 21{,}600$, $l_e/l_d \geq 12{,}000$ and $\lambda/l_d \geq 90$. Clearly, $L_e > l_e > \lambda > l_d$.

Finally, the relation between the two "turbulent Reynolds numbers" is

$$Re_T = \tfrac{3}{20} Re_\lambda^2 \tag{3.9}$$

For the values of Re_λ just cited, namely, 20, 100 and 2000, the corresponding values of Re_T are 60, 1500 and 600,000.

The spectra of turbulence. For isotropic turbulence, the *spectral distributions* in wave number space are *spherical*. Not to be confused with the kinetic energy, let k represent the wave number of turbulence in this space.[2] Then the *three-dimensional energy spectrum* of turbulence is approximately given by

$$E(k) = \alpha\left(\varepsilon\nu^5\right)^{1/4}\left(k_d/k_e\right)^{5/3}\left(k/k_e\right)^4 \exp\left[-\tfrac{3}{2}r\alpha\left(k/k_d\right)^{4/3}\right]\left[1+\left(k/k_e\right)^2\right]^{-17/6} \tag{3.10}$$

where $\alpha \approx 1.7$ is Kolmogorov's constant,[3] k_e and k_d are the wave numbers of the large and small eddies, that is, $1/l_e$ and $1/l_d$ respectively, and r (which depends on the Reynolds number) approximately equals one. This expression, which was originally obtained by Mayle et al. [1998], matches von Kármán's [1948] formula for small wave numbers, Pao's [1965] solution for large wave numbers, and Kolmogorov's [1941] expression for wave numbers within the *inertial subrange*. Kolmogorov's result, which is valid only when $k_e << k << k_d$, is given by

$$E(k) = \alpha\left(\varepsilon\nu^5\right)^{1/4}\left(k/k_d\right)^{-5/3} \tag{3.11}$$

The intensity of turbulence is obtained by integrating the energy spectrum over all wave numbers. Hence,

$$\hat{u}^2 = \frac{2}{3}\int_0^\infty E(k)\,dk \tag{3.12}$$

where the integral itself equals the kinetic energy of turbulence. Since the relation between the dissi-

[1] For those familiar with Hinze, $\lambda = \lambda_g$.

[2] For anisotropic turbulence, the components of the three-dimensional wave number k_i with $i = 1,2,3$ are different.

[3] See Screenivasan [1995] for a discussion of Kolmogorov's constant.

pation and energy spectra is given by[1]

$$\hat{\varepsilon} = 2\nu \int_0^{\infty} k^2 E(k)\,dk \qquad (3.13)$$

the quantity $k^2E(k)$ represents the dissipation spectrum.

Equations (3.10) and (3.11), as well as $k^2E(k)$ are plotted in Fig. 3.1 for the ratio $k_e/k_d = l_d/l_e = 0.002$ ($Re_\lambda = 250$, $Re_T = 9400$). For smaller values of k_e/k_d, the maxima of these spectrums move farther apart, and Kolmogorov's inertial subrange occupies a greater portion of the energy spectrum. For larger values, the maxima move closer together until, for k_e/k_d larger than about 0.01 ($Re_\lambda < 85$, $Re_T < 1500$) no inertial subrange exists and the curve for $E(k)$ begins to fall below Kolmogorov's relation. Therefore, referring to the discussion following Eq. (3.8), few, if any, of the spectra obtained experimentally using turbulence grids will exhibit a true inertial subrange. Despite this, however, there is evidence that Eq. (3.10) still remains a viable expression for $E(k)$ at these larger k_e/k_d values (smaller Reynolds numbers).

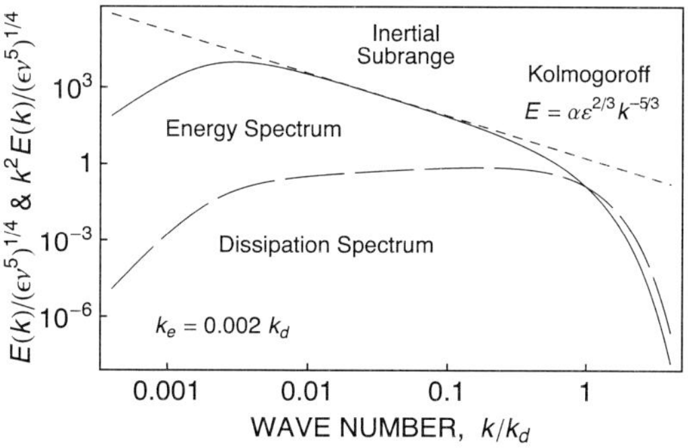

Figure 3.1.
Energy and dissipation spectral distributions for isotropic turbulence and Kolmogorov's result for the inertial subrange.

A length frequently cited in literature is the *integral length scale* of turbulence. Corresponding to the average dimension of the eddies and roughly equal to the size of the large eddies, it is defined as

$$\Lambda = \frac{\pi}{2\hat{u}^2} \int_0^{\infty} \frac{E(k)}{k}\,dk \qquad (3.14)$$

An approximate relation between the integral length scale and l_e is provided in Table 3.1. For large Reynolds numbers, Λ/l_e is virtually equal to the generally accepted value of 0.75. Approximate expressions for the quantities A and r in Eqs. (3.6) and (3.10) are also provided in this table.

Table 3.1
Approximate expressions for $Re_\lambda > 25$.

Λ/l_e	$0.74(1 + 90/Re_\lambda)^{2/9}$
A	$0.80(1 + 24/Re_\lambda)$
r	$1 - 15\,Re_\lambda^{-3/2}$

Numerical tables for these quantities at various Reynolds numbers can be found in Appendix B, to-

[1] See Hinze [1959] for a detailed derivation of this expression and others in this section.

gether with tables for the *one-dimensional energy spectrum* of the streamwise velocity fluctuations. The latter, which is the spectrum most commonly reported in the literature, can be obtained (Hinze [1959]) from

$$E_1(k_1) = \int_{k_1}^{\infty} \frac{E(k)}{k}\left(1 - \frac{k_1^2}{k^2}\right) dk \tag{3.15}$$

Integrating this distribution over all wave numbers yields

$$\hat{u}^2 = \int_0^{\infty} E_1(k_1)\, dk_1 \tag{3.16}$$

which should be compared with Eq. (3.12). An approximate relation for $E_1(k_1)$ that is good up to the third moment of accuracy was also obtained by Mayle et al., namely,

$$\frac{E_1(k_1)}{\left(\varepsilon \nu^5\right)^{1/4}} = \frac{18}{55}\alpha \frac{r\left(k_d/k_e\right)^{5/3}}{1 + r\left(k_1/k_e\right)^{5/3} \exp\left[\frac{14}{3}\left(k_1/k_d\right)\right]} \tag{3.17}$$

An equivalent expression, given in terms of the intensity and integral length scale, is

$$\frac{2\pi E_1(k_1)}{\hat{u}^2 \Lambda} = \frac{4}{1 + r\left(k_1/k_e\right)^{5/3} \exp\left[\frac{14}{3}\left(k_1/k_d\right)\right]} \tag{3.18}$$

This equation is compared to data in Fig. 3.2. In addition, von Kármán's formula for large Reynolds numbers, namely,

$$\frac{2\pi E_1(k_1)}{\hat{u}^2 \Lambda} = 4\left[1 + \left(k_1/k_e\right)^2\right]^{-5/6}$$

is also plotted. In this case, even though an inertial subrange doesn't exist, the agreement between Eq. (3.18) and the data is good.

According to Eq. (3.18), it is natural to use Λ as the length scale for the ordinate in this figure and k_e in the abscissa. For large Reynolds numbers, Λ and $1/k_e$ may be exchanged freely since Λk_e equals a constant. But for small Reynolds numbers, when Λk_1 rather than k_1/k_e is used for the abscissa, which is a common practice, the spectral curves for various Reynolds numbers will overlap near $k_1/k_e = 1$.

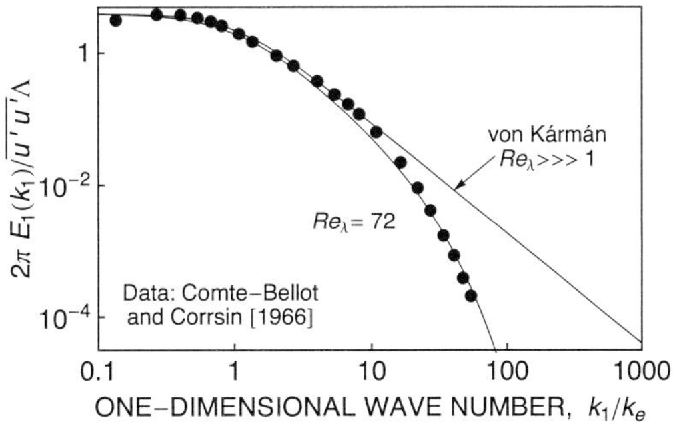

Figure 3.2.
A measured and calculated one-dimensional energy spectrum of isotropic turbulence.

Finally, the one-dimensional energy spectrum of the *transverse* fluctuations is related to that for the streamwise fluctuations (Hinze [1959]) by

$$E_2(k_1)=\frac{1}{2}\left[E_1(k_1)-k_1\frac{\partial E_1(k_1)}{\partial k_1}\right] \tag{3.19}$$

while the three-dimensional energy spectrum can be determined from

$$E(k_1)=\frac{k_1}{2}\left[k_1\frac{\partial^2 E_1(k_1)}{\partial k_1^2}-\frac{\partial E_1(k_1)}{\partial k_1}\right] \tag{3.20}$$

Since spectra are often presented as a function of frequency, it is also worth noting that the relations between the wave number and frequency, and their corresponding spectral functions are

$$k_1=\frac{2\pi}{\bar{u}}f \quad \text{and} \quad E_1(k_1)=\frac{\bar{u}}{2\pi}E_1(f) \tag{3.21}$$

where f is the circular frequency. Note that the dimension of $E_1(f)$ is a length squared divided by time.

3.2 A Phenomenological Approach.

The law of the wall. The law of the wake. An approximate turbulent boundary layer solution.

By "a phenomenological approach" we mean the strategy of solving for turbulent flow based on experimental evidence and some fundamental considerations. During the first half of the 1900s, it eventually became obvious from increasing amounts of data that turbulent boundary layers generally consist of three overlapping layers. These are the *viscous constant-shear layer* immediately next to the wall, the *turbulent constant-shear layer* slightly farther away, and the *intermittent turbulent layer* much farther away in the outer region of the boundary layer. Generally, the thickness of these layers increase in size and the relative change of the mean velocity across them decrease the farther they are from the wall. That is, the greatest change in the mean velocity occurs in the thinnest layer immediately next to the wall. The mean-velocity distribution in the first two layers is generally referred to as the "law of the wall." It only applies to the inner ten percent or less of the boundary layer. The distribution in the remaining ninety-some percent of the boundary layer is referred to as either the "velocity-defect law" or the "law of the wake."

In this section we will develop the mean-velocity distribution within a turbulent boundary layer using a phenomenological approach and then obtain an approximate solution for a turbulent boundary layer on a flat plate with a zero-pressure-gradient. The intent here is simply to review classical thinking, provide its analytical and empirical results for future comparisons, and establish the relevant scales for a turbulent boundary layer.

3.2.1 The law of the wall.

Constant-shear-stress layer. Viscous sublayer. Logarithmic layer. The law of the wall. Near-wall dissipation.

Let's first consider the layer near the wall. According to Eq. (1.49), the total shear stress there varies as

$$\tau=\mu\frac{\partial\bar{u}}{\partial y}-\rho\overline{u'v'}=\tau_0+\frac{dp_\infty}{dx}y+\frac{1}{6}\frac{\rho\tau_0}{\mu^2}\frac{d\tau_0}{dx}y^3+\ldots \tag{3.22}$$

where the time-average bars on $\bar{\tau}$ and $\bar{\tau}_0$ have been dropped. Introducing the friction velocity given by Eq. (2.14), namely,

$$u_\tau=\sqrt{\tau_0/\rho}=\bar{u}_\infty\sqrt{c_f/2} \tag{3.23}$$

and the dimensionless "wall coordinates," defined by

$$u^+ = \frac{\bar{u}}{u_\tau} \quad \text{and} \quad y^+ = \frac{u_\tau y}{\nu}$$

the shear stress distribution for flow with a zero pressure gradient may be written as

$$\frac{\tau}{\tau_0} = 1 + \frac{1}{6}\sqrt{\frac{2}{c_f}}\frac{d\ln c_f}{dRe_x} y^{+3} + \ldots \ ; \quad (d\bar{p}_\infty / dx = 0)$$

which, incidentally, is valid whether the flow is laminar or turbulent. Since c_f for a turbulent boundary layer varies approximately with the inverse of x raised to one-fifth power, as we will see, this expression may be written as

$$\frac{\tau}{\tau_0} \approx 1 - \frac{1}{30}\sqrt{\frac{2}{c_f}}\frac{1}{Re_x} y^{+3} + \ldots \approx 1 - \frac{4}{5}\frac{1}{Re_x} y^{+3} + \ldots \ ; \quad (\text{turbulent, } d\bar{p}_\infty / dx = 0) \tag{3.24}$$

The last approximation follows from the experimental fact that $30\sqrt{(c_f/2)}$ is roughly equal to four-fifths for turbulent boundary layers. Consequently, the second term can be neglected within the layer immediately next to the wall for very large Reynolds numbers yielding

$$\tau = \mu \frac{\partial \bar{u}}{\partial y} - \rho \overline{u'v'} \approx \tau_0 \ ; \quad (y^+ << Re_x^{1/3})$$

Known as the *Couette-flow approximation*, this is equivalent to neglecting the convective terms near the wall in the momentum equation and integrating the result with respect to y.

The thickness of this constant-shear layer can be directly estimated from Eq. (3.24). Supposing that the second term in the expansion cannot be neglected once its value exceeds one-tenth, the shear stress for a Reynolds number equal to 10^7 remains nearly constant within the layer $y^+ < 100$ next to the wall. This corresponds to a thickness of roughly 0.02δ, as we will see, or a very small fraction of the boundary layer. For larger Reynolds numbers, because the boundary-layer thickness increases at a rate greater than that of the constant-shear layer, the constant-shear layer becomes an even smaller fraction of the boundary-layer thickness.

As a first step in solving the closure problem, Boussinesq [1877] introduced an *eddy viscosity* μ_T, defined by the relation

$$-\rho \overline{u'v'} = \mu_T \frac{\partial \bar{u}}{\partial y} \tag{3.25}$$

Unlike its molecular counterpart μ, however, μ_T depends on the motion of the fluid and is therefore unknown. Substituting this into the above equation yields

$$(\nu + \nu_T)\frac{\partial \bar{u}}{\partial y} \approx \frac{\tau_0}{\rho}$$

where ν_T is the *kinematic eddy viscosity* or *eddy diffusivity* of momentum. Clearly, the velocity distribution in the constant-shear layer depends on the molecular viscosity, eddy viscosity, and wall shear stress. Introducing the wall coordinates, this equation transforms into

$$(1 + \nu_T/\nu)\frac{\partial u^+}{\partial y^+} \approx 1 \ ; \quad (d\bar{p}_\infty / dx = 0) \tag{3.26}$$

Near-wall velocity distribution. As unpretentious as the above equation appears, it cannot be solved until something is said about ν_T. From Eqs. (1.47) and (1.64) we see that very near the wall the mean velocity and the turbulent shear stress vary according to the first and third power of y respectively. It then follows that near the wall

$$\nu_T = -\overline{u'v'}\Big/\left(\partial\bar{u}/\partial y\right) \propto y^3 \tag{3.27}$$

This was first pointed out by Reichardt [1951]. Therefore, very near the wall, $\nu_T/\nu << 1$ and Eq. (3.26) reduces to $\partial u^+/\partial y^+ = 1$ with the solution

$$u^+ = y^+ \quad \text{(constant shear, small } y^+\text{)} \tag{3.28}$$

which is simply a restatement of Eq. (1.47). This distribution is associated with that portion of the constant-shear layer commonly referred to as the "viscous sublayer," that is, the layer where momentum is diffused across it by molecular motion. Experimentally, it is found to be valid for $y^+ < 5$.

Farther away from the wall, $\nu_T/\nu >> 1$, that is, momentum is diffused by turbulent motion, and Eq. (3.26) reduces to

$$\frac{\nu_T}{\nu}\frac{\partial u^+}{\partial y^+} \approx 1$$

This is the first time that we must seriously think about the eddy viscosity. Presumably, it reflects the diffusion of momentum by the turbulent eddies. It has the same dimensions as the kinematic viscosity, that is, a length squared divided by a time, which in turn is the same as a velocity times a length. From a dimensional standpoint, we might expect that the characteristic velocity near the wall (based on the previous solution) is u_τ. We might also expect that the characteristic length associated with turbulent momentum transfer is the size of the large eddies. Supposing that the size of these eddies beyond the viscous sublayer, yet still within the constant-shear layer, varies in proportion to the distance away from the wall, dimensional analysis yields

$$\nu_T = \kappa u_\tau y \tag{3.29}$$

where κ is a proportionality constant. Substituting this expression into the preceding equation yields

$$\frac{du^+}{dy^+} = \frac{1}{\kappa y^+}$$

the solution of which is

$$u^+ = \frac{1}{\kappa}\ln y^+ + B \quad \text{(constant shear, large } y^+\text{)} \tag{3.30}$$

where B is a constant of integration. Comparisons with numerous experimental results over the years suggest that the best values for κ, known as *von Kármán's constant*, and B are 0.41 and 5.0 respectively. It is noteworthy that this profile can also be obtained using either Taylor's [1915], Prandtl's [1925, 1927] or Kármán's [1930] mixing-length theories. Furthermore, the profile can also be obtained by considering only the functional forms of the mean-velocity profiles associated with both the *inner* and *outer* regions of the boundary layer using functional and dimensional analysis. In this manner Millikan [1938] showed that the *logarithmic law* arises naturally as the only profile that can exist within the *overlap* region between them. Consequently, as we will see, this law is empirically observed even beyond the region of constant shear.

Equating Eqs. (3.28) and (3.30) using the values of κ and B given above and solving for y^+, we obtain $y^+ = 10.8$ as the position where the two velocity profiles match. If we considered only these two profiles within the constant-shear layer, this distance then represents the thickness of the viscous sublayer δ_V^+, that is, $\delta_V^+ = 10.8$. For a distance Reynolds number of 10^7, this thickness is roughly one-tenth the thickness of the constant-shear layer or an incredibly small fraction of the boundary-layer thickness, namely, 0.002δ. As a consequence, it is nearly impossible to accurately measure the drastic changes of the velocity and Reynolds stresses occurring within it, which in turn hampers any Reynolds-stress modeling within the layer.

Nevertheless, to smoothly match the velocity profiles given by Eqs. (3.28) and (3.30), Reichardt [1951] proposed using

$$\frac{\nu_T}{\nu} = \kappa y^+ - \kappa\delta_V^+ \tanh\frac{y^+}{\delta_V^+} \tag{3.31}$$

which is easily shown to match the behavior for ν_T given by Eq. (3.27) for small values of y^+ and Eq. (3.29) for the large. Substituting this expression into Eq. (3.26) yields

$$u^+ = \int_0^{y^+}\left[1+\kappa z - \kappa\delta_V^+ \tanh\left(z/\delta_V^+\right)\right]^{-1} dz$$

which has no analytical solution. However an approximate solution given by Reichardt is

$$u^+ = \frac{1}{\kappa}\ln\left(1+\kappa y^+\right) + \left(B - \ln\kappa/\kappa\right)\left[1-\exp\left(-y^+/\delta_V^+\right) - \frac{y^+}{\delta_V^+}\exp\left(-y^+/3\right)\right] \tag{3.32}$$

As shown in Fig. 3.3, this expression agrees very well with data in the constant-shear layer of a turbulent boundary layer and even somewhat beyond. Far away from the wall, depending on the Reynolds number, the deviations become significant. The solutions for the viscous sublayer and logarithmic constant-shear layer are shown in the figure. In addition, the dashed line corresponds to a commonly assumed form for a turbulent velocity profile, namely, $u^+ = Cy^{+1/n}$. In this case, the best fit to the law of the wall is obtained when $C = 8.4$ and $n = 7$. As seen in the figure, this *power-law profile* also happens to follow the data farther away from the wall. We will return to this observation later.

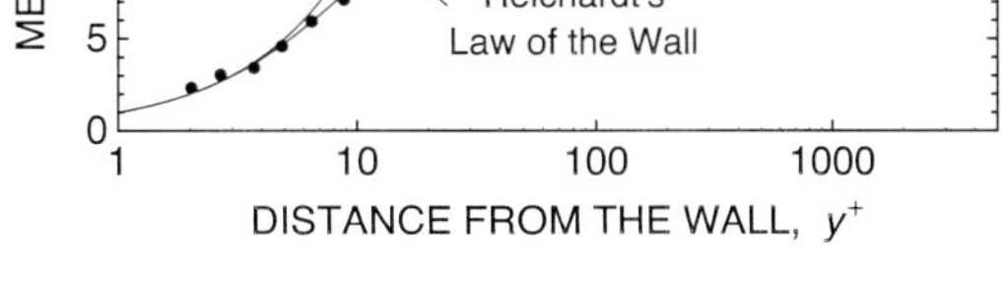

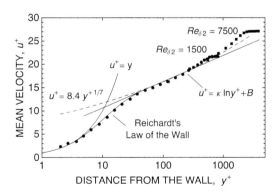

Figure 3.3.
Reichardt's law of the wall compared with measured turbulent boundary-layer profiles plotted in the wall-coordinate system. Data from Klebanoff [1955] and Johnson and Johnston [1989].

Near-wall dissipation. Recalling that the convective terms in the momentum equation are negligible within the constant-shear layer, it follows that they can be neglected in the turbulent-kinetic-energy equation as well. Consequently, Eq. (1.51) reduces to

$$-\overline{u'v'}\frac{\partial\bar{u}}{\partial y} + \frac{\partial}{\partial y}\left[\nu\frac{\partial k}{\partial y} - \overline{v'\left(k+p'/\rho\right)}\right] - \hat{\varepsilon} = 0 \tag{3.33}$$

where the second term in brackets represents the transport of kinetic energy by the turbulence. Within the viscous sublayer, this term can be neglected compared to the viscous term together with the production term since both, according to Eqs. (1.63) and (1.64), vary as y^3. Therefore, within the viscous sublayer, the viscous diffusion of turbulent energy is balanced by its dissipation.

Farther away from the wall, yet within the logarithmic constant-shear layer where the viscous diffusion of turbulent energy can be neglected, the turbulent transport term may also be neglected since the turbulent kinetic energy (as we will see) is nearly a maximum there. Therefore, the production of turbulent energy is balanced by its dissipation. Consequently, the dissipation is given by

$$\hat{\varepsilon} \approx -\overline{u'v'}\frac{\partial \bar{u}}{\partial y} = \nu_T\left(\frac{\partial \bar{u}}{\partial y}\right)^2 = \frac{u_\tau^3}{\kappa y} \quad \text{(logarithmic layer)} \tag{3.34}$$

Since the dissipation within the viscous sublayer near the wall varies linearly with respect to y according to Eq. (1.66), drawing a tangent to the curve given by Eq. (3.34) at $y = \delta_V$ and evaluating it at the wall provides a rough approximation that

$$\hat{\varepsilon}_0 \approx 2\frac{u_\tau^3}{\kappa\delta_V} = 2\frac{u_\tau^4}{\kappa\nu\delta_V^+}$$

Using κ = 0.41 and δ_V^+ = 10.8 yields $\hat{\varepsilon}_0 \approx u_\tau^4/2\nu$, and from Eq. (3.33) we obtain $k \approx \frac{1}{4}\, u_\tau^2 y^{+2}$ through the viscous sublayer. It then follows from Eqs. (3.1) and (3.4) that the characteristic size of the large and small eddies in the sublayer are $L_e \approx \frac{1}{4}(\nu/u_\tau)y^{+3}$ and $l_d \approx \nu/u_\tau$, and their characteristic velocities are $V_e \approx \frac{1}{2} u_\tau y^+$ and $v_d \approx u_\tau$. Therefore, when $y^+ \approx 3/2$, the large and small eddies are indistinguishable and the wave number of the maxima in the energy and dissipation spectra (not surprisingly) virtually correspond.

3.2.2 The law of the wake.

The defect law. Cole's wake function and wake parameter. Outer-layer shear stress and intermittency.

The results presented in this section are primarily empirical because no satisfactory analytical solution for the mean-velocity profile beyond the constant-shear layer has been obtained. Perhaps the best attempt in this regard is that provided by Townsend [1956] who assumed a "wake-like" similarity-flow solution in the outer portion of the boundary layer.

Velocity distribution. It quickly became known from early measurements that the outer portion of turbulent velocity profiles could be correlated reasonably well when plotted in the format

$$\frac{\bar{u}_\infty - \bar{u}}{u_\tau} = f(y/\delta)$$

which is known as the *velocity-defect law*.

Later, after a series of thorough investigations into turbulent boundary layers with favorable and adverse pressure gradients, Clauser [1954] suggested that turbulent boundary layers having outer-flow similarity be called "equilibrium boundary layers," and showed that it was better to replace δ in the above expression with the quantity $\delta_1\bar{u}_\infty/u_\tau$.

Taking a slightly different tack, and after a critical examination of the data, Coles [1956] determined that the mean-velocity in this region is well correlated by the expression

$$u^+ = \frac{1}{\kappa}\ln y^+ + B + \frac{\Pi}{\kappa}w(y/\delta) \tag{3.35}$$

where $w(y/\delta)$ is a universal function called the "wake function" and Π is Coles' wake parameter. Known as the *law of the wake*, Coles normalized the wake function by arbitrarily setting $w(1) = 2$ and

$\int_0^1 w(\eta)d\eta = 1$. Hence, the skin-friction law becomes

$$u_\infty^+ = \frac{1}{\kappa}\ln\delta^+ + B + \frac{2\Pi}{\kappa} \tag{3.36}$$

when we recognize that $u_\infty^+ = \sqrt{(2/c_f)}$. Moreover, the velocity-defect law becomes

$$\frac{\bar{u}_\infty - \bar{u}}{u_\tau} = -\frac{1}{\kappa}\ln\left(y/\delta\right) + \frac{\Pi}{\kappa}\left[2 - w(y/\delta)\right]$$

For an analytical approximation to his empirical results, Coles proposed using

$$w = 2\sin^2\left(\pi y/2\delta\right) = 1 - \cos\left(\pi y/\delta\right) \tag{3.37}$$

for the law of the wake. A quick check will confirm that it meets his normalizing criteria.

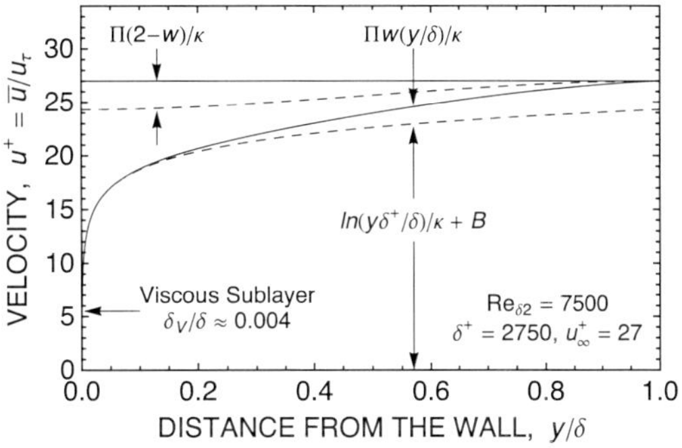

Figure 3.4.
The turbulent mean-velocity profile and its various components plotted in the standard-coordinate system.

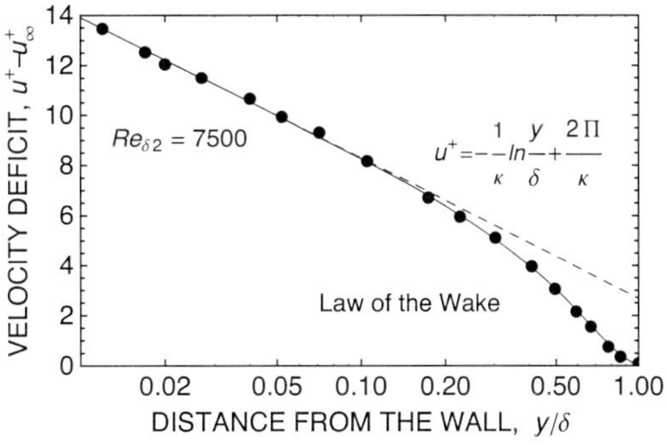

Figure 3.5.
The law of the wake for a constant free-stream velocity plotted in the velocity-defect format. Data from Klebanoff [1955].

The velocity profile of Eq. (3.35) is plotted in Fig. 3.4 for $\delta^+ = 2750$ and $u_\infty^+ = 27$ along with its logarithmic component. In addition, the quantity $\Pi(2\text{–}w)/\kappa$, which is the wake-like component of the defect law, is shown. The same profile, plotted in the velocity-defect format, is shown in Fig. 3.5 compared with the data of Klebanoff [1955]. The agreement is excellent, but it should be kept in mind that this data set was one of many used by Coles to obtain his correlation.

Since Eq. (3.35) with Coles' wake function Eq. (3.37) correlates the mean-velocity through most of the boundary layer, the displacement and momentum thicknesses for a turbulent boundary layer can be obtained from

$$\frac{\delta_1}{\delta} = \int_0^\infty \left(1 - \frac{u}{\bar{u}_\infty}\right) d\frac{y}{\delta} = \frac{1}{\kappa}(1+\Pi)\sqrt{\frac{c_f}{2}}$$

and

$$\frac{\delta_2}{\delta} = \int_0^\infty \frac{u}{\bar{u}_\infty}\left(1 - \frac{u}{\bar{u}_\infty}\right) d\frac{y}{\delta} = \frac{\delta_1}{\delta} - \frac{1}{\kappa^2}\left(2+3.18\Pi+1.52\Pi^2\right)\frac{c_f}{2}$$

As it turns out, however, the wake parameter Π depends on the momentum-thickness Reynolds number, pressure gradient and free-stream turbulence level. For a fully-developed, large Reynolds-number turbulent boundary layer with both the pressure gradient and free-stream turbulence level equal to zero, $\Pi = 0.55$. By fitting Coles' empirical results for various Reynolds numbers, the wake parameter for smaller Reynolds numbers varies approximately as

$$\Pi = 0.55\left\{1 - \exp\left[-0.4\left(Re_{\delta 2}/400 - 1\right)\right]\right\}; \quad \left(Re_{\delta 2} \geq 400,\ d\bar{p}_\infty/dx = 0\right)$$

Below a Reynolds number of about 400, no wake can be discerned and $\Pi = 0$. For the effects of pressure gradients and free-stream turbulence level, the reader is referred respectively to White [1974] and Dyban and Epik [1985]; but now it is just as easy to calculate these profiles using one of the many numerical computational programs.

Outer-layer shear stress. Near the wall, we obtained the mean velocity from a known shear-stress distribution. Farther away from the wall, we can obtain the shear stress from a known mean-velocity distribution. From the x-momentum equation with $dp/dx = 0$, Eq. (1.46), it follows that

$$-\frac{\overline{u'v'}}{u_\tau^2} = 1 + \frac{1}{u_\tau^2}\int_0^y \left(\bar{u}\frac{\partial \bar{u}}{\partial x} + \bar{v}\frac{\partial \bar{u}}{\partial y}\right) dy$$

Since the mean velocity normal to the wall can be calculated by means of the continuity equation, the above expression may be evaluated. The result of Coles' calculation using Klebanoff's [1955] velocity profile with $Re_{\delta 2} = 7500$ is shown in Fig. 3.6. Its agreement with data is excellent but expected for the reasons just pointed out. The solid line near the wall is the turbulent shear stress calculated from Eq. (3.25) using Eqs. (3.26) and (3.31). Its rapid rise to almost the wall value underscores the thinness of this layer and the difficulty of modeling the Reynolds-stress equations within it.

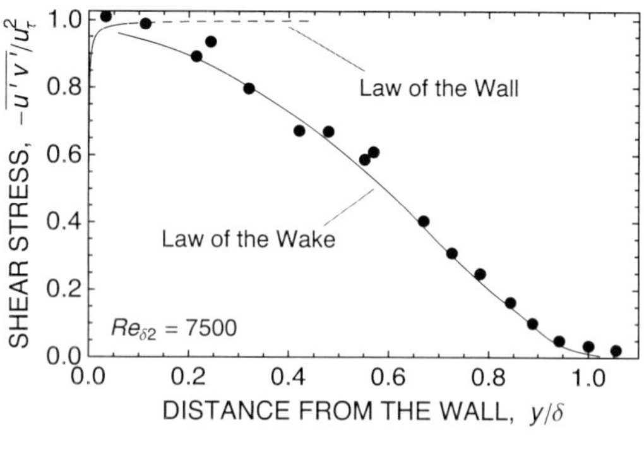

Figure 3.6.
Turbulent shear-stress distribution in a boundary layer for a constant free-stream velocity. Data from Klebanoff [1955].

Intermittency. Because the larger eddies in a turbulent boundary layer protrude into the free stream, the flow near its outer edge randomly alternates between a near-wall turbulent and farther out free-stream flow. Defining the *intermittency*, γ, as the fraction of time the flow is turbulent, its measured variation through the boundary layer is as plotted in Fig. 3.7. As we will learn in §5.2.5, intermittency measurements near zero and one are difficult to accurately obtained. Nevertheless, it is obvious that only about half of the boundary layer may be considered fully turbulent. In some early computational programs (see Cebeci and Smith's [1974] for example) this is taken into account by substituting the quantity ν_T/γ for the eddy viscosity since it is experimentally found to be nearly constant in the outer layer (see Schlichting [1979]). The effect of intermittent flow in the more recent programs, however, is usually absorbed within the constants of their eddy-diffusivity model. More about this later.

A correlation of this data, which is also shown in the figure, is given by

$$\gamma = \frac{1}{2}\,\mathrm{erfc}\left[5\left(y/\delta - 0.8\right)\right] \tag{3.38}$$

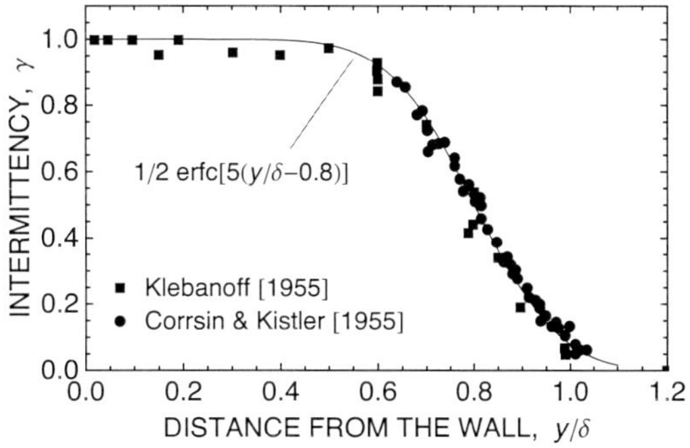

Figure 3.7.
Intermittency distribution in a turbulent boundary layer on a smooth flat plate with a constant free-stream velocity.

3.2.3 An approximate turbulent boundary layer solution.

The following analysis can be traced back to Prandtl [1927, 1932], Kármán [1930], and the measurements of turbulent flow in a pipe by Nikuradse [1926]. As shown in Fig. 3.3, the best fit to the logarithmic law of the wall using a one-seventh-power law is

$$\bar{u} = 8.4 u_\tau \left(u_\tau y/\nu\right)^{1/7} \;;\quad (0 \le y \le \delta)$$

The corresponding skin-friction law, obtained by evaluating this expression at $y = \delta$, is

$$c_f/2 = 0.0241 Re_\delta^{-1/4}$$

Since the velocity distribution may also be written as $\bar{u}/\bar{u}_\infty = (y/\delta)^{1/7}$, the displacement and momentum thicknesses for this profile are

$$\frac{\delta_1}{\delta} = \frac{1}{8} \quad \text{and} \quad \frac{\delta_2}{\delta} = \frac{7}{72} \tag{3.39}$$

The corresponding shape factor is $H = \delta_1/\delta_2 = 1.29$, compared to 2.59 for laminar flow.

The integral-momentum equation for a constant free-stream velocity, Eq. (1.78), may be written as

$$\frac{dRe_{\delta 2}}{dRe_x} = \frac{c_f}{2}$$

Eliminating c_f and δ in this equation by using the foregoing equations, and integrating, yields

$$Re_{\delta 2} = 0.038\,Re_x^{4/5}$$

where the momentum thickness Reynolds number has been arbitrarily set equal to zero at $x = 0$. The skin-friction coefficient is now easily obtained by taking the derivative of this expression with respect to Re_x. This yields $c_f/2 = 0.030\,Re_x^{-1/5}$. A somewhat better fit to skin-friction data is obtained, however, by changing the coefficient from 0.030 to 0.0287. As a result, the corrected expression

$$c_f/2 = 0.0287\,Re_x^{-1/5}\ ; \quad (Re_x \leq 2\ 10^6) \tag{3.40}$$

correlates the skin-friction data when $Re_x \leq 2\ 10^6$. The corrected version for the momentum thickness Reynolds number is then

$$Re_{\delta 2} = 0.036\,Re_x^{4/5}\ ; \quad (Re_x \leq 2\ 10^6) \tag{3.41}$$

For larger Reynolds numbers, the empirical correlation given by Schultz-Grunow [1940] should be used. This skin-friction law, which is valid from $Re_x \approx 10^5$ to very high numbers, is given by

$$c_f/2 = 0.185\left(\log_{10} Re_x\right)^{-2.584}\ ; \quad (Re_x \geq 10^5) \tag{3.42}$$

It is plotted in Fig. 3.8 together with the previous result and the skin-friction law for laminar flow. These results were sketched as the theories in Fig. A of the Introduction, but in a linear-linear rather than the standard log-log format. Presently, c_f is most easily computed numerically.

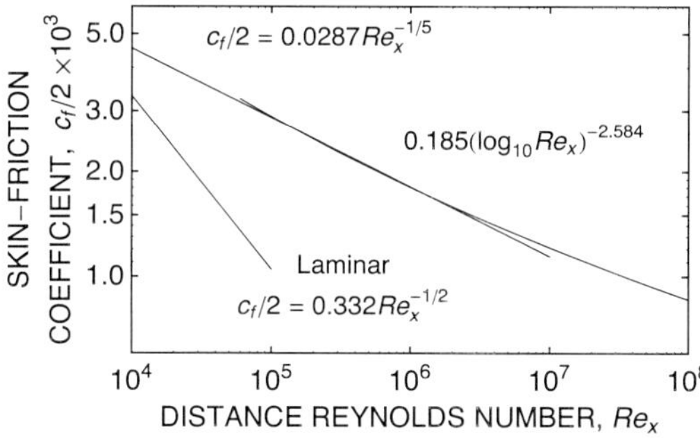

Figure 3.8.
The laminar and turbulent skin-friction laws.

3.3 A Differential Approach.

The relevant equations. Closing the Reynolds-stress equation. A dissipation transport equation.

In this section we will present a popular method for calculating turbulent boundary layers. This method involves solving the equations of motion, the Reynolds-stress transport equations, and a transport equation for the viscous dissipation of turbulent energy. Since the Reynolds-stress and dissipation equations contain unknown third and fourth order terms of the velocity and pressure fluctuations, these must be modeled as functions of the first and second order terms to form a closed set of equations. Modeling the turbulence at this level is generally referred to as the "method of second-moment closure."

Although it is not the author's intent to support any particular model here, the closure discussed in the following pages is essentially that developed by Launder, Hanjalić, Rodi, and their coworkers simply because it is well documented and, although dated, still widely used. The only difference between their models and that presented herein is the treatment of the *near-wall* dissipation component. As a result, some of the functions and constants differ from theirs.

The relevant equations. The basic equations were given in § 1.2 and are repeated here for convenience. The equations for the mean motion, Eqs. (1.18) and (1.22), are

$$\frac{\partial \bar{u}_i}{\partial x_i}=0 \tag{3.43}$$

and

$$\bar{u}_j\frac{\partial \bar{u}_i}{\partial x_j}=-\frac{1}{\rho}\frac{\partial p}{\partial x_i}+\frac{\partial}{\partial x_j}\left[\nu\left(\frac{\partial \bar{u}_i}{\partial x_j}+\frac{\partial \bar{u}_j}{\partial x_i}\right)-\overline{u_i'u_j'}\right] \tag{3.44}$$

where we have set the body forces equal to zero. The Reynolds-stress equation, Eq. (1.32), is

$$\bar{u}_k\frac{\partial \overline{u_i'u_j'}}{\partial x_k}=P_{ij}+\frac{\partial}{\partial x_k}\left(\nu\frac{\partial \overline{u_i'u_j'}}{\partial x_k}\right)-\frac{\partial}{\partial x_k}\left(\overline{u_i'u_j'u_k'}\right)-\left(\overline{u_i'\frac{\partial p'/\rho}{\partial x_j}}+\overline{u_j'\frac{\partial p'/\rho}{\partial x_i}}\right)-\hat{\varepsilon}_{ij}$$

where P_{ij} is the production-rate tensor

$$P_{ij}=-\overline{u_i'u_k'}\frac{\partial \bar{u}_j}{\partial x_k}-\overline{u_j'u_k'}\frac{\partial \bar{u}_i}{\partial x_k} \tag{3.45}$$

The typical approach to modeling the Reynolds-stress equation is to rewrite the pressure-force production terms using the velocity-pressure correlation $\overline{u_i'p'}$. Then the equation takes the form

$$\bar{u}_k\frac{\partial \overline{u_i'u_j'}}{\partial x_k}=P_{ij}+\frac{\partial}{\partial x_k}\left(\nu\frac{\partial \overline{u_i'u_j'}}{\partial x_k}\right)-\frac{\partial}{\partial x_k}\left(\overline{u_i'u_j'u_k'}+\frac{1}{\rho}\overline{u_i'p'}\delta_{jk}\right)+\varphi_{ij}-\hat{\varepsilon}_{ij} \tag{3.46}$$

where

$$\phi_{ij}=\overline{\frac{p'}{\rho}\left(\frac{\partial u_i'}{\partial x_j}+\frac{\partial u_j'}{\partial x_i}\right)}\quad\text{and}\quad \hat{\varepsilon}_{ij}=2\nu\overline{\frac{\partial u_i'}{\partial x_k}\frac{\partial u_j'}{\partial x_k}} \tag{3.47}$$

are the *pressure-strain-rate* and *dissipation* tensors. By contracting Eq. (3.46), the corresponding turbulent-kinetic-energy equation is obtained, namely,

$$\bar{u}_j\frac{\partial k}{\partial x_j}=P+\frac{\partial}{\partial x_j}\left(\nu\frac{\partial k}{\partial x_j}-\overline{u_j'\left(\tfrac{1}{2}u_k'u_k'\right)}-\frac{1}{\rho}\overline{u_i'p'}\delta_{ij}\right)-\hat{\varepsilon} \tag{3.48}$$

where $k=\overline{u_i'u_i'}/2$, $P=P_{ii}/2$, and $\hat{\varepsilon}=\hat{\varepsilon}_{ii}/2$.

To close this set of equations, approximate forms for the last three terms in Eq. (3.46) must be developed. The basic premise underlying all second-moment closures is that these terms should be expressed as functions of the mean-flow quantities, the Reynolds stresses, and the dissipation of kinetic energy. Since the dissipation is unknown, however, a transport equation for $\hat{\varepsilon}$ must also be developed without introducing any other unknown quantities. This modeling is briefly described in the next two subsections.

3.3.1 Modeling terms in the Reynolds stress equation.

Turbulent transport. Pressure-strain modeling. Dissipation.

The turbulent-transport term. There are two components in Eq. (3.46) to the turbulent transport term; namely, the *triple-velocity correlation* representing the transport of the various Reynolds-stress components by the turbulent velocity fluctuations, and the *pressure-diffusion* component. Of these, the triple-velocity correlation can be approximated by

$$-\overline{u_i'u_j'u_k'} = c_s \frac{k}{\hat{\varepsilon}} \left[\overline{u_i'u_l'} \frac{\partial \overline{u_j'u_k'}}{\partial x_l} + \overline{u_j'u_l'} \frac{\partial \overline{u_k'u_i'}}{\partial x_l} + \overline{u_k'u_l'} \frac{\partial \overline{u_i'u_j'}}{\partial x_l} \right] \tag{3.49}$$

where c_s is an empirical constant. Proposed by Hanjalić and Launder [1972], this expression implies a gradient driven diffusion with an eddy viscosity proportional to the product of a stress and the time scale $T_e = k/\hat{\varepsilon}$. Unlike other models, this expression remains unchanged under the permutation of its indices and yields an *anisotropic eddy-viscosity tensor*. This anisotropy is quickly seen by considering a boundary layer where x_2 is taken as the direction normal to the wall. The eddy viscosity in the x_1, x_2 and x_3 directions are then given by

$$c_s \frac{k}{\hat{\varepsilon}} \overline{u_2'u_2'}\,, \quad 3c_s \frac{k}{\hat{\varepsilon}} \overline{u_2'u_2'} \quad \text{and} \quad c_s \frac{k}{\hat{\varepsilon}} \overline{u_2'u_2'}$$

respectively. That is, the turbulent diffusion normal to the wall is *three* times that parallel to the wall.

Another model for this term is that proposed by Daly and Harlow [1970],

$$-\overline{u_i'u_j'u_k'} = c_s \overline{u_k'u_l'} \frac{k}{\hat{\varepsilon}} \frac{\partial \overline{u_i'u_j'}}{\partial x_l} \tag{3.50}$$

Also presuming an eddy viscosity proportional to $\overline{u_i'u_j'}T_e$, this expression yields an isotropic eddy-viscosity tensor even when the turbulence is known to be highly anisotropic, such as near a wall. Despite this, it still remains the "standard" for many turbulent-flow computer programs. As we will see in §3.5, Daly and Harlow's model appears to be the best suited for turbulent boundary layers where the intermittent flow in its outer region is on average "mixed," whereas Hanjalić and Launder's model appears to be the best suited for turbulent flows where the turbulence is highly anisotropic.

In either of these models, since k approaches zero while the dissipation remains finite, T_e and consequently the eddy viscosity approach zero at a wall. Considering this a problem, some investigators[1] have suggested replacing $k/\hat{\varepsilon}$ in these models with the timescale

$$T = k/\hat{\varepsilon} + \sqrt{v/\hat{\varepsilon}}$$

such that the lower bound on the timescale becomes that associated with the dissipative eddies.

The second component of the turbulent-transport term is the pressure-velocity component. Compared with all the other terms in Eq. (3.46), however, it is generally considered small and quickly dispatched by neglecting it.[2] More about this term can be found in Launder [1989].

The pressure-strain-rate term. Without going into details, the pressure-strain term Eq. (3.47) is usually taken to be the sum of three components. Guided by the integral solution[3] to the Poisson equation for ϕ_{ij}, one component accounts solely for interactions between turbulent velocity fluctuations, the second accounts for interactions between the velocity fluctuations and the mean velocity gradients, and the third accounts for wall effects. Denoted by $\phi_{ij,1}$, $\phi_{ij,2}$, and $\phi_{ij,w}$ respectively, these are sometimes referred to as the "slow, rapid, and wall-reflection" components. The distinction between the first two refers to the rate at which they affect the flow. Introducing this notation, the decomposition becomes

[1] Durbin [1991, 1993], Yang and Shih [1993].

[2] Mansour, Kim and Moin [1988], Perot and Moin [1995].

[3] See Chou [1945] and Launder, Reece and Rodi [1975].

$$\phi_{ij} = \overline{\frac{p'}{\rho}\left(\frac{\partial u_i'}{\partial x_j} + \frac{\partial u_j'}{\partial x_i}\right)} = \phi_{ij,1} + \phi_{ij,2} + \phi_{ij,w}$$

Clearly ϕ_{ij} is a symmetric tensor and since $\partial u_k'/\partial x_k = 0$, its trace is also zero. Therefore, it acts only to *redistribute* energy between the various Reynolds-stress components.

Considered initially by Rotta [1951], the first component $\phi_{ij,1}$ promotes a return to isotropy. Consequently, a common strategy for modeling the component is to relate it to the *anisotropic stress tensor*

$$a_{ij} = \frac{\overline{u_i'u_j'}}{k} - \frac{2}{3}\delta_{ij} \tag{3.51}$$

which vanishes if the turbulence is isotropic.[1] This tensor has five independent elements (as $\phi_{ij,1}$). It also has two invariants that remain unaffected by any change in the coordinate system, namely,

$$A_2 = a_{ij}a_{ji} \quad \text{and} \quad A_3 = a_{ij}a_{jk}a_{ki} \tag{3.52}$$

For isotropic turbulence, $A_2 = A_3 = 0$. For *two-dimensional turbulence* where one fluctuating velocity component is negligibly small, such as turbulence very near a wall, $A_2 – A_3 = 8/9$. And for *axisymmetric turbulence*, such as turbulence near the centerline of a variable area circular duct, $A_2^3 = 6A_3^2$. As it turns out, all possible states of turbulence are confined to a region in an A_2A_3-plane defined by these lines.[2] By definition, $A_2 > 0$, while A_3 can be either positive or negative depending on whether A_2 is greater or less than 8/9. The points where the turbulence is both two-dimensional and axisymmetric are $(A_2, A_3) = (2/3, –2/9)$ and $(8/3, 16/9)$. The first is attained near a wall where the axial component of turbulence approaches zero, or on the centerline of a duct under very high acceleration. The second corresponds to *one-dimensional turbulence* where the axial component of turbulence becomes significantly larger than the radial.

Since $A_2 – A_3 = 8/9$ for two-dimensional turbulence, the invariant

$$A = 1 - \tfrac{9}{8}\left(A_2 - A_3\right) \tag{3.53}$$

vanishes in the limit of two-dimensional turbulence. For isotropic turbulence, since $A_2 = A_3 = 0$, $A = 1$. Hence A varies between zero and one. The *degree of two-dimensionality* is measured by the quantity $1–A$. An equivalent expression for A is

$$A = \det\left(\overline{u_i'u_j'}\right) \Big/ \left(2k/3\right)^3 \tag{3.54}$$

which is easily checked to be correct for isotropic turbulence. Since $A_2^3 = 6A_3^2$ for axisymmetric turbulence, the *degree of axial symmetry* is measured by the quantity

$$A_1 = \sqrt{6}\, A_3 \Big/ A_2^{3/2}$$

which varies between plus and minus one.

Returning to the problem of relating $\phi_{ij,1}$ and a_{ij}, it can be shown that the most general relation between two second-rank symmetric tensors with zero trace, say F_{ij} and G_{ij}, is

$$F_{ij} = f_1 G_{ij} + f_2\left(G_{ik}G_{kj} - \tfrac{1}{3}\delta_{ij}G_2\right) \tag{3.55}$$

where the quantities f_1 and f_2 are considered to be functions of the invariants G_2 and G_3 of the tensor G_{ij}, that is, $G_2 = G_{ij}\,G_{ji}$ and $G_3 = G_{ij}\,G_{jk}\,G_{ki}$, and any scalar parameter. Consequently, the most general relation that can be formed between the dimensionless pressure-strain-rate and a_{ij} is

---

[1] See "Supplement on Cartesian Tensors."

[2] See Lumley [1978] for details.

$$\frac{\phi_{ij,1}}{\hat{\varepsilon}} = f_1 a_{ij} + f_2\left(a_{ik}a_{kj} - \tfrac{1}{3}\delta_{ij}A_2\right)$$

where f_1 and f_2 are presumed to be functions of A_2, A_3 and various scalars such as the turbulent Reynolds number Re_T. It follows then that the simplest *nonlinear* model for $\phi_{ij,1}$ is

$$\phi_{ij,1} = f_1\hat{\varepsilon}\left(\frac{\overline{u_i'u_j'}}{k} - \frac{2}{3}\delta_{ij}\right)$$

where $f_1 = fnc(A_2, A_3, Re_T)$.

As pointed out by Launder [1989], most models now use Lumley and Newman's [1977] method of requiring the diagonal component of $\phi_{ij,1}$ to vanish when the same component of the Reynolds stress vanishes. This is usually accomplished by considering f_1 to have the form $f_1 = A^n fnc(A_2, A_3, Re_T)$, which causes *all* components of $\phi_{ij,1}$ to vanish when the turbulence is two-dimensional. Choosing $n =$ 1 yields

$$\phi_{ij,1} = -c_1 f_{\phi 1} A\hat{\varepsilon}\left(\frac{\overline{u_i'u_j'}}{k} - \frac{2}{3}\delta_{ij}\right) \tag{3.56}$$

where c_1 is a positive constant and the sign has been chosen to promote a return to isotropy. For the turbulence model presented herein, a good expression for $f_{\phi 1}$ seems to be

$$f_{\phi 1} = 1 - \exp\left[-\left(Re_T/150\right)^2\right]$$

which is the function suggested by Hanjalić et al.[1]

Equation (3.56) with $f_{\phi 1}A = 1$ was first obtained by Rotta [1951], who also determined from the experiments conducted by Uberoi [1957] on the decay of axisymmetric, anisotropic turbulence that $c_1 \approx 2.5$. It also turns out that this value is reasonable for Uberoi's data using $f_{\phi 1}A$ not equal to unity except near the entrance of the test section where $c_1 \approx 3$ is a better choice.

The second component of the pressure-strain-rate $\phi_{ij,2}$ accounts for interactions between the velocity fluctuations and the mean velocity gradients. The simplest model is obtained by relating it to the anisotropic *production* tensor which, if we consider only the first term in Eq. (3.55), provides

$$\phi_{ij,2} = f_1\left(P_{ij} - \tfrac{2}{3}\delta_{ij}P\right)$$

Using arguments similar to those leading to Eq. (3.56) and using $n = 1/2$, as suggested by Hanjalić et al., we obtain

$$\phi_{ij,2} = -c_2\sqrt{A}\left(P_{ij} - \tfrac{2}{3}\delta_{ij}P\right) \tag{3.57}$$

For nearly isotropic distortion, Launder et al. [1975] point out that c_2 should be about 0.6, while for turbulent boundary layers Hanjalić et al. suggest using $c_2 \approx 0.8$. As discussed by Launder et al. and Launder [1989], the above model is a much too simplified approximation of the term.

Following Launder et al. [1975] on this matter, consider equilibrium, nearly-homogeneous turbulence in uniform shear flow as an example. For this case it can be shown that the production of turbulence by the mean flow is balanced by its dissipation such that $P = \hat{\varepsilon}$, and that its solution for large Reynolds numbers using Eqs. (3.56) and (3.57) reduces to

[1] Most of the empirical functions in the following are either those recently suggested by Hanjalić et al. [1997, 1999], good approximations to them, or previously accepted and widely used variations thereof.

$$a_{ij} = \frac{\overline{u_i'u_j'}}{k} - \frac{2}{3}\delta_{ij} = \frac{1-c_2\sqrt{A}}{1+c_1 A}\left(\frac{P_{ij}}{\hat{\varepsilon}} - \frac{2}{3}\delta_{ij}\right) = n\left(\frac{P_{ij}}{\hat{\varepsilon}} - \frac{2}{3}\delta_{ij}\right)$$

where n is a number. Taking x_1 in the direction of the flow and x_3 in the direction of the velocity gradient, the only nonzero components of the production tensor are P_{11} which equals twice the dissipation, and P_{12} which equals $(a_{22}+2/3)/a_{12}$ times the dissipation. Consequently, $a_{11} = 4n/3$, $a_{22} = a_{33} = -2n/3$, and $a_{12} = -\sqrt{(a_{22}+2/3)}n$. The important point here is that the above model for $\phi_{ij,2}$ yields the same value for both a_{22} and a_{33}, which, as shown in Table 3.2, is not found experimentally. For reference, the values of a_{ij} using Eq. (3.57) and the average experimental value of A are given in the last row. Clearly, the values of a_{22} and a_{12} are reasonably reproduced while the partition of energy between the x_1 and x_3 stress components are grossly in error. This disagreement will again be seen in §3.5.2 when we use this model to calculate the turbulent boundary layer. Although more complex models[1] have been developed to correct this deficiency, they are beyond the scope of this book. Therefore, we will use (as others) the expression given in Eq. (3.57) for $\phi_{ij,2}$.

Table 3.2

Results for equilibrium, nearly-homogeneous turbulence in uniform shear flow.

Investigator	a_{11}	a_{22}	a_{33}	a_{12}	A
Rose [1966]	0.24	–0.19	–0.06	–0.33	0.66
Champagne et al. [1970]	0.25	–0.17	–0.08	–0.38	0.63
Eq. (3.57): $c_1 = 2.5$, $c_2 = 0.6$	0.35	–0.17	–0.17	–0.36	0.645

The third component of the pressure-strain-rate $\phi_{ij,w}$ accounts for the near-wall effect on the large scale eddies (see §3.5.1 for a physical description, and Launder, et al. [1975] and Gibson and Launder [1978] for the model details). Assuming what appears to be a rather obscure form to guaranty its proper behavior regardless the direction the wall faces, it is given by

$$\phi_{ij,w} = c_{1w} f_{\phi,w} \frac{\hat{\varepsilon}}{k}\left(\overline{u_k'u_l'}n_k n_l \delta_{ij} - \tfrac{3}{2}\overline{u_k'u_i'}n_k n_j - \tfrac{3}{2}\overline{u_k'u_j'}n_k n_i\right)$$
$$+ c_{2w} f_{\phi,w}\left(\phi_{kl,2} n_k n_l \delta_{ij} - \tfrac{3}{2}\phi_{ki,2} n_k n_j - \tfrac{3}{2}\phi_{kj,2} n_k n_i\right) \quad (3.58)$$

where the c's are again empirical constants and n_i is the unit vector normal to the wall. Containing two groups of terms, the first increases the anisotropy of the normal stresses as the wall is approached while the second reduces the shear stress.

The function $f_{\phi,w}$ used to scale this effect is generally assumed to have the form

$$f_{\phi,w} = \left(L_e / x_n\right)^m$$

where L_e is the length scale of the large eddies defined in Eq. (3.1) and x_n is the coordinate normal to the wall. This function is of course singular at the wall and sometimes yields values of $\phi_{ij,w}$ that are too large. As a consequence, the function is usually limited to some maximum value.[2] Launder, et al., Gibson and Launder, and Gibson and Rodi [1989] use the exponent $m = 1$ for their boundary-layer calculations, while Demuren and Rodi [1984] use $m = 2$ for their channel calculations. For the turbu-

[1] See Launder [1989] for a description of these.

[2] See Hanjalić et al. [1999] for example.

lence model presented herein, the better value appears to be $m = 1$. In any case, $\phi_{ij,w}$ with the above expression for $f_{\phi,w}$ is clearly important only when the distance from the wall is of the order L_e and less.

The dissipation term. There are basically two paths that can be pursued here. One is to use the dissipation $\hat{\varepsilon}$ for scaling the turbulence and develop a transport equation for it. The other is to use the dissipation associated with the transfer of energy from the large to small eddies, say $\tilde{\varepsilon}$, for scaling and develop a transport equation for it. Taking the latter path[1] and presuming that the difference between the two dissipation scales is found primarily in the viscous layer near a wall, we may express the dissipation as

$$\hat{\varepsilon}_{ij} = \tilde{\varepsilon}_{ij} + \tilde{\varepsilon}_{ij,w} \tag{3.59}$$

where $\tilde{\varepsilon}_{ij}$ represents the dissipation component associated with the transfer process and $\tilde{\varepsilon}_{ij,w}$ represents the dissipation not accounted for by $\tilde{\varepsilon}_{ij}$ as a wall is approached. The contraction of this yields $\hat{\varepsilon} = \tilde{\varepsilon} + \tilde{\varepsilon}_w$ where $\tilde{\varepsilon} = \tilde{\varepsilon}_{ii}/2$ and $\tilde{\varepsilon}_w = \tilde{\varepsilon}_{ii,w}/2$.

Splitting the dissipation tensor into the sum of its anisotropic and isotropic components, we obtain

$$\tilde{\varepsilon}_{ij} = \tilde{\varepsilon}\left(\frac{\tilde{\varepsilon}_{ij}}{\tilde{\varepsilon}} - \frac{2}{3}\delta_{ij}\right) + \frac{2}{3}\delta_{ij}\tilde{\varepsilon}$$

where the quantity in parentheses, usually denoted by e_{ij}, is the *anisotropic dissipation tensor*. Since the trace of the anisotropic component is zero, the first term simply acts to *redistribute* the dissipation among its various components. As a consequence, only the isotropic portion of $\tilde{\varepsilon}_{ij}$ is "dissipative."

If we assume, as we did for the pressure-strain-rate tensor, that e_{ij} is a function of the Reynolds-stress anisotropy tensor a_{ij} and consider only the simplest nonlinear model, it follows that

$$\tilde{\varepsilon}_{ij} = \tilde{\varepsilon} f_{\varepsilon 1}\left(\frac{\overline{u_i' u_j'}}{k} - \frac{2}{3}\delta_{ij}\right) + \frac{2}{3}\delta_{ij}\tilde{\varepsilon} \tag{3.60}$$

where in general $f_{\varepsilon 1} = fnc(A_2, A_3, Re_T)$. Clearly, when the turbulence is isotropic $\tilde{\varepsilon}_{ij} = \frac{2}{3}\delta_{ij}\tilde{\varepsilon}$ irrespective of $f_{\varepsilon 1}$. When $f_{\varepsilon 1}$ equals one, however, the above expression reduces to

$$\tilde{\varepsilon}_{ij} = \frac{\overline{u_i' u_j'}}{k}\tilde{\varepsilon}$$

which is Rotta's [1951] result for the dissipation of large scale eddies in anisotropic turbulent flow.

In the past, $f_{\varepsilon 1}$ was considered a function of Reynolds number, but now it is normally considered a function of the anisotropic Reynolds-stress-tensor invariants (and sometimes the invariants of the anisotropic dissipation tensor). Recalling that A vanishes in the limit of two-component turbulence, the simplest function that yields Rotta's expression on approach to a wall is $f_{\varepsilon 1} = 1 - A^n$. Choosing $n = 1$ for now, we obtain

$$f_{\varepsilon 1} = 1 - A$$

That is, $f_{\varepsilon 1}$ equals the degree of two-dimensionality.

Returning to Eq. (3.59), the component of the dissipation near the wall, after substituting Eq. (3.60) and setting $f_{\varepsilon 1} \approx 1$, becomes

$$\tilde{\varepsilon}_{ij,w} \approx \hat{\varepsilon}_{ij} - \frac{\overline{u_i' u_j'}}{k}\tilde{\varepsilon} \tag{3.61}$$

[1] The other, which is incorporated into most computational boundary-layer programs, is described in Appendix C.

Using the expansions given in Eqs. (1.64)–(1.67) for $\hat{\varepsilon}_{ij}$, $\overline{u_i'u_j'}$ and k, the near-wall-dissipation components in terms of the distance away from the wall, say x_2, become

$$\tilde{\varepsilon}_{11,w} = 2\nu\overline{a_1a_1} + \ldots - 2\frac{\overline{a_1a_1}}{(\overline{a_1a_1} + \overline{c_1c_1})}\tilde{\varepsilon}\,, \quad \tilde{\varepsilon}_{22,w} = 8\nu\overline{b_2b_2}x_2^2 + \ldots - 2\frac{\overline{b_2b_2}}{(\overline{a_1a_1} + \overline{c_1c_1})}\tilde{\varepsilon}x_2^2\,, \quad \text{etc.}$$

where a_n, b_n and c_n are the coefficients of the x_2^n terms in the expansions for u_1', u_2' and u_3', that is, $u_1' = a_1x_2 + a_2x_2^2 + \ldots$, etc. If we presumed that the trace of the large-scale-eddy dissipation $\tilde{\varepsilon}$ vanishes at the wall, then

$$\tilde{\varepsilon}_{11,w} \approx 2\nu\overline{a_1a_1} + \cdots, \quad \tilde{\varepsilon}_{22,w} \approx 8\nu\overline{b_2b_2}x_2^2 + \cdots, \quad \tilde{\varepsilon}_{12,w} \approx 4\nu\overline{a_1b_2}x_2 + \cdots, \quad \text{etc.}$$

Since $\overline{u_1'u_1'} = \overline{a_1a_1}x_2^2 + 2\overline{a_1a_2}x_2^3 + \cdots$, the first of these can be expressed in terms of the normal stress by either

$$\tilde{\varepsilon}_{11,w} = 2\nu\frac{\partial\sqrt{\overline{u_1'u_1'}}}{\partial x_2}\frac{\partial\sqrt{\overline{u_1'u_1'}}}{\partial x_2}, \quad \tilde{\varepsilon}_{11,w} = 2\nu\frac{\overline{u_1'u_1'}}{x_2^2} \quad \text{or} \quad \tilde{\varepsilon}_{11,w} = \nu\sqrt{\left|\frac{\partial\overline{u_1'u_1'}}{x_2\partial x_2}\right|}\sqrt{\left|\frac{\partial\overline{u_1'u_1'}}{x_2\partial x_2}\right|} \tag{3.62}$$

The first relation, or a version of it written in terms the kinetic energy k, was introduced by Jones and Launder [1972] and became the standard near-wall correction for the early k–ε models. The second was introduced by Chien [1982]. As it turns out, however, only the third relation yields $\overline{u_1'u_1'} \propto x_2^2$ as the solution to the near-wall Reynolds-stress equation given by

$$\nu\frac{\partial^2\overline{u_1'u_1'}}{\partial x_2^2} \approx \tilde{\varepsilon}_{11,w}$$

without specifying any condition at the wall other than $\overline{u_1'u_1'} = 0$. Consequently, this becomes the natural choice of the three models for the near-wall dissipation component $\tilde{\varepsilon}_{11,w}$.

Relations similar to Eq. (3.62) are found for the other near-wall dissipation components as well, but to satisfy each component of the near-wall Reynolds-stress equation, the coefficients in the relations containing the velocity normal to the wall must be different. In this case, it's not too difficult to show that a general expression for all the components when one of the coordinates is always taken normal to the wall is given by

$$\frac{\tilde{\varepsilon}_{ij,w}}{\nu} = \mathrm{sgn}\left(\overline{u_i'u_j'}\right)f_{\varepsilon,w}\left[\left|\frac{\partial\overline{u_i'u_j'}}{x_n\partial x_n}\right| + n_in_k\left|\frac{\partial\overline{u_k'u_j'}}{x_n\partial x_n}\right| + n_kn_j\left|\frac{\partial\overline{u_i'u_k'}}{x_n\partial x_n}\right|\right] \tag{3.63}$$

where the *sgn* function properly accounts for the dissipation of the off-diagonal stress components. The function $f_{\varepsilon,w}$ in the above expression, which must equal one at the wall, has been introduced to dampen the contribution of this component away from the wall. Whereas the wall component of the pressure-strain-rate function $f_{\phi,w}$ depends on L_e, we should expect that $f_{\varepsilon,w}$ depends on the dissipation length scale l_d. Based on very limited experience and comparisons with data, it presently appears that a good approximation for $f_{\varepsilon,w}$ is

$$f_{\varepsilon,w} = \exp\left[-0.5\sqrt{x_n/l_d}\right]$$

A contraction of Eq. (3.63) neglecting the smaller $(\partial\overline{u_n'u_n'}/\partial x_n)^2$ terms yields

$$\tilde{\varepsilon}_w = \nu f_{\varepsilon,w}\left|\frac{1}{x_n}\frac{\partial k}{\partial x_n}\right| \quad \text{or} \quad \tilde{\varepsilon}_w = 2\nu f_{\varepsilon,w}\sqrt{\left|\frac{\partial k}{\partial x_n^2}\right|}\sqrt{\left|\frac{\partial k}{\partial x_n^2}\right|}$$

depending on one's predisposition toward symmetry.

A gathering of terms. Substituting all the above models into Eq. (3.46), an approximation to the Reynolds-stress equation is given by

$$\bar{u}_k \frac{\partial \overline{u_i' u_j'}}{\partial x_k} = P_{ij} + \frac{\partial}{\partial x_k}\left[\nu \frac{\partial \overline{u_i' u_j'}}{\partial x_k} + c_s \overline{u_k' u_l'} \frac{k}{\hat{\varepsilon}} \frac{\partial \overline{u_i' u_j'}}{\partial x_l} \right]$$
$$-\left(c_1 f_{\phi 1} A + f_{\varepsilon 1}\right) \tilde{\varepsilon} \left[\frac{\overline{u_i' u_j'}}{k} - \frac{2}{3}\delta_{ij} \right] - c_2 \sqrt{A}\left(P_{ij} - \tfrac{2}{3}\delta_{ij} P\right) - \tfrac{2}{3}\delta_{ij}\tilde{\varepsilon} + \varphi_{ij,w} - \tilde{\varepsilon}_{ij,w}$$

where P_{ij}, $\phi_{ij,w}$ and $\tilde{\varepsilon}_{ij,w}$ are given by Eqs. (3.45), (3.58) and (3.63). Note that the timescale associated with turbulent diffusion is $k/\hat{\varepsilon}$ and not $k/\tilde{\varepsilon}$, otherwise the diffusivity near the wall would be over estimated since $\tilde{\varepsilon} \to 0$. Also note that $k/\hat{\varepsilon}$ approaches zero near the wall since k and $\hat{\varepsilon}$ vary according to x_n^2 and x_n^0 there.

3.3.2 A dissipation transport equation.

Equal to one-half the trace of the dissipation tensor in Eq. (3.47), $\hat{\varepsilon}$ is given by

$$\hat{\varepsilon} = \nu \overline{\frac{\partial u_i'}{\partial x_j} \frac{\partial u_i'}{\partial x_j}}$$

As shown by Davidov [1961], Harlow and Nakayama [1968], and Daly and Harlow [1970], an exact equation for the transport of $\hat{\varepsilon}$ can be obtained by differentiating the equation of motion for the fluctuating velocity Eq. (1.21) with respect to x_k, multiplying through by $2\nu(\partial u_i'/\partial x_k)$, and taking the time-average of the result. This yields

$$\bar{u}_j \frac{\partial \hat{\varepsilon}}{\partial x_j} = -2\nu \left(\overline{\frac{\partial u_i'}{\partial x_k}\frac{\partial u_j'}{\partial x_k}} + \overline{\frac{\partial u_k'}{\partial x_i}\frac{\partial u_k'}{\partial x_j}} \right) \frac{\partial \bar{u}_i}{\partial x_j} - 2\nu \overline{u_j' \frac{\partial u_i'}{\partial x_k}} \frac{\partial^2 \bar{u}_i}{\partial x_j \partial x_k}$$
$$- \frac{\partial}{\partial x_i}\left(\overline{u_i' \varepsilon'} + \frac{2\nu}{\rho} \overline{\frac{\partial u_i'}{\partial x_j}\frac{\partial p'}{\partial x_j}} - \nu \frac{\partial \hat{\varepsilon}}{\partial x_i} \right) - 2\nu \overline{\frac{\partial u_i'}{\partial x_j}\frac{\partial u_i'}{\partial x_k}\frac{\partial u_j'}{\partial x_k}} - 2\overline{\left(\nu \frac{\partial^2 u_i'}{\partial x_j \partial x_k} \right)^2} \tag{3.64}$$

The first two terms represent the production of $\hat{\varepsilon}$ by the mean strain field. It can be shown that these terms are smaller than the other terms by a factor proportional to $Re_T^{1/2}$ and Re_T respectively, and hence, together with the viscous diffusion term, can be neglected for large turbulent-Reynolds-number flows. The next term represents the turbulent transport of $\hat{\varepsilon}$ by the velocity fluctuations, pressure fluctuations, and viscous diffusion. Finally, the last two terms represent the generation and destruction of $\hat{\varepsilon}$ by vortex stretching and viscosity. For large Reynolds numbers, these are the largest terms on the right hand side, but their difference is generally the same order of magnitude as the other terms in the equation.

Clearly, all the terms on the right hand side of this equation (except the viscous diffusion term) must be modeled. In the early 1970s, these terms were modeled using large scale eddy thinking since "wall functions" were used and only the turbulent motion beyond them had to be modeled. Consequently, the earlier approximations of the dissipation equation are more appropriate for $\tilde{\varepsilon}$ than for $\hat{\varepsilon}$ except that they lack a near-wall viscous term.[1] Because this is exactly what's needed for the present model, Hanjalić and Launder's [1972] modeling of the dissipation equation will be used for $\tilde{\varepsilon}$ except near

---

[1] In this regard, the interested reader should compare the dissipation-transport equation in Hanjalić and Launder [1972] with those in Launder, Reece and Rodi [1975], Hanjalić and Launder [1976], Launder [1989] and Hanjalić et al. [1999].

the wall where it must be modified accordingly.

Since the quantity multiplying the mean strain rate in the first term of Eq. (3.64) is related to the components of the dissipation, it is reasonable to assume that its product with the mean-strain rate can be approximated according to

$$2\nu\left(\overline{\frac{\partial u_i'}{\partial x_k}\frac{\partial u_j'}{\partial x_k}}+\overline{\frac{\partial u_k'}{\partial x_i}\frac{\partial u_k'}{\partial x_j}}\right)\frac{\partial \bar{u}_i}{\partial x_j}=c_{\varepsilon 1}\left(\frac{\overline{u_i'u_j'}}{k}\tilde{\varepsilon}\right)\frac{\partial \bar{u}_i}{\partial x_j}=-c_{\varepsilon 1}\frac{\tilde{\varepsilon}}{k}P$$

where $c_{\varepsilon 1}$ is an empirical constant and P is the production of turbulent energy. The generally accepted value for $c_{\varepsilon 1}$ is 1.44.

Using Taylor's [1915] vorticity transport theory, Hanjalić and Launder [1976] proposed that the second term in Eq. (3.64) might be approximated by

$$-2\nu\overline{u_j'\frac{\partial u_i'}{\partial x_k}}\frac{\partial^2 \bar{u}_i}{\partial x_j\partial x_k}=c_{\varepsilon 3}\nu\frac{k\overline{u_l'u_j'}}{\tilde{\varepsilon}}\left(\frac{\partial^2 \bar{u}_i}{\partial x_l\partial x_k}\right)\left(\frac{\partial^2 \bar{u}_i}{\partial x_j\partial x_k}\right)$$

where the generally accepted value for $c_{\varepsilon 3}$ is 2.

The common procedure in modeling the turbulent transport of the dissipation, $\overline{u_i'\varepsilon'}$, is to assume a gradient-driven process for that portion caused by the velocity fluctuations and neglect that caused by the pressure fluctuations. Thus, using Daly and Harlow's diffusion model, Eq. (3.50), the turbulent transport term is approximated by

$$\overline{u_i'\varepsilon'}=-c_\varepsilon\overline{u_i'u_j'}\frac{k}{\hat{\varepsilon}}\frac{\partial\tilde{\varepsilon}}{\partial x_j}$$

Rodi [1971] argued that the last two terms in Eq. (3.64) should be considered together. When the Reynolds number is large enough for an inertial subrange to exist, the sum of these terms depend on the turbulent transport of energy from the low to high wave numbers, which in turn is controlled by the large scale eddies and therefore independent of viscosity. From dimensional considerations, it is then reasonable to expect that the sum of these terms will be proportional to the dissipation divided by the relevant timescale, which in this case is $k/\tilde{\varepsilon}$. Hence, these terms are approximated by

$$2\nu\overline{\frac{\partial u_i'}{\partial x_j}\frac{\partial u_i'}{\partial x_k}\frac{\partial u_j'}{\partial x_k}}+2\overline{\left(\nu\frac{\partial^2 u_i'}{\partial x_j\partial x_k}\right)^2}=c_{\varepsilon 2}f_{\varepsilon 2}\frac{\tilde{\varepsilon}^2}{k}$$

where $c_{\varepsilon 2}$ is a proportionality constant, and $f_{\varepsilon 2}$ is a function that accounts for the viscous effect on the turbulent eddies at low Reynolds numbers. From experiments on the decay of isotropic turbulence, it is found that $c_{\varepsilon 2} = 1.92$ and

$$f_{\varepsilon 2}=1-\frac{c_{\varepsilon 2}-1.4}{c_{\varepsilon 2}}\exp\left[-\left(Re_T/7\right)^2\right]$$

Clearly $f_{\varepsilon 2}$ differs from unity only when Re_T is of the order of seven or less, that is, extremely small. The details regarding $c_{\varepsilon 2}$ and $f_{\varepsilon 2}$ will be fully discussed in §3.4.1.

Finally, the viscous dissipation is that portion of the term

$$2\overline{\left(\nu\frac{\partial^2 u_i'}{\partial x_j\partial x_k}\right)^2}$$

near the wall not taken into account in the previous expression. Supposing that it can be modeled in a fashion similar to the dissipation of turbulent energy, we will consider that

$$2\left[\overline{\left(\nu\frac{\partial^2 u_i'}{\partial x_j \partial x_k}\right)^2}\right]_w = \nu f_{\varepsilon,w}\left|\frac{\partial \tilde{\varepsilon}}{x_n \partial x_n}\right|$$

This differs from the near-wall component proposed by Hanjalić and Launder [1976] in their low-Reynolds number model, namely,

$$2\nu c_{\varepsilon 2} f_{\varepsilon 2}\varepsilon\left(\partial k^{1/2}/\partial x_n\right)/k$$

Substituting all the above expressions into Eq. (3.64), our approximate equation for $\tilde{\varepsilon}$ is given by

$$\bar{u}_j\frac{\partial \tilde{\varepsilon}}{\partial x_j} = c_{\varepsilon 1}\frac{\tilde{\varepsilon}}{k}P + \frac{\partial}{\partial x_i}\left[\nu\frac{\partial \tilde{\varepsilon}}{\partial x_i} + c_\varepsilon \overline{u_i'u_j'}\frac{k}{\hat{\varepsilon}}\frac{\partial \tilde{\varepsilon}}{\partial x_j}\right]$$
$$- c_{\varepsilon 2} f_{\varepsilon 2}\frac{\tilde{\varepsilon}^2}{k} + c_{\varepsilon 3}\nu\frac{k\overline{u_l'u_j'}}{\tilde{\varepsilon}}\left(\frac{\partial^2 \bar{u}_i}{\partial x_l \partial x_k}\right)\left(\frac{\partial^2 \bar{u}_i}{\partial x_j \partial x_k}\right) - \nu f_{\varepsilon,w}\left|\frac{\partial \tilde{\varepsilon}}{x_n \partial x_n}\right| \quad (3.65)$$

where the empirical constants remain to be determined.

3.3.3 A gathering of equations.

The equations developed in the two previous subsections, together with all the model's functions and constants are summarized below. Here it should be noted that the Reynolds-stress and kinetic-energy equations have been written using Hanjalić and Launder's turbulent diffusion model. For Daly and Harlow's model, use only the *last* term of the turbulent diffusion model in these equations.

$$\bar{u}_k\frac{\partial \overline{u_i'u_j'}}{\partial x_k} = P_{ij} + \frac{\partial}{\partial x_k}\left[\nu\frac{\partial \overline{u_i'u_j'}}{\partial x_k} + c_s\frac{k}{\hat{\varepsilon}}\left(\overline{u_i'u_l'}\frac{\partial \overline{u_j'u_k'}}{\partial x_l} + \overline{u_j'u_l'}\frac{\partial \overline{u_k'u_i'}}{\partial x_l} + \overline{u_k'u_l'}\frac{\partial \overline{u_i'u_j'}}{\partial x_l}\right)\right]$$
$$-\left(c_1 f_{\phi 1}A + f_{\varepsilon 1}\right)\tilde{\varepsilon}\left[\frac{\overline{u_i'u_j'}}{k} - \frac{2}{3}\delta_{ij}\right] - c_2\sqrt{A}\left[P_{ij} - \frac{2}{3}\delta_{ij}P\right] - \frac{2}{3}\delta_{ij}\tilde{\varepsilon} + \phi_{ij,w} - \tilde{\varepsilon}_{ij,w} \quad (3.66)$$

$$\bar{u}_j\frac{\partial k}{\partial x_j} = P + \frac{\partial}{\partial x_j}\left[\nu\frac{\partial k}{\partial x_j} + c_s\frac{k}{\hat{\varepsilon}}\left(\overline{u_k'u_l'}\frac{\partial \overline{u_j'u_k'}}{\partial x_l} + \overline{u_j'u_l'}\frac{\partial k}{\partial x_l}\right)\right] - \tilde{\varepsilon} - \nu f_{\varepsilon,w}\left|\frac{1}{x_n}\frac{\partial k}{\partial x_n}\right| \quad (3.67)$$

$$\bar{u}_j\frac{\partial \tilde{\varepsilon}}{\partial x_j} = c_{\varepsilon 1}\frac{\tilde{\varepsilon}}{k}P + \frac{\partial}{\partial x_i}\left[\nu\frac{\partial \tilde{\varepsilon}}{\partial x_i} + c_\varepsilon \overline{u_i'u_j'}\frac{k}{\hat{\varepsilon}}\frac{\partial \tilde{\varepsilon}}{\partial x_j}\right]$$
$$- c_{\varepsilon 2} f_{\varepsilon 2}\frac{\tilde{\varepsilon}^2}{k} + c_{\varepsilon 3}\nu\frac{k\overline{u_l'u_j'}}{\tilde{\varepsilon}}\left(\frac{\partial^2 \bar{u}_i}{\partial x_l \partial x_k}\right)\left(\frac{\partial^2 \bar{u}_i}{\partial x_j \partial x_k}\right) - \nu f_{\varepsilon,w}\left|\frac{\partial \tilde{\varepsilon}}{x_n \partial x_n}\right| \quad (3.68)$$

where

$$P_{ij} = -\overline{u_i'u_k'}\frac{\partial \bar{u}_j}{\partial x_k} - \overline{u_j'u_k'}\frac{\partial \bar{u}_i}{\partial x_k}, \qquad P = \frac{P_{kk}}{2} \quad (3.69)$$

$$\phi_{ij,w} = f_{\phi,w}\left[c_{1w}\frac{\tilde{\varepsilon}}{k}\left(\overline{u'_k u'_l}n_k n_l \delta_{ij} - \frac{3}{2}\overline{u'_k u'_i}n_k n_j - \frac{3}{2}\overline{u'_k u'_j}n_k n_i\right)\right.$$
$$\left. + c_{2w}\left(\phi_{kl,2}n_k n_l \delta_{ij} - \frac{3}{2}\phi_{ki,2}n_k n_j - \frac{3}{2}\phi_{kj,2}n_k n_i\right)\right] \qquad (3.70)$$

$$\frac{\tilde{\varepsilon}_{ij,w}}{\nu} = \mathrm{sgn}\left(\overline{u'_i u'_j}\right) f_{\varepsilon,w}\left[\left|\frac{\partial\overline{u'_i u'_j}}{x_n \partial x_n}\right| + n_i n_k\left|\frac{\partial\overline{u'_k u'_j}}{x_n \partial x_n}\right| + n_k n_j\left|\frac{\partial\overline{u'_i u'_k}}{x_n \partial x_n}\right|\right] \qquad (3.71)$$

with

$$f_{\phi 1} = 1 - \exp\left[-\left(Re_T/150\right)^2\right], \quad f_{\phi,w} = L_e/x_n$$
$$f_{\varepsilon 1} = 1 - A, \quad f_{\varepsilon 2} = 1 - \frac{c_{\varepsilon 2} - 1.4}{c_{\varepsilon 2}}\exp\left[-\left(Re_T/7\right)^2\right], \quad f_{\varepsilon,w} = \exp\left[-0.5\sqrt{x_n/l_d}\right] \qquad (3.72)$$

and

$$A = \det\left(\overline{u'_i u'_j}\right)/\left(2k/3\right)^3 \qquad (3.73)$$

As a reminder, n_i and x_n are the unit normal vector and the direction normal to a wall. In addition, for isotropic turbulence, the anisotropic Reynolds-stress invariant A equals one, whereas near the wall for two-dimensional turbulence it equals zero.

The constants used to correlate various empirical data sets are given in Table 3.3. While most of these constants nearly equal the values found in the literature, the value of c_s in the second row is significantly different from that in the third. The reason for this is that Hanjalić and Launder's diffusion model Eq. (3.49) was used to correlate the shear-free turbulent boundary-layer data, while Daly and Harlow's model Eq. (3.50) was used to correlate the turbulent boundary-layer data. Neither of these models correlated both of these sets using the same constants. We will return to this point in §3.5.

Table 3.3
Empirical constants for Eqs. (3.66)–(3.72).

Obtained by correlating:	c_1	c_2	$c_{\varepsilon 1}$	$c_{\varepsilon 2}$	$c_{\varepsilon 3}$	c_s	c_ε	c_{1w}	c_{2w}
Free-stream turbulence data	2.5	0.6	1.44	1.92	–	–	–	–	–
Shear-free turbulent boundary-layer data	2.5	–	–	1.92	–	0.03	0.18	0.14	–
Turbulent boundary-layer data	2.5	0.8	1.44	1.92	2	0.22	0.18	0.08	0.06

3.4 Free-Stream Turbulence.

Decay of isotropic and anisotropic turbulence.

We briefly considered two cases concerning free-stream turbulence in the previous section. Namely, the decay of anisotropic turbulence in a uniform flow with a constant mean velocity, where the advection, redistribution and dissipation of turbulence prevailed,[1] and equilibrium, anisotropic turbulence

[1] See the discussion regarding Eq. (3.56) about c_1.

in a uniform shear flow, where the production, redistribution and dissipation of turbulence prevailed.[1]

In this section we will consider two other cases of free-stream turbulence. First, we will consider the decay of isotropic turbulence in a uniform flow with a constant mean velocity, where only the advection and dissipation of turbulence are significant, and then the decay of anisotropic turbulence with a variable mean velocity, where all the terms in the Reynolds-stress equation except the diffusion term are important.

3.4.1 Decay of isotropic turbulence.

Consider turbulence in a uniform flow of constant velocity downstream of a grid placed perpendicular to the flow. Without any energy sources to sustain it, the turbulence simply decays with distance. Sufficiently downstream of the grid, providing the grid is uniform, the normal-stress components of turbulence become equal and the shear-stress components become zero. That is, the turbulence becomes isotropic. If the turbulence decays "slowly" such that its diffusion can be neglected compared to its dissipation, then the turbulence may be considered *quasi-homogeneous* in the direction of the flow and homogeneous in planes perpendicular to the flow. Comparing the diffusion and dissipation terms in the kinetic energy equation, Eq. (3.67), quasi-homogeneity is attained when the distance associated with the decay process is much larger than the size of the large eddies.

Then for flow in the x direction with a uniform mean velocity $\bar{u}_\infty$, Eqs. (3.67) and (3.68) reduce to

$$\bar{u}_\infty \frac{\partial k}{\partial x} = -\tilde{\varepsilon} \quad \text{and} \quad \bar{u}_\infty \frac{\partial \tilde{\varepsilon}}{\partial x} = -c_{\varepsilon 2} f_{\varepsilon 2} \frac{\tilde{\varepsilon}^2}{k} \tag{3.74}$$

where

$$c_{\varepsilon 2} f_{\varepsilon 2} = c_{\varepsilon 2} - \left(c_{\varepsilon 2} - 1.4\right)\exp\left[-(Re_T/7)^2\right]$$

as initially proposed by Hanjalić and Launder [1976].[2]

When the turbulent Reynolds number is either very large or very small, $c_{\varepsilon 2} f_{\varepsilon 2}$ is constant[3] and the solutions to Eq. (3.74) are

$$T_e = \frac{k}{\tilde{\varepsilon}} = \frac{1}{n}\frac{(x - x_0)}{\bar{u}_\infty} \quad \text{and} \quad k = C\left[\frac{1}{n}\frac{(x - x_0)}{\bar{u}_\infty}\right]^{-n}$$

where $n = 1/(c_{\varepsilon 2} f_{\varepsilon 2} - 1)$, x_0 is the *virtual origin of turbulence*, and C is an integration constant. Thus in both cases the timescale of turbulence increases in direct proportion to the elapsed time from a virtual origin, but not at the same rate nor from the same virtual origin. In addition, the length scales of turbulence as defined in Eqs. (3.1) and (3.4) vary according to

$$L_e \propto \left(x - x_0\right)^{(2-n)/2} \quad \text{and} \quad l_d \propto \left(x - x_0\right)^{(1+n)/4}$$

From experiments on decaying grid-generated turbulence, the best fit to data indicate that $n = 1.09$ in the initial stage of decay when the Reynolds number is large. That is, the turbulent kinetic energy decays approximately as the inverse of distance or the elapsed time. It also implies that L_e and l_d increase approximately with the square root of distance, and, since $Re_T = k^2/\nu\tilde{\varepsilon}$ is proportional to $(x - x_0)^{1-n}$, that the turbulent Reynolds number is nearly constant. Furthermore, for this value of n, after realizing that $f_{\varepsilon 2} = 1$ for large Re_T, we obtain $c_{\varepsilon 2} = 1.92$, which is the value given in Table 3.3.

[1] See the discussion regarding Eq. (3.57) about c_2.

[2] The value of seven in the exponential of this expression differs from their value of six to account for the change in $c_{\varepsilon 2}$ from their value of 1.8 to the presently more accepted value of 1.92.

[3] See Hinze [1959].

Theoretically, in the final stage of decay when the Reynolds number is small and even the largest eddies are strongly affected by viscosity, it is found that $n = 2.5$.[1] Consequently, the turbulent kinetic energy decays as the inverse of the five-halves power of distance, the length scale L_e decreases by the one-quarter power of distance, while l_d increases by the seven-eighths power, and the Reynolds number decreases by the three-halves power of the distance or elapsed time. In this case, $c_{\varepsilon 2} f_{\varepsilon 2} = (1+n)/n = 7/5$ which is the origin of the number 1.4 in the previous expression for $c_{\varepsilon 2} f_{\varepsilon 2}$.

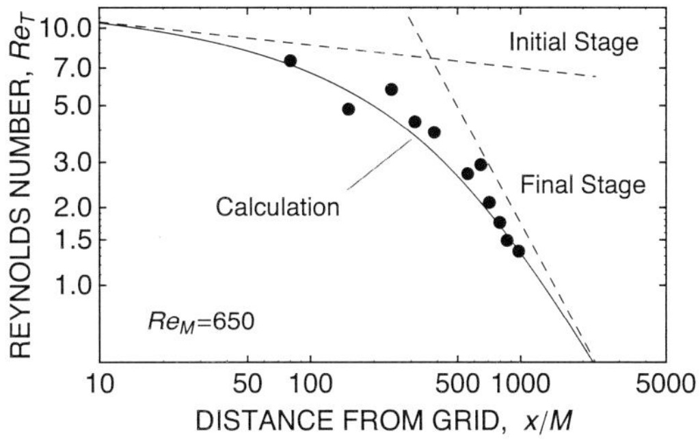

Figure 3.9.
Calculated and measured variations of turbulent Reynolds number downstream of a grid. Data from Batchelor and Townsend [1948].

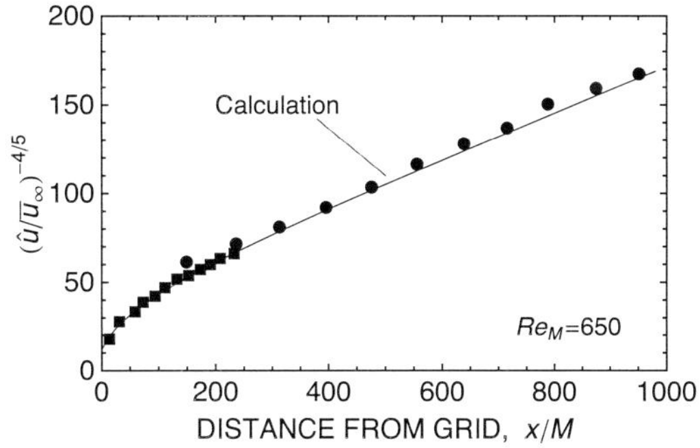

Figure 3.10.
Calculated and measured variations of the intensity of turbulence downstream of a grid. Data from Batchelor and Townsend [1948].

The solutions to Eq. (3.74) are plotted in Figs. 3.9 and 3.10. The data in these figures are from Batchelor and Townsend [1948] for a grid Reynolds number $Re_M = 650$ where M is the grid's mesh size. In the first figure the Reynolds number variation with distance from the grid is plotted. In addition, the Reynolds number variation for both the initial and final stages of decay are shown. The variation of the intensity of turbulence with distance is presented in the second figure, where the intensity is given by

$$\hat{u} = \sqrt{2k/3}$$

This plot is shown using the typical format for the final stage of decay where the intensity should vary according to the inverse four-fifths power of x. The correlation is reasonably good, as should be expected, since this is the data from which the factor $c_{\varepsilon 2} f_{\varepsilon 2}$ was determined.

In this case, that is, for isotropic free-stream turbulence, the turbulence level is defined by

[1] See Hinze [1959] for a detailed analysis of this problem.

$$Tu_\infty = \hat{u}_\infty / \bar{u}_\infty = \sqrt{(\overline{u'u'})_\infty} / \bar{u}_\infty \tag{3.75}$$

which is easily obtained from measurements using a single hot-wire anemometer.

3.4.2 Anisotropic turbulence with a variable streamwise velocity.

Now consider the decay of turbulence in a two-dimensional duct of variable height. Again the turbulence may be considered quasi-homogeneous in the direction of the flow and homogeneous near the central plane of the duct in planes perpendicular to the flow. For flow in the x-direction with a mean velocity of $\bar{u}_\infty(x)$, Eqs. (3.66)–(3.68) reduce to

$$\bar{u}_\infty \frac{d\overline{u'u'}}{dx} = -2\overline{u'u'}\frac{d\bar{u}_\infty}{dx} - \left(c_1 f_{\phi 1} A + f_{\varepsilon 1}\right)\frac{\tilde{\varepsilon}}{k}\left[\overline{u'u'} - \frac{2}{3}k\right] + \frac{2}{3}c_2\sqrt{A}\left(2\overline{u'u'} + \overline{v'v'}\right)\frac{d\bar{u}_\infty}{dx} - \frac{2}{3}\tilde{\varepsilon} \tag{3.76}$$

$$\bar{u}_\infty \frac{d\overline{v'v'}}{dx} = 2\overline{v'v'}\frac{d\bar{u}_\infty}{dx} - \left(c_1 f_{\phi 1} A + f_{\varepsilon 1}\right)\frac{\tilde{\varepsilon}}{k}\left[\overline{v'v'} - \frac{2}{3}k\right] - \frac{2}{3}c_2\sqrt{A}\left(\overline{u'u'} + 2\overline{v'v'}\right)\frac{d\bar{u}_\infty}{dx} - \frac{2}{3}\tilde{\varepsilon} \tag{3.77}$$

$$\bar{u}_\infty \frac{dk}{dx} = \left(\overline{v'v'} - \overline{u'u'}\right)\frac{d\bar{u}_\infty}{dx} - \tilde{\varepsilon} \tag{3.78}$$

$$\bar{u}_\infty \frac{d\tilde{\varepsilon}}{dx} = c_{\varepsilon 1}\frac{\tilde{\varepsilon}}{k}\left(\overline{v'v'} - \overline{u'u'}\right)\frac{d\bar{u}_\infty}{dx} - c_{\varepsilon 2} f_{\varepsilon 2}\frac{\tilde{\varepsilon}^2}{k} \tag{3.79}$$

where the expressions for $f_{\varepsilon 1}$, $f_{\varepsilon 2}$ and $f_{\phi 1}$ are given in Eq. (3.72).

The equations for the shear-stress components are all functionally the same. In particular, the equation for $\overline{u'v'}$ is

$$\bar{u}_\infty \frac{d\overline{u'v'}}{dx} = -\left(c_1 f_{\phi 1} A + f_{\varepsilon 1}\right)\frac{\tilde{\varepsilon}}{k}\overline{u'v'}$$

Since this equation is linear and the function multiplying $\overline{u'v'}$ on the right-hand side of the equation is well behaved, its solution is $\overline{u'v'} = 0$ when $\overline{u'v'}$ initially equals zero. Consequently, if we suppose that the turbulence is initially isotropic in the duct, all the Reynolds shear-stress components will remain zero even as the mean-flow velocity varies along the duct. In this case, only Eqs. (3.76)–(3.79) need to be solved.

From Eq. (3.73), the invariant A for this flow is

$$A = \frac{27}{8}\frac{\overline{u'u'}\,\overline{v'v'}\,\overline{w'w'}}{k^3} \tag{3.80}$$

where $\overline{w'w'} = 2k - \overline{u'u'} - \overline{v'v'}$.

The data from one set of Roach and Brierley's [1990] tests with a variable free-stream velocity are shown in Fig. 3.11. The turbulence entering the duct was nearly isotropic and the shear-stress components along the duct were virtually zero. The initial turbulence level for this test was about three percent and the turbulent Reynolds number Re_T was about a hundred. The measured variation of the mean-flow velocity and the sixth order polynomial fit used for the calculations are shown in the upper portion of the figure, whereas the variation of the turbulent components with distance are plotted in the lower. These data were obtained in the free stream just outside the viscous boundary layer. The calculations were performed using the constants given in the first row of Table 3.3. It turns out that the invariant A for this data is nearly unity along the duct implying that the turbulence is nearly isotropic. More specifically, the greatest deviation from isotropy for this data is about four percent, which one might find surprising considering the difference between the various components.

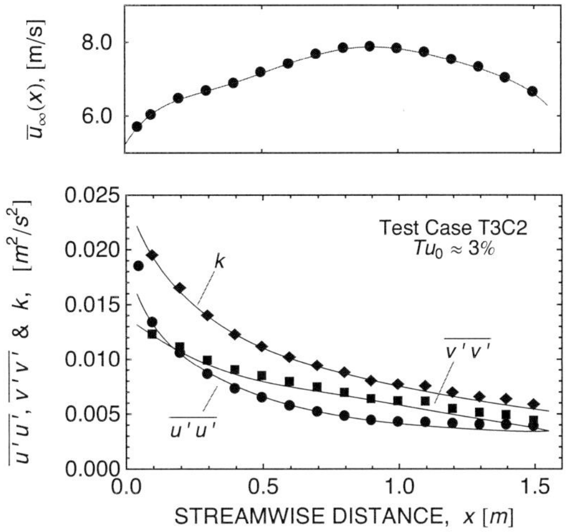

Figure 3.11.
Calculated and measured variations of turbulent stresses downstream of a grid with a variable free-stream velocity. Data from Roach and Brierley [1990].

For anisotropic free-stream turbulence, the turbulence level is defined by

$$Tu_\infty = \sqrt{2k_\infty/3}\Big/\overline{u}_\infty = \sqrt{(\overline{u'u'}+\overline{v'v'}+\overline{w'w'})_\infty/3}\Big/\overline{u}_\infty \qquad (3.81)$$

For the data plotted in Fig. 3.11, the maximum difference between using all three normal-stress components to calculate the free-stream turbulence level versus just the streamwise component as in Eq. (3.75) is about ten percent and occurs near the streamwise distance of one meter.

3.5 Near-Wall Turbulence.

Shear-free turbulent boundary layers. Turbulent boundary layers.

In this section we will examine two cases of near-wall turbulence. First we will consider the turbulent boundary layer that develops along a wall moving parallel to and at the same speed as the turbulent free stream. This flow, which is an important ingredient to understanding laminar flow with free-stream turbulence, is traditionally called a “shear-free” turbulent boundary layer. Second, we will consider the standard zero-pressure-gradient turbulent boundary layer on a flat plate. In both cases we will present the relevant equations and compare the calculated results with data.

As we will see, these two flows are fundamentally different. The first, the shear-free flow, simply concerns the interaction of a turbulence field with a solid wall, whereas the second concerns the interaction of a turbulent flow with a solid wall and its non-turbulent free stream. In the first, since the wall moves at the same speed as the free stream, the mean-flow velocity gradient and hence production of turbulent energy is zero. Thus, this flow concerns only the diffusion, redistribution, and decay of turbulent energy, whereas that in the standard turbulent boundary layer concerns in addition the production of turbulent energy and the boundary layer’s interaction with its free stream. As might be expected, the production, or lack thereof for the shear-free flow, significantly affects the turbulence in the inner region of these flows, while the free-stream interaction or lack thereof significantly affects the turbulent diffusivity in the outer regions. In particular, regarding the latter, and as previously mentioned, the author found that Hanjalić and Launder’s anisotropic model correlated the shear-free results best, Daly and Harlow’s isotropic model correlated the turbulent-boundary-layer results best, while neither reproduced the observed partitioning of turbulent energy in the outer portions of both without changing some of the constants in the model. Since the outer portion of a turbulent boundary layer is intermittent and on average well mixed, this might be expected.

Finally, all the following results were calculated using a modified Patankar-Spalding [1970] algorithm. To capture the details near the wall, one hundred and twenty nodes were taken across the flow about half of which for the turbulent boundary layer were within the distance of $y^+ = 50$.

3.5.1 Shear-free turbulent boundary layers.

Consider turbulence near a wall where both the wall and the mean velocity of the turbulence have the same speed. This is best achieved experimentally[1] by investigating wind-tunnel turbulence near a belt that slides along a flat wall at the same velocity as the mean flow. Physically, by changing to a coordinate system moving with the flow, the problem is the same as investigating the temporal change to a large field of turbulence after suddenly inserting a large, thin flat plate into the field.[2] In this case, the turbulence far away from the plate simply decays with time, while, in response to the boundary conditions at the plate, a boundary layer of turbulence develops near it. If the turbulence far from the plate is homogeneous, then the turbulence near the plate will be axisymmetric about an axis perpendicular to the plate. And if the distant turbulence is isotropic, then the shear-stress components of turbulence will be zero everywhere. In this case then, we only need to consider the diffusion and dissipation of the normal stresses, and the redistribution or "repartitioning" of energy between them near the plate.

Without resorting to any mathematics, first consider what must happen to the large eddies as they approach the surface of the moving belt or inserted plate. If we treat an eddy as a separate fluid entity, it's not too difficult to imagine that the eddy will begin to flatten when it moves to within one diameter from the surface and continue to flatten as it moves closer. Since this distortion transfers the turbulent energy within the eddy from that associated with fluctuations *normal* to the surface to the energy associated with fluctuations *parallel* to the surface, it follows that an anisotropic layer of turbulence must develop near the surface roughly equal to the size of the larger free-stream or distant eddies L_∞; where we will henceforth differentiate between the local size of the turbulent eddies and its free-stream value by using L_e and L_∞. This distortion forms what may be called an "inertial boundary layer of turbulence."

It is also not difficult to imagine what happens when the dissipative eddies approach the surface. They will not only distort, but also dissipate. Using a similar notation, the thickness of this layer will be about the size of the smaller free-stream or distant dissipative eddies l_∞. This forms the *viscous layer of turbulence*. If the turbulent Reynolds number is large, more precisely, if $Re_T^{3/4} >> 1$ according to Eqs. (3.8) and (3.9), then $l_\infty << L_\infty$ and the thickness of these two layers of turbulence will be drastically different. For smaller Reynolds numbers, however, the two length scales and, therefore, the inertial and viscous boundary-layer thicknesses will be nearly the same.

This has important implications for pre-transitional flow with free-stream turbulence. Namely, if the turbulent Reynolds number is large, then the normal component of turbulent energy just outside the laminar boundary layer will be much smaller than either its free-stream value or streamwise component even when the free-stream turbulence is isotropic. On the other hand, if the Reynolds number is small, all components will be nearly the same.

Realizing that either the steady or unsteady viewpoint of this flow can be taken, consider the unsteady version of a plate suddenly "inserted" on the $y = 0$ plane at the time $t = 0$. Also consider before the plate's insertion that the turbulence is everywhere both isotropic and homogeneous. Then Eqs. (3.66)–(3.68) using the transformation $\partial/\partial t = \bar{u}\partial/\partial x$ with $\bar{v} = 0$, and $\overline{u_i'u_j'} = 0$ for $i \neq j$, reduce to

[1] Uzkan and Reynolds [1967], Thomas and Hancock [1977] and Aaronson, Johansson, and Löfdahl [1997].

[2] Hunt and Graham [1978] and Perot and Moin [1995].

$$\frac{\partial\overline{u'u'}}{\partial t}=\frac{\partial}{\partial y}\left[\left(\nu+c_s\overline{v'v'}\frac{k}{\hat{\varepsilon}}\right)\frac{\partial\overline{u'u'}}{\partial y}\right]$$

$$-\left(c_1 f_{\phi 1}A+f_{\varepsilon 1}\right)\frac{\tilde{\varepsilon}}{k}\left[\overline{u'u'}-\frac{2}{3}k\right]+c_{1w}f_{\phi,w}\frac{\tilde{\varepsilon}}{k}\overline{v'v'}-\frac{2}{3}\tilde{\varepsilon}-\nu f_{\varepsilon,w}\left|\frac{\partial\overline{u'u'}}{y\partial y}\right| \tag{3.82}$$

$$\frac{\partial\overline{v'v'}}{\partial t}=\frac{\partial}{\partial y}\left[\left(\nu+3c_s\overline{v'v'}\frac{k}{\hat{\varepsilon}}\right)\frac{\partial\overline{v'v'}}{\partial y}\right]$$

$$-\left(c_1 f_{\phi 1}A+f_{\varepsilon 1}\right)\frac{\tilde{\varepsilon}}{k}\left[\overline{v'v'}-\frac{2}{3}k\right]-2c_{1w}f_{\phi,w}\frac{\tilde{\varepsilon}}{k}\overline{v'v'}-\frac{2}{3}\tilde{\varepsilon}-3\nu f_{\varepsilon,w}\left|\frac{\partial\overline{v'v'}}{y\partial y}\right| \tag{3.83}$$

$$\frac{\partial k}{\partial t}=\frac{\partial}{\partial y}\left[\nu\frac{\partial k}{\partial y}+c_s\overline{v'v'}\frac{k}{\hat{\varepsilon}}\frac{\partial(k+\overline{v'v'})}{\partial y}\right]-\tilde{\varepsilon}-\nu f_{\varepsilon,w}\left|\frac{\partial k}{y\partial y}\right| \tag{3.84}$$

$$\frac{\partial\tilde{\varepsilon}}{\partial t}=\frac{\partial}{\partial y}\left[\left(\nu+c_\varepsilon\overline{v'v'}\frac{k}{\hat{\varepsilon}}\right)\frac{\partial\tilde{\varepsilon}}{\partial y}\right]-c_{\varepsilon 2}f_{\varepsilon 2}\frac{\tilde{\varepsilon}^2}{k}-\nu f_{\varepsilon,w}\left|\frac{\partial\tilde{\varepsilon}}{y\partial y}\right| \tag{3.85}$$

where Hanjalić and Launder's turbulent-diffusion model is used to account for the anisotropic nature of the flow near the plate.

Since the turbulent fluctuations normal to the plate are instantly reduced to zero, that is, in a time too short for viscosity to have an effect, and since we are considering an incompressible fluid, that is, the speed of sound is infinite, the flow at time $t = 0^+$ can be determined from the inviscid, incompressible equations of motion. The solution to this problem, which provides the initial conditions for our problem, is given by Hunt and Graham [1978].

In the free stream, Eqs. (3.84) and (3.85) reduce to those given in Eq. (3.74). Since the turbulence remains isotropic within planes parallel to the plate, the equation for $\overline{w'w'}$ is identical to that for $\overline{u'u'}$, and since the boundary conditions for both are the same, $\overline{w'w'} = \overline{u'u'}$. Consequently,

$$k=\overline{u'u'}+\overline{v'v'}/2 \tag{3.86}$$

and only two of the first three equations above need be solved together with Eq. (3.85). In addition, from Eq. (3.73) we obtain

$$A=\frac{27}{8}\frac{(\overline{u'u'})^2\overline{v'v'}}{k^3} \tag{3.87}$$

It also follows since $a_{ii} = 0$, that the anisotropic stress components are related by $a_{11} = a_{33} = -a_{22}/2$. Hence the other invariants for this flow are

$$A_1=\left(\frac{\overline{v'v'}}{k}-\frac{2}{3}\right)\Bigg/\left|\frac{\overline{v'v'}}{k}-\frac{2}{3}\right|,\quad A_2=\frac{3}{2}\left(\frac{\overline{v'v'}}{k}-\frac{2}{3}\right)^2\ \text{and}\ A_3=\frac{3}{4}\left(\frac{\overline{v'v'}}{k}-\frac{2}{3}\right)^3$$

Since we expect the eddies to "flatten" as they approach the plate, $\overline{v'v'}$ will be less than $2k/3$. As a result, $A_1 = -1$, A_2 of course will be positive, and A_3 will be negative.

One also obtains $\tilde{\varepsilon}_{11}=\tilde{\varepsilon}_{33}=\tilde{\varepsilon}-\tilde{\varepsilon}_{22}/2$, and similarly for the components of $\hat{\varepsilon}$. According to Eq. (3.60),

$$\tilde{\varepsilon}_{22} = f_{\varepsilon 1}\tilde{\varepsilon}\left(\frac{\overline{v'v'}}{k} - \frac{2}{3}\right) + \frac{2}{3}\tilde{\varepsilon}$$

While according to Eqs. (3.59) and (3.63),

$$\hat{\varepsilon} = \tilde{\varepsilon} + \nu f_{\varepsilon,w}\left|\frac{\partial k}{y\partial y}\right| \quad , \quad \hat{\varepsilon}_{22} = \tilde{\varepsilon}_{22} + 3\nu f_{\varepsilon,w}\left|\frac{\partial \overline{v'v'}}{y\partial y}\right| \tag{3.88}$$

and

$$\hat{\varepsilon}_{11} = \hat{\varepsilon}_{33} = \left(\tilde{\varepsilon} + \nu f_{\varepsilon,w}\left|\frac{\partial k}{y\partial y}\right|\right) - \frac{1}{2}\left(\tilde{\varepsilon}_{22} + 3\nu f_{\varepsilon,w}\left|\frac{\partial \overline{v'v'}}{y\partial y}\right|\right) \tag{3.89}$$

The calculated results for the rather low initial turbulence Reynolds number of $(Re_T)_0 = 150$ ($Re_\lambda = 32$) are plotted in Figs. 3.12 and 3.13, where the dashed lines represent the initial profiles as determined by Hunt and Graham. These figures illustrate the development of the kinetic-energy and normal-stress profiles with respect to time, where time has been normalized by the initial time scale of turbulence, namely, $T_0 = (k/\hat{\varepsilon})_0$. The turbulence far from the surface of the plate in these figures, as well as the next, is in its initial state of decay (see §3.4.1). As a result, the turbulence Reynolds number far from the surface barely changes with time. Furthermore, the ratio of turbulence length scales l_∞/L_∞, which varies inversely with the three-quarters power of Reynolds number, remains nearly constant even though each length scale increases approximately as the square root of $1 + t/T_0$. Therefore, when $(Re_T)_0 = 150$, we obtain $Re_T = k_\infty^2/\nu\hat{\varepsilon}_\infty \approx 150$, $l_\infty/L_\infty \approx 0.025$, and $\lambda_\infty/L_\infty \approx 0.26$.

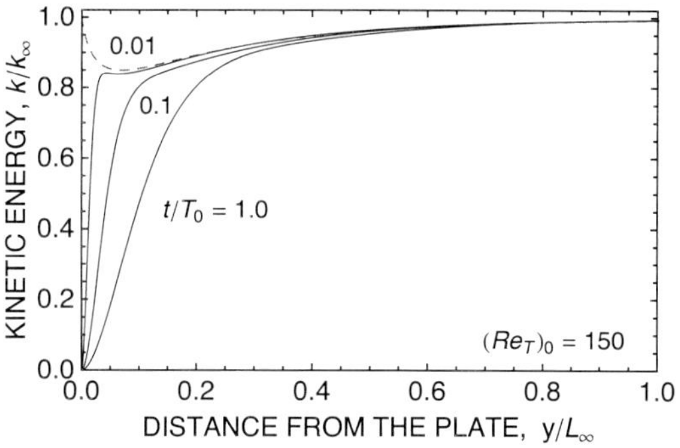

Figure 3.12.
Temporal development of the kinetic energy profile for a turbulent shear-free mean flow with an initial turbulence Reynolds number $(Re_T)_0 = 150$. The initial (dashed) profile is from Hunt and Graham [1978].

The temporal development of the kinetic-energy profile is presented in Fig. 3.12. According to Hunt and Graham (dashed line), the effect of inserting the plate immediately "blocks" the inviscid motion of the eddies normal to its surface reducing the kinetic-energy most effectively near $y/L_\infty = 0.1$.[1] This position depends on the spectrum of turbulence. After insertion, and according to Eq. (3.84), the kinetic-energy of turbulence simply diffuses and dissipates. Since the coordinates of the plot are scaled by the conditions far away from the plate, the effect displayed in this figure reflect only the diffusion and "near-wall" dissipation of turbulence even though k_∞ and L_∞ change with time. For $Re_T \approx 150$, viscous diffusion is limited to a distance of about $0.008L_\infty$ away from the plate. In other words, all temporal development roughly beyond the time $t/T_0 = 0.01$ are dominated by *turbulent* diffusion

[1] As it turns out, and as we will see in §4.3.1, Eq. (4.27), this distance corresponds to the size of the most "effective" eddies l_{eff}, which in this case is approximately equal to $0.08L_\infty$.

and near-wall dissipation.

The temporal development of the normal-stress profiles is presented in Fig. 3.13. Upon inserting the plate, the energy is immediately "repartitioned" between the turbulent components parallel and normal to it. At the plate, the $\overline{v'v'}$ stress component immediately reduces to zero, while the $\overline{u'u'}$ component immediately increases to three-halves its initial value. As expected, this repartitioning is confined to a thickness roughly equal to the size of the large eddies. Following insertion, profile development is controlled by the "return to isotropy" and "wall" components of the pressure-strain-rate terms in Eqs. (3.82) and (3.83), namely $\phi_{ij,1}$ and $\phi_{ij,w}$, as well as by turbulent diffusion and near-wall dissipation. The thickness of this developing layer of turbulence, which is more obvious in the $\overline{v'v'}$ stress profiles, constitutes the inertial boundary layer of turbulence.

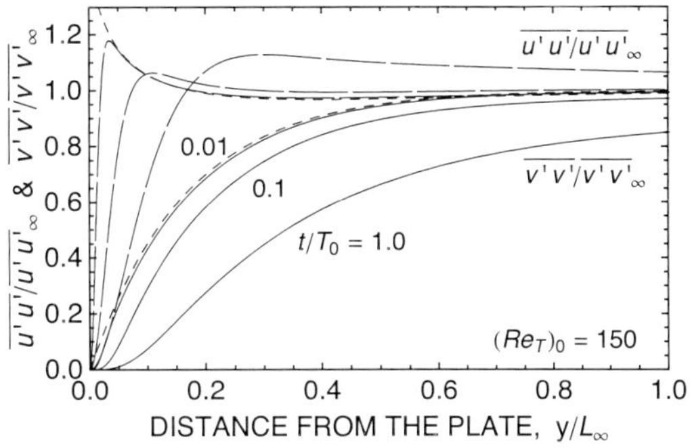

Figure 3.13.
Temporal development of the normal stress profiles for a turbulent shear-free mean flow with an initial turbulence Reynolds number $(Re_T)_0 = 150$. The initial (dashed) profiles are from Hunt and Graham [1978].

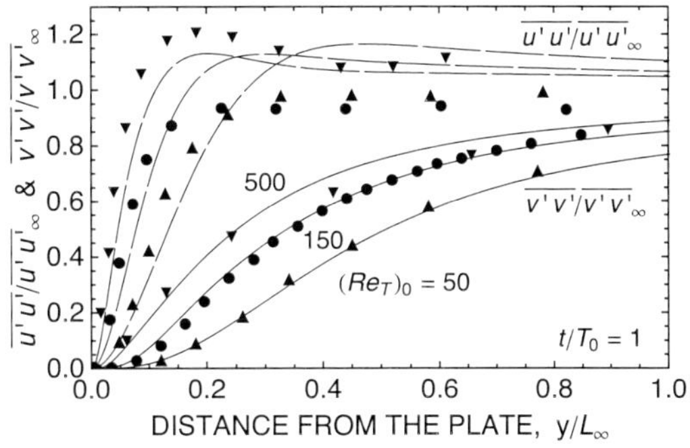

Figure 3.14.
Comparison between the normal-stress profiles for a turbulent shear-free mean flow at $t/T_0 = 1$ as calculated by a second moment closure of the Reynolds-stress equations and a DNS analysis. The DNS "data" are from Perot and Moin [1995] for $Re_T = 54$, 134 and 374.

A comparison between the results calculated at $t/T_0 = 1$ from the above set of equations and a DNS analysis is provided in Fig. 3.14. In this figure, the DNS "data" was obtained by Perot and Moin [1995] for initial turbulent Reynolds numbers of 54, 134 and 374. As it turns out, since the differences between the results for these Reynolds numbers and those used for the present calculations are hardly noticeable, the comparison shown in this figure is valid. Furthermore, since the experimental data obtained by Aaronson, Johansson, and Löfdahl [1997] for $Re_T = 400$ agree quite well with Perot and Moin's large Reynolds number results, the formers' data are not shown. Upon comparison, the $\overline{v'v'}$ calculations and DNS results agree reasonably well, which indicates that the wall component of the pressure-strain-rate term is correctly modeled. However, the agreement between the $\overline{u'u'}$ results at large distances from the plate are rather poor, particularly for the two lower Reynolds numbers.

Since the $\overline{u'u'}$ component should approach its final value from above, these two $\overline{u'u'}$ DNS results are suspect.

Finally, it should be noted that the invariant A for this flow increases monotonically from zero at the surface to unity within roughly one eddy diameter L_∞ away from it. It is also found that A_2 decreases from the value of two-thirds and A_3 increases from the negative value of two-ninths at the surface, both asymptotically approaching zero within roughly one-third of an eddy diameter. Compared to the free-stream turbulent flow discussed in §3.4.2, this is truly an anisotropic flow.

3.5.2 Turbulent Boundary Layers.

Now consider a time-averaged, two-dimensional turbulent flow over a flat plate aligned parallel to the flow. As before, we will measure x along the plate from its leading edge and y normal to the plate. Then the corresponding mean-flow velocity components are $\bar{u}$ and $\bar{v}$, hence the continuity and momentum equations, Eqs. (3.43) and (3.44), reduce to

$$\frac{\partial \bar{u}}{\partial x}+\frac{\partial \bar{v}}{\partial y}=0$$
$$\bar{u}\frac{\partial \bar{u}}{\partial x}+\bar{v}\frac{\partial \bar{u}}{\partial y}=\bar{u}_\infty\frac{\partial \bar{u}_\infty}{\partial x}+\frac{\partial}{\partial y}\left(\nu\frac{\partial \bar{u}}{\partial y}-\overline{u'v'}\right) \tag{3.90}$$

The various components of the Reynolds-stress equation, Eq. (3.66), using Daly-Harlow's diffusion model become

$$\bar{u}\frac{\partial \overline{u'u'}}{\partial x}+\bar{v}\frac{\partial \overline{u'u'}}{\partial y}=-2\left(1-2c_2\sqrt{A}/3\right)\overline{u'v'}\frac{\partial \bar{u}}{\partial y}+\frac{\partial}{\partial y}\left[\left(\nu+c_s\overline{v'v'}\frac{k}{\hat{\varepsilon}}\right)\frac{\partial \overline{u'u'}}{\partial y}\right]$$
$$-\left(c_1 f_{\phi 1}A+f_{\varepsilon 1}\right)\frac{\tilde{\varepsilon}}{k}\left[\overline{u'u'}-\frac{2}{3}k\right]+f_{\phi,w}\left[c_{1w}\frac{\tilde{\varepsilon}}{k}\overline{v'v'}-\frac{2}{3}c_{2w}c_2\sqrt{A}\,\overline{u'v'}\frac{\partial \bar{u}}{\partial y}\right]-\frac{2}{3}\tilde{\varepsilon}-\nu f_{\varepsilon,w}\left|\frac{\partial \overline{u'u'}}{y\partial y}\right| \tag{3.91}$$

$$\bar{u}\frac{\partial \overline{v'v'}}{\partial x}+\bar{v}\frac{\partial \overline{v'v'}}{\partial y}=-\frac{2}{3}c_2\sqrt{A}\,\overline{u'v'}\frac{\partial \bar{u}}{\partial y}+\frac{\partial}{\partial y}\left[\left(\nu+c_s\overline{v'v'}\frac{k}{\hat{\varepsilon}}\right)\frac{\partial \overline{v'v'}}{\partial y}\right]$$
$$-\left(c_1 f_{\phi 1}A+f_{\varepsilon 1}\right)\frac{\tilde{\varepsilon}}{k}\left[\overline{v'v'}-\frac{2}{3}k\right]-2f_{\phi,w}\left[c_{1w}\frac{\tilde{\varepsilon}}{k}\overline{v'v'}-\frac{2}{3}c_{2w}c_2\sqrt{A}\,\overline{u'v'}\frac{\partial \bar{u}}{\partial y}\right]-\frac{2}{3}\tilde{\varepsilon}-3\nu f_{\varepsilon,w}\left|\frac{\partial \overline{v'v'}}{y\partial y}\right| \tag{3.92}$$

$$\bar{u}\frac{\partial \overline{w'w'}}{\partial x}+\bar{v}\frac{\partial \overline{w'w'}}{\partial y}=-\frac{2}{3}c_2\sqrt{A}\,\overline{u'v'}\frac{\partial \bar{u}}{\partial y}+\frac{\partial}{\partial y}\left[\left(\nu+c_s\overline{v'v'}\frac{k}{\hat{\varepsilon}}\right)\frac{\partial \overline{w'w'}}{\partial y}\right]$$
$$-\left(c_1 f_{\phi 1}A+f_{\varepsilon 1}\right)\frac{\tilde{\varepsilon}}{k}\left[\overline{w'w'}-\frac{2}{3}k\right]+f_{\phi,w}\left[c_{1w}\frac{\tilde{\varepsilon}}{k}\overline{v'v'}-\frac{2}{3}c_{2w}c_2\sqrt{A}\,\overline{u'v'}\frac{\partial \bar{u}}{\partial y}\right]-\frac{2}{3}\tilde{\varepsilon}-\nu f_{\varepsilon,w}\left|\frac{\partial \overline{w'w'}}{y\partial y}\right| \tag{3.93}$$

$$\bar{u}\frac{\partial \overline{u'v'}}{\partial x}+\bar{v}\frac{\partial \overline{u'v'}}{\partial y}=-\left(1-c_2\sqrt{A}\right)\overline{v'v'}\frac{\partial \bar{u}}{\partial y}+\frac{\partial}{\partial y}\left[\left(\nu+c_s\overline{v'v'}\frac{k}{\hat{\varepsilon}}\right)\frac{\partial \overline{u'v'}}{\partial y}\right]$$
$$-\left(c_1 f_{\phi 1}A+f_{\varepsilon 1}\right)\frac{\tilde{\varepsilon}}{k}\overline{u'v'}-\frac{3}{2}f_{\phi,w}\left[c_{1w}\frac{\tilde{\varepsilon}}{k}\overline{u'v'}+c_{2w}c_2\sqrt{A}\,\overline{v'v'}\frac{\partial \bar{u}}{\partial y}\right]+2\nu f_{\varepsilon,w}\left|\frac{\partial \overline{u'v'}}{y\partial y}\right| \tag{3.94}$$

where the plus sign on the dissipation term in the last equation reflects the fact that the shear stress is negative. Finally from Eqs. (3.67) and (3.68), we obtain

$$\bar{u}\frac{\partial k}{\partial x}+\bar{v}\frac{\partial k}{\partial y}=-\overline{u'v'}\frac{\partial \bar{u}}{\partial y}+\frac{\partial}{\partial y}\left[\nu\frac{\partial k}{\partial y}+c_s\overline{v'v'}\frac{k}{\hat{\varepsilon}}\frac{\partial k}{\partial y}\right]-\tilde{\varepsilon}-\nu f_{\varepsilon,w}\left|\frac{\partial k}{y\partial y}\right| \tag{3.95}$$

and

$$\bar{u}\frac{\partial \tilde{\varepsilon}}{\partial x}+\bar{v}\frac{\partial \tilde{\varepsilon}}{\partial y}=-c_{\varepsilon 1}\frac{\tilde{\varepsilon}}{k}\overline{u'v'}\frac{\partial \bar{u}}{\partial y}+\frac{\partial}{\partial y}\left[\left(\nu+c_{\varepsilon}\overline{v'v'}\frac{k}{\hat{\varepsilon}}\right)\frac{\partial \tilde{\varepsilon}}{\partial y}\right]$$

$$-c_{\varepsilon 2}f_{\varepsilon 2}\frac{\tilde{\varepsilon}^2}{k}+c_{\varepsilon 3}\nu\overline{v'v'}\frac{k}{\tilde{\varepsilon}}\left(\frac{\partial^2 \bar{u}}{\partial y^2}\right)^2-\nu f_{\varepsilon,w}\left|\frac{\partial \tilde{\varepsilon}}{y\partial y}\right| \tag{3.96}$$

Following others, the maximum of the wall function for this flow is limited by using

$$f_{\phi,w}=\min\left(L_e/x_n, 3.0\right)$$

In addition, the intermittent flow in the outer portions of the boundary layer was taken into account by using an eddy viscosity in the region equal to eighty-seven percent of its maximum value. As previously mentioned, this feature will be considered more thoroughly in §7.1.

The invariant A for this flow is

$$A=\frac{27}{8}\frac{\overline{w'w'}(\overline{u'u'}\,\overline{v'v'}-\overline{u'v'}\,\overline{u'v'})}{k^3} \tag{3.97}$$

The other invariants are

$$A_2=\left(\frac{\overline{u'u'}}{k}-\frac{2}{3}\right)^2+\left(\frac{\overline{v'v'}}{k}-\frac{2}{3}\right)^2+\left(\frac{\overline{w'w'}}{k}-\frac{2}{3}\right)^2+2\left(\frac{\overline{u'v'}}{k}\right)^2$$

which is obviously positive, and

$$A_3=\left(\frac{\overline{u'u'}}{k}-\frac{2}{3}\right)^3+\left(\frac{\overline{v'v'}}{k}-\frac{2}{3}\right)^3+\left(\frac{\overline{w'w'}}{k}-\frac{2}{3}\right)^3+3\left(\frac{\overline{u'v'}}{k}\right)^2\left(\frac{\overline{u'u'}}{k}+\frac{\overline{v'v'}}{k}-\frac{4}{3}\right)$$

The calculated results are plotted in Figs. 3.15, 3.16 and 3.17. In the first figure, the calculated mean-velocity and turbulent shear-stress distributions are compared with the data of Klebanoff [1955]. The agreement between the calculated and measured results is excellent. The important aspect of this plot, however, is the thinness of the viscous layer adjacent to the wall, where both the velocity and turbulent shear stress change dramatically.

To emphasize the changes taking place in this layer, the calculated mean-velocity and large-scale-eddy dissipation profiles are plotted in Fig. 3.16 using the wall coordinates. Here the mean-velocity profile is shown by using every other calculated point, while the dissipation profile is shown by the a dashed line. In addition, for the sake of comparison, some of the relations presented in §3.2.1 are also plotted. The agreement between the calculated mean velocity, the viscous sublayer law Eq. (3.28), and Reichardt's law of the wall, Eq. (3.32), is excellent, as is the agreement between the calculated dissipation profile and the dissipation based on the law of the wall, Eq. (3.34). The interesting feature of this figure is, however, the dramatic change that occurs in the calculated dissipation function $\tilde{\varepsilon}$ and where it occurs. Everything happens in the layer sandwiched between the viscous sublayer and the logarithmic layer where the production of turbulent energy is nearly equal to its dissipation. In the past, this layer was commonly referred to as the "buffer layer." To correctly capture the behavior in this layer and appropriately model it, a large number of points must be calculated and measured within it. Carrying out such calculations were difficult in the past, but with the arrival of multiprocessor computers and terabytes of memory this is no longer a problem. The same cannot be said about measurements this close to the wall, however, until some new device such as a *nano-anemometer*[1] becomes available. As a consequence, DNS "experiments," as we saw in the previous subsection, are beginning to play a more important role in modeling turbulent boundary layers.

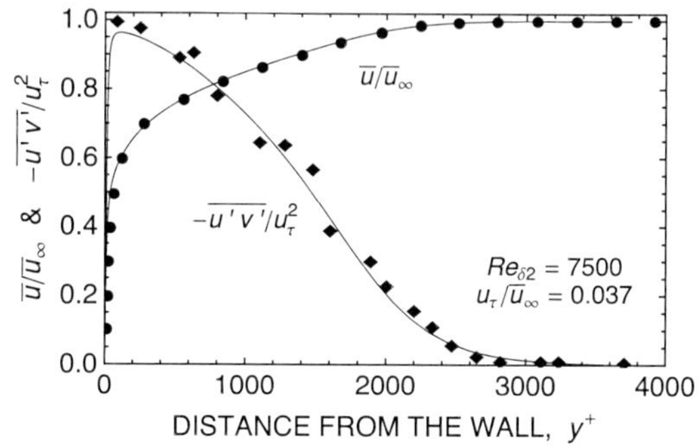

Figure 3.15.
Calculated and measured mean-velocity and shear-stress profiles in a turbulent boundary layer. Data from Klebanoff [1955].

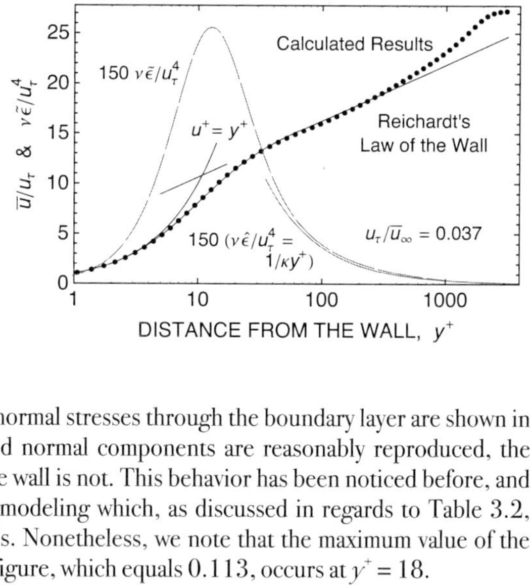

Figure 3.16.
Non-dimensional mean velocity and dissipation function plotted in wall coordinates. Calculated for Klebanoff's [1955] data.

The calculated and measured variation of the normal stresses through the boundary layer are shown in Fig. 3.17. Although both the streamwise and normal components are reasonably reproduced, the component transverse to the flow very near the wall is not. This behavior has been noticed before, and can be attributed to the wall-pressure-strain modeling which, as discussed in regards to Table 3.2, can be "fixed" by using more complex models. Nonetheless, we note that the maximum value of the dimensionless streamwise component in this figure, which equals 0.113, occurs at $y^+ = 18$.

In conclusion, we note that the invariant A for this flow increases monotonically from zero at the wall to about ninety-five percent of its free-stream value at $y^+ \approx 500$, and from there asymptotically approaches unity. It should also be noted that the calculated variations of the skin-friction coefficient and momentum-thickness Reynolds number with distance are virtually the same as that given by Eqs. (3.40) and (3.41).

---

[1] The author knows of no such device.

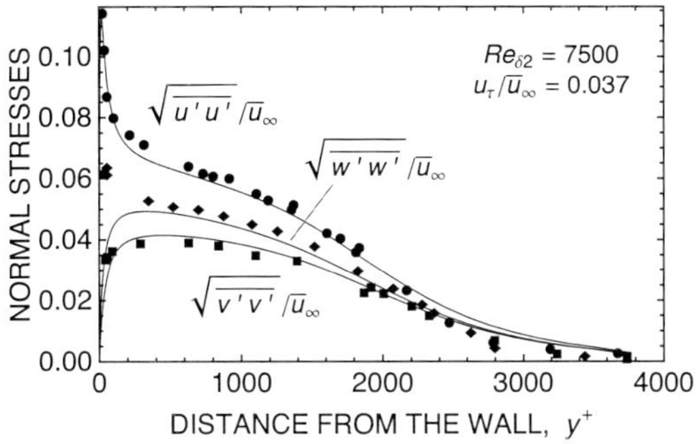

Figure 3.17.
Calculated and measured Reynolds normal-stress profiles in a turbulent boundary layer. Data from Klebanoff [1955].

References

Aaronson, D., Johansson, A.V., and Löfdahl, L., 1997, "Shear-Free Turbulence Near a Wall," J. Fluid Mech., 338, pp. 363–385.

Batchelor, G.K., 1953, The Theory of Homogeneous Turbulence, Cambridge University Press, Cambridge.

Batchelor, G.K., and Townsend, A.A., 1948, "Decay of Isotropic Turbulence in the Initial Period," Proc. Roy. Soc. London A, 193, pp. 539–558; also, "Decay of Isotropic Turbulence in the Final Period," Proc. Roy. Soc. London A, 194, pp. 527–543.

Boussinesq, J., 1877, "Essai sur la théorie des eaux courantes," Mém. prés. par div. savants á l'acad. sci., Paris, 23, p. 46.

Cebeci, T., and Smith, A.M.O., 1974, Analysis of Turbulent Boundary-Layers, Academic Press, New York.

Champagne, F.H., Harris, V.G., and Corrsin, S., 1970, "Experiments on Nearly Homogeneous Turbulent Shear Flow," J. Fluid Mech., 41, pp. 81–141.

Chien, K.Y., 1982, "Predictions of Channel and Boundary Flows with a Low-Reynolds-Number Turbulence Model," AIAA Journal, 20, pp. 33–38.

Chou, P.Y., 1945, "On Velocity Correlations and the Solution of the Equations of Turbulent Fluctuation," Quart. Appl. Math., 3, p. 38.

Clauser, F.H., 1954, "Turbulent Boundary Layers in Adverse Pressure Gradients," J. Aero. Sci., 21, pp. 91–108.

Coles, D., 1956, "The Law of the Wake in the Turbulent Boundary Layer," J. Fluid Mech., 1, pp. 191–226.

Comte-Bellot, G., and Corrsin, S., 1966, "The Use of a Contraction to Improve the Isotropy of Grid-Generated Turbulence," J. Fluid Mech., 25, pp. 657–682.

Corrsin, S., and Kistler, A.L., 1955, "Free-Stream Boundaries of Turbulent Flows," NACA TR 1244; also 1954, NACA TN 3133.

Daly, B.J., and Harlow, F.H., 1970, "Transport Equations in Turbulence," Phys. Fluids, 13, pp. 2634–2649.

Davidov, B.I., 1961, "On the Statistical Dynamics of an Incompressible Turbulent Flow," Dokl. Acad. Nauk., SSSR, 136, pp. 47–50.

Demuren, O., and Rodi, W., 1984, "Calculation of Turbulence-Driven Secondary Motion in Non-

Circular Ducts," J. Fluid Mech., 140, p. 189.

Durbin, P.A., 1991, "Near-Wall Turbulence Modeling Without Damping Functions," Comput. Fluid Dyn., 3, pp. 1–13.

Durbin, P.A., 1993, "A Reynolds Stress Model for Near-Wall Turbulence," J. Fluid Mech., 249, pp. 465–498.

Dyban, E., and Epik, E., 1985, Тепломссобмен и Гидродинамика Тубулизированных Потоков, Киев Наукоа Думка (in Russian).

Gibson, M.M., and Launder, B.E., 1978, "Ground Effects on Pressure Fluctuations in the Atmospheric Boundary Layer," J. Fluid Mech., 86, pp. 491–511.

Gibson, M.M., and Rodi, W., 1989, "Simulation of Free Surface Effect on Turbulence with a Reynolds Stress Model," J. Hydraulic Research, 27, pp. 233–244.

Hanjalić, K., 1994, "Advanced Turbulence Closure Models: A View of Current Status and Future Prospects," Int. J. Heat and Fluid Flow, 15, pp. 178–203.

Hanjalić, K., and Launder, B.E., 1972, "A Reynolds-Stress Model of Turbulence and its Application to Thin Shear Flows," J. Fluid Mech., 52, pp. 609–638.

Hanjalić, K., and Launder, B.E., 1976, "Contribution Towards a Reynolds-Stress Closure for Low-Reynolds-Number Turbulence," J. Fluid Mech., 74, pp. 593–610.

Hanjalić, K., Jakirlić, S., and Hadzić, I., 1997, "Expanding the Limits of 'Equilibrium' Second-Moment Turbulence Closures," Fluid Dyn. Res., 20, pp. 25–41.

Hanjalić, K., Hadzić, I., Jakirlić, S, and Basara, B., 1999, "Modeling the Turbulent Wall Flows Subjected to Strong Pressure Variations," Modeling Complex Turbulent Flows, (eds., M.D. Salas et al.), Kluwer Acad. Publ., pp. 203–222.

Harlow, F.H., and Nakayama, P.I., 1968, "Transport of Turbulence Energy Decay Rate," Los Alamos Sci. Lab. University of California Report LA 3854.

Hinze, J.O., 1959, Turbulence, McGraw-Hill, New York.

Hunt, J.C.R., and Graham, J.M.R., 1978, "Free-Stream Turbulence Near Plane Boundaries," J. Fluid Mech., 84, pp. 209–235.

Johnson, P.L., and Johnston, J.P., 1989, Seventh Symposium on Turbulent Shear Flows, Dept. of Mech. Engng., Stanford University, pp. 20.2.1–20.2.6.

Jones, W.P., and Launder, B.E., 1972, "The Prediction of Laminarization With a Two-Equation Model of Turbulence," Int. J. Heat and Mass Transfer, 15, pp. 301–314.

Kármán, Th. von, 1930, "Mechanische Ähnlichkeit und Turbulenz," Nachr. Ges. Wiss. Göttingen, Math. Phys. Klasse, 58; also 1931, NACA TM 611.

Kármán, Th. von, 1948, "Progress in the Statistical Theory of Turbulence," Proc. Natl. Acad. Science, 34, p. 530.

Klebanoff, P.S., 1955, "Characteristics of Turbulence in a Boundary Layer with Zero Pressure Gradient," NACA TR 1247.

Kolmogorov, A.N., 1941, "The Local Structure of Turbulence in Incompressible Viscous Fluid for Very Large Reynolds numbers," C.R. Acad. Sci. U.R.S.S., 30, pp. 301–305; also 1941, "Energy Dissipation of Locally Isotropic Turbulence," C.R. Acad. Sci. U.R.S.S., 32, pp. 16–18.

Launder, B.E., 1989, "Second-Moment Closure: Present ... and Future," Int. J. Heat and Fluid Flow," 10, pp. 282–300.

Launder, B.E., and Spalding, D.B., 1974, "The Numerical Computation of Turbulent Flows," Computer Methods in Applied Mechanics and Engineering, 3, pp. 269–289.

Launder, B.E., Reece, G.J. and Rodi, W., 1975, "Progress in the Development of a Reynolds-Stress Turbulence Closure," J. Fluid Mech., 68, pp. 537–566.

Lumley, J.L., 1978, "Computational Modeling of Turbulent Flows," Adv. App. Mech., 18, pp. 123–176.

Lumley, J.L., and Newman, G.R., 1977, "The Return to Isotropy of Homogeneous Turbulence," J. Fluid Mech., 82, pp. 161–178.

Mansour, N.N., Kim, J., and Moin, P., 1989, "Near-Wall k-ε Turbulence Modeling," AIAA Journal, 27, pp. 1068–1073.

Mayle, R.E., Dullenkopf, K., and Schulz, A., 1998, "The Turbulence That Matters," J. Turbomachinery, 120, pp. 402–409.

Millikan, C.B., 1938, Proc. 5th Intern. Congr. Appl. Mech., Cambridge, Mass., p.386.

Nikuradse, J., 1926, "Untersuchungen über die Geschwindigkeitsverteilung in turbulenten Strömungen," Diss. Göttingen; VDI-Forschungsheft 281.

Pao, Y.H., 1965, "Structure of Turbulent Velocity and Scalar Fields at Large Wavenumbers," Phys. Fluids, 8, pp. 1063–1075.

Patankar, S.V., and Spalding, D.B., 1970, Heat and Mass Transfer in Boundary Layers, Intertext, London.

Perot, B., and Moin, P., 1995, "Shear-Free Turbulent Boundary Layers. Part 1. Physical Insights into Near-Wall Turbulence," J. Fluid Mech., 295, pp. 199–227; also Part 2. "New Concepts for Reynolds Stress Equation Modelling of Inhomogeneous Flows," pp. 229–245.

Prandtl, L., 1925, "Über die ausgebildete Turbulenz," ZAMM, 5, pp. 136–139.

Prandtl, L., 1927, "Über den Reibungswiderstand strömender Luft," Ergebnisse AVA Göttingen, III Lfg., pp. 1–5; also 1932, "Zur turbulenten Strömung in Rohren und längs Platte," Ergebnisse AVA Göttingen, Ergebnisse der AVA Göttingen, IV Lfg., pp. 18–29.

Roach, P.E., and Brierley, D.H., 1990, "The Influence of a Turbulent Free-Stream on Zero Pressure Gradient Transitional Boundary Layer Development Including the T3A & T3B Test Case Conditions," Proc. 1st ERCOFTAC Workshop on Numerical Simulation of Unsteady Flows, Transition to Turbulence and Combustion, Lausanne; also 1992, Numerical Simulation of Unsteady Flows and Transition to Turbulence, (eds., O. Perinea et al.), Cambridge University Press, pp. 319–347, and 1993, data transmitted by J. Coupland from the Rolls-Royce Applied Science Laboratory.

Reichardt, H., 1951, ZAMM, 31, p. 208.

Rodi, W., 1971, "On the Equation Governing the Rate of Turbulent Energy Dissipation," Mech. Engng. Dept. Imperial College Report TM/TN/A/14.

Rodi, W., 1978, "Turbulence Models and Their Application in Hydraulics – A State of the Art Review," SFB 80/T/127, University of Karlsruhe, Germany.

Rose, W.G., 1966, "Results of an Attempt to Generate a Homogeneous Turbulent Shear Flow," J. Fluid Mech., 25, pp. 97–120.

Rotta, J.C., 1951, "Statistische Theorie nichthomogener Turbulenz I," Z. Phys., 129, pp. 547–573; also 1951, "Statistische Theorie nichthomogener Turbulenz II," Z. Phys., 131, pp. 51–77.

Rotta, J.C., 1972, Turbulente Strömungen, B.G. Teubner, Stuttgart (in German).

Schlichting, H., 1979, Boundary-Layer Theory, McGraw-Hill, New York.

Schultz-Grunow, F., 1940,"Neues Widerstandsgesetz für glatte Platten," Luftfahrtforschung, 17, p. 239; also 1941, NACA TM 986.

Screenivasan, K.R., 1995, "On the Universality of the Kolmogorov Constant," Phys. Fluids, 7, pp.

2778–2784.

Speziale, Ch.G., Abid, R., and Anderson, E.C., 1992, "Critical Evaluation of Two-Equation Models for Near-Wall Turbulence," AIAA Journal, 30, pp. 324–331.

Taylor, G.I., 1915 "The Transport of Vorticity and Heat Through Fluids in Turbulent Motion," Phil. Trans., A215, pp. 1–26.

Taylor, G.I., 1935, "Statistical Theory of Turbulence, Parts 1–4," Proc. Roy. Soc. London A 151, pp. 421–478; also 1938, "The Spectrum of Turbulence," Proc. Roy. Soc. London A, 164, pp. 476–490.

Tennekes, H., and Lumley, J.L., 1972, A First Course in Turbulence, MIT Press.

Thomas, N.H., and Hancock, P.E., 1977, "Grid Turbulence Near a Moving Wall," J. Fluid Mech., 82, pp. 481–496.

Townsend, A.A., 1956, The Structure of Turbulent Shear Flow, Cambridge University Press, New York.

White, F.M., 1974, Viscous Fluid Flow," McGraw-Hill, New York.

Uberoi, M.S., 1957, "Equipartition of Energy and Local Isotropy in Turbulent Flows," J. Appl. Phys., 28, pp. 1165–1170.

Uzkan, T., and Reynolds, W.C., 1967, "A Shear-Free Turbulent Boundary Layer," J. Fluid Mech., 28, pp. 803–821.

Yang, Z., and Shih, T.H., 1993, "New Time Scale Based k-ε Model for Near-Wall Turbulence," AIAA Journal, 31, pp. 1191–1198; also 1992, "A k, ε Calculation of Transitional Boundary Layers," NASA TM 105604.

Part II

In Between

4 Pre-Transitional Flow

Controlling mechanisms. Path to natural transition. Path to by-pass transition.

In the broadest sense, a pre-transitional boundary layer begins at the leading edge of a surface and extends downstream to where turbulent spots first appear, and from there intermittently between the spots until the end of transition. As a rule, its extent lengthens with either a favorable pressure gradient or a reduced disturbance level and shortens with either an adverse gradient or increased disturbance level. A disturbance as referred to here is anything that causes either a momentary or continual unsteadiness within the boundary layer, such as free-stream turbulence, pressure waves with either a broad frequency spectrum (free-stream turbulence, noise) or a distinct frequency (sound), isolated surface imperfections or distributed roughness, and surface vibrations in any direction. In many practical situations, however, free-stream turbulence is the primary disturbance.

Pre-transitional flow is characterized by an unsteadiness within the laminar boundary layer that amplifies with distance via a mechanism that is peculiar to either the disturbance or group of disturbances. If we consider only the effect of free-stream turbulence, two mechanisms are involved. For *low* turbulence levels, the mechanism is primarily associated with the natural *instability* of the flow within the boundary layer. For *high* turbulence levels, the mechanism is primarily associated with the work done on fluctuations within the boundary layer by the impressed unsteady pressure forces in the free stream. Historically, transition following these paths are respectively called "natural" and "bypass" transition. The latter is used to indicate that transition by this mechanism completely bypasses the mechanism associated with the instability of the flow. For *moderate* turbulence levels lying within the relatively narrow band from about one-half to two percent, both mechanisms can cause the unsteady flow within the boundary layer to amplify depending on the amplitude and spectrum of the turbulence as well as its rate of decay. Thus, the effect of turbulence on transition is by itself rather complex.

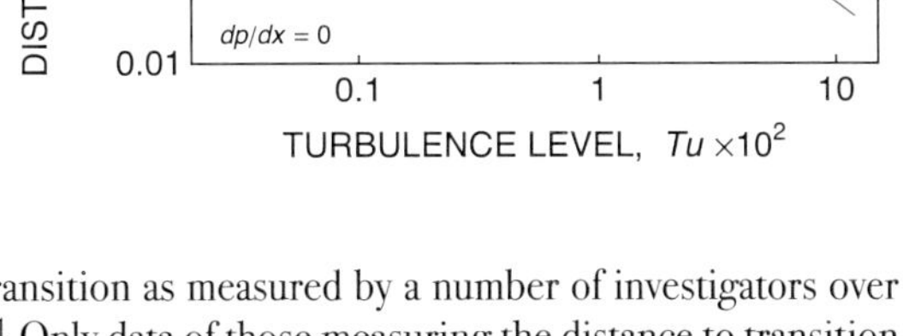

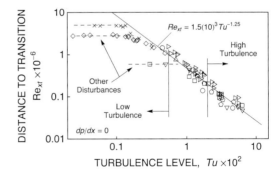

Figure 4.1.
The effect of free-stream turbulence on the onset of transition without a pressure gradient. (See Fig. 6.4 for legend.)

The effect of turbulence on the "onset" of transition as measured by a number of investigators over the last half of a century is shown in Fig. 4.1.[1] Only data of those measuring the distance to transition from the plate's leading edge for flow with a zero pressure gradient is presented in this figure. Plotted as the distance Reynolds number to transition from the plate's leading edge against turbulence level,

[1] The turbulence levels for all but Abu-Ghannam and Shaw's (upward pentagons) and Roach and Brierley's (circles) data in this figure were measured at the leading edge of the plate. Their values were measured halfway between the leading edge and onset, and at onset, respectively. In most cases, Tu is based on all three of the fluctuating velocity components. Finally, the onset of transition for all but Wells' (times symbols), and Gostelow and Ramachandran's and Gostelow and Blunden's (squares) data was determined from wall-shear stress measurements (see §6.1.1). Wells determined onset from the average breakup position of laminar smoke filaments near the surface, while Gostelow and his associates used intermittency measurements (see §6.1.2). See Fig. 6.4 for legend.

the majority of the data lies in a wide downward sloping swath with increasing turbulence level.[1] That is, the distance to transition or the length of pre-transitional flow decreases with increasing turbulence. As indicated in the figure, low turbulence levels are considered to be less than about one-half percent, while high levels are greater than about two percent. For low turbulence levels, the data independent of the turbulence level indicate transition affected by disturbances other than turbulence. Usually, as reported by Schubauer and Skramstad [1943], these disturbances can be attributed to tunnel noise, sound waves, and the experimental setup. For high turbulence levels, the effect of these disturbances are generally negligible.

In this chapter, we will first consider the mechanisms responsible for amplifying the unsteady components of the flow in a pre-transitional boundary layer. We will then consider the instabilities associated with a laminar boundary layer and their growth as natural transition is approached, the effect of free-stream turbulence on the turbulent-like unsteadiness in a laminar boundary layer and their growth as bypass transition is approached, and finally evaluate some old and new criteria regarding the onset of transition.

Some of the important results in this chapter are:

1. Natural transition is the process by which relatively *small-amplitude unstable* disturbances within the boundary layer grow and breakdown into turbulent spots. In this case, a practical method for calculating pre-transitional flow from its beginning to end is still beyond us.
2. Bypass transition is the process by which relatively *large-amplitude* disturbances within the boundary layer are *forced* to grow and breakdown into turbulent spots. In this case, a practical method for calculating pre-transitional flow is presently available when the disturbance is free-stream turbulence.
3. In either case, natural or bypass transition, the conditions when amplified laminar unsteadiness break down into turbulent eddies are still unknown. That is, we still can't predict when laminar flow breaks down.

Beside the references at the end of this chapter, other relevant articles may be found in the Bibliography.

4.1 Seeds of Transition.

Integral kinetic-energy equation. Unstable oscillations. Forced oscillations.

Consider laminar flow on a flat plate perturbed by a small-amplitude oscillation in velocity. If we suppose that the wavelength of the oscillation is small compared to distance over which the mean flow changes, we can presume the mean velocity component normal to the wall is essentially zero over the distance of several wavelengths such that the kinetic-energy equation, Eq. (1.51), reduces to

$$\bar{u}\frac{\partial k}{\partial x} = -\overline{u'v'}\frac{\partial \bar{u}}{\partial y} - \frac{1}{\rho}\overline{u'\frac{\partial p'}{\partial x}} + \nu\frac{\partial^2 k}{\partial y^2} - \hat{\varepsilon} \tag{4.1}$$

where both the mean-velocity profile and boundary-layer thickness may now be considered independent of x.

Furthermore, consider integrating this equation with respect to y from the wall into the free stream. First suppose that the amplitude of the oscillation is proportional to the root-mean-square of u' in the boundary layer, $\hat{u}(x)$. Second, assume that the time-averaged profiles at any distance x constitute a one-parameter family of curves using the free-stream turbulence level as the parameter. Then, $k = \hat{u}^2\, fnc_1(y/\delta, \hat{u}_\infty/\bar{u}_\infty)$, $\overline{u'v'} = \hat{u}^2\, fnc_2(y/\delta, \hat{u}_\infty/\bar{u}_\infty)$, etc., and integrating the kinetic-energy equation

[1] Correlation for these results will be discussed in Chapter 6.

yields

$$\frac{d\hat{u}^2}{d(x/\delta)} \approx c_1\hat{u}^2 - c_2\frac{\delta}{\rho\bar{u}_\infty}\int_0^1 \overline{u'\frac{\partial p'}{\partial x}}d(\frac{y}{\delta}) - c_3\frac{1}{Re_\delta}\hat{u}^2 \tag{4.2}$$

where the constants c_n are the numerical results of the integration. Here, the last term on the right-hand side is the integrated dissipation term. The integral of the diffusion term in Eq. (4.1) is zero because $\partial k/\partial y = 0$ at both limits.

The mechanism associated with natural transition. Suppose in the limit of very low free-stream turbulence that $\hat{u}_\infty$ equals zero. Then the pressure fluctuations p' must be proportional to $\rho\bar{u}_\infty u'$ and Eq. (4.2) reduces to

$$\frac{d\hat{u}^2}{d(x/\delta)} \approx a_1\hat{u}^2 - a_2\frac{1}{Re_\delta}\hat{u}^2$$

Here a_1 can be either positive or negative while $a_2 = c_3$ is always positive. The first term on the right-hand side is the sum of the first two terms in Eq. (4.2). Since, as we will see later, the integral of the pressure term is nearly zero in this case, the right-hand side of this equation is primarily the difference between the *energy generated by the shear-stress* $(-\overline{u'v'})$ and the *energy dissipated.* Clearly, the development of $\hat{u}$ depends on whether this difference is greater or less than zero.

The solution to the above equation is

$$\hat{u} \approx \hat{u}_0 \exp\left[\frac{1}{2}\left(a_1 - a_2/Re_\delta\right)\frac{(x-x_0)}{\delta}\right] \tag{4.3}$$

where $\hat{u}_0$ is the root-mean-square of the oscillation within the boundary layer at x_0. Obviously, when $\hat{u}_0 = 0$ there is no response. However, when $\hat{u}_0$ is finite, even infinitesimal, the response either grows or decays exponentially depending on the sign of the quantity in parenthesis. In particular, when $a_1 < 0$, which requires the integral of the Reynolds shear-stress term to be *negative*, the oscillation decays independent of the Reynolds number. When $a_1 > a_2/Re_\delta > 0$, the oscillation grows. This is the behavior associated with the well-known stability theory of Orr and Sommerfeld. Whether the oscillation continues to grow or not depends on the frequency of the initial oscillation; something this simple analysis cannot provide. Nevertheless, when $a_1 = a_2/Re_\delta$, the oscillation, known as a *neutral disturbance*, neither grows nor decays. Rearranging this condition provides $Re_\delta = a_2/a_1$, that is, a Reynolds number above which the oscillation may *initially* grow. This is called the "critical Reynolds number."

The mechanism associated with bypass transition. Now consider oscillations in a boundary layer with free-stream turbulence. In this case, the pressure fluctuations are proportional to $\rho\bar{u}_\infty u'_\infty$ instead of $\rho\bar{u}_\infty u'$, and unsteady changes in the pressure occur over a distance equal to the characteristic size of the turbulent eddy l_∞. When the initial Reynolds' shear stress equals zero, the first production term can be neglected and Eq. (4.2) reduces to

$$\frac{d\hat{u}^2}{d(x/\delta)} \approx b_1\frac{\omega_\infty\delta\hat{u}_\infty}{\bar{u}_\infty}\hat{u} - b_2\frac{1}{Re_\delta}\hat{u}^2 \tag{4.4}$$

where the eddy size was eliminated by introducing the characteristic frequency $\omega_\infty = \bar{u}_\infty/l_\infty$. In contrast to the energy equation for natural transition, the right-hand side of this equation is the difference between the *energy generated by the pressure fluctuations* and the *dissipation*; and it contains the free-stream turbulent intensity. The ratio of these terms is proportional to $\omega_\infty\delta^2/\nu$, which is the square of the ratio between the steady and unsteady boundary-layer thicknesses associated with transverse viscous oscillations as we learned in §2.3.

The solution to Eq. (4.4) is

$$\hat{u} = \hat{u}_0 \exp\left[-b_2 \frac{1}{2Re_\delta} \frac{(x-x_0)}{\delta}\right] + \frac{b_1}{b_2} \frac{\omega_\infty \delta^2}{\nu} \hat{u}_\infty \left\{1 - \exp\left[-b_2 \frac{1}{2Re_\delta} \frac{(x-x_0)}{\delta}\right]\right\}$$

The first term represents the *decay* of the initial disturbance, while the second represents the *growth* of the oscillations within the boundary layer caused by the turbulence. In contrast to the solution found above for an unstable disturbance, this solution grows even when $\hat{u}_0 = 0$. In fact, we can easily show that these oscillations initially grow according to

$$\hat{u} \approx \frac{1}{2} b_2 \hat{u}_\infty \frac{\omega_\infty (x - x_0)}{\bar{u}_\infty} + \cdots$$

Hence, the initial growth of the oscillations is linear when $\hat{u}_0$ is infinitesimal compared to the exponential growth associated with an unstable boundary layer.

As a matter of fact, the pressure fluctuations in this case are proportional to $\rho u'_\infty(2\bar{u}_\infty + u'_\infty) - \rho\overline{u'_\infty u'_\infty}$ which when substituted into Eq. (4.2) and evaluated leads to a production term that equals that shown in Eq. (4.4) multiplied by the factor $(1 + \frac{1}{2}\hat{u}_\infty / u_\infty)$. The second term in this factor represents the fractional increase in the production of kinetic energy caused by the imposed *mean* pressure associated with the free-stream turbulence. Clearly, this contribution, which was determined by Taylor [1936], has a second order effect on the growth of the oscillations compared to the contribution caused by the imposed pressure fluctuations.

4.2 Path to Natural Transition.

Initial stage. Later stages.

The physics of pre-transitional flow with low free-stream turbulence is now reasonably well understood. If we consider a laminar boundary layer on a flat plate subjected to a small amplitude unsteadiness, the development of the flow within the boundary layer en route to transition can be divided into four consecutive regions. In the first, which begins at the leading edge of the plate, any unsteadiness is viscously damped. In the second, which begins when the critical Reynolds number is reached, two and three-dimensional unstable wave components of the unsteadiness grow. This growth continues into the third region, where the "most-amplified" wave component of the unsteadiness affects the mean flow and a secondary system of streamwise vortices develops within the boundary layer. Finally, in the fourth region, a secondary high-frequency instability develops within the "upwellings" of the vortex system which in turn rapidly breaks down into turbulent eddies.

Although a practical method for calculating the continuous progression of this flow is not available, short of running a DNS experiment, we can solve various parts of it in a piecemeal fashion. Therefore, in this section, we will first examine the stability of a laminar boundary layer, which covers the first two regions of pre-transitional flow, and then briefly discuss the flow associated with the last two regions.

4.2.1 The initial stage.

Laminar stability theory. Tollmien-Schlichting waves. The neutral stability curve. Perturbation velocities and Reynolds stresses. Effects of pressure gradient.

The earliest analytical investigations into shear-flow stability date back to the late 1800s when Rayleigh [1880, 1887, 1892] examined the stability of parallel flows. Neglecting the effect of viscosity, he established that all mean-velocity profiles with an inflection point are unstable. Although he also showed that profiles without an inflection point are always stable, this turns out not to be necessarily true when viscosity is taken into account.

The earliest investigations including the effect of viscosity on stability are those described by Reynolds [1895] and Lorentz [1907]. Both used an integral of the kinetic-energy equation to examine the variation of the disturbance's energy with time. Even though their method provides an insight into the mechanism responsible for boundary-layer instability, the theory admits any disturbance that satisfies the continuity without necessarily satisfying the momentum equation. Consequently, the theory yields the unlikely result of different critical Reynolds numbers for the same mean-velocity profile. A rudimentary version of their analysis was just presented in the introduction of this chapter.

The classic approach to evaluating the stability of laminar boundary layers was developed by Orr [1907] and Sommerfeld [1908]. It is a *linear* theory and unlike that developed by Reynolds and Lorentz, it is a differential approach capable of providing a *stability criterion* for any disturbance that satisfies the equations of motion. Since their analysis is limited to *small-amplitude* disturbances, it is only valid within the first stage of pre-transition when the free-stream turbulence is very low. Their equation was first solved by Tollmien [1929] for Blasius' flow, and later by Schlichting [1933] for boundary layers with nonzero free-stream pressure gradients.

In the following we will develop the Orr-Sommerfeld equation and examine various aspects of its solution. For those interested in all the details regarding the theory of hydrodynamic stability, the books written by Lin [1955], Chandrasekhar [1961] and Betchov and Criminale [1967] should be consulted.

Laminar stability theory. Let's first consider the mean flow. As discussed in §1.2, if we presume that the amplitude of the unsteadiness divided by the free-stream velocity is very much smaller than Re_δ^{-1}, we can neglect the Reynolds stress terms in the momentum equation compared to the viscous terms such that the mean-flow equations reduce to those for steady flow. Consequently, the solution for a mean flow with $dp/dx = 0$ perturbed by a small amplitude unsteadiness will be exactly that given by Blasius. Furthermore, by presuming that the characteristic wavelengths of the unsteadiness are not too large, the boundary-layer thickness will remain almost unchanged over several wavelengths such that we can use the parallel-flow assumption described in §2.3.3. Then the mean-flow velocity components are given by $\bar{u} = \bar{u}_\infty f'(\eta)$ and $\bar{v} = 0$ where $f(\eta)$ is the Blasius function and η is the similarity variable given by Eq. (2.3).

Now consider the unsteady motion. If we restrict ourselves to two-dimensional perturbations such that u', v' and p' are functions of x, y and t only, with $w' = 0$,[1] then the linearized versions of Eqs. (1.57)–(1.59) become

$$\frac{\partial u'}{\partial x} + \frac{\partial v'}{\partial y} = 0 \tag{4.5}$$

$$\frac{\partial u'}{\partial t} + \bar{u}\frac{\partial u'}{\partial x} = -\frac{1}{\rho}\frac{\partial p'}{\partial x} + \nu\frac{\partial^2 u'}{\partial x^2} + \nu\frac{\partial^2 u'}{\partial y^2} - v'\frac{d\bar{u}}{dy} \tag{4.6}$$

$$\frac{\partial v'}{\partial t} + \bar{u}\frac{\partial v'}{\partial x} = -\frac{1}{\rho}\frac{\partial p'}{\partial y} + \nu\frac{\partial^2 v'}{\partial x^2} + \nu\frac{\partial^2 v'}{\partial y^2} \tag{4.7}$$

As for any set of linear differential equations, a single equation for any one of the dependent variables may be obtained. This is most easily done by first cross-differentiating the last two equations and subtracting to eliminate the pressure. The result is a perturbation vorticity equation, which could have been obtained directly from Eq. (1.70). Expressing the velocity components in terms of a perturbation stream function $\psi'(x, y, t)$, the continuity equation is identically satisfied. This leaves only the

---

[1] As shown by Squire [1933], the critical Reynolds number for perturbations traveling in a direction skewed to the mean flow is higher than perturbations traveling in the same direction. Therefore, this analysis provides the *lowest* limit of instability.

vorticity equation, rewritten in terms of the stream function to be solved.

Following standard practice, but using slightly different notation, we consider a periodic perturbation of the form

$$\psi'(x,y,t)=\hat{u}\delta_1\phi(y)e^{i(kx-\omega t)} \tag{4.8}$$

where $\hat{u}$ is the amplitude of the perturbation at some position x, δ_1 is the boundary layer's displacement thickness at that position, ϕ is a dimensionless perturbation stream function, and k and ω are the perturbation's wave number and frequency respectively. This perturbation is essentially the same traveling wave we considered in Eq. (2.42), which in this case is known as a *Tollmien-Schlichting wave*. The corresponding perturbation velocities are

$$u'=\frac{\partial\psi'}{\partial y}=\hat{u}\delta_1\frac{d\phi}{dy}e^{i(kx-\omega t)} \quad \text{and} \quad v'=-\frac{\partial\psi'}{\partial x}=-ik\hat{u}\delta_1\phi e^{i(kx-\omega t)} \tag{4.9}$$

Using Eqs. (4.6)–(4.9) to eliminate u', v', and p', and introducing the dimensionless variables

$$\eta=y/\delta_1,\quad \alpha=k\delta_1,\quad \beta=\omega\delta_1/\bar{u}_\infty \quad \text{and} \quad U(\eta)=\bar{u}/\bar{u}_\infty$$

the perturbation vorticity equation may be written in its standard form after much algebra as

$$\frac{i}{Re_{\delta 1}}\left(\frac{d^2}{d\eta^2}-\alpha^2\right)^2\phi+\left(\alpha U-\beta\right)\left(\frac{d^2}{d\eta^2}-\alpha^2\right)\phi-\alpha\phi\frac{d^2U}{d\eta^2}=0 \tag{4.10}$$

This fourth-order linear differential equation for the dimensionless stream function $\phi(\eta)$ is the well-known *Orr-Sommerfeld equation*.

The boundary conditions on ϕ follow from the fact that the perturbation velocities must vanish both at the surface and far out in the free stream when considering the stability of the flow.[1] The first condition leads to

$$\phi(0)=\phi'(0)=0 \tag{4.11}$$

where the prime now refers to differentiation with respect to η. In the free stream, we have U equal to unity such that Eq. (4.10) reduces to

$$\left(\frac{d^2}{d\eta^2}-\alpha^2\right)\left[\frac{i}{Re_{\delta 1}}\left(\frac{d^2}{d\eta^2}-\alpha^2\right)\phi+\left(\alpha-\beta\right)\phi\right]=0$$

This equation has the solution

$$\phi=Ae^{-\alpha\eta}+Be^{-\gamma\eta}$$

where only the solutions decaying with respect to η have been retained, A and B are arbitrary constants, and $\gamma^2=\alpha^2+i\,Re_{\delta 1}(\alpha-\beta)$. For the relevant values of α, β and $Re_{\delta 1}$, one finds $|\gamma| >> |\alpha|$ such that the second term in the above expression may always be neglected compared to the first for large values of η. Therefore, the asymptotic form of the solution in the free stream is

$$\phi\propto e^{-\alpha\eta}\to 0 \quad \text{as} \quad \eta\to\infty \tag{4.12}$$

which provides the second set of boundary conditions on ϕ and ϕ'.[2]

It is well known that a solution to a homogeneous differential equation with homogeneous boundary conditions is only possible for particular values of its parameters, called "eigenvalues." Historically,

---

[1] That is, there are no external disturbances.

[2] In other words, the disturbance must also satisfy the equations of motion.

temporally-amplified solutions of Eq. (4.10) were sought. This involves finding the complex eigenvalue for the frequency β given real values for the wave number and Reynolds number, α and $Re_{\delta 1}$. Since it has been shown that the temporally and spatially-amplified solutions are virtually the same,[1] we will discuss the spatially-amplified solutions. In this case, β and $Re_{\delta 1}$ are assumed real and the complex eigenvalues for $\alpha = \alpha_r + i\alpha_i$ with its corresponding *eigenfunctions* ϕ are sought. It then follows from Eq. (4.8) that the perturbations decay when $\alpha_i = \mathrm{Im}\{k\delta_1\}$ is greater than zero, amplify when α_i is less than zero, and neither decay nor amplify when α_i equals zero. These conditions correspond respectively to a mean flow that is *initially* stable, unstable, and neutrally stable to the boundary-layer perturbation.

The solution is not easy to obtain. Since neither the eigenvalues nor eigenfunctions are known a priori, the method for solving Eq. (4.10) is to begin with a guess. Generally, values for α_i, β and $Re_{\delta 1}$ are presumed and a solution sought by systematically guessing α_r. But because Eq. (4.10) has a singularity where $U = \beta/\alpha$, and because one of its four independent solutions grow exponentially in η, this process is particularly toilsome. The solution near the singularity has been extensively studied by Lin [1955], and one of the latest numerical solution methods has been presented by Osborne [1967]. The eigenvalues reported in the following are those obtained by Jordinson [1970, 1971], while the eigenfunctions, unless otherwise stated, were obtained by the author using Jordinson's eigenvalues as a first guess and a standard numerical package[2] for solving ordinary differential equations. All the results in this section pertain to boundary-layer flow over a semi-infinite plate with a constant free-stream velocity. Consequently, in Eq. (4.10), the velocity profile is given by $U(\eta) = f'(\eta_B)$, where f is Blasius' solution with $\eta_B = 1.72\eta$.

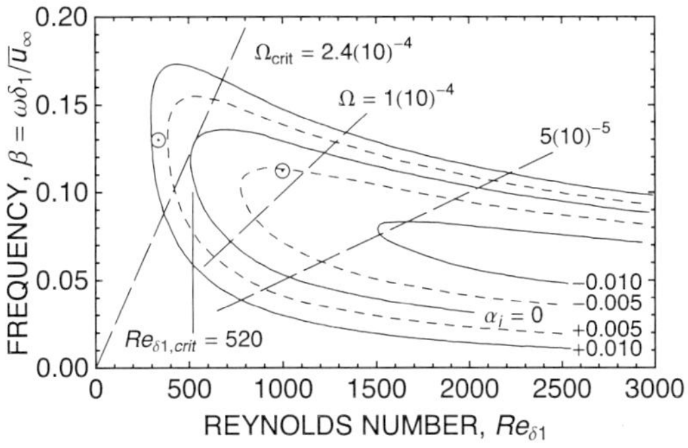

Figure 4.2.
Curves of constant spatial amplification and attenuation for a boundary layer on a flat plate at zero incidence. After Jordinson [1970].

The curves of constant growth and decay are shown in Fig. 4.2. The curve with $\alpha_i = 0$ is the neutral stability loop. The coordinates of this curve are given in Appendix D, Table D.1. All values of β and $Re_{\delta 1}$ within this loop have $\alpha_i < 0$ and represent conditions for which the perturbations will initially grow with distance. Outside the loop, all perturbations will decay. Since they all decay for Reynolds numbers to the left of the loop, a necessary and sufficient condition for the stability of a parallel flow with a Blasius' profile is

$$Re_{\delta 1} \leq 520 \quad \text{(stable)} \tag{4.13}$$

[1] Jordinson [1970] determined this by comparing his numerical spatially-amplified solution with the numerical temporally-amplified solutions of Kurtz and Crandall [1962], Kaplan [1964], and Wazzan et al. [1968]. In addition, see Gaster [1962, 1965]. Furthermore, note that Wazzan et al. also provide the spatially-amplified solutions for the Falkner-Skan class of flows discussed at the end of this section.

[2] Mathematica™.

Thus the *critical Reynolds number* is $Re_{\delta 1,crit} = 520$.[1] The corresponding boundary layer thickness and distance Reynolds numbers, $Re_{\delta,crit}$ and $Re_{x,crit}$, are about 1500 and 91,400 respectively. The former number has been cited several times before. For larger Reynolds numbers, a perturbation will either decay or amplify depending on its frequency.

Since $\beta = \omega\delta_1/\bar{u}_\infty = (\omega\nu/\bar{u}_\infty^2)Re_{\delta 1}$, lines of constant frequency in Fig. 4.2 are straight lines passing through the origin, and their slope is equal to the frequency parameter first introduced in §2.3.1, namely,

$$\Omega = \omega\nu/\bar{u}_\infty^2$$

Any one of these lines also represents the path of events traced by a perturbation of *constant frequency* with increasing distance along the plate. Hence, any of them to the left of the neutral stability loop represent a path for decaying perturbations, whereas those intersecting the loop represent paths for growing and decaying perturbations. The line tangent to the loop is the path where the perturbation just becomes neutrally stable before decaying, which provides another necessary and sufficient condition for viscous stability, namely,

$$\Omega \geq 2.4(10)^{-4} \quad \text{(stable)} \tag{4.14}$$

That is, the *critical frequency* is $\Omega_{crit} = 2.4(10)^{-4}$, and the corresponding critical value of $(\omega\delta^2/\nu)_{crit} = (\Omega Re_\delta^2)_{crit}$ is 540. Therefore, according to the criterion given in Eq. (2.25), the *critical Tollmien-Schlichting wave* is a high-frequency perturbation.[2]

At this position on the neutral stability curve, the values obtained from Table D.1 for α_r and β are approximately 0.35 and 0.14 respectively. Therefore, the wavelength of the disturbance corresponding to the critical frequency is

$$\lambda = 2\pi/k_r = 2\pi\delta_1/\alpha_r \approx 18\delta_1$$

which is about six times the boundary-layer thickness. The corresponding propagation velocity can be obtained from the relation[3]

$$c_r = \omega/k_r = \beta\bar{u}_\infty/\alpha_r \tag{4.15}$$

which yields $c_{r,crit} = 0.4\bar{u}_\infty$. Thus, the critical Tollmien-Schlichting wave is a *high-frequency* perturbation traveling at a speed *less* than half of the free-stream velocity.

The position where the mean velocity equals c_r is known as the *critical layer*. For the critical Tollmien-Schlichting wave, this layer occurs where $y/\delta_1 \approx 0.7$. At this location, as noted earlier, the inviscid form of the Orr-Sommerfeld equation becomes singular, and just as we discovered in §2.3.3 for oscillations traveling slower than the free-stream velocity, the viscous forces must damp the large inertia forces that would exist there without viscosity.

A comparison between the calculated and experimentally determined neutral-stability curves is shown in Fig. 4.3. The experimental data were obtained by Schubauer and Skramstad [1943, 1948], in their historic confirmation of Tollmien's theory, and later by Ross et al. [1970] by placing a vibrating rib-

---

[1] The critical Reynolds number for Blasius' *flow*, in contrast to a *parallel flow* with a Blasius' profile, was calculated by Barry and Ross [1970] to be about 420. Thus, the boundary layer is slightly less stable when its growth is taken into account.

[2] In fact, a form of the high-frequency perturbation equation given by Eq. (2.44) can be obtain directly from the Orr-Sommerfeld equation by neglecting the α^2 term in the differential operators of Eq. (4.10), integrating once, and applying the free-stream boundary conditions.

[3] This expression is only valid on the neutral stability curve. Elsewhere, the propagation velocity is obtained by using $\beta = [(\alpha_r + i\alpha_i)(c_r + ic_i)/u_\infty]$.

bon within the boundary layer and observing the growth and decay of the oscillations downstream of the ribbon for various frequencies.[1] The scatter in the data, particularly near the critical Reynolds number and frequency, is inherent to the process of finding the minimum and maximum in the amplitude of the oscillation with distance (that is, the displacement-thickness Reynolds number).

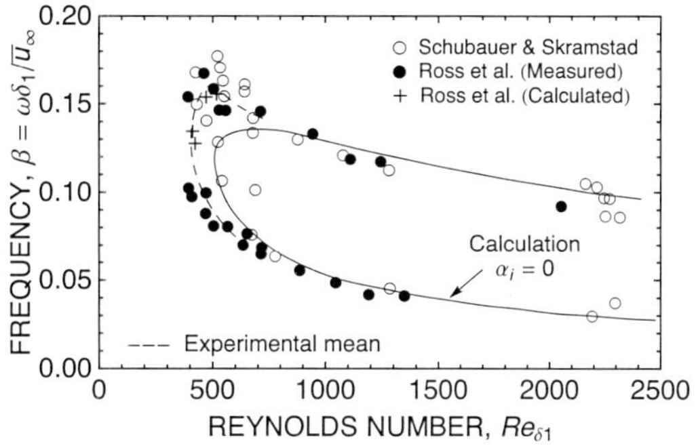

Figure 4.3.
Numerical solution of Jordinson [1970] for the neutral-stability curve compared with the data.

Notwithstanding, it is obvious that the critical Reynolds number and frequency are substantially different from those calculated using the Orr-Sommerfeld equation. As shown by Ross et al., this discrepancy is a consequence of neglecting the growth of the boundary layer and, subsequently, the variation of the eigenfunction with distance in the analysis; an error that becomes substantial near the critical Reynolds number. To confirm this, the neutral stability curve obtained by Barry and Ross [1970] for Blasius' flow was used to determine four points on the "true" neutral stability curve. These points are marked "+" in the figure, and indeed lie near the mean *measured* neutral stability curve shown by the dashed line. Considering these results and taking the minimum value of the measured curve and slope of the line tangent to it passing through the origin, the *actual* values of $Re_{\delta 1,crit}$ and Ω_{crit} are respectively about 400 and $3.4(10)^{-4}$ for Blasius' flow.

Perturbation velocities and Reynolds stresses. The dimensionless perturbation-velocity components are obtained by taking the real parts of Eq. (4.9). This yields

$$u'/\hat{u} = e^{-\alpha_i(x-x_0)}\left(\phi_r' \cos\beta t^* + \phi_i' \sin\beta t^*\right)$$

and (4.16)

$$v'/\hat{u} = e^{-\alpha_i(x-x_0)}\left[\left(\alpha_r\phi_i + \alpha_i\phi_r\right)\cos\beta t^* + \left(\alpha_i\phi_i - \alpha_r\phi_r\right)\sin\beta t^*\right]$$

where ϕ_r and ϕ_i are the real and imaginary parts of the eigenfunction $\phi(\eta)$, $\beta t^* = \beta t - \alpha_r(x - x_0)$ is a modified time interval, and x_0 is the position where the displacement-thickness Reynolds number equals $Re_{\delta 1}$.

The solution obtained by Jordinson for $Re_{\delta 1} = 998$ and $\beta = 0.1122$ is given in Table 4.1. This corresponds to his Case 3 perturbation, which, as shown by the circled point lying within the neutral-stability loop in Fig. 4.2, is unstable. The values in the table have been normalized by the maximum value of ϕ_r which occurs near $y/\delta_1 = 1.92$. The real and imaginary components of the complex eigenvalue for this case are $\alpha_r = 0.3086$ and $\alpha_i = -0.0057$, which yield $c_r = 0.3635\bar{u}_\infty$ and $c_i = 0.0067\bar{u}_\infty$ as the components of the complex wave speed. Consequently, the critical layer appears approximately

---

[1] As demonstrated by Schubauer and Skramstad, results for boundary-layer oscillations with and without a vibrating ribbon and from several sources of sound are all about the same. The vibrating ribbon, however, provides a well controlled disturbance.

where $y/\delta_1 = 0.64$. This solution was also reproduced by the author to obtain the more detailed results shown in the following figures.

Table 4.1

Real and imaginary parts of the eigenfunction $\phi(\eta)$ for $Re_{\delta 1} = 998$ and $\beta = 0.1122$. After Jordinson [1970].

y/δ_1	Re{$\phi(\eta)$}	Im{$\phi(\eta)$}	y/δ_1	Re{$\phi(\eta)$}	Im{$\phi(\eta)$}	y/δ_1	Re{$\phi(\eta)$}	Im{$\phi(\eta)$}
0	0	0	1.5	0.94942	–0.00549	3.9	0.62961	0.00654
0.15	0.05296	–0.02852	1.8	0.99600	–0.00144	4.2	0.57403	0.00693
0.3	0.16979	–0.0513	2.1	0.99221	0.00111	4.5	0.52329	0.00720
0.45	0.30310	–0.05076	2.4	0.95270	0.00270	4.8	0.47701	0.00738
0.6	0.43501	–0.03921	2.7	0.89272	0.00379	5.1	0.43482	0.00746
0.75	0.55890	–0.02728	3.0	0.82447	0.00466	5.4	0.39636	0.00747
0.9	0.66994	–0.01918	3.3	0.75568	0.00540	5.7	0.36130	0.00743
1.2	0.84269	–0.01104	3.6	0.69024	0.00603	6.0	0.32934	0.00733

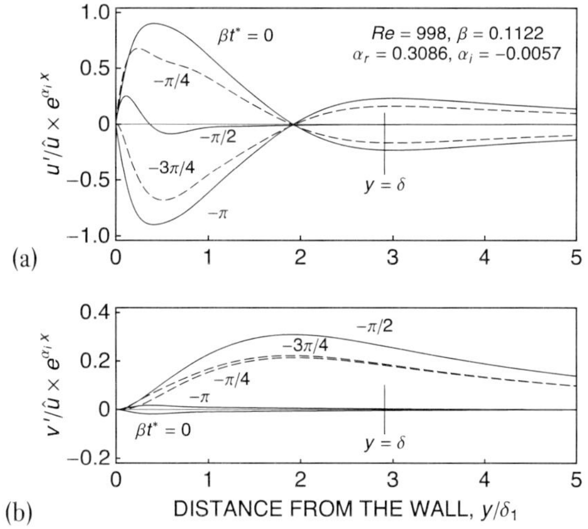

Figure 4.4a,b.
Streamwise perturbation distributions through one-half of a cycle for an unstable Tollmien-Schlichting wave. Calculated for Jordinson's unstable Case 3.

The instantaneous perturbation profiles are shown in Fig. 4.4. Profiles for the other half of the cycle at $\beta t^* = \pi/4$, $\pi/2$ and $3\pi/4$ can be obtained by flipping the $-3\pi/4$, $-\pi/2$ and $-\pi/4$ curves about the horizontal axis. In part (a) of the figure, we see that the streamwise perturbation in the outer portion of the boundary layer simply oscillates about a nodal plane where ϕ_r is a maximum, while that between the wall and critical layer oscillates as an outwardly propagating viscous shear wave in response to harmonic motion farther away. This behavior is similar to the response for a free-stream perturbation traveling at a speed less than the free-stream velocity as shown in Fig. 2.11. In both cases, the unsteady component of the streamwise velocity changes phase within the boundary layer. In this case though, the outer portion oscillates about a common node. Furthermore, comparing the profiles in parts (a) and (b), we see that the streamwise and normal components are *negatively correlated* throughout most of the boundary layer and cycle. Hence, as initially pointed out by Prandtl [1921], a positive Reynolds shear stress ($-\overline{u'v'}$) is produced within the boundary layer.

The Reynolds stresses can be calculated by integrating the time-dependent stress components with respect to time over the period of one oscillation. Thus, with

$$\overline{u'u'}/\hat{u}^2 = \frac{\beta}{2\pi}\int_{t_0}^{t_0+2\pi/\beta} u'(x,\eta,t)u'(x,\eta,t)\,dt$$

etc., and the expressions given in Eq. (4.16), we obtain

$$\overline{u'u'}/\hat{u}^2 = \frac{1}{2}e^{-2\alpha_i x}\left(\phi_r'^2+\phi_i'^2\right), \quad \overline{v'v'}/\hat{u}^2 = \frac{1}{2}e^{-2\alpha_i x}\left(\alpha_r^2+\alpha_i^2\right)\left(\phi_r^2+\phi_i^2\right)$$

and

$$-\overline{u'v'}/\hat{u}^2 = \frac{1}{2}e^{-2\alpha_i x}\left[\alpha_r\left(\phi_r\phi_i'-\phi_i\phi_r'\right)-\alpha_i\left(\phi_r\phi_r'+\phi_i\phi_i'\right)\right]$$

These profiles are plotted in Fig. 4.5. The $\overline{u'u'}$ profile attains its maximum near the wall slightly below the critical layer, decreases to zero where u' changes phase at $y/\delta_1 = 1.92$, rises to another maximum near the edge of the boundary layer, and finally trails off to zero in the free stream. The $\overline{v'v'}$ profile, on the other hand, varying with η^4 at the wall, simply rises to its maximum at $y/\delta_1 = 1.92$ before trailing off. The shear stress attains its *maximum* near the critical layer. As noted above, it is positive throughout the boundary layer providing an energy transfer *from* the mean flow *to* the perturbations. This is characteristic of an unstable perturbation. Finally, the $\overline{u'u'}$ profile in this figure should be compared to that in Fig. 2.12. The difference between them of course is that the oscillation in the present case is self-excited and therefore decays with an increasing distance from the wall, while that in Fig. 2.12 is externally excited.

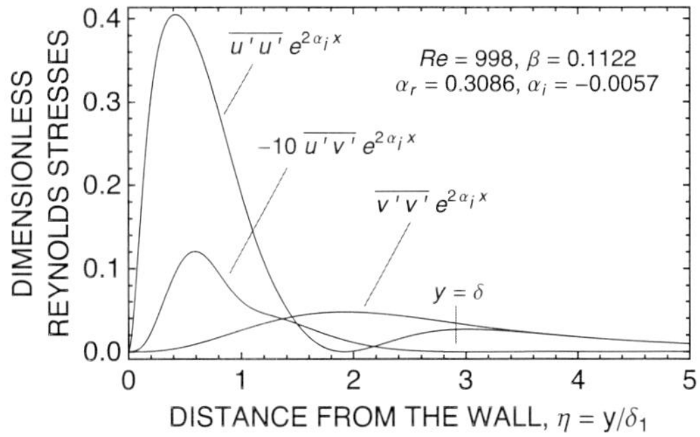

Figure 4.5.
Reynolds-stress profiles for an unstable perturbation. Calculated for Jordinson's Case 3.

A comparison between the shear-stress distributions for a stable and unstable perturbation is given in Fig, 4.6. The stable distribution corresponds to Jordinson's Case 1 perturbation which is shown by the circled point to the left of the neutral-stability loop in Fig. 4.2 at $Re_{\delta 1} = 336$ and $\beta = 0.1297$. The corresponding eigenvalues are $\alpha_r = 0.3084$ and $\alpha_i = 0.0079$. The Reynolds stress for this case is positive near the wall and *negative* elsewhere. Consequently, energy is transferred *from* the mean flow *to* the perturbations near the wall and *from* the perturbations *to* the mean flow in the outer portion of the boundary layer. Since this perturbation is stable, the *net* energy gain by the perturbations is less than that dissipated.

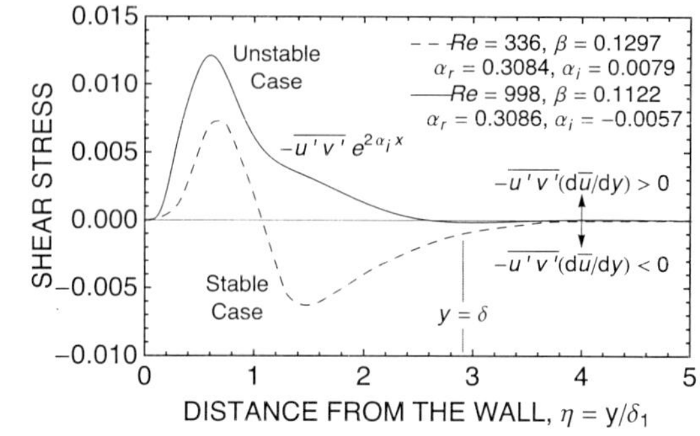

Figure 4.6.
Reynolds shear-stress profiles for a stable (dashed) and unstable (solid) perturbation. Calculated for Jordinson's Cases 1 and 3.

Reynolds-stress terms. The dimensionless form of the Reynolds-stress equations for a parallel mean flow are

$$
\begin{aligned}
U\frac{\partial\overline{u'u'}}{\partial x} &= -2\overline{u'v'}\frac{dU}{d\eta}-2\overline{u'\frac{\partial p'}{\partial x}}+\frac{1}{Re_{\delta 1}}\left(\frac{\partial^2\overline{u'u'}}{\partial x^2}+\frac{\partial^2\overline{u'u'}}{\partial\eta^2}\right)-\frac{1}{Re_{\delta 1}}\hat{\varepsilon}_{11}\\
U\frac{\partial\overline{v'v'}}{\partial x} &= -2\overline{v'\frac{\partial p'}{\partial\eta}}+\frac{1}{Re_{\delta 1}}\left(\frac{\partial^2\overline{v'v'}}{\partial x^2}+\frac{\partial^2\overline{v'v'}}{\partial\eta^2}\right)-\frac{1}{Re_{\delta 1}}\hat{\varepsilon}_{22}\\
U\frac{\partial\overline{u'v'}}{\partial x} &= -\overline{v'v'}\frac{dU}{d\eta}-\overline{u'\frac{\partial p'}{\partial\eta}}-\overline{v'\frac{\partial p'}{\partial x}}+\frac{1}{Re_{\delta 1}}\left(\frac{\partial^2\overline{u'v'}}{\partial x^2}+\frac{\partial^2\overline{u'v'}}{\partial\eta^2}\right)-\frac{1}{Re_{\delta 1}}\hat{\varepsilon}_{12}
\end{aligned}
\tag{4.17}
$$

where the pressure fluctuation and dissipation have been rendered dimensionless using $\rho\bar{u}_\infty\hat{u}$ and $\nu\hat{u}^2/\delta_1^2$ respectively. Since the solution is known, all the terms in these equations can be determined. The calculated results for Jordinson's unstable Case 3 are shown in Fig. 4.7. Each term, multiplied by the exponential of $(2\alpha_i x)$, is shown as it appears in the above equations. Thus, when the term (with its sign) is positive, it causes an exponential increase of the stress component with distance. The dashed line represents the advection of the stress component with distance and hence is the sum of all the terms on the right-hand side of these equations. For clarity, only the largest terms are labeled. Note that the terms associated with the wall-normal-stress component are shown at a twenty-to-one scale of the others. The smaller terms in the shear-stress equation are also scaled by twenty to one.

Clearly, everything but the effect of the pressure fluctuation on the wall-normal component occurs within the boundary layer. Since the gradients of the wall-normal and shear-stress components are relatively small near the wall where viscosity plays a major role, the diffusion and dissipation of these stresses are orders of magnitude smaller than the diffusion and dissipation of the streamwise component. As a consequence, the advection of the wall-normal stress nearly equals its production by the fluctuating pressure force, and since the production of the Reynolds shear stress by the mean-velocity strain rate and the pressure force nearly balance one another, its advection is nearly zero. The situation is much different for the streamwise component where all the terms are important, but within their individual regions. Here, the production by the mean-velocity strain rate, which is most important near the critical layer and beyond, is almost balanced by the pressure-force production term, whereas near the wall the pressure-force production, diffusion, and dissipation terms are important.

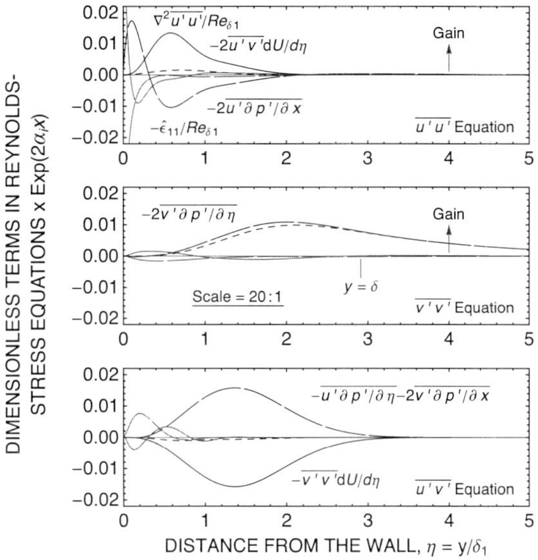

Figure 4.7.

Calculated distributions of each term in the Reynolds-stress equations for an unstable Tollmien-Schlichting perturbation. The production term in each plot is represented by the solid line, the velocity-pressure-gradient term by the long-dashed line, the advection by the short-dashed line, the viscous diffusion by the solid gray line, and the dissipation by the gray-dashed line. The vertical scales for the plots are $\{1, 20, 1\}$ times the actual values. The advection, diffusion and dissipation terms in the shear-stress plot are shown twenty times their actual value. Calculated for Jordinson's unstable Case 3.

The various components of the pressure-velocity correlation are shown in Fig. 4.8. While none of them are particularly easy to evaluate, they can be obtained from either

$$-\frac{\partial p'}{\partial x}=\beta\frac{\partial u'}{\partial t}+U\frac{\partial u'}{\partial x}+v'\frac{dU}{d\eta}-\frac{1}{Re_{\delta 1}}\left(\frac{\partial^2 u'}{\partial x^2}+\frac{\partial^2 u'}{\partial \eta^2}\right)$$

and

$$-\frac{\partial p'}{\partial \eta}=\beta\frac{\partial v'}{\partial t}+U\frac{\partial v'}{\partial x}-\frac{1}{Re_{\delta 1}}\left(\frac{\partial^2 v'}{\partial x^2}+\nu\frac{\partial^2 v'}{\partial \eta^2}\right)$$

or, equivalently by solving

$$\frac{\partial^2 p'}{\partial x^2}+\frac{\partial^2 p'}{\partial \eta^2}=-\frac{\partial v'}{\partial x}\frac{dU}{d\eta}=-\alpha^2\phi f'' e^{i(\alpha x-\beta t)} \tag{4.18}$$

for p' and then calculating the relevant time-averaged quantities. In either case, the boundary conditions are $\partial p'/\partial\eta = 0$ at $\eta = 0, p'(x,\infty) = 0$, and p' is periodic in x with a period of $2\pi/\alpha_r$.

Note that all the correlations in the figure equal zero at the wall. This is forced by the boundary conditions on the velocity fluctuations, not the pressure. Furthermore, note through differentiation that the sum of these correlations must equal zero, and since the derivative of the velocity-pressure correlation is relatively small compared to the others that

$$-\overline{u'\frac{\partial p'}{\partial x}}\approx\overline{p'\frac{\partial u'}{\partial x}}$$

The large peaks very near the wall in these terms are caused by the outwardly propagating viscous wave, which might be surprising since the viscosity does not explicitly appear in equation Eq. (4.18). Nevertheless, a calculation of the correlation on the left-hand side of the above relation neglecting the viscous terms yields the dashed line in the figure. It follows then that the secondary peaks in each of these correlations near the critical layer result from the essentially inviscid-harmonic motion away from the wall. However, unlike the inwardly propagating transverse wave we found in §2.3.3, these peaks result from a simple harmonic oscillation of the flow beyond the critical layer, not in the free stream.

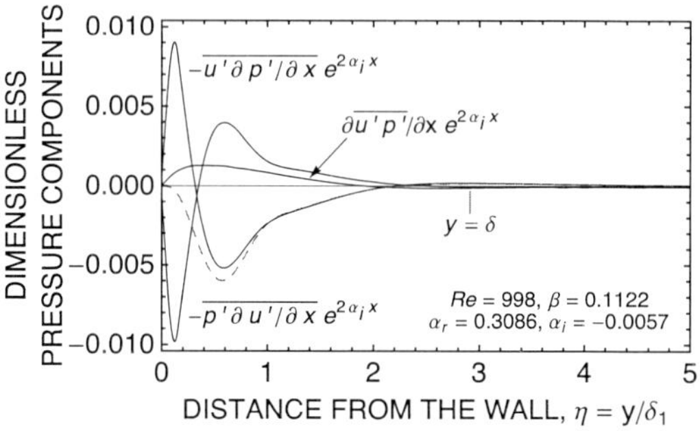

Figure 4.8.
Profiles of the pressure components. Calculated for Jordinson's unstable Case 3.

A very interesting consequence of the above result follows from the continuity equation, Eq. (4.5), namely,

$$\overline{u'\frac{\partial p'}{\partial x}} \approx -\overline{p'\frac{\partial u'}{\partial x}} = \overline{p'\frac{\partial v'}{\partial \eta}} = \frac{\partial \overline{v'p'}}{\partial \eta} - \overline{v'\frac{\partial p'}{\partial \eta}}$$

Since the magnitude of the last term in this series of equalities, according to Fig. 4.7, is much less than the most left-hand expression, we obtain by reversing the order of the above sequence that

$$\frac{\partial \overline{v'p'}}{\partial \eta} \approx \overline{p'\frac{\partial v'}{\partial \eta}} = -\overline{p'\frac{\partial u'}{\partial x}}$$

In other words, the diffusion of the wall-normal velocity pressure correlation nearly equals its pressure-strain-rate companion and both are at least an order of magnitude larger than any of the other terms in the $\overline{v'v'}$ equation. This is counter the usual assumption made for turbulent flow as discussed in §3.3.1.

Effects of pressure gradient on stability. Before continuing with the remaining stages associated with an unstable perturbation in pre-transitional flow, we will briefly consider the effects of a pressure gradient on laminar stability. In particular, let's consider the stability results for what are known as *wedge flows*. Anyone wanting to learn the details about the following results should see Schlichting [1979] (or any text on laminar boundary layers) for the mean-velocity solutions and Wazzan et al. [1968] for the stability solutions. Stability results for other mean-velocity profiles may be found in Ombrewski et al. [1969].

Here, we consider a laminar boundary layer developing on a surface with a free-stream velocity varying as

$$\overline{u}_\infty = ax^m \tag{4.19}$$

where a is a *dimensional* proportionality constant. This class of flows correspond to potential flow past a wedge with an included angle of $2\pi m/(m+1)$ symmetrically aligned to the incident velocity with x measured along its surface from the apex. Clearly, the pressure gradient in the flow is favorable or unfavorable depending on whether m is greater or less than zero.

The solution to the boundary-layer equations for the mean velocity, Eqs. (1.45) and (1.46), is a one parameter set of similarity profiles known as the *Falkner-Skan profiles*.[1] The streamwise component of these profiles are plotted in Fig. 4.9, where the similarity parameter η is given by

$$\eta = \sqrt{\frac{m+1}{2}\frac{\bar{u}_\infty}{\nu x}}\,y = \sqrt{\frac{m+1}{2}\frac{a x^{m-1}}{\nu}}\,y \tag{4.20}$$

The profile for $m = 0$ corresponds to Blasius', while those for $m = 1$ and $m = -0.091$ are for stagnating and separating flow respectively. As discussed in §1.4, velocity profiles in an adverse pressure gradient ($m < 0$) exhibit an inflection point, while those in a favorable pressure gradient ($m > 0$) do not. For $m = 0$, the inflection point resides at the wall.

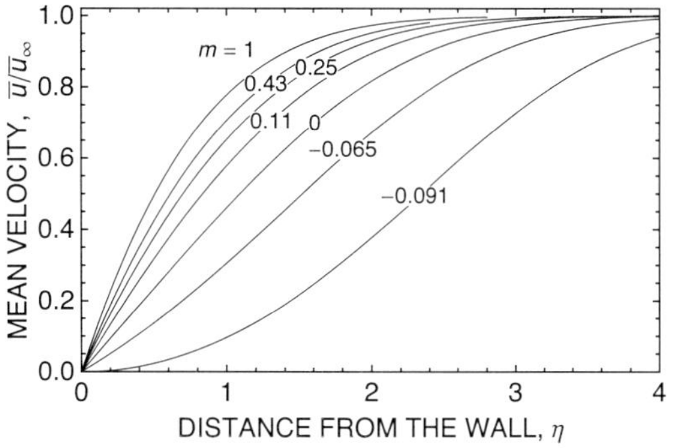

Figure 4.9.
The velocity profiles for Falkner-Skan flows past a wedge with an included angle of $2\pi m/(m+1)$.

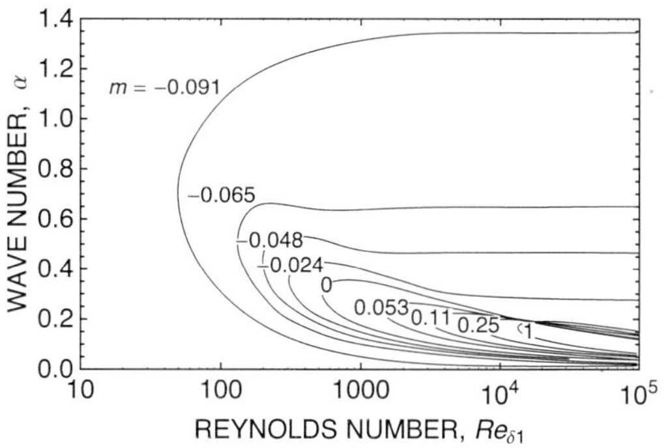

Figure 4.10.
The neutral stability curves for Falkner-Skan flows past a wedge with an included angle of $2\pi m/(m+1)$. After Wazzan et al. [1968].

The neutral stability curves for this class of flows are shown in Fig. 4.10. These are the time-amplified solutions of Eq. (4.10), not the space-amplified as presented in Fig. 4.2. Therefore, the wave number and Reynolds number are the natural independent parameters rather than the frequency and Reynolds number. For very large Reynolds numbers with $m < 0$, the upper branch of the neutral stability curve approaches the solution to Rayleigh's equation for inviscid flow; that is, the flow is stable above a cer-

[1] Falkner and Skan [1931].

tain wave number and unstable below. When $m \geq 0$, both branches of the curve approach zero for indefinitely large Reynolds numbers. In any case, the critical Reynolds number clearly decreases for an adverse pressure gradient and increases for a favorable one.

The critical Reynolds number is plotted as a function of the exponent m in Fig. 4.11. For separated flow, the critical Reynolds number is about seventy, and for stagnation flow it's about 12,500. Here we recall that the critical value for a zero pressure gradient is about 520. Lists of the critical values for the Falkner-Skan profiles and their corresponding temporal and spatial amplification rates are provided in Appendix D, Table D.3.

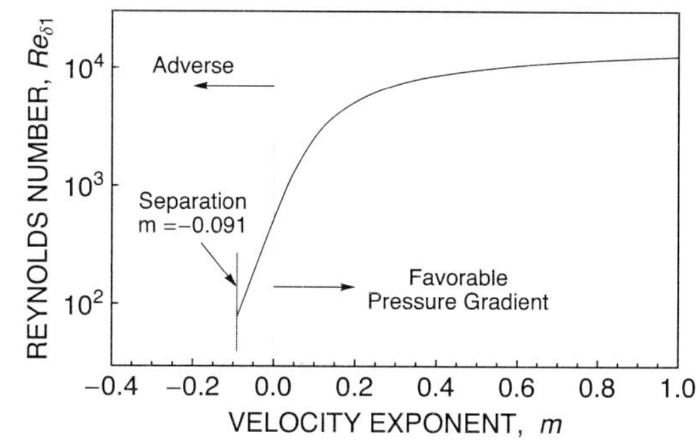

Figure 4.11.
The critical Reynolds number for Falkner-Skan flows past a wedge with an included angle of $2\pi m/(m+1)$. After Wazzan et al. [1968].

Two of the most common parameters used to characterize the local free-stream change in the flow are the *acceleration parameter K* and *pressure-gradient parameter* $\lambda_{\delta 2}$.[1] These are defined as

$$K = \frac{\nu}{\bar{u}_\infty^2}\frac{d\bar{u}_\infty}{dx} \quad \text{and} \quad \lambda_{\delta 2} = -\frac{\delta_2^2}{\mu \bar{u}_\infty}\frac{d\bar{p}_\infty}{dx} \tag{4.21}$$

where $\lambda_{\delta 2}$ is a ratio of the pressure-to-viscous forces within the boundary layer. It is not too difficult to show that $\lambda_{\delta 2} = Re_{\delta 2}^2 K$. For wedge flows, that is, a free-stream velocity varying according to Eq. (4.19), these parameters are given by

$$K = \frac{m}{Re_x} \quad \text{and} \quad \lambda_{\delta 2} = \frac{2m}{m+1}\eta_{\delta 2}^2$$

where Re_x is the *local* streamwise Reynolds number and $\eta_{\delta 2}$ is the value of the similarity variable η given in Eq. (4.20) with y equal to the momentum thickness δ_2. Clearly, the acceleration parameter varies inversely with distance from the leading edge of the wedge, whereas the pressure-gradient parameter remains constant. Consequently, all Falkner-Skan profiles correspond to solutions with constant $\lambda_{\delta 2}$. The values of $\lambda_{\delta 2}$ for different m values are also provided in Table D.3.

4.2.2 The later stages.

Initial amplification. Three-dimensional instability and mean flow distortion. Secondary instability and laminar breakdown.

Initial amplification. The amplitude of a Tollmien-Schlichting wave can be calculated by integrat-

[1] Pohlhausen [1921] originally introduced a parameter similar to the latter of these in his integral boundary-layer theory. Known as *Pohlhausen's pressure-gradient parameter*, it represents the ratio of pressure to viscous forces and is defined using the boundary-layer thickness. The quantity defined in Eq. (4.21) was first introduced by Holstein and Bohlen [1940] in their simplified version of Pohlhausen's theory. Subsequently, Thwaites [1949] used their version in his correlation of laminar boundary-layer data.

ing the amplification rate along its path line. Since a straight line passing through the origin in Fig. 4.2 represents the path line for a wave of any given frequency, its amplitude is obtained from

$$\frac{A}{A_{crit}} = \exp\left[-\int_{Re_{x,crit}}^{Re_x} \alpha_i \, dRe_x\right] = \exp\left[-\frac{2}{(1.72)^2}\int_{Re_{\delta 1,crit}}^{Re_{\delta 1}} \alpha_i \, dRe_{\delta 1}\right]$$

where A_{crit} is the oscillation's amplitude and $Re_{x,crit}$ is the distance Reynolds number when the path crosses the *lower* branch of the stability curve.

The results of this calculation, carried out by Jordinson using the long dashed lines in Fig. 4.2 for $\Omega = 10^{-4}$ and $5(10)^{-5}$, are presented in Fig. 4.12.[1] The minimum of these curves correspond to where the path line intersects the lower branch of the loop, while the maximum corresponds to their intersection with the upper branch. The decay upstream of the minimum corresponds to the decay of the wave before reaching its critical Reynolds number. Although not stressed earlier when discussing the curves of constant amplification in Fig. 4.2, it is easy to see in this figure that the higher frequencies of a disturbance amplify first; the highest being nearest the critical frequency of $\Omega_{crit} = 2.4(10)^{-4}$. However, whether they intensify more than their low frequency counterparts depends on the disturbance's energy spectrum and the decay of the various frequency components before reaching their corresponding critical Reynolds number. As long as its amplitude remains small, all of this can be calculated from a Fourier analysis once the spectral details of the disturbance are known.

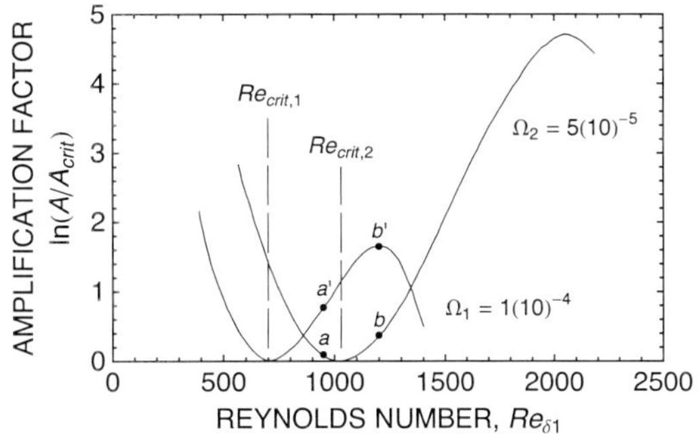

Figure 4.12.
Calculated amplitude curves for two unstable Tollmien-Schlichting waves in a laminar boundary layer on a flat plate. After Jordinson [1970].

Of the two results shown in the figure, the wave with the lower frequency amplifies the most. This follows from Fig. 4.2 where we see that the path for the lower frequency cuts through a larger portion of the unstable region. Its maximum value is almost e^5 or 150. Even larger values have been calculated by Jaffe et al. [1970] who obtained a maximum of about e^{12} or $1.6\ (10)^5$ at a Reynolds number of about 4000 when $\Omega = 2(10)^{-5}$. As we will see, however, using a linear analysis to calculate amplification factors up to this order is questionable. Readers interested in methods for predicting the onset of natural transition using linear stability theory are referred to White [1974] or Reshotko [1976].

Three-dimensional instability. It is well known that two-dimensional Tollmien-Schlichting waves are inherently unstable to spanwise disturbances. As investigated experimentally by Schubauer [1957], Klebanoff and Tidstrom [1959], Klebanoff et al. [1962] and Kovasznay et al. [1962], and theoretically by Benney and Lin [1960], Benney [1961], Lin and Benney [1962], Greenspan and Benney [1963], and Benney [1964], a wave subjected to small spanwise irregularities in boundary-

[1] Similar calculations were carried out by Shen [1954] who obtained temporally-amplified solutions to the Orr-Sommerfeld equation and Jaffe et al. [1970] who obtained spatially-amplified solutions.

layer thickness soon develops spanwise variations in amplitude that intensify into streamwise streaks of low and high amplitude oscillations. This behavior is illustrated in Fig. 4.13 where the spanwise variation of intensity at several streamwise distances downstream of a vibrating ribbon is plotted for roughly a twenty-five percent periodic irregularity in the boundary-layer thickness. In this experiment, segments of tape were placed on the surface beneath the ribbon to produce a disturbance with a spanwise wavelength equal to about five boundary-layer thicknesses at the ribbon. This period was chosen to reproduce to the average wavelength of the "natural" spanwise irregularities observed earlier by both Schubauer, and Klebanoff and Tidstrom in their experiments.

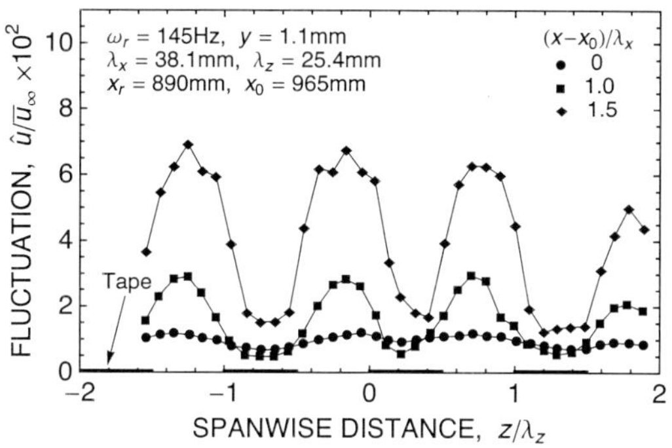

Figure 4.13.
Spanwise variations of intensity in a disturbed boundary layer for a disturbance introduced near the upper branch of the neutral-stability curve. Data from Klebanoff et al. [1962].

The initial response of the wave to a spanwise irregularity can be determined by considering its local streamwise growth rate for a particular wave frequency as shown in Fig. 4.12. For example, consider the spanwise positions where the boundary layer is nominally thin and thick, which, for a constant free-stream velocity, correspond to where the Reynolds number $Re_{\delta 1}$ is low and high. In addition, consider for the sake of argument that these positions correspond to points a and b in the figure which are near the lower branch of the neutral-stability curve. Then, since the amplification rate is larger at point b, the intensity of the oscillations in the *thicker* portion of the boundary layer will grow faster than that in the thinner. On the other hand, if we suppose that the frequency of the disturbance is higher such that these positions correspond to points a' and b' near the upper branch of the neutral-stability curve, then the intensity of the oscillations in the *thinner* portion of the boundary layer will grow faster. In other words, *any* spanwise irregularity in boundary-layer thickness produces an unstable three-dimensional response.

The results shown in Fig. 4.13 are for a disturbance (the vibrating ribbon) introduced at $Re_{\delta 1} = 1625$ with $\beta = \omega\delta_1/\bar{u}_\infty = 0.1$ which is near the upper branch of the neutral-stability curve. As a consequence, the maximum of the oscillations downstream of the tape occur in the streaks with the thinner boundary layer, while their minimum occur in the streaks with the thicker boundary layer. According to the reasons just discussed, this behavior was experimentally found by Klebanoff et al. to reverse when the frequency of the ribbon was halved.

In either case, the growth of a Tollmien-Schlichting wave beyond its initial response depends on the intensity of the disturbance. This is shown in Fig. 4.14 where the streamwise variation of maximum and minimum intensities within the streamwise streaks are plotted for a weak and strong disturbance, again with $Re_{\delta 1} = 1625$ and $\beta = 0.1$. For the weak disturbance, both the maximum and minimum intensities grow and decay according to linear-stability theory as they first pass through the upper region of the neutral-stability loop, where all oscillations amplify, and then into the region above the loop where they all decay. On the other hand, after a distance of about three-and-a-half wavelengths from the ribbon, the responses to the strong disturbance deviate from this path. In particular, the maximum in intensity begins to grow faster than that for a weak disturbance while the minimum first decays and then grows, both becoming no longer "small." A few wavelengths downstream from its departure, the growth of the maximum intensity typically levels off and sharp spikes begin to appear in the instantaneous velocity with each passing wave. These spikes are the precursor of high frequency laminar oscillations, laminar "breakdown" into small scale turbulence, and subsequent turbulent spots. As observed by Schubauer, Klebanoff et al., and Kovasznay et al., this phenomenon only occurs on the streaks of maximum intensity. Following breakdown, the intensity increases for another two-to-three wavelengths to its maximum at which point it decreases to a level commensurate with a turbulent boundary layer. The distance between breakdown and maximum intensity on this path is approximately that required for separate spots (which occur one wavelength apart) to grow and begin interfering with one another. As it turns out, the initial decrease of the minimum intensity curve is a result of energy transferred from the streaks with a minimum intensity to those with a maximum. Following that, the intensity increases at a rapid rate as the flow in these streaks becomes disturbed by spots expanding into it until they merge to form a turbulent boundary layer. A quick calculation will show that the final intensity when scaled by the free-stream velocity is about 0.08 for both strong disturbance curves.

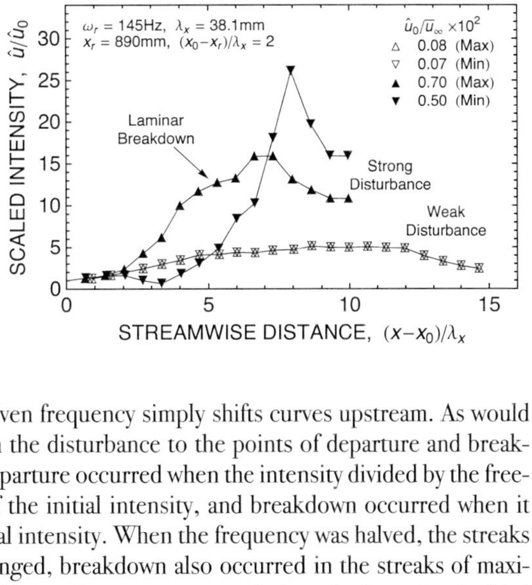

Figure 4.14.
Variation in intensity in the peaks and valleys of a distorted wave. Data from Klebanoff et al. [1962].

Increasing the disturbance's intensity for a given frequency simply shifts curves upstream. As would be expected, this decreases the distance from the disturbance to the points of departure and breakdown. For the frequency used in Fig. 4.14, departure occurred when the intensity divided by the free-stream velocity reached 0.01 independent of the initial intensity, and breakdown occurred when it reached 0.074, again independent of the initial intensity. When the frequency was halved, the streaks of maximum and minimum intensity interchanged, breakdown also occurred in the streaks of maximum spanwise intensity when it reached 0.074, but departure occurred when the intensity doubled to 0.02.

Mean-flow distortion. The rapid increase in intensity is also accompanied by the development of secondary flows within the original two-dimensional boundary layer. As shown in Fig. 4.15a, this flow consists of an alternating pattern of clockwise and counterclockwise streamwise vortices. The measurements, taken by Klebanoff et al., were obtained six wavelengths downstream from the ribbon, which is about one wavelength before breakdown (see Fig. 4.14). The spanwise variation in the mean streamwise and spanwise velocity components at this location are shown in parts (b) and (c) of the figure, while the corresponding variation in streamwise intensity is presented in part (d). The data shown by the filled circles were obtained about three-tenths of the boundary-layer thickness away from the surface which is *above* the critical layer, while those shown by the empty circles in part (c) of the figure were obtained at about one-tenth of the thickness, which is *below* the critical layer.

The maximum mean velocity in the streamwise direction is greater than that for Blasius' flow where the streamwise intensity is at its minimum, and less where the intensity is a maximum. The spanwise component of the mean velocity, in comparison, is a maximum on either side of the peak in streamwise intensity, and changes sign from one side of the peak to the other. It also changes sign across the boundary layer. The arrows in this part of the figure represent the direction of the spanwise flow nearest the surface which corresponds to the data represented by the empty circles. Alternating direction, they indicate a converging and diverging secondary flow near the surface. Farther away from the surface, the direction of the secondary flow is exactly opposite which is consistent with the secondary flow system of alternating clockwise and counterclockwise streamwise vortices shown in part (a) of the figure.[1] The deficits and excesses in the streamwise velocity are also consistent with this pattern, since the converging spanwise flow near the surface transports the slower moving fluid away from the surface while the diverging flow draws in the faster moving fluid from above. When the frequency is halved, such that the disturbance is introduced near the lower branch of the neutral-stability curve, the whole system as demonstrated by Klebanoff et al. simply shifts spanwise by half a wavelength.

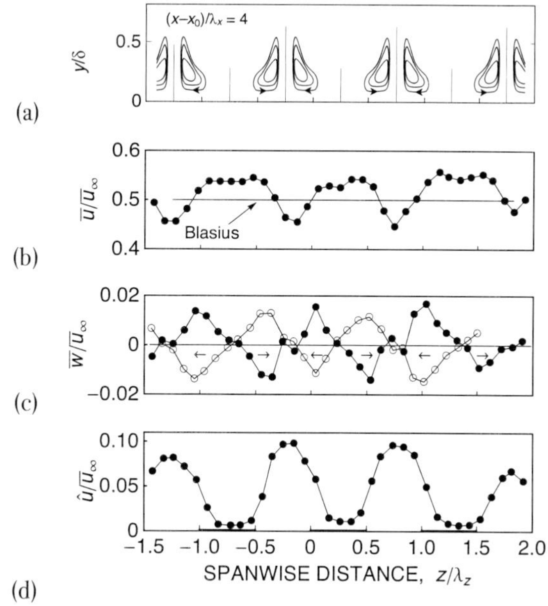

Figure 4.15a,b,c,d.
Spanwise variations in the streamwise and spanwise mean velocity components and intensities in a boundary layer disturbed near the upper branch of the neutral-stability curve. Data from Klebanoff et al. [1962].

The birth of this system is coincident with the departure of the intensity curves from the *weak* disturbance curve in Fig. 4.14 and linear-stability theory. It is also consistent with the calculations of Benney and Lin who considered the nonlinear behavior of a primary disturbance whose amplitude also varies periodically in the spanwise direction. That is, once the intensity of the primary wave reaches a certain level, a secondary vortical flow with rapidly amplifying oscillations develops. As previously noted, this level depends on the frequency of the disturbance.

The subsequent breakdown along the streaks of peak intensity is best understood by considering the

[1] The streamlines shown here are reflected and repeated versions of those determined by Klebanoff et al. for the half-wavelength between $z/\lambda_z = -0.25$ and -0.75.

instantaneous velocity profiles shown in Fig. 4.16. Measured by Kovasznay et al. [1962], who also used tape segments to produce a periodic varying boundary-layer thickness and a vibrating ribbon, these profiles were obtained about six wavelengths downstream from the ribbon. In addition, the test was conducted with $Re_{\delta 1} = 1100$ at the ribbon, $\beta = 0.08$, and the intensity of the ribbon adjusted to yield a "spike" in the trace of the instantaneous velocity at the measurement position. Therefore, these results are for a disturbance introduced near the maximum rate of amplification within the neutral-stability curve and for a boundary layer just before laminar breakdown.

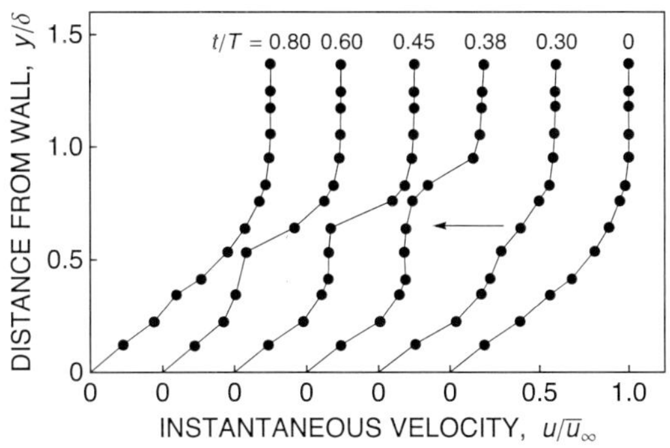

Figure 4.16.
Instantaneous velocity profiles at the location of first spike. Data from Kovasznay et al.[1962].

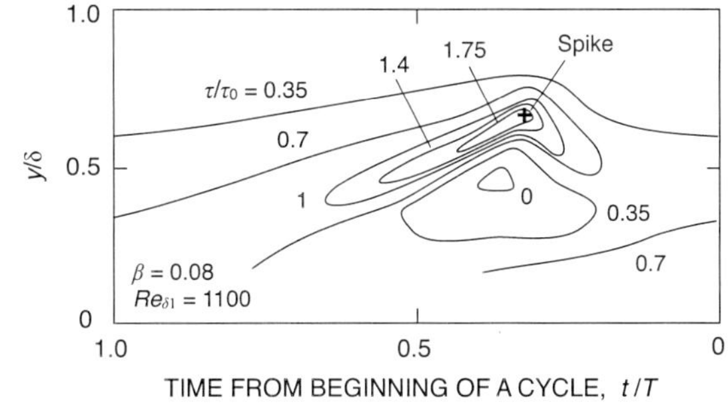

Figure 4.17.
Shear-stress contours at the peak position in the one spike stage. Data from Kovasznay, et al. [1962].

The profiles in this figure are plotted for consecutive instants of time during the period T of the disturbance, but in reverse order to illustrate how the profiles would appear if measured instantaneously at consecutive distances downstream. The most dramatic and important feature of the profiles is the formation of an intense shear layer in the outer portion of the boundary layer. During the rest of the cycle the shear layer practically disappears and the profiles appear more Blasius-like. The position of the shear layer is shown more precisely in Fig. 4.17 where contours of the local shear stress divided by Blasius' wall-shear stress are plotted. As it turns out, the shear layer at this position is "kinked" and its maximum strength is almost twice that at the wall. In addition, the layer is accompanied by a mild shear region below, which when grouped together forms an extended region of highly distorted flow about a third of the way through the cycle in the middle of the boundary layer. As shown in the figure, the spike occurs at the peak of this region. Its coordinates are roughly $t/T = 0.32$ and $y/\delta = 0.65$, which corresponds with the position shown by the arrow in the previous figure where the velocity changes dramatically from high to low and back. Consequently, the spike, which is normally considered a precursor of laminar breakdown for natural transition, is simply a result of the rapidly changing velocity occurring in the kink of the intense shear layer. Upstream of this position, the shear layer is developing and neither as intense nor kinked. Downstream, however, it develops a waviness or additional kinks that are associated with a secondary instability.

Secondary instability and laminar breakdown. It is not difficult to conclude that the internal shear layer shown in Fig. 4.16 is unstable. Since the effects of viscosity on the relatively low frequency oscillations away from the wall can be neglected, all mean-velocity profiles with an inflection point according to Rayleigh [1880, 1887, 1892], such as those shown in the figure, are unstable. As shown by Greenspan and Benney [1963], who examined the stability of an inviscid internal shear layer along the streak of peak oscillations using Kelvin-Helmholtz' stability theory,[1] the development of this instability is exceedingly rapid. In particular, using the velocity profile measured by Kovasznay et al. they calculated a tenfold increase in amplitude for the secondary instability within one half the wavelength of the primary disturbance. In addition, they calculated a wavelength and frequency for the secondary instability approximately equal to one-fifth and eight times that of the disturbance, which are in reasonable agreement with the measurements of Klebanoff et al. and Kovasznay et al. Consequently, it appears that the sudden amplification of high-frequency oscillations and, subsequently, breakdown within each cycle of the disturbance as it travels downstream are caused by a secondary instability of the distorted mean flow boundary-layer profile along the streak of peak oscillations.[2]

The situation is complicated by the fact that the intense shear layer is shaped more like a highly sweptback flying wing with a span equal to about one-third the spanwise wavelength of the disturbance and a kinked leading-edge tip, than a two-dimensional version of the layer shown in Fig. 4.17. Nevertheless, a picture of the final stage along this path to transition is that the flow distorted by the growing three-dimensional wave develops an unstable shear layer as it propagates downstream, which in turn rapidly breaks down after forming its first kink.

Several aspects of this process are illustrated in the *xt*-diagram of Fig. 4.18, which is a modified version of that presented by Greenspan and Benney. Although we should expect a somewhat irregular pattern for natural transition (both in space and time), the pattern shown in this figure is idealized and conforms more with the controlled experiments conducted by Klebanoff et al. and Kovasznay et al. Furthermore, this is only a picture of the situation along the streak of peak oscillations. Nevertheless, the abscissa is the streamwise distance arbitrarily measured from the position where the spike and first secondary wave appear, divided by the wavelength of the primary (Tollmien-Schlichting) instability. The ordinate is the time it takes a free-stream particle to transverse one wavelength. Therefore, since the primary instability in the figure is shown propagating at a speed equal to 0.4 times the free-stream velocity, the dimensionless period of a cycle is equal to 2.5.

The trajectories of the primary and secondary waves are shown in the lowest diagram of the figure. In this case, the secondary instability is shown propagating at twice the primary speed[3] with roughly one-eighth the wavelength. Consequently, the secondary waves would begin to interfere with the primary wave after two primary wavelengths (point "a" in the figure) if nothing else happened. As shown by Schubauer and Klebanoff et al., however, breakdown occurs about three to four Kelvin-Helmholtz wavelengths downstream once the spikes (kinks in the shear layer) appear. This position is marked by the circles in the lower central diagram of the figure. Called an "incipient" turbulent spot, its leading and trailing edges follow the trajectories plotted by the heavy lines, which we assume in this example propagate at velocities equal to 0.9 and 0.5 times the free-stream velocity, respectively. Although

[1] See Appendix E and a related discussion of separated-flow transition in §8.3.

[2] Again see §8.3 where similar results are found for separated-flow transition.

[3] As calculated by Greenspan and Benny.

these values are typical, we will see in Chapter 5 that they depend on the transition Reynolds number and pressure gradient. We will also see that the initial growth of a spot is not linear. Nevertheless, projecting these trajectories *upstream*, we determine that the virtual origin for the spot lies shortly downstream from the appearance of the first spike as indicated by the heavy crosses. This position, as we will see in §5.2.3, generally but not always coincides with the onset of transition.

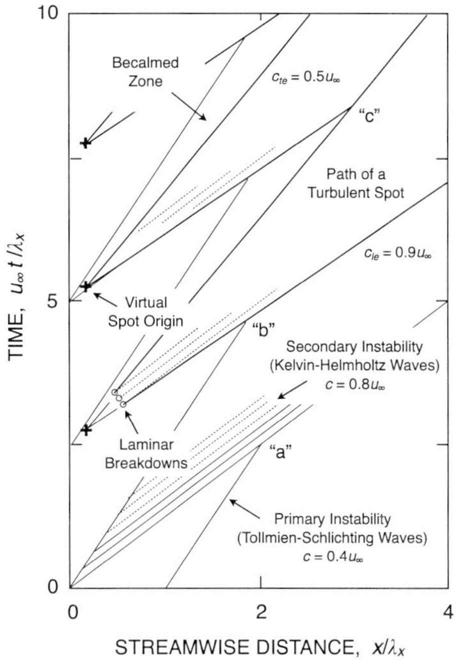

Figure 4.18.
Space-time diagram of the primary and secondary instability waves, turbulent spots, and becalmed regions.

Propagating faster than the primary Tollmien-Schlichting wave ahead of it, the leading edge of the spot catches up with the wave within two of its wavelengths (point "b" in the figure). As we will learn in Chapter 5, the region between the trailing edge of the spot and following primary wave is called the "becalmed zone," since any instabilities in the flow there are subdued by a velocity profile that is fuller than a Blasius profile. Finally, we see that the leading edge of a spot catches up to the trailing edge of the previous spot about three wavelengths downstream from its origin (point "c" in the figure). This roughly corresponds, as shown in Fig. 4.14, to where the intensity of the disturbance reaches its maximum along the streak of peak oscillations.

More recent analyses and scenarios of these stages have been presented by Cohen et al. [1991] and Breuer et al. [1997]. While helpful to understanding the problem, their results again provide only part of the solution for a pre-transitional boundary layer with a very low free-stream turbulence level. Clearly, this unsteady flow with its multiple regions of two and three-dimensional instabilities is a complex problem to which we have only a few piecewise solutions. In other words, we still lack a practical method to calculate it. In this regard, direct numerical simulation (DNS) appears to be a promising method of attack, even though it is presently far from being practical.

4.2.3 Laminar breakdown.

In 1957, Schubauer reported that "breakdown" to turbulent flow in his experiments on natural transition without a pressure gradient occurred when the maximum intensity $\sqrt{u'u'}/\bar{u}_\infty$ reached a value of 0.074. For their experiments, Klebanoff et al. [1962] determined that the Strouhal number for laminar breakdowns in a laminar boundary layer was independent of their disturbance frequency. In particular, they found that the average Strouhal number based on the free-stream velocity and displacement thickness was about 0.13, that is,

$$S_{\delta 1,b} = \left(f\delta_1/u_\infty\right)_b \approx 0.13$$

They also determined that the breakdowns propagated independent of the disturbance frequency with a velocity equal to about 0.68 times the free-stream velocity, that is, a velocity $c = 0.68u_\infty$.

As a consequence, since $f = \omega/2\pi = ck/2\pi$, where k is the wave number of the breakdowns, it is relatively easy to obtain

$$\left(\omega\delta_1/u_\infty\right)_b \approx 0.82 \quad \text{and} \quad \left(k\delta_1\right)_b \approx 1.2 \tag{4.22}$$

Since the maximum dimensionless frequency for the neutral stability curve in Fig. 4.2 is about 0.14,[1] we readily determine that laminar breakdown in natural transition occurs at a frequency or wave number about five times higher than that associated with a Tollmien-Schlichting instability and are, therefore, probably associated with the secondary "Kelvin-Helmholtz-like" instability.

4.3 Path to Bypass Transition.

Some preliminaries. Mayle-Schulz model. Calculations and experimental results. Laminar breakdown and onset of transition.

The first serious measurements taken in a pre-transitional boundary layer with free-stream turbulence were reported by Dyban et al. [1976], who obtained both mean-flow and Reynolds-stress profiles. A sample of their data showing the Reynolds-stress components near the wall is given in Fig. 4.19. Clearly, this behavior cannot result from turbulence being diffused into the boundary layer from the free stream as many investigators modeling pre-transitional flow presume. While numerous experimental investigations have been carried out since, some of which are referenced in the Bibliography, the most extensive data set to date is that obtained by Roach and Brierley [1990]. We will examine their data later.

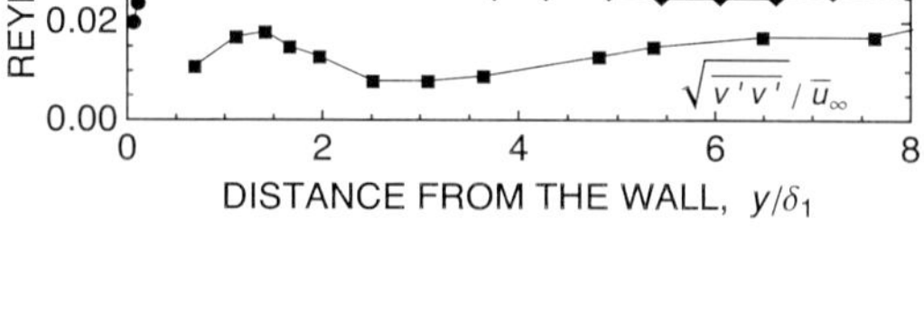

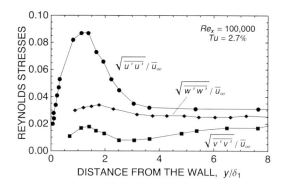

Figure 4.19.
Reynolds-stress profiles in a pre-transitional boundary layer with zero pressure gradient. Data from Dyban et al. [1976].

[1] See also Appendix D, Table D.1.

The first attempt to evaluate the effect of turbulence on a laminar boundary layer was made by Taylor [1936]. He developed an expression for the time-averaged contribution to the pressure gradient for an isotropic turbulent free stream and suggested that this be used in addition to the mean pressure gradient to calculate the mean-velocity profile. In the 1950s it was popular to examine the effects of small amplitude, unsteady, free-stream perturbations on laminar boundary layers.[1] Unlike the traveling wave solutions examined in §2.3.3, however, none of these analyses produced intensities of the size shown in Fig. 4.19, nor large changes in the time-averaged velocity. As a consequence, it was generally thought that the effect of free-stream turbulence on a laminar boundary layer was small until in the 1960s and 70s, it was found that turbulence significantly affected laminar heat transfer and Dyban et al. conducted their experiments.

In 1966, Smith and Kuethe presented a unique "exact" solution for laminar stagnation flow with free-stream turbulence by using an eddy-viscosity model. Since the eddy viscosity for stagnation flow reduces to a constant, it simply augments the molecular viscosity in proportion to the increased turbulence level. Although other algebraic eddy-viscosity models were proposed to calculate the effect of turbulence on laminar boundary-layer flow,[2] this approach was quickly abandoned in the 1970s and 80s when turbulent flow models were modified to *mimic* transition. In these models, as mentioned above, the kinetic-energy in the free stream was simply allowed to diffuse (generally with a diffusivity equal to ν) and dissipate through the pre-transitional boundary layer until the onset of transition, where it then most often suddenly "jumped" into a transitional mode.

In 1997, Mayle and Schulz proposed using Lin's [1957] production term for high-frequency oscillations (§2.3.1) and Dullenkopf and Mayle's [1995] idea of an "effective" turbulence to evaluate the production of turbulent kinetic energy within a pre-transitional boundary layer. Their analysis was unique in that it introduced a new mechanism based on the free-stream pressure fluctuations for amplifying the velocity fluctuations within the boundary layer, which then provided a practical differential approach for calculating the effect of free-stream turbulence on a laminar boundary layer. Soon after, an integral approach using Taylor's expression for the time-averaged pressure gradient was presented by Roach and Brierley [2000]. Both of these analyses are capable of calculating the effect of free-stream turbulence on both the laminar mean flow and *streamwise* fluctuating velocity component as shown in Fig. 4.19. However, since the ideas and analysis of Mayle and Schulz are easily extended to calculating the other Reynolds-stress components as well, we will only consider their approach and the extension thereof in the following.

In this section, after considering some necessary preliminaries, we will develop a set of Reynolds-stress equations that accounts for the effects of turbulence level and length scales on pre-transitional laminar boundary layers, compare some calculated and experimental results, and then consider some old and new criteria for the onset of bypass transition.

4.3.1 Mayle-Schulz model.

The other production term. The turbulence that matters. The role of turbulent length scales.

We saw in §4.1 how free-stream turbulence can directly affect the production of turbulent-like fluctuations within a laminar boundary layer. In this section we will develop an expression for the relevant production term, determine the scale of *free-stream* turbulence that most affects a laminar boundary layer, and consider the general role of turbulent length scales on the boundary layer.

The other production term. Let's first consider the Reynolds-stress equation for the streamwise

---

[1] See introduction to §2.3.

[2] Here the reader is referred to Dyban and Epik [1985], Ames et al. [1999], and Steelant and Dick [2001]. However, since they determined the eddy viscosities directly from their measurements, only Dyban and Epik's analysis predicts an increase in streamwise-normal intensity comparable to that shown in Fig. 4.19.

stress component. Given by Eq. (1.52), this equation after neglecting the smallest terms reduces to

$$\bar{u}\frac{\partial\overline{u'u'}}{\partial x}+\bar{v}\frac{\partial\overline{u'u'}}{\partial y}=-2\overline{u'v'}\frac{\partial\bar{u}}{\partial y}-2\overline{u'\frac{\partial p'/\rho}{\partial x}}+\frac{\partial}{\partial y}\left(\nu\frac{\partial\overline{u'u'}}{\partial y}-\overline{v'u'u'}\right)-2\nu\overline{\left(\frac{\partial u'}{\partial y}\right)^2}$$

Since free-stream oscillations with the lowest frequency produce a quasi-steady laminar flow within the boundary layer (§2.3.4), we should expect that the largest turbulent eddies will have little effect on the time-averaged flow. And since the highest frequency oscillations are dominated by viscosity and quickly dissipate within the boundary layer, we should expect that the smallest turbulent eddies will also have little effect. Therefore, the fluctuations observed growing within the boundary layer must be produced by free-stream eddies with a "most effective" size which is neither too large nor too small.

Following the reasoning given in §2.3.1, if we suppose that these eddies produce fluctuations in the boundary layer with a frequency $\omega >> \nu/\delta^2$, we may then use Lin's theory for high-frequency oscillations to eliminate the fluctuating pressure in the above equation. That is, from Eq. (2.35) we have

$$-\frac{1}{\rho}\frac{\partial p'}{\partial x}\approx\frac{\partial u'_\infty}{\partial t}; \quad (\omega\delta^2/\nu >> 1)$$

As a consequence, the velocity/pressure-gradient correlation in the above equation is given by

$$-\overline{u'\frac{\partial p'/\rho}{\partial x}}\approx\overline{u'\frac{\partial u'_\infty}{\partial t}}$$

Furthermore, presuming, as we did in §2.3.2, that changes in the unsteady components occur in the y direction over the distance δ, which in turn is of order $\sqrt{(\nu L/u_\infty)}$, changes with respect to time are of order ω, u' is of order u'_∞, and v' is of order $u'_\infty(\delta/L)$. As a result we obtain

$$\overline{u'\frac{\partial p'/\rho}{\partial x}}\Big/\varepsilon=O(\omega\delta^2/\nu) \quad \text{and} \quad \overline{u'\frac{\partial p'/\rho}{\partial x}}\Big/\overline{u'v'}\frac{\partial\bar{u}}{\partial y}=O(\omega\delta^2/\nu)$$

This not only implies that the velocity/pressure-gradient term overwhelms the dissipation when Lin's high-frequency criterion is met, causing the fluctuations to grow, but that it also overwhelms the standard production term, which then can initially be neglected by comparison. Using the same order of magnitude analysis, it can also be shown that the diffusion by the wall-normal fluctuating velocity component can be neglected compared to the viscous diffusion.

Consequently, if we use Lin's theory for high-frequency oscillations, the above Reynolds-stress equation for a laminar boundary layer with a turbulent free stream reduces to

$$\bar{u}\frac{\partial\overline{u'u'}}{\partial x}+\bar{v}\frac{\partial\overline{u'u'}}{\partial y}=2\overline{u'\frac{\partial u'_\infty}{\partial t}}+\nu\frac{\partial^2\overline{u'u'}}{\partial y^2}-2\nu\overline{\left(\frac{\partial u'}{\partial y}\right)^2}; \quad (\omega\delta^2/\nu >> 1) \tag{4.23}$$

which explicitly includes the free-stream fluctuation. As shown in §4.1, the integral solution to this equation yields a fluctuation intensity that initially grows linearly with distance. Of course, the relevant frequency still remains to be determined.

The turbulence that matters. First introduced by Dullenkopf and Mayle [1995] to explain leading-edge heat transfer results, later extended by Mayle and Schulz [1997] to pre-transitional boundary layers, and afterward formalized by Mayle et al. [1998], the turbulence that affects a laminar boundary layer depends on the mechanism involved. For pre-transitional flow, the mechanism is revealed in the first term on the right-hand side of Eq. (4.23). Introducing a correlation coefficient $R_{\phi 11}$, which generally depends on both x and y, this term may be written as

$$\overline{u'\frac{\partial u'_\infty}{\partial t}} = R_{\phi 11}\sqrt{\overline{u'u'}}\sqrt{\overline{\frac{\partial u'_\infty}{\partial t}\frac{\partial u'_\infty}{\partial t}}} \tag{4.24}$$

The fact that $R_{\phi 11}$ depends on y is easily seen by realizing that any fluctuation and its temporal derivative is ninety degrees out of phase. Consequently, the coefficient must decrease across the boundary layer from a finite value at the wall to zero in the free stream. Furthermore, since $R_{\phi 11}$ is dimensionless, the relation must be a function of y divided by some characteristic boundary-layer or turbulent length scale, both of which depend on x.

Guided by dimensional considerations, Mayle and Schulz introduced an *effective intensity and frequency* of turbulence by proposing that

$$\sqrt{\overline{\frac{\partial u'_\infty}{\partial t}\frac{\partial u'_\infty}{\partial t}}} \propto \omega_{eff}\sqrt{\overline{u'^2_{eff}}} \tag{4.25}$$

Since the energy contained in these "effective" eddies is only a fraction of the total turbulent energy, we should expect that the "most effective" turbulence level is also only a fraction of the "reported" turbulence level which in turn is essentially that associated with the large eddies.

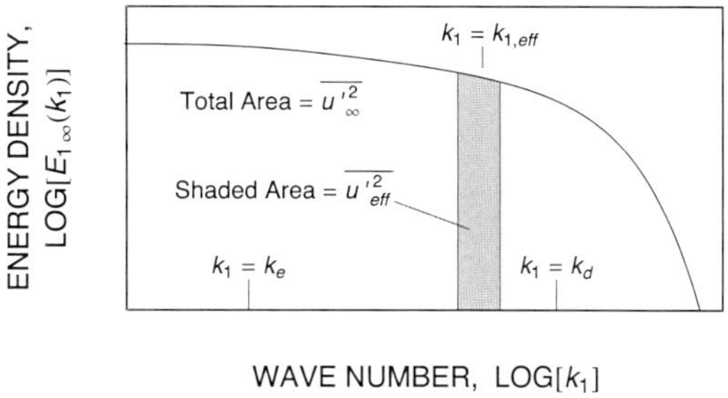

Figure 4.20.
The effective turbulence as a fraction of the total turbulent energy.

Recalling the various definitions in §3.1, the integral of the one-dimensional energy spectrum $E_1(k_1)$ over all wave numbers k_1 equals the mean-square of turbulence. This equals the "Total Area" in Fig. 4.20. The effective intensity of turbulence on the other hand is the integral of $E_1(k_1)$ over only a *band of wave numbers* centered about the effective wave number as shown by the "Shaded Area" in the figure. If we use $k_{1,eff}$ to denote the characteristic wave number of this band, then the characteristic size and frequency of the eddies are $l_{eff} = 1/k_{1,eff}$ and $\omega_{eff} = \bar{u}_\infty k_{1,eff}$. Hence, the right-hand side of the previous relation may be expressed as

$$\omega_{eff}\sqrt{\overline{u'^2_{eff}}} = \bar{u}_\infty k_{1,eff}\sqrt{\overline{u'^2_{eff}}} \propto \bar{u}_\infty k_{1,eff}\sqrt{k_{1,eff}E_1(k_{1,eff})} \tag{4.26}$$

where we have tacitly assumed the bandwidth of effective turbulence to be some fraction of the effective wave number.

Furthermore, if we plot a dimensionless form of the last term in Eq. (4.26) against wave number using the isotropic-turbulence spectrum given by Eq. (3.17), we obtain the distributions presented in Fig. 4.21. Clearly, not only does the function have a maximum, but the maximum is virtually independent of the turbulent Reynolds number. From Eq. (3.17) we determine that this maximum occurs when $k_{1,eff} \approx 0.3k_{d,\infty}$. Hence, the length scale and frequency of free-stream turbulence most effective in producing pre-transitional boundary layer fluctuations are

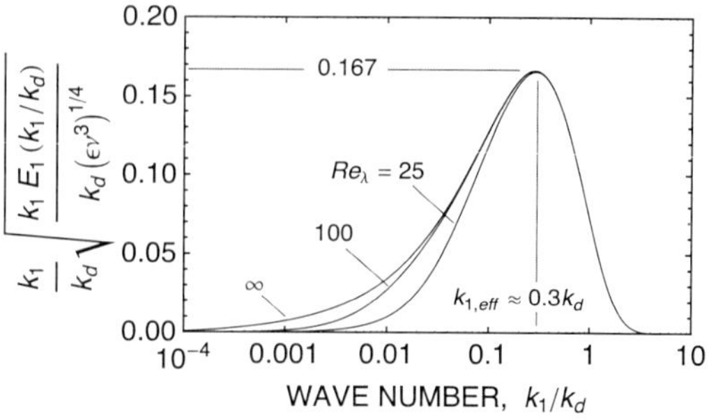

Figure 4.21.
The effective turbulence for a pre-transitional boundary layer.

$$l_{eff} \approx 3.3 l_{d,\infty} = 3.3\left(\nu^3/\varepsilon_\infty\right)^{1/4} \quad \text{and} \quad \omega_{eff} = 0.3\bar{u}_\infty\left(\varepsilon_\infty/\nu^3\right)^{1/4} \tag{4.27}$$

In addition, we can determine from the maximum that

$$k_{1,eff}\sqrt{k_{1,eff}E_1(k_{1,eff}/k_{d,\infty})} = 0.167k_{d,\infty}\sqrt{k_{d,\infty}(\varepsilon_\infty\nu^5)^{1/4}}$$

Substituting this into Eq. (4.26) and recalling that $k_{d,\infty} = (\varepsilon_\infty/\nu^3)^{1/4}$, we obtain

$$\omega_{eff}\sqrt{\overline{u'^2_{eff}}} \propto \bar{u}_\infty\sqrt{\varepsilon_\infty/\nu} \tag{4.28}$$

Surprisingly this doesn't contain the free-stream turbulence level, but only its rate of dissipation. In hindsight this result follows directly from Eq. (4.25) using $\partial/\partial t = \bar{u}_\infty \partial/\partial x$ and the definition of the dissipation. However, if this tack had been initially taken, neither the fact that the effective turbulence is independent of the turbulent Reynolds number nor the value of the effective wave number would have been obtained.[1]

Nevertheless, substituting Eq. (4.28) into Eq. (4.25), and its result into Eq. (4.24) yields

$$\overline{u'\frac{\partial u'_\infty}{\partial t}} = R_{\phi 11}\bar{u}_\infty\sqrt{\varepsilon_\infty/\nu}\sqrt{\overline{u'u'}} \tag{4.29}$$

where the coefficient $R_{\phi 11}$ now differs by a constant from that originally defined in Eq. (4.24). Furthermore, by eliminating ω_{eff} between Eqs. (4.27) and (4.28) we obtain

$$\sqrt{\overline{u'^2_{eff}}} \propto \left(\nu\varepsilon_\infty\right)^{1/4} = \upsilon_{d,\infty}$$

where υ_d is Kolmogorov's velocity scale. Hence the scales associated with the most effective turbulence for pre-transitional flow are all related to the Kolmogorov scales of turbulence. Note that if we had plotted Fig. 4.21 using axes based on the wave number associated with the large eddies, k_e, we would have found that the maximum, and hence all the scales of the effective turbulence, would not only depend on the scales associated with the large eddies, but *also* on the turbulent Reynolds number.[2] However, after eliminating these larger scales in terms of the dissipative scales using the expressions given in §3.1, we would again obtain the results given by Eqs. (4.27) and (4.28).

From the above expressions, the frequency parameter associated with the most effective free-stream eddies is easily determined to be

---

[1] See Mayle and Schulz [1997], who simply reasoned that the effective scale was proportional to Kolmogorov's length scale.

[2] Compare the papers by Mayle and Schulz [1997] and Mayle et al. [1998].

$$\Omega_{eff} = \frac{\omega_{eff}\,\nu}{\overline{u}_\infty^2} = 0.3\frac{\left(\nu\varepsilon_\infty\right)^{1/4}}{\overline{u}_\infty} = 0.3\frac{\upsilon_{d,\infty}}{\overline{u}_\infty} \tag{4.30}$$

These quantities are rather straightforward to evaluate given sufficient free-stream turbulence data. If the turbulence is isotropic, either Eq. (3.74) or its equivalent Eq. (3.75) can be used to determine the required free-stream dissipation, otherwise the equations given in §3.4.2 must be used. The results of these calculations for several zero-pressure-gradient tests are listed in Table 4.2. Except where noted, the transition Reynolds number based on the boundary-layer thickness $Re_{\delta,t}$ is "as best as can be determined." For reference, the critical Tollmien-Schlichting wave length and frequency are also given. These quantities are given at the point of instability, which is always upstream of where natural transition begins. Here it is interesting that while the *effective* dimensional circular frequencies for these tests vary in no particular order between 600 and 3000 at the onset of transition, the effective Strouhal number at onset is $S_{\delta,t} \approx 0.58$ for all the experiments.

Table 4.2

Estimates of the *most effective* free-stream eddy size and frequency *controlling* the growth of fluctuations in a laminar boundary layer evaluated at the plate's leading edge and beginning of transition. Data with $dp_\infty/dx = 0$ from Roach and Brierley [1990], Dyban and Epik [1985], and Westin et al. [1994].

Investigators	$Tu_0(\%)$	$\Omega_{eff,0}$	$Re_{\delta,t}$	$Tu_t(\%)$	$\Omega_{eff,t}$
Roach & Brierley	5.3	$3.6(10)^{-3}$	1220	4.7	$3.2(10)^{-3}$
Roach & Brierley	2.7	$3.1(10)^{-3}$	1840	1.7	$2.0(10)^{-3}$
Dyban & Epik †	–	–	>2800	<1.2	$<1.4(10)^{-3}$
Westin et al.	1.4	$1.5(10)^{-3}$	3650	1.0	$1.0(10)^{-3}$
Roach & Brierley	0.9	$1.1(10)^{-3}$	6000	0.5	$6.0(10)^{-4}$
Tollmien-Schlichting	0	–	1510	0	$2.4(10)^{-4}$

† These results were obtained from Figs. 52 and 68 of their book. An extrapolation of their data upstream yields an initial turbulence level of roughly two percent. The extrapolation, however, is too uncertain to provide even a rough estimate of the initial frequency parameter.

Clearly, the effective frequency parameter decreases with decreasing turbulence level, and near the onset of transition it approaches the critical Tollmien-Schlichting frequency. This behavior, together with the fact that the order of magnitude of these frequencies are not outrageously different from Tollmien-Schlichting's value even though they were calculated from a completely different outlook, strongly reinforces our view concerning the physics associated with the effect of turbulence on a laminar boundary layer.

Moreover, the results in this table also suggest that natural transition will not occur when the turbulence levels are greater than about one-to-two percent. This we deduce from the listed transition Reynolds numbers and the fact that the *transition* Reynolds number for zero free-stream turbulence must be larger than the critical value of $Re_{\delta,crit} \approx 1500$. For turbulence levels less than one-to-two percent, "natural" transition is still possible because these instabilities grow exponentially while the fluctuations caused by free-stream turbulence grow linearly. This behavior has been observed and commented on by Roach and Brierley [1990], Mayle and Schulz [1997], and Westin et al. [1998]. Therefore, the general picture emerging for pre-transitional flow in a turbulent free stream is one where the unsteadiness within the boundary layer is amplified by the free-stream turbulence unless its intensity is so low that the exponentially growing Tollmien-Schlichting instabilities take over. From Morkovin's [1969] viewpoint, we would say that the boundary layer becomes more "receptive" to bypass transition as the turbulence level increases.

The role of turbulent length scales. It is impossible to understand the effect of turbulence on a laminar boundary layer without considering the size of the turbulent eddies in the free stream relative to the thickness of the boundary layer. From Eqs. (3.1) and (3.4), the ratio of the two for the larger (energetic) and smaller (dissipative) eddies are

$$\frac{L_e}{\delta} = \frac{\bar{u}_\infty}{k_\infty^{1/2}} \frac{Re_T}{Re_\delta} \quad \text{and} \quad \frac{l_d}{\delta} = \frac{\bar{u}_\infty}{\upsilon_{d,\infty}} \frac{1}{Re_\delta}$$

The variation of these ratios through pre-transitional, transitional, and into post-transitional flow for a moderate free-stream turbulence level are shown in Fig. 4.22. The lines were calculated assuming a laminar boundary layer. The limits of Tu and Re_T given in the figure correspond to those at the leading edge of the test plate and the onset of transition. Since the turbulent Reynolds number is relatively high, it and the ratio L_e/l_d (equal to $Re_T^{3/4}$) change little with distance through the test region. Consequently, the curves and data for the two scales run virtually parallel to one another in the figure. Also, since the turbulent length scales themselves change little over the very forward portion of the plate, the decrease of their ratio with the boundary-layer thickness up to a Reynolds number of about 10,000 is caused by the growing boundary layer. Hence their ratios initially vary according to the inverse of the square root of distance. As the length scales grow, their rate of decrease with distance slows until transition where the boundary layer thickness rapidly increases.

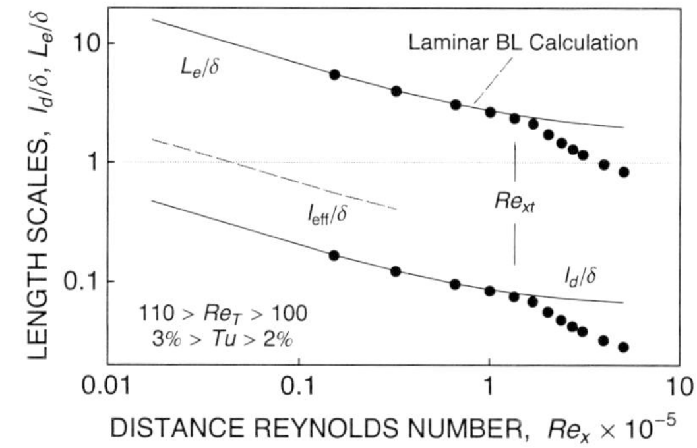

Figure 4.22.
Calculated variations of turbulent length scales to boundary-layer thickness ratios for Roach and Brierley's [1990] T3A test.

For this test case, as with the other tests listed in Table 4.2, not only are the large eddies an order of magnitude or so larger than the small eddies, but they are also larger than the boundary layer before transition. Consequently, the *inertial boundary layer of turbulence* must be taken into account when considering pre-transitional flow. As explained in §3.5.1, this requires one to consider the distortion of the large eddies and the associated redistribution of turbulent energy within them as they approach the plate. Since the smaller dissipative eddies are always a fraction of the boundary-layer thickness, they will generally be entrained, distorted and dissipated within the boundary layer. As shown by the dashed line in the figure, the most effective eddies are the same order of magnitude as the boundary-layer thickness. If we calculate $\Omega_{eff} Re_\delta^2$ for this test at the distance Reynolds number of 1000, which is just downstream of our lower limit for using the boundary-layer approximation (see §2.1), we obtain a value of 75. That is, $\Omega_{eff} Re_\delta^2 >> 1$, which, in turn according to Eq. (2.25), implies that the most effective eddies produce high-frequency disturbances within the boundary layer. Hence, our initial supposition in deriving Eq. (4.23) is correct.

Some other calculated length-scale ratios and Reynolds numbers for the tests listed in Table 4.2 are given in Table 4.3. At the leading edge, the estimated lengths-to-leading edge diameter ratios are given.[1] From these values, we may conclude that the large eddies begin distorting and stretching around the plate once they are about one "eddy-diameter" away from it. This not only produces a "dancing" stagnation line with unsteady leading-edge separation bubbles, but also a "streaky" turbulence on the plate as the eddies stretch downstream from around its leading edge. As one might expect, this spanwise inhomogeneity will affect the growth of the pre-transitional fluctuations within and between the streaks differently and, consequently, the onset of transition. On the other hand, the dissipative eddies, being smaller than the leading edge, "pass" to either side of it with much less inhibition. For a discussion of their effect on the leading-edge boundary layer, the reader is referred to Hunt [1973].

Table 4.3

Approximate length-scale ratios and turbulent Reynolds numbers for several zero-pressure-gradient tests. Data from Roach and Brierley [1990] and Westin et al. [1994].

Investigators	$Tu_0(\%)$	$L_{e,0}/d_{le}$	$l_{d,0}/d_{le}$	$(\delta/d)_{le}$	$(L_e/\delta)_t$	$(\lambda/\delta)_t$	$Re_{T,t}$	$Re_{\lambda,t}$	$(l_d/\delta)_t$
Roach & Brierley	0.9	3.1	0.14	0.028	1.7	0.67	58	20	0.083
Westin et al.	1.4	~7	~0.2	~0.04	3.0	0.77	122	28	0.081
Roach & Brierley	2.7	6.4	0.19	0.054	2.6	0.75	100	26	0.081
Roach & Brierley	5.3	14.6	0.09	0.039	13	1.27	870	76	0.078

At transition, the turbulent length scale L_e is still greater than the boundary-layer thickness, but varies significantly with the turbulence level. Taylor's length scale, which nearly equals the boundary-layer thickness, also varies with turbulence level. The variation in their corresponding turbulent Reynolds numbers follows suit. In contrast, however, the dissipation length scale at transition is an order of magnitude less than, but nearly a constant ratio of the boundary-layer thickness. We will discuss this point again in §4.3.4.

4.3.2 Modeling pre-transitional flow with free-stream turbulence.[2]

Inner and outer equations. A Reynolds-stress model for pre-transitional flow.

If we begin by considering the boundary-layer thickness to be much smaller than the large-scale free-stream eddies, then the inner viscous and outer inertial layer of turbulence can be treated almost independently. According to our discussion in the previous section, the equation for the streamwise component of the Reynolds stress in the inner layer, Eq. (4.23), becomes

$$\bar{u}\frac{\partial\overline{u'u'}}{\partial x}+\bar{v}\frac{\partial\overline{u'u'}}{\partial y}=R_{\phi 11}\bar{u}_\infty\sqrt{\frac{\varepsilon_\infty}{\nu}}\sqrt{\overline{u'u'}}+\nu\frac{\partial^2\overline{u'u'}}{\partial y^2}-\tilde{\varepsilon}_{11,w} \tag{4.31}$$

where Eq. (4.29) has been used, and only the wall component of the dissipation tensor need be considered. By neglecting the turbulent diffusion and stress-strain-rate components of the production

---

[1] The values shown here are easily calculated using the result for stagnation flow, namely, $\delta/d \approx 1.2\,Re_d^{-1/2}$ (see Schlichting [1979]). For Roach and Brierley's tests, the reported leading-edge diameter is $d_{le} = 1.5$ mm, while that for Westin's et al. is roughly 2 mm.

[2] Although developed for isotropic free-stream turbulence herein, an equivalent, albeit more complex and unsubstantiated model for anisotropic turbulence is proposed in Appendix F.

terms in Eqs. (1.53)–(1.55), and introducing expressions similar to Eq. (4.29) for the velocity-pressure-gradient correlations, the remaining Reynolds-stress equations for the inner layer are

$$\bar{u}\frac{\partial\overline{v'v'}}{\partial x}+\bar{v}\frac{\partial\overline{v'v'}}{\partial y}=R_{\phi 22}\bar{u}_\infty\sqrt{\frac{\varepsilon_\infty}{\nu}}\sqrt{\overline{v'v'}}+\nu\frac{\partial^2\overline{v'v'}}{\partial y^2}-\tilde{\varepsilon}_{22,w} \tag{4.32}$$

$$\bar{u}\frac{\partial\overline{w'w'}}{\partial x}+\bar{v}\frac{\partial\overline{w'w'}}{\partial y}=R_{\phi 33}\bar{u}_\infty\sqrt{\frac{\varepsilon_\infty}{\nu}}\sqrt{\overline{w'w'}}+\nu\frac{\partial^2\overline{w'w'}}{\partial y^2}-\tilde{\varepsilon}_{33,w} \tag{4.33}$$

$$\bar{u}\frac{\partial\overline{u'v'}}{\partial x}+\bar{v}\frac{\partial\overline{u'v'}}{\partial y}=-R_{\phi 12}\bar{u}_\infty\sqrt{\frac{\varepsilon_\infty}{\nu}}\sqrt{\overline{v'v'}}+\nu\frac{\partial^2\overline{u'v'}}{\partial y^2}-\tilde{\varepsilon}_{12,w} \tag{4.34}$$

The first of these four equations is essentially that solved by Mayle and Schulz. The first and last are essentially those solved by Mayle et al. [2008]. However, in their analysis they used the near-wall expansion given by Eq. (1.64) to determine $\overline{v'v'}$ instead of Eq. (4.32). In the following analysis, all the Reynolds-stress equations are considered.

In the outer layer, the appropriate equations are essentially those for a free-shear turbulent boundary layer. Given in §3.5.1, these are Eqs. (3.82)–(3.85). In the present case, however, the derivative with respect to time is replaced by $D/Dt=\bar{u}_i\partial/\partial x_i$, the viscous-diffusion and wall-dissipation terms can be neglected, and the mean velocity is nearly uniform and approximately equal to the free-stream velocity. If the free-stream pressure gradient is not zero, then the terms containing the streamwise variation of the mean velocity should be considered. These terms, which can be obtained from Eq. (3.66), are generally small, but they provide the appropriate behavior as the free stream is approached and are easy to include. As a result, the equations for the outer layer are

$$\begin{aligned}\bar{u}_\infty\frac{\partial\overline{u'u'}}{\partial x}+\bar{v}_\infty\frac{\partial\overline{u'u'}}{\partial y}=&\frac{\partial}{\partial y}\left[c_s\overline{v'v'}\frac{k}{\hat{\varepsilon}}\frac{\partial\overline{u'u'}}{\partial y}\right]-\left(c_1 f_{\phi 1}A+f_{\varepsilon 1}\right)\frac{\tilde{\varepsilon}}{k}\left[\overline{u'u'}-\frac{2}{3}k\right]\\&+c_{1w}\overline{v'v'}\frac{k^{1/2}}{y}-\frac{2}{3}\tilde{\varepsilon}-\left\langle\left\{2\overline{u'u'}-\frac{2}{3}c_2\sqrt{A}\left[2\overline{u'u'}+\overline{v'v'}\right]\right\}\frac{\partial\bar{u}_\infty}{\partial x}\right\rangle\end{aligned} \tag{4.35}$$

$$\begin{aligned}\bar{u}_\infty\frac{\partial\overline{v'v'}}{\partial x}+\bar{v}_\infty\frac{\partial\overline{v'v'}}{\partial y}=&\frac{\partial}{\partial y}\left[3c_s\overline{v'v'}\frac{k}{\hat{\varepsilon}}\frac{\partial\overline{v'v'}}{\partial y}\right]-\left(c_1 f_{\phi 1}A+f_{\varepsilon 1}\right)\frac{\tilde{\varepsilon}}{k}\left[\overline{v'v'}-\frac{2}{3}k\right]\\&-2c_{1w}\overline{v'v'}\frac{k^{1/2}}{y}-\frac{2}{3}\tilde{\varepsilon}+\left\langle\left\{2\overline{v'v'}-\frac{2}{3}c_2\sqrt{A}\left[\overline{u'u'}+2\overline{v'v'}\right]\right\}\frac{\partial\bar{u}_\infty}{\partial x}\right\rangle\end{aligned} \tag{4.36}$$

$$\begin{aligned}\bar{u}_\infty\frac{\partial\overline{w'w'}}{\partial x}+\bar{v}_\infty\frac{\partial\overline{w'w'}}{\partial y}=&\frac{\partial}{\partial y}\left[c_s\overline{v'v'}\frac{k}{\hat{\varepsilon}}\frac{\partial\overline{w'w'}}{\partial y}\right]-\left(c_1 f_{\phi 1}A+f_{\varepsilon 1}\right)\frac{\tilde{\varepsilon}}{k}\left[\overline{w'w'}-\frac{2}{3}k\right]\\&+c_{1w}\overline{v'v'}\frac{k^{1/2}}{y}-\frac{2}{3}\tilde{\varepsilon}-\left\langle\frac{2}{3}c_2\sqrt{A}\left[\overline{u'u'}-\overline{v'v'}\right]\frac{\partial\bar{u}_\infty}{\partial x}\right\rangle\end{aligned} \tag{4.37}$$

$$\bar{u}_\infty\frac{\partial\tilde{\varepsilon}}{\partial x}+\bar{v}_\infty\frac{\partial\tilde{\varepsilon}}{\partial y}=\frac{\partial}{\partial y}\left[\left(\nu+c_\varepsilon\overline{v'v'}\frac{k}{\hat{\varepsilon}}\right)\frac{\partial\tilde{\varepsilon}}{\partial y}\right]-c_{\varepsilon 2}f_{\varepsilon 2}\frac{\tilde{\varepsilon}^2}{k} \tag{4.38}$$

where we have chosen to use the equation for the transverse normal-stress component instead of that for the turbulent kinetic energy.

Presumably, Eqs. (4.31)–(4.34) should be solved in the inner layer, Eqs. (4.35)–(4.38) solved in the outer layer, and the solutions somehow combined. One way to combine them is to introduce a function $\gamma(x,y)$ that varies from zero near the wall where the flow is laminar and the inner-layer equations apply, to unity near the free stream where the flow is turbulent and the outer-layer equations apply. Then the solution for $\overline{u'u'}$ would be given by

$$\overline{u'u'} = (1-\gamma)(\overline{u'u'})_i + \gamma(\overline{u'u'})_o \tag{4.39}$$

where the subscripts refer to the inner and outer solutions, with similar expressions for the other stresses. It's not clear, at least to the author, whether or not "laminar" and "turbulent" fluctuations can be distinguished in a laminar boundary layer. Hence, we will simply consider the function γ as the fraction of time that the outer solutions apply, and not the intermittency we will introduce in the following chapter.

Reynolds-stress model. Rather than solving the full set of equations given above, it is possible to develop an "equivalent" set of equations after realizing that the wall-normal stress component is most affected by the wall in the outer inertial layer, and that it is small compared to the other components near the wall (see Fig. 3.13). Consequently, it appears that we only need to solve Eqs. (4.36) and (4.38) in the outer layer and suitably modified versions of Eqs. (4.31)–(4.34) in the inner layer. Considering the physical nature of each layer, one such equivalent set of equations might be

$$\bar{u}\frac{\partial \overline{u'u'}}{\partial x} + \bar{v}\frac{\partial \overline{u'u'}}{\partial y} = R_{\phi 11}\bar{u}_\infty \sqrt{\frac{\varepsilon_\infty}{\nu}}\sqrt{\overline{u'u'}} + \nu \frac{\partial^2 \overline{u'u'}}{\partial y^2} - \tilde{\varepsilon}_{11,w}$$
$$+ c_{1w}(\overline{v'v'})_o \frac{k^{1/2}}{y} - 2\overline{u'u'}\frac{\partial \bar{u}}{\partial x} - \gamma\left\{c_1 f_{\phi 1} A \frac{\tilde{\varepsilon}}{k}\left[\overline{u'u'} - \frac{2}{3}k\right] + \overline{u'u'}\frac{\tilde{\varepsilon}}{k}\right\} \tag{4.40}$$

$$\bar{u}\frac{\partial (\overline{v'v'})_i}{\partial x} + \bar{v}\frac{\partial (\overline{v'v'})_i}{\partial y} = R_{\phi 22}\bar{u}_\infty \sqrt{\frac{(\overline{v'v'})_\infty}{(\overline{u'u'})_\infty}}\sqrt{\frac{L_e}{\delta_1}}\sqrt{\frac{\varepsilon_\infty}{\nu}}\sqrt{(\overline{v'v'})_i} + \nu \frac{\partial^2 (\overline{v'v'})_i}{\partial y^2} - \tilde{\varepsilon}_{22,w} \tag{4.41}$$

$$\bar{u}\frac{\partial \overline{w'w'}}{\partial x} + \bar{v}\frac{\partial \overline{w'w'}}{\partial y} = R_{\phi 33}\bar{u}_\infty \sqrt{\frac{(\overline{w'w'})_\infty}{(\overline{u'u'})_\infty}}\sqrt{\frac{L_e}{\delta_1}}\sqrt{\frac{\varepsilon_\infty}{\nu}}\sqrt{\overline{w'w'}} + \nu \frac{\partial^2 \overline{w'w'}}{\partial y^2} - \tilde{\varepsilon}_{33,w}$$
$$+ c_{1w}(\overline{v'v'})_o \frac{k^{1/2}}{y} - \gamma\left\{c_1 f_{\phi 1} A \frac{\tilde{\varepsilon}}{k}\left[\overline{w'w'} - \frac{2}{3}k\right] + \overline{w'w'}\frac{\tilde{\varepsilon}}{k}\right\} \tag{4.42}$$

$$\bar{u}\frac{\partial \overline{u'v'}}{\partial x} + \bar{v}\frac{\partial \overline{u'v'}}{\partial y} = -R_{\phi 12}\bar{u}_\infty \sqrt{\frac{\varepsilon_\infty}{\nu}}\sqrt{\overline{v'v'}} + \nu \frac{\partial^2 \overline{u'v'}}{\partial y^2} - \tilde{\varepsilon}_{12,w} \tag{4.43}$$

$$\bar{u}\frac{\partial (\overline{v'v'})_o}{\partial x} + \bar{v}\frac{\partial (\overline{v'v'})_o}{\partial y} = \frac{\partial}{\partial y}\left[3c_s(\overline{v'v'})_o \frac{k}{\hat{\varepsilon}}\frac{\partial (\overline{v'v'})_o}{\partial y}\right] - \left(c_1 f_{\phi 1} A + f_{\varepsilon 1}\right)\frac{\tilde{\varepsilon}}{k}\left[(\overline{v'v'})_o - \frac{2}{3}k\right]$$
$$-2c_{1w}(\overline{v'v'})_o \frac{k^{1/2}}{y} - \frac{2}{3}\tilde{\varepsilon} - \gamma\left\{2(\overline{v'v'})_o \frac{\partial \bar{u}}{\partial x} + \frac{2}{3}c_2\sqrt{A}\left[\overline{u'u'} + 2(\overline{v'v'})_o\right]\frac{\partial \bar{u}}{\partial x}\right\} \tag{4.44}$$

$$\bar{u}\frac{\partial \tilde{\varepsilon}}{\partial x} + \bar{v}\frac{\partial \tilde{\varepsilon}}{\partial y} = \frac{\partial}{\partial y}\left[c_\varepsilon(\overline{v'v'})_o \frac{k}{\hat{\varepsilon}}\frac{\partial \tilde{\varepsilon}}{\partial y}\right] - c_{\varepsilon 2} f_{\varepsilon 2}\frac{\tilde{\varepsilon}^2}{k} \tag{4.45}$$

where the subscripts "i" and "o" refer to the inner and outer components, and

$$\hat{\varepsilon}=\tilde{\varepsilon}+\tilde{\varepsilon}_w,\quad \tilde{\varepsilon}=\tilde{\varepsilon}_{ii}/2,\quad \tilde{\varepsilon}_w=\tilde{\varepsilon}_{ii,w}/2$$

In addition, we use $f_{\varepsilon 1}=1-A$ with A defined by Eq. (3.54), and $f_{\phi 1}$ and $f_{\varepsilon 2}$ as defined in Eq. (3.72). In the present case, the latter can be set equal to one because the turbulent Reynolds number in the outer layer is generally larger than ten.

The first four of these equations are the modified equations for the inner layer. The modifications, which represent the influence of the wall on the stresses lying in the plane parallel to the wall, are found in the second lines of Eqs. (4.40) and (4.42). Included are Hanjalić and Launder's wall-distortion term and Rotta's return-to-isotropy and anisotropic-dissipation terms. Generally, these terms are smaller than the terms in the first line of each equation within the inner layer, yet provide the proper behavior in the outer layer particularly near the free stream. Note here that the outer solution $\overline{(v'v')}_o$ is used in the wall-distortion term. Also note that Rotta's terms are multiplied by the factor γ to restrict their influence to the outer layer. Initial calculations using the production terms given in Eqs. (4.32) and (4.33) indicated the growth of $\overline{(v'v')}_i$ and $\overline{w'w'}$ relative to $\overline{u'u'}$ was too slow, especially for high turbulence levels. Multiplying their rates by the ratio of free-stream intensities and the factor $(L_e/\delta_1)^{1/2}$ seems to help, although this correction is certainly too simple to account for the anisotropy and distortion of free-stream turbulence on their production. In addition, it was found that some of the turbulent anisotropies in the free stream could be taken into account by multiplying the normal and transverse components of the production terms by the ratio of the normal and transverse components of free-stream intensities to the streamwise component. These modifications, being far from certain, require further investigation.

The last two equations concern the wall effect on the wall-normal stress component. In general, these equations are the same as Eqs. (3.83) and (3.85) for a free-shear turbulent boundary layer, except for the last term in Eq. (4.44) which accounts for a variable streamwise velocity. Note that Hanjalić and Launder's turbulent diffusion model given by Eq.(3.49), which best correlated the free-shear turbulent boundary layer data, is used here and not Daly and Harlow's model.[1]

The wall-normal stress component can be obtained by combining its inner and outer solutions according to

$$\overline{v'v'}=(1-\gamma)\overline{(v'v')}_i+\gamma\overline{(v'v')}_o \tag{4.46}$$

Presuming an exponential behavior, it is found that the function

$$\gamma=1-\exp\left[-0.08\left(y/\delta_1\right)^2\right] \tag{4.47}$$

best correlates the data. Note that $\gamma\approx 1/2$ at the edge of the viscous boundary layer, that is, where $y/\delta_1\approx 3$.

We could use either Eq. (3.63) for the wall-dissipation rates or, following Mayle and Schulz, the second expression in Eq. (3.62). Choosing the latter we obtain

$$\tilde{\varepsilon}_{11,w}=2\nu\frac{\overline{u'u'}}{y^2},\quad \tilde{\varepsilon}_{22,w}=12\nu\frac{\overline{(v'v')}_i}{y^2},\quad \tilde{\varepsilon}_{33,w}=2\nu\frac{\overline{w'w'}}{y^2}\quad\text{and}\quad \tilde{\varepsilon}_{12,w}=6\nu\frac{\overline{u'v'}}{y^2} \tag{4.48}$$

where the coefficients have been determined by equating the dissipation and diffusion of each stress component at the wall to yield the near-wall behavior described by Eq. (1.64).

As previously noted the correlation coefficients must account for a change of phase through the boundary layer. Mayle et al. [2008] successfully used a coefficient of the form

[1] See the discussion in §3.3.3 regarding Table 3.3.

$$R_{\phi,ij} = c_{ij} \exp\left(-a\, y/\delta_1\right) \tag{4.49}$$

where the constants a and c_{ij} are determined by optimizing the fit between the calculations and pre-transitional data.

The solution of these equations are also subject to initial and boundary conditions. At some predetermined streamwise position, initial profiles for all the dependent variables must be provided.[1] The results are rather insensitive to the initial profiles of $\overline{u'u'}$ and $\overline{w'w'}$ as long as they are physically reasonable. The initial values for both $\overline{(v'v')}_i$ and $\overline{u'v'}$ can be set equal to zero, while either the solution obtained by Hunt and Graham [1978], or a good approximation to it, can be used for the initial $\overline{(v'v')}_o$ profile.[2] For the calculations shown in the following figures, the approximation

$$\frac{\overline{(v'v')}_o}{\overline{(v'v')}_\infty} = \left[1 - \exp\left(-6\,y/L_{e,0}\right)\right]$$

was used, where $L_{e,0}$ is the initial free-stream turbulent length scale.

For the boundary conditions, all the dependent variables must equal zero at the wall. In the free stream, $\overline{(v'v')}_i$ equals zero while all the other variables must satisfy the free-stream conditions given by Eqs. (3.76)–(3.79). Here we note that "in the free stream" means beyond the *outer* layer. In the present case then, since only the wall-normal stress component is being modeled in this layer, we *must* use its asymptotic behavior for large values of y to determine the "edge" of the computational thickness.

The "best" values for the constants in Eqs. (4.40)–(4.45) and (4.49) are given in Table 4.4. These have been determined by optimizing the fits between mean-flow velocity and Reynolds-stress profiles both with and without a pressure gradient. The values of c_{11} and a are nearly the same as determined by Mayle et al. [2008], while the constants pertaining to the outer layer are identical to those listed in Table 3.3 for near-wall shear-free turbulence.

Table 4.4
Empirical constants in Eqs. (4.40)–(4.45) and Eq. (4.49).

c_{11}	c_{22}	c_{33}	c_{12}	a	c_1	c_2	c_s	c_ε	$c_{\varepsilon 1}$	$c_{\varepsilon 2}$	c_{1w}
0.066	0.011	0.003	0.024	0.65	2.5	0.6	0.03	0.18	1.44	1.92	0.14

4.3.3 Calculated and experimental results.

Flows without a pressure gradient. Flows with a pressure gradient.

Although a lot of data is available on pre-transitional flow in a turbulent free stream, surprisingly few investigations are complete enough to be useful. Unlike a turbulent boundary layer that is largely "insulated" by its intermittent wake-like flow from the free stream, a laminar boundary layer is not. This is particularly problematic for analyzing laminar flows with free-stream turbulence since the inertial boundary layer of turbulence that extends beyond the viscous layer is usually not included in the measurements.

As a result, the most useful pre-transition data to date is that obtained by Roach and Brierley [1990].

---

[1] For the calculations shown in the following figures, this position was taken to be $Re_x = 1000$ which is just downstream of the lower limit for using the boundary-layer approximation.

[2] See Fig. 3.13 for Hunt and Graham's solution.

Their experiments are exceptional in that the free-stream turbulence, mean-velocity, and Reynolds-stress profiles are documented well beyond the viscous boundary layer in most cases. Collectively called the "T3" series, they are distinguished by the suffix "A," "AM," "B," and "C1" through "C5." The first three experiments were conducted with a nominally constant free-stream velocity, whereas the remaining five were conducted with an accelerating/decelerating free-stream flow. The inlet free-stream velocity for these tests varied between 1.5 and 19m/s. The nominal free-stream turbulence level was either 1, 3 or 6 percent. The experiments by Westin et al. [1994] are also useful even though they only report the mean-velocity and streamwise-normal stress profiles. The free-stream turbulence for their case with a mean velocity of 8m/s is also well documented. In this test, the nominal free-stream turbulence level is about 1.5 percent, but they have conducted tests with turbulence levels up to 6.6 percent and free-stream velocities from 1.2 to 12m/s.

Regarding the other data, much of it is for turbulence levels near or less than one percent which appears to be affected by some other mechanism such as laminar instabilities, vibrations, or sound. On the other hand, much of the data for high turbulence levels were obtained with turbulence-generating grids placed upstream of the inlet nozzle in contrast to the above mentioned experiments where the grids were placed downstream of the nozzle directly upstream of the test plate. As a result, the incident turbulence in the former tests was highly anisotropic, which in turn appears to affect the growth of the pre-transitional fluctuations in a manner not yet understood. Therefore, the following comparisons are restricted to those tests with nearly isotropic turbulence at initial turbulence levels greater than one percent.

The zero-pressure-gradient results. The data from Roach and Brierley's [1990] T3A test are shown in Figs. 4.23 to 4.27. The data were obtained from an experiment conducted on a flat plate aligned parallel to a steady stream with a free-stream velocity of about 5.2m/s and a nominal pressure gradient of zero. The free-stream turbulence level varied from about three to about one percent over the length of the plate. The turbulent Reynolds number Re_T and degree of turbulence isotropy A, see §3.4.2, Eq. (3.80), are about 100 and 0.96 respectively for this flow. The streamwise variations of the free-stream turbulent length scales have been plotted in Fig. 4.22. The measured decay of turbulence is shown in Fig. 4.23. The lines in the figure show the calculated results using Eqs. (3.76)–(3.79). These results were obtained after choosing an initial position to match the data and adjusting the value of the dissipation at that position to provide the best overall fit to the measured distributions. Other information regarding this data can be found in Tables 4.2 and 4.3. Transition occurs somewhere between the measurement stations at Re_x = 100,600 and 134,800 (somewhat closer to the latter) where the free-stream turbulence level is about two percent. This position is indicated by the downward pointing arrow in the figure.

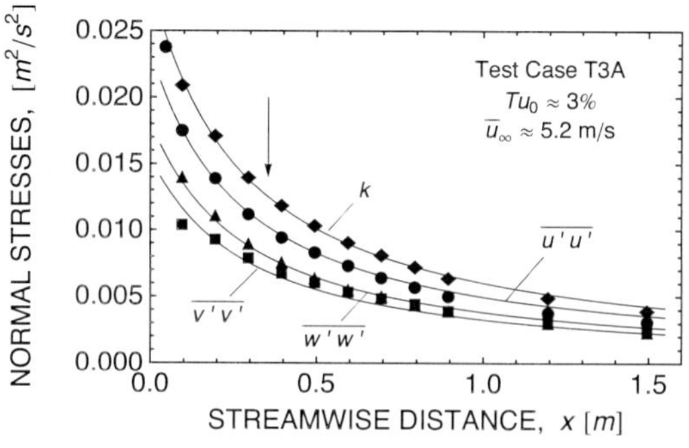

Figure 4.23.
The measured and calculated free-stream decay of Reynolds normal stresses downstream from the leading edge of the test plate. Data from Roach and Brierley [1990].

The intensity profiles at $Re_x = 100{,}600$ just before transition are shown in Fig. 4.24. At this location $L_e/\delta_1 \approx 8.4$ and $l_d/\delta_1 \approx 0.27$. These data, which qualitatively agree with Dyban's et al. [1976] measurements shown earlier in Fig. 4.19, are characteristic of all pre-transitional data with free-stream turbulence. The lines in the figure show the calculated results using the pre-transition model outlined in the previous section. The agreement with data is good, including the rise, fall, and final rise of the wall-normal intensity to its free-stream value. Although not obvious in the figure, but in agreement with expected behavior, the calculated near-wall variation of the streamwise and transverse intensity components with respect to the distance away from the wall are linear, while that of the wall-normal component is quadratic.

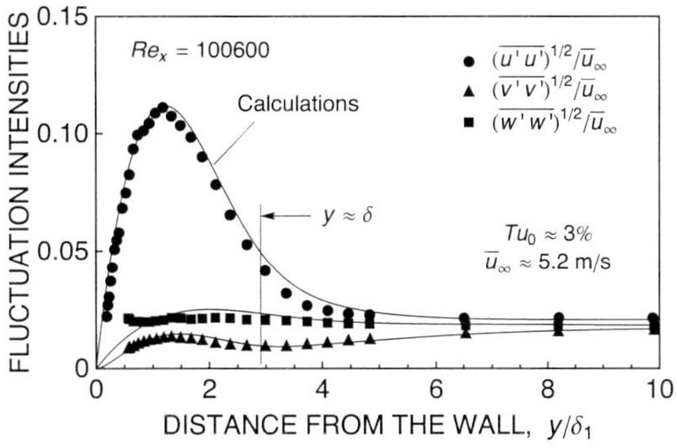

Figure 4.24.
Measured and calculated intensity profiles at $Re_x = 100{,}600$. Data from Roach and Brierley's [1990] T3A test.

Clearly the effect of the wall extends beyond the edge of the viscous boundary layer and, as expected from our analysis of a free-shear turbulent boundary layer, the streamwise and wall-normal stress components are most affected. In general, it may be said that the effect of turbulence on the fluctuations within the inner layer is most dramatic for the streamwise component and less so for the transverse component. An important point, however, is that neither of these components play a major role in the development of the Reynolds shear stress. As seen in Eq. (4.43), this role is allocated to the wall-normal stress component which is typically an order of magnitude less than the streamwise component within the inner layer.

The development of these profiles with distance is shown in Fig. 4.25. As previously mentioned, the profiles at $Re_x = 134{,}800$ are somewhat downstream from the onset of transition. The discrepancy between the measured and calculated shear-stress profiles in Fig. 4.25c at this Reynolds number appears to be typical for a flow near transition. In fact, an investigation into the behavior of these profiles show that the measured shear-stress profiles downstream of transition not only shift toward the wall together with the intensity profiles, but spread and increase in magnitude relative to that calculated. As noted by both Matsubara and Alfredsson [2001] and Mayle et al. [2008], a shift in the streamwise intensity profiles toward the wall also occurs, but this is barely detectable in Fig. 4.25a. However, comparing the profiles near the onset of transition in these figures with those for turbulent flow in Figs. 3.15 and 3.17 shows that the maximum of the streamwise intensity is about the same as that in a fully-developed turbulent boundary layer while the intensity of the wall-normal component and shear stress is only about a third.

The effect of free-stream turbulence on the mean-velocity profile and its deviation from Blasius' profile is shown in Fig. 4.26. At this measurement station, since the reported data appears shifted slightly away from the wall, a plot of the deviation using data corrected for this shift is also provided.[1] Since the calculated difference solely depends on the Reynolds shear-stress term in the momentum equation, we can conclude that the main effect of turbulence on the mean velocity profile, and hence the wall shear stress, is through the production of Reynolds shear stress in Eq. (4.43) and the development of the wall-normal stress component. Thus, the model for the wall-normal stress component is crucial for predicting pre-transitional flow.[2] For the calculations shown in these and the following figures, this model is represented by Eqs. (4.41), (4.44), (4.46) and (4.47).

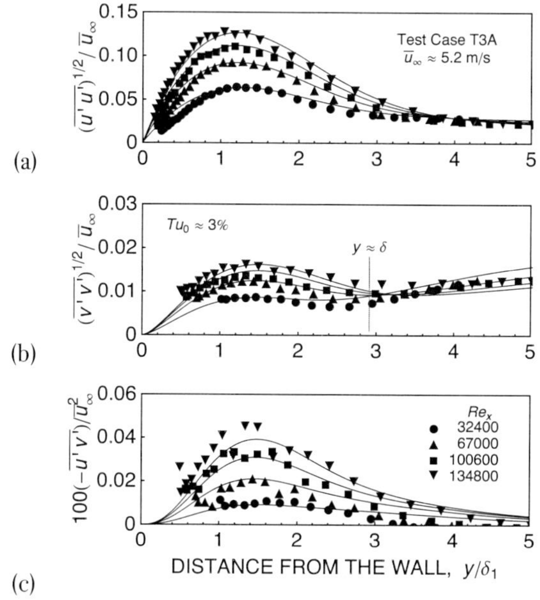

Figure 4.25a,b,c.
Measured and calculated intensity and shear-stress profiles at various downstream distances. Data from Roach and Brierley [1990].

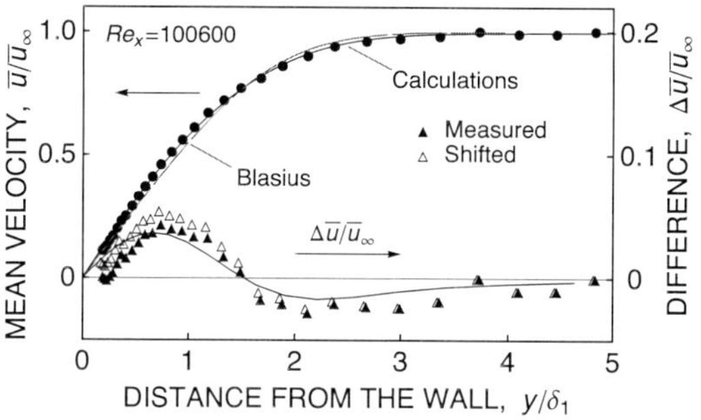

Figure 4.26.
Measured and calculated mean-velocity profile at $Re_x = 100{,}600$. Data from Roach and Brierley's [1990] T3A test.

Since the mean velocity near the wall is higher, the skin-friction coefficient for pre-transitional flows in a turbulent free stream will be higher than that predicted for a non-turbulent free stream. This has

[1] The shift by the way amounts to a distance of about 0.03mm, which is well within the accuracy of positioning the probe.
[2] As seen in Eq. (3.94), this is also true for predicting turbulent flow. See Durbin [1993].

been well documented by Dyban et al. [1976], Roach and Brierley [1990], Blair [1992], Zhou and Wang [1995], Gibbings and Al-Shukri [1997], and Wang and Keller [1999], among others.

Table 4.5
Conditions where the maximum deviation in the mean velocity profile is about five percent.

Investigators	Re_x	Tu_x (%)	$\Delta c_f / c_{f,Blasius}$
Westin et al. (1.4%)	533,000	1.0	0.21
Roach and Brierley (3%)	134,800	2.0	0.16
Roach and Brierley (6%)	59,000	5.0	0.26

Correlations for the effect of free-stream turbulence on the skin-friction coefficient usually take the form $\bar{c}_f = (1+aTu)c_{f,Blasius}$, where a is a constant. If we suppose that the correction to the skin-friction coefficient is additive, it is easy to show that the correction must take the form $\Delta c_f / c_{f,Blasius} \propto \Delta\bar{u}_{max}/\bar{u}_\infty$. In other words, the percentage change in the laminar skin-friction coefficient is proportional to the maximum deviation in the mean velocity. Table 4.5 lists the values for some tests where the maximum deviation in the mean-velocity profile is about five percent, that is, where $\Delta\bar{u}_{max}/\bar{u}_\infty = 0.05$. Clearly, the correction does not vary in proportion to the turbulence level as commonly supposed, but is roughly constant. Thus, any correlation for the pre-transitional skin-friction coefficient in terms of free-stream turbulence level alone is essentially useless. This is because the effect of turbulence on the mean flow is cumulative. That is, the same maximum in $\Delta\bar{u}/\bar{u}_\infty$ can be obtained with smaller free-stream turbulence levels as long as the fluctuations have enough time (distance) to develop. As a consequence, a more appropriate expression for the effect of turbulence on the skin-friction coefficient is

$$\bar{c}_f = \left(1 + 4.2\,\Delta\bar{u}_{\max}/\bar{u}_\infty\right)c_{f,Blasius}$$

where the average of the corrections in the table has been used to determine the coefficient. However, since a relation between the maximum deviation and turbulence level is presently unknown, this expression is also useless for practical purposes. In other words, it is best to calculate the effect of free-stream turbulence on pre-transitional flow by some algorithm similar to that presented in the previous section.

The streamwise variation of the maxima in the intensities and shear-stress within the inner-viscous layer are shown in Fig. 4.27. As previously mentioned, transition begins somewhere between the indicated measurement stations. Accordingly, the calculations are shown as a solid/dashed line, where the solid line represents the pre-transitional portion of the flow. Despite the flow being different, we see that the growth of the streamwise intensity before and after the onset of transition is about the same. The same cannot be said of the wall-normal intensity and shear stress, however, since their growth rate changes dramatically with the onset of transition.

The data for different free-stream turbulence levels are plotted in Fig. 4.28. As in the previous figure, the calculations are shown as solid/dashed lines with the break near the onset of transition. In these and the following figures, the data are distinguished by using black symbols before transition and gray after. A zone where transition most likely begins is shaded gray in part (b) of the figure (see following section for the discussion of this band). The maximum intensity for a fully-established turbulent boundary layer, as measured by Klebanoff [1955], is shown in part (a) of the figure.

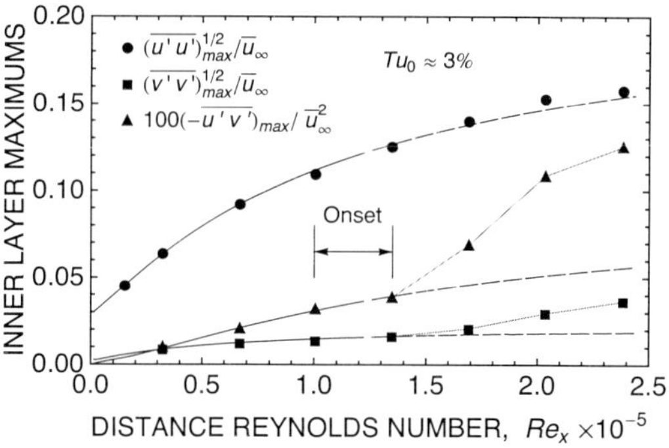

Figure 4.27.
Streamwise variations of the maximum intensity and shear-stress profiles within the inner-viscous layer. Data from Roach and Brierley's [1990] T3A test.

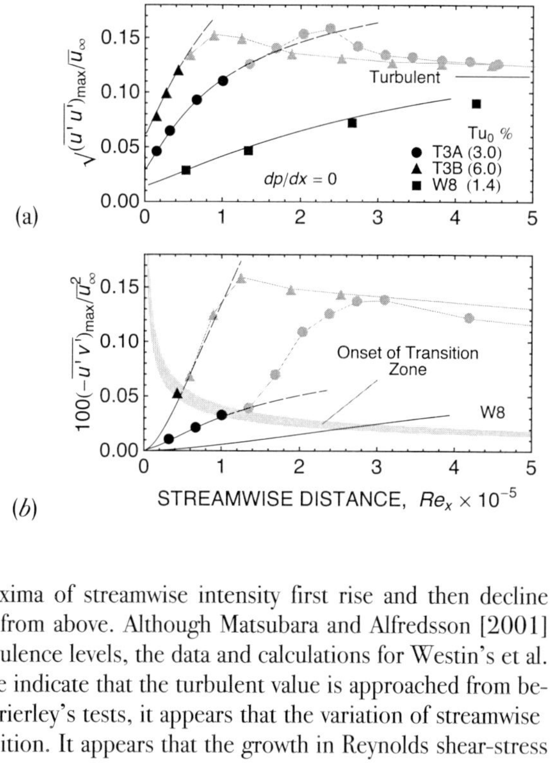

Figure 4.28a,b.
Streamwise variations of the maximum intensity and shear stress for pre-transitional flow without a free-stream pressure gradient. Data from Roach and Brierley's [1990] T3A and T3B test cases, and Westin's et al. [1994] test case W8.

For the higher turbulence levels, the maxima of streamwise intensity first rise and then decline through transition to approach this value from above. Although Matsubara and Alfredsson [2001] also show a similar behavior for lower turbulence levels, the data and calculations for Westin's et al. one-and-a-half percent turbulence test case indicate that the turbulent value is approached from below. Nevertheless, in both of Roach and Brierley's tests, it appears that the variation of streamwise intensities can be calculated through transition. It appears that the growth in Reynolds shear-stress can also be calculated for the higher free-stream turbulence level since its growth rate is the same before and after, while that for the lower level cannot.

The nonzero pressure-gradient results. Similar comparisons for flows with a pressure gradient are shown in Fig. 4.29. The variations in free-stream velocity are shown in part (a) of the figure. For three of the tests, T3C1, T3C2, and T3C5, transition occurs in an accelerating flow. For the other two, T3C3 and T3C4, transition occurs in a decelerating flow. The positions of the maximum free-stream velocity for these two tests are $Re_x \approx 350{,}000$ and $125{,}000$, respectively. The calculated results were obtained by multiplying the production term in Eq. (4.40) by the factor

$$1+12\frac{\delta_2^2}{\nu}\frac{d\bar{u}_\infty}{dx}$$

to approximate the effect of the free-stream strain rate on production.[1] This correction, which uses the pressure gradient parameter, could be interpreted as the first two terms in a series expansion of the presumably complicated function describing the effect.

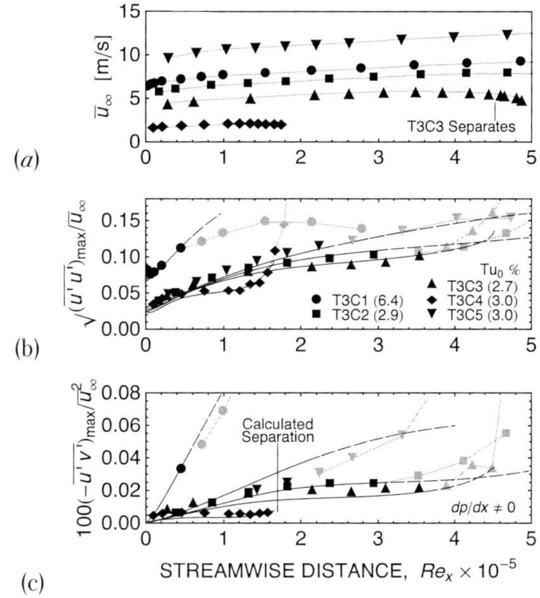

Figure 4.29a,b,c.
Calculated and measured streamwise variations for pre-transitional flow with free stream pressure gradients. Data from Roach and Brierley [1990].

For the most part, the intensity calculations in part (b) of the figure agree with the data throughout pre-transition and sometimes beyond. The amplified intensities for the T3C3 (triangles) and T3C4 (diamonds) test cases are caused by the deceleration of the free stream following the position of maximum velocity. For these tests, separation is calculated at $Re_x \approx 450{,}000$ and 170,000 respectively. But it appears from the mean-velocity profiles (not shown), the Reynolds shear-stress results shown in Fig. 4.29c, and skin-friction coefficient distributions shown in Fig. 4.30 that the laminar boundary layer for the T3C3 test transitions near the position of maximum free-stream velocity and therefore avoids separation, whereas the boundary layer for the T3C4 test separates and quickly reattaches as a turbulent boundary layer.[2] For design purposes, this behavior must not only be completely understood but calculable. In Fig. 4.30, it is noteworthy that the skin-friction coefficient for laminar flow with free-stream turbulence is larger than that without, and calculated surprising well by the present model. In some cases, however, as observed also by Roach and Brierley [2000] using their model, it over predicts the skin-friction coefficient.

[1] A similar factor had to be used by Mayle et al. [2008] to modify the production terms for both the streamwise normal-stress component and the shear stress.

[2] See §8.3 about separated-flow transition.

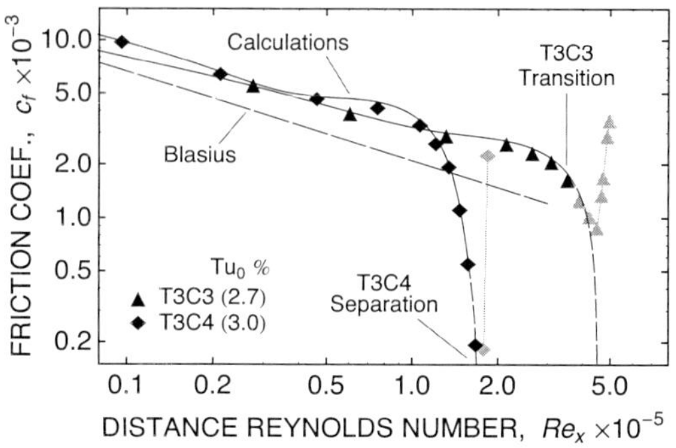

Figure 4.30.
Calculated and measured streamwise variations of the skin-friction coefficient. Data from Roach and Brierley's [1990] T3C3 and T3C4 tests.

4.3.4 Onset of transition.

Something old. Something new.

Since a theory of laminar breakdown for boundary layers with high free-stream turbulence has yet to be developed, "postulates" have been proposed instead. These generally take the form of statements concerning conditions that might occur at the onset of transition.[1] In this section, we will examine some of the older postulates and then a new one. Since data on either laminar breakdown or the onset of transition is sparse, not complete, and not particularly accurate, it is impossible at this time to determine whether any of the postulates are correct, although efforts to do so without any new data continue.

Something old. In 1943, Liepmann postulated that transition begins where the maximum Reynolds shear stress within the boundary layer equals the wall shear stress. Using the data of Roach and Brierley [1990], the maximum values of $\overline{u'v'}/u_\tau^2$ measured at the stations bracketing transition are plotted in Fig. 4.31. The measurement stations upstream and downstream of transition (open and solid symbols respectively) were determined by picking out where a notable shift in the intensity profile toward the wall, accompanied by an increase in the Reynolds shear stress occurred. While this is somewhat subjective, it turns out that all stations downstream of transition were either before or at the position of minimum skin-friction coefficient.

The dashed horizontal lines in the figure, corresponding to the values of 0.28 and 0.38, are the average of the maxima at the upstream and downstream locations. Therefore, the onset of transition apparently occurs when

$$(-\overline{u'v'})_{\max} \big/ u_\tau^2 \approx 0.33 \pm 0.05 \tag{4.50}$$

which is one-third the value originally proposed by Liepmann. Although not shown, the data of Acharya [1985] and Zhou and Wang [1995], who measured Reynolds stress distributions within transitioning boundary layers with free stream turbulence, also fit this criterion.

The region corresponding to $0.28 < (-\overline{u'v'})_{\max}/u_\tau^2 < 0.38$ is shown by the gray band in Fig. 4.28*b*. For this plot, since $dp/dx = 0$, Blasius' solution was used to determine the variation of the friction velocity with distance Reynolds number, not the solution for various turbulence levels.[2] Here, the

---

[1] Here it is worth emphasizing the fact that the onset of transition does not necessarily correspond with the first breakdown of laminar flow. To be specific, breakdown is a single random physical occurrence, while onset is the mean of the probability distribution of these occurrences. This distinction will become evident when we examine the theory of turbulent spots in the next chapter and particularly their source distributions in §5.2.3.

[2] See Eqs. (2.14) and (2.10).

band cuts across the calculated pre-transition curves about where Roach and Brierley's data indicate transition occurs, as it should, and near Westin's et al. third measurement station (see part (a) of the figure). Although Westin et al. did not measure Reynolds shear-stress distributions, they did detect laminar breakdowns at their fifth and last measurement station corresponding to $Re_x = 533{,}000$.

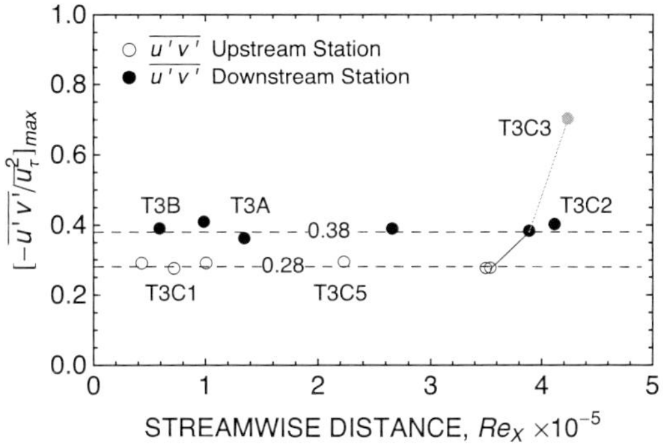

Figure 4.31.
Maximum Reynolds shear stress before and after the onset of transition. Data from Roach and Brierley [1993].

Other observations and postulates have also been made. As previously mentioned, Schubauer [1957] reported that "breakdown" to turbulent flow in natural transition occurred when the maximum intensity $\sqrt{\overline{u'u'}}/\overline{u}_\infty$ reached a value of 0.074. For the tests presently being considered, the average of this value at the positions bracketing transition is about 0.12 (±9%). Sharma et al. [1982] suggested that transition might begin where the maximum value of $\sqrt{\overline{u'u'}}$ within the boundary layer equals $3u_\tau$. This was examined by Mayle and Schulz [1997], as well as by Roach and Brierley [2000], and found not always true. In addition, Johnson [1993], and later Roach and Brierley [2000], proposed that transition occurs where the local intensity $\sqrt{\overline{u'u'}}/\overline{u}$ within the boundary layer first attains a value of about 0.3. The author found that the average of this quantity for the positions bracketing transition is about 0.35 (±34%). Clearly, more experimental work is needed before the onset of transition can be linked to the fluctuations.

Something new. Upon preparing Table 4.3, the author was struck by the fact that $l_d/\delta \approx 0.081$ at transition for all four tests listed. Considering the vastly different turbulence grids used in these tests (either a parallel or square array of round or square bars or wires of different mesh), and subsequent differences in turbulence levels and length scales, this result appears significant.

The data in Table 4.3 is plotted in Fig. 4.32 against the transition Reynolds number using the dimensionless wave number $k_{eff}\delta_1$ at transition. The effective wave number was determined from Eq. (4.27) with $k_{eff} = 1/l_{eff}$, namely,

$$k_{eff} = 0.3\left(\varepsilon_\infty / \nu^3\right)^{1/4}$$

after fitting the free-stream turbulence data as explained in §4.3.3 to obtain the dissipation as a function of the streamwise distance. For *isotropic* turbulence with $dp/dx = 0$, the dissipation can be obtained from Eq. (3.74), which in terms of the free-stream turbulence level Tu_∞ is given by

$$\varepsilon_\infty = -\frac{3}{2}\overline{u}_\infty^3 \frac{dTu_\infty^2}{dx} \tag{4.51}$$

where, according to Eq. (3.75),

$$Tu_\infty = \sqrt{(\overline{u'u'})_\infty} \Big/ \overline{u}_\infty$$

For *anisotropic* turbulence, the free-stream dissipation and turbulence level are given by Eqs. (3.78) and (3.81) respectively.

In addition, the results of Klebanoff et al. [1962] for laminar breakdown in natural transition (discussed in §4.2.3), and the neutral stability curve for a Blasius boundary layer (see Fig. 4.2 or Appendix D) are plotted. Furthermore, the data of Jonas et al. [2000] are plotted.[1] Their test conditions are about the same as Roach and Brierley's T3A test with a *nominal* turbulence level (given in parentheses in the figure) of about three percent, but for different turbulent length scales. In all cases, the transition Reynolds numbers plotted are the author's best estimates.

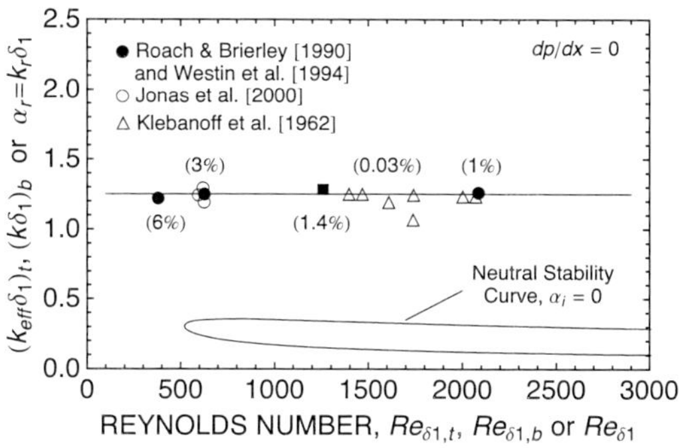

Figure 4.32.
Onset criterion for natural and bypass transition based on the effective wave number.

Apparently, transition for these tests occurs when

$$\left(k_{eff}\delta_1\right)_t \approx 1.25 \tag{4.52}$$

which is not only independent of the transition Reynolds number, but nearly identical to the result given by Eq. (4.22) for laminar breakdown. To the author's knowledge, this coincidence is something new. If not fortuitous, it could infer that similar breakdown mechanisms are involved in both cases. But this remains to be investigated. Nevertheless, the wave number of the disturbance responsible for transition in a turbulent free stream, $k_{eff,t}$, is about five times larger than that responsible for laminar instability, k_r. This agrees with the hypothesis initially proposed by Mayle and Schulz [1997] and Mayle et al. [1998] which is described in §4.3.1. However, as shown in Appendix F.2, the most effective wave number in the *boundary layer* is about one-fifth that in the *free stream*, and its value there is about the same for all components of the wave number. In other words, the most "effective" turbulence *within* the boundary layer, which is produced by the free-stream turbulence with wave numbers near $k_{eff,t}$, has about the same wave number as those which cause a laminar boundary layer to become unstable.

The criterion given by Eq. (4.52) indicates that the onset of boundary-layer transition in a turbulent free stream doesn't depend on the turbulence level at all, but on its dissipation. As might be recalled,[2] a similar situation was found regarding the production of fluctuations in pre-transitional laminar flow. Thus these two results agree, whereas correlations of the onset of transition with respect to the free-

[1] These results are the best estimates obtained by the author after spending much effort reducing their data. Determining the onset of transition for this data, which the author found to be significantly different from that reported by them, is particularly difficult because both the distances between measurement stations and the errors in measuring the skin-friction coefficients are relatively large. Nevertheless, since the experimental setup was essentially the same as Roach and Brierley's, it was deemed worthwhile to include these best estimates.

[2] See §4.3.2.

stream turbulence level, such as shown in Fig. 4.1, does not. As we will see in §6.2.2, however, this apparent disagreement is reconciled once the length scale of turbulence is considered.

The form of the above criterion is not new. Deduced originally from stability considerations and dimensional analysis, Reshotko [1969, 1976] proposed that the transition Reynolds number for bypass transition be given by

$$Re_{\delta 1,t} \propto \left(u_\infty^2/\omega\nu\right)^n \quad \text{or equivalently} \quad Re_{\delta 1,t} \propto \left(u_\infty^2/\nu c_r k\right)^n$$

where ω and k are the characteristic frequency and wave number of the disturbance, and c_r is its phase velocity. As pointed out by Reshotko, the coefficient and exponent may depend on other dimensionless quantities as well. Nevertheless, comparing the second of these expressions with Eq. (4.52) using $c_r = u_\infty$, it appears that $n = 1$. As a consequence, since $\Omega = \omega\nu/u_\infty^2$, the first of the above relations yields

$$\left(\Omega_{eff} Re_{\delta 1}\right)_t = 1.25 \tag{4.53}$$

where the constant of proportionality was set equal to 1.25 in compliance with the condition given by Eq. (4.52).

When either Eq. (4.52) or (4.53) is used as a transition criterion, it should be noted that the range over which calculations must be performed or data must be gathered to accurately determine onset increases significantly as the free-stream turbulence level decreases. This follows from the fact that the decay rates and therefore the effective wave numbers decrease significantly with distance for the lower turbulence levels, which in turn reduces the slope of the $k_{eff}\delta_1$ curve as it approaches the value of 1.25. This behavior is illustrated in Fig. 4.33. The solid lines in this figure correspond to the author's calculated results for Roach and Brierley's [1990] T3MA, T3A and T3B test cases, whereas the dashed line is given by Eq. (4.52). The range of distance Reynolds numbers corresponding to a ±5 percent error in either calculating or measuring the transition onset parameter $(k_{eff}\delta_1)_t$ is about $\Delta Re_x = 1.1(10)^6$ for the one percent test case, whereas for the three percent test case it is about $\Delta Re_x = (10)^5$.

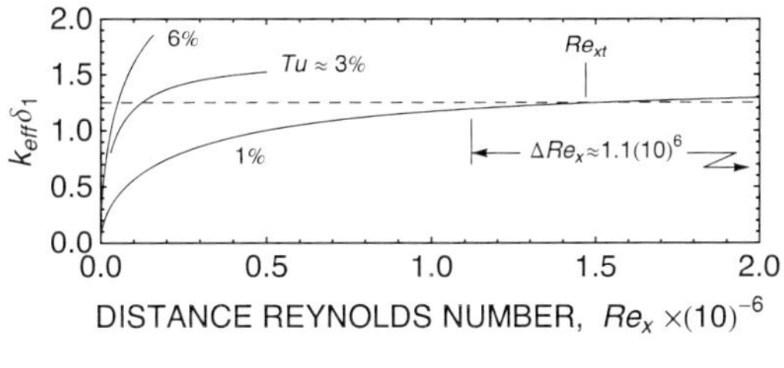

Figure 4.33.
Variation of the transition onset parameter with distance for three different free-stream turbulence levels. Calculations for Roach and Brierley's [1990] T3MA, T3A and T3B test cases.

References

Ames, F.E., Kwon, O., and Moffat, R.J., 1999, "An Algebraic Model for High Intensity Large Scale Turbulence," ASME Paper 99-GT-160.

Barry, M.D.J., and Ross, M.A.S., 1970, "The Flat Plate Boundary Layer. Part 2. The Effect of Increasing Thickness on Stability," J. Fluid Mech., 43, pp. 813–818.

Benney, D.J., 1961, "A Non-Linear Theory for Oscillations in Parallel Flow," J. Fluid Mech., 10, pp. 209–236.

Benney, D.J., 1964, "Finite Amplitude Effects in an Unstable Laminar Boundary Layer," Phys. Fluids, 7, pp. 319–326.

Benney, D.J., and Lin, C.C., 1960, "On the Secondary Motion Induced by Oscillations in a Shear Flow," Phys. of Fluids, 3, pp. 656–657.

Blair, M.F., 1992, "Boundary-Layer Transition in Accelerating Flows with Intense Freestream Turbulence: Part 1–Disturbances Upstream of Transition Onset," J. Fluids Engng., 114, pp. 313–321.

Betchov, R., and Criminale, W.O., 1967, Stability of Parallel Flow, Academic Press, New York.

Breuer, K.S., Cohen, J., and Haritonidis, J.H., 1997, "The Late Stages of Transition Induced by a Low-Amplitude Wavepacket in a Laminar Boundary Layer," J. Fluid Mech., 340, pp, 395–411.

Chandrasekhar, S., 1961, Hydrodynamic and Hydromagnetic Stability, Oxford University Press; also 1981, Dover Publications.

Cohen, J., Breuer, K.S., and Haritonidis, J.H., 1991, "On the Evolution of a Wave Packet in a Laminar Boundary Layer," J. Fluid Mech., 225, pp. 575–606.

Dullenkopf, K., and Mayle, R.E., 1995, "An Account of Free-Stream Turbulence Length Scale on Laminar Heat Transfer," J. Turbomachinery, 117, pp. 410–406.

Durbin, P.A., 1993, "A Reynolds Stress Model for Near-Wall Turbulence," J. Fluid Mech., 249, pp. 465–498.

Dyban, E., Epik, E., and Suprun, T.T., 1976, "Characteristics of the Laminar Boundary Layer in the Presence of Elevated Free-Stream Turbulence," Fluid Mech.-Soviet Res., 5, pp. 30–36.

Dyban, E., and Epik, E., 1985, Тепломссобмен и Гидродинамика Тубулизированных Потоков, Киев Наукоа Думка (in Russian).

Falkner, V.M., and Skan, S.W., 1931, "Some Approximate Solutions of the Boundary Layer Equation," Phil. Mag.,12, pp. 865–896; also 1930, ARC RM 1314.

Gaster, M., 1962, "A Note on the Relation Between Temporally-Increasing and Spatially-Increasing Disturbances in Hydrodynamic Stability," J. Fluid Mech., 14, pp. 222–224.

Gaster, M., 1965, "The Role of Spatially Growing Waves in the Theory of Hydrodynamic Stability," Progress Aero. Sciences (ed., D. Küchemann), 6, pp. 251–270.

Gibbings, J.C., and Al-Shukri, S.M., 1997, "Effect of Sandpaper Roughness and Stream Turbulence on the Laminar Layer and Its Transition," Aeronautical J., 101, pp. 17–24.

Greenspan, H.P., and Benney, D.J., 1963, "On Shear Instability, Breakdown and Transition," J. Fluid Mech., 15, pp. 133–153.

Holstein, H., and Bohlen, T., 1940, "Ein einfaches Verfahren zur Berechnung laminarer Reibungsschichten," Lilienthal-Bericht, 10, pp. 5–16.

Hunt, J.C.R., 1973, "A Theory of Turbulent Flow Round Two-Dimensional Bluff Bodies," J. Fluid Mech., 61, p. 625.

Hunt, J.C.R., and Graham, J.M.R., 1978, "Free-Stream Turbulence Near Plane Boundaries," J. Fluid Mech., 84, pp. 209–235.

Jaffe, N.A., Okamura, T.T., and Smith, A.M.O., 1970, "Determination of Spatial Amplification Factors and Their Application to Predicting Transition," AIAA Journal, 8, pp. 301–308.

Johnson, M.W., 1993, "A Bypass Transition Model for Boundary Layers," ASME Paper 93-GT-9.

Jonàs, P., Mazur, O., and Uruba, V., 2000, "On the Receptivity of the By-Pass Transition to the Length Scale of the Outer Stream Turbulence," European J. Mech. B/Fluids, 19, pp. 707–722.

Jordinson, R., 1970, "The Flat Plate Boundary Layer. Part I: Numerical Integration of the Orr-Sommerfeld Equation," J. Fluid Mech., 43, pp. 801–811.

Jordinson, R., 1971, "Spectrum of Eigenvalues of the Orr-Sommerfeld Equation for Blasius Flow," Phys. Fluids, 14, pp. 2535–2537.

Kaplan, R.E., 1964, "The Stability of Laminar Incompressible Boundary Layers in the Presence of Compliant Boundaries," Ph.D Thesis, Massachusetts Institute of Technology; also Aero-Elastic and Structures Research Lab. Report ASRL-TR-116-1,

Klebanoff, P.S., 1955, "Characteristics of Turbulence in a Boundary Layer with Zero Pressure Gradient," NACA TR 1247.

Klebanoff, P.S., and Tidstrom, K.D., 1959, "Evolution of Amplified Waves Leading to Transition in a Boundary Layer with Zero Pressure Gradient," NASA TN D-195.

Klebanoff, P.S., Tidstrom, K.D., and Sargent, L.M.,1962, "The Three-Dimensional Nature of Boundary-Layer Instability," J. Fluid Mech., 12, pp. 1–34.

Kovasznay, L.S.G., Komoda, H., and Vasudeva, B.R., 1962, "Detailed Flow Field in Transition," Proc. 1962 Heat Transfer and Fluid Mech. Inst., Stanford University Press, Stanford, pp. 1–26.

Kurtz, E.F., and Crandall, S.H., 1962, "Computer-Aided Analysis of Hydrodynamic Stability," J. Math. Phys., 44, pp. 264–279.

Liepmann, H.W., 1943, "Investigations on Laminar Boundary-Layer Stability and Transition on Curved Boundaries," NACA ACR 3H30; also NACA WR-W-107.

Lin, C.C., 1955, The Theory of Hydrodynamic Stability, Cambridge Univ. Press.

Lin, C.C., 1957, "On the Instability of Laminar Flow and Its Transition to Turbulence," Grenzschlichtforschung Symposium Freiburg im Brieslau (ed., H. Görtler, 1958), Springer-Verlag, Berlin, pp. 144–160.

Lin, C.C., and Benney, D.J., 1962, "On the Instability of Shear Flows," Proc. Symp. Applied Math., 13, Hydrodynamic Instability, pp. 1–24.

Lorentz, H.A., 1907, "Über die Entstehung turbulenter Flüssigkeitsbewegungen und über den Einfluss dieser Bewegungen bei der Strömung durch Rohren," Abh. theor. Phys., 1, pp. 43–71 (new version of earlier paper: Akad.v. Wet. Amsterdam, 6, p. 28, 1897).

Matsubara,M., and Alfredsson, P.H., 2001, "Disturbance Growth in Boundary Layers Subjected to Free Stream Turbulence," J. Fluid Mech., 430, pp. 149–168.

Mayle, R.E., and Schulz, A., 1997, "The Path to Predicting Bypass Transition," J. Turbomachinery, 119, pp. 405–411.

Mayle, R.E., Dullenkopf, K., and Schulz, A., 1998, "The Turbulence That Matters," J. Turbomachinery, 120, pp. 402–409.

Mayle, R.E., Schulz, A., and Bauer, H.-J., 2008, "Reynolds Stress Calculations for Pre-transitional Boundary Layers with Turbulent Free Streams," ASME Paper GT2008-50109.

Morkovin, M.V., 1969, "On the Many Faces of Transition," Viscous Drag Reduction (ed., C.S. Wells), Plenum Press, New York, pp. 1–31.

Osborne, M.R., 1967, "Numerical Methods for Hydrodynamic Stability Problems," SIAM J. Appl. Math., 15, pp. 539–557.

Ombrewski, H.G., Morkovin, M.V., and Landahl, M., 1969, "A Portfolio of Stability Characteristics of Incompressible Boundary Layers," AGARDograph No. 134.

Orr, W.M.F., 1907, "The Stability or Instability of the Steady Motions of a Perfect Liquid and of a Viscous Liquid. Part I: A Perfect Liquid. Part II: A Viscous Liquid," Proc. Roy. Irish Acad., 27, pp. 9–68 and 69–138.

Pohlhausen, K., 1921, "Zur näherungsweisen Integration der Differentialgleichung der laminaren

Reibungsschicht," ZAMM, 1, pp. 252–268.

Prandtl, L., 1921, "Bemerkungen über die Entstehung der Turbulenz," ZAMM, 1, pp. 431–436; also 1922, Phys. Z., 23, pp. 19–25.

Rayleigh, J.W.S., Lord, 1880, "On the Stability, or Instability, of Certain Fluid Motion," Scientific Papers, I, Cambridge University Press, pp. 474–487.

Rayleigh, J.W.S., Lord, 1887, "On the Stability, or Instability, of Certain Fluid Motion," Scientific Papers, II, Cambridge University Press, pp. 2–23.

Rayleigh, J.W.S., Lord, 1892, "On the Question of the Stability of the Flow of Fluids," Scientific Papers, III, Cambridge University Press, pp. 575–584.

Reshotko, E., 1969, "Stability Theory as a Guide to the Evaluation of Transition Data," AIAA Journal, 7, pp. 1086–1091.

Reshotko, E., 1976, "Boundary Layer Stability and Transition," Ann. Rev. Fluid Mech., 8, pp. 311–349.

Reynolds, O., 1895, "On the Dynamical Theory of Incompressible Viscous Fluids and the Determination of the Criterion," Phil. Trans. Roy. Soc. A, 186, pp. 123–164.

Roach, P.E., and Brierley, D.H., 1990, "The Influence of a Turbulent Free-Stream on Zero Pressure Gradient Transitional Boundary Layer Development Including the T3A & T3B Test Case Conditions," Proc. 1st ERCOFTAC Workshop on Numerical Simulation of Unsteady Flows, Transition to Turbulence and Combustion, Lausanne; also 1992, Numerical Simulation of Unsteady Flows and Transition to Turbulence, (eds., O. Perinea et al.), Cambridge University Press, pp. 319–347, and 1993, data transmitted by J. Coupland from the Rolls-Royce Applied Science Laboratory.

Roach, P.E., and Brierley, D.H., 2000, "Bypass Transition Modeling: A New Method which Accounts for Free-Stream Turbulence and Length Scale," ASME Paper 2000-GT-278.

Ross, J.A., Barnes, F.H., Burns, J.G., and Ross, M.A.S., 1970, "The Flat Plate Boundary Layer. Part 3. Comparison of Theory with Experiment," J. Fluid Mech., 43, pp. 819–832.

Schlichting, H., 1933, "Zur Entstehung der Turbulenz bei der Plattenströmung," Nachr. Ges. Wiss. Göttingen, Math. Phys. Klasse, pp. 182–208; also 1933, ZAMM, 13, pp. 171–174.

Schlichting, H., 1979, Boundary-Layer Theory, McGraw-Hill, New York.

Schubauer, G.B., 1957, "Mechanism of Transition at Subsonic Speeds," Grenzschlichtforschung Symposium Freiburg im Brieslau (ed., H. Görtler, 1958), Springer-Verlag, Berlin, pp. 85–109.

Schubauer, G.B., and Skramstad, H.K., 1943, "Laminar Boundary-Layer Oscillations and Stability of Laminar Flow," NACA WR-W-8.

Schubauer, G.B., and Skramstad, H.K., 1948, "Laminar Boundary-Layer Oscillations and Transition on a Flat Plate," NACA Rept. 909.

Sharma, O.P., Wells, R.A., Schlinker, R.H., and Bailey, D.A., 1982, "Boundary Layer Development on Turbine Airfoil Suction Surfaces," Jour. Engng. for Power, 104, pp. 698–706.

Shen, S.F., 1954, "Calculated Amplified Oscillations in Plane Poiseuille and Blasius Flows," JAS, 21, pp. 62–64.

Smith, M.C., and Kuethe, A.M., 1966, "Effects of Turbulence on Laminar Skin Friction and Heat Transfer," Phys. Fluids, 9, pp. 2337–2344.

Sommerfeld, A., 1908 "Ein Beitrag zur hydrodynamischen Erklärung der turbulenten Flüssigkeitsbewegungen," Atti der 4. Congr. Internat. dei Mat. III, Roma, pp. 116–124.

Squire, H.B., 1933, "On the Stability of Three-Dimensional Distribution of Viscous Fluid Between Parallel Walls," Proc. Roy. Soc. London A, 142, pp. 621–628.

Steelant, J., and Dick, E., 2001, "Modelling of Laminar-Turbulent Transition for High Freestream Turbulence," J. Fluids Engng., 123, pp. 22–30.

Taylor, G.I., 1936, "Effect of Turbulence on Boundary Layer," Proc. Roy. Soc., London A, 156, pp. 307–317.

Tollmien, W., 1929, "Über die Entstehung der Turbulenz," 1. Mitt. Nachr. Ges. Wiss. Göttingen, Math. Phys. Klasse, pp. 21–44; also 1931, NACA TM 609.

Thwaites, B., 1949, "Approximate Calculation of the Laminar Boundary Layer," Aeronautical Quart., 7, pp. 245–280.

Wang, T., and Keller, 1999, "Intermittent Flow and Thermal Structures of Accelerated Transitional Boundary Layers, Part 1: Mean Quantities, and Part 2: Fluctuation Quantities," J. Turbomachinery, 121, pp. 98–105 and 106–112.

Wazzan, A.R., Okamura, T.T, and Smith, A.M.O., 1968, "Spatial and Temporal Stability Charts for the Falkner-Skan Boundary-Layer Profiles," Douglas Aircraft Report No. DAC-67086.

Westin, K.J.A., Boiko, A.V., Klingmann, B.G.B., Kozlov, V.V., and Alfredsson, P.H., 1994, "Experiments in a Boundary Layer Subjected to Free Stream Turbulence. Part 1. Boundary Layer Structure and Receptivity," J. Fluid Mech., 281, pp.193–218.

Westin, K.J.A., Bakchinov, A.A., Kozlov, V.V., and Alfredsson, P.H., 1998, "Experiments on Localized Disturbances in a Flat Plate Boundary Layer. Part 1. The Receptivity and Evolution of a Localized Free Stream Disturbance," European J. Mech. B/Fluids, 17, pp. 823–846.

White, F.M., 1974, Viscous Fluid Flow, McGraw-Hill, New York.

Zhou, D., and Wang, T., 1995, "Effects of Elevated Free-Stream Turbulence on Flow and Thermal Structures in Transitional Boundary Layers," J. Turbomachinery, 117, pp. 407–417.

5 Turbulent Spots

Anatomy of a spot. Emmons' turbulent spot theory. Spot Kinematics. Emmons' spot parameters.

At first transition was thought to be a more or less abrupt two-dimensional breakdown of laminar to turbulent flow.[1] Then Dryden [1934–39] demonstrated that the flow within the "transition region" was part of the time laminar and part of the time turbulent – something like that shown in Fig. 5.1. Maintaining the historical view, however, he concluded that transition occurs on a spanwise-irregular but abrupt front that oscillates back and forth over a streamwise distance between the regions of fully-laminar and fully-turbulent flow. This remained the popular view of transitional flow until Emmons [1951] discovered turbulent spots.

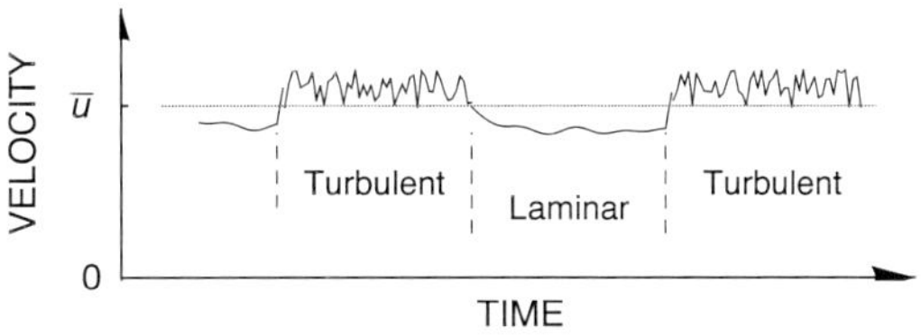

Figure 5.1.
Typical hot-wire signal in transitional flow.

Historically, Emmons' discovery was a major breakthrough in both our perception and treatment of transitional flow. In particular, he observed that transition begins with "tiny spots of turbulence" arising within the laminar boundary layer that grow and coalesce as they propagate downstream until they eventually cover the surface. Consequently, the turbulent portions of the signal shown in Fig. 5.1 correspond to turbulent spots passing over the probe, and not the advance and retreat of a turbulent front. Furthermore, since he observed that the spots randomly appeared over the surface both temporally and spatially, he concluded that transition is a stochastic, three-dimensional, unsteady process that must be analyzed accordingly.

There are three major aspects of transition that must be understood – its onset, its spots, and its randomness. Our primary interest in this chapter are the spots. First, we will consider the anatomy of a turbulent spot and its evolution. We will then examine Emmons' turbulent spot theory, and consider variations in the production of spots. After this, we will consider the kinematics of a single turbulent spot in flows with a zero, favorable, and adverse pressure gradient. Then we will consider the effect of "becalmed" zones on transition and finally Emmons' propagation parameter.

Some of the important results in this chapter are:

1. A turbulent spot is a very flat arrowhead shaped patch of turbulence.
2. Emmons' turbulent-spot theory easily accounts for the stochastic intermittent flow of turbulent spots in transition as long as his spot production and propagation rates are known a priori.
3. Since most of the turbulent spot data are for fully-developed spots propagating in flows with very low free-stream turbulence, they are of very limited practical use when the turbulence is high.
4. After more than a half-century of work, we have yet to *measure* the birth rate of turbulent spots and have barely assessed the physical characteristics of the becalmed zone following a spot.

Beside the references at the end of this chapter, other relevant articles may be found in the Bibliography.

---

[1] See the experimental results of Burgers [1924] and Hegge Zijnen [1924], and the theoretical analysis of Prandtl[1927] and Prandtl and Schlichting [1934]. Discussions of these investigations may also be found in Rotta [1972] and Schlichting [1979].

5.1 Anatomy of a Spot.

General features. Spot mean flow. Reynolds-stresses. Becalmed zones.

When a laminar flow breaks down, the "newborn" turbulent eddies, variously called "incipient," "embryonic," "nascent," or "immature" turbulent spots, pass through several stages of development on their way downstream. In the early stages, the spots, also known as *Emmons' spots*, grow and develop isolated from one another, while in the later stages farther downstream, where more of the surface is covered with them, the spots grow, interact, and coalesce into irregular shaped turbulent patches that eventually engulf all of the laminar flow. Usually, three distinct stages can be observed in which the spots may be described as *developing* or *immature* turbulent spots, *fully-developed* or *mature* spots, and *merging spots* or *turbulent patches*. For flows with free-stream turbulence above about two percent, transition is generally complete before any of the spots are fully developed. To the author's knowledge, virtually all experiments on turbulent spots examine their development before they merge with one another, and almost all of them examine spots in a flow with low free-stream turbulence. Furthermore, virtually all the investigations examine the development of a single spot. This is presumably a result of the earlier experiments conducted on two laterally displaced spots by Schubauer and Klebanoff [1955] and Elder [1960], who observed that spot interference hardly affects their growth rates. More recently, however, Makita and Nishizawa [1998, 1999], who conducted similar experiments, observed that merging interactions do affect their growth rates, and that the effect becomes more complex when a spot merges off-center with another from behind. The interaction between a turbulent spot and an unstable Tollmien-Schlichting wave was investigated by Schubauer and Klebanoff, Amini and Lespinard [1982], and Chambers and Thomas [1983], while the interaction between a turbulent spot and a turbulent boundary layer was examined by Zilberman et al. [1977]. In the later case, it was found that a turbulent spot not only retains its own identity as it enters a turbulent boundary layer, but that it also suffers a negligible loss of intensity.

In most turbulent spot experiments, the growth and propagation of a single spot generated by a device causing a momentary three-dimensional disturbance to the boundary layer is considered. Some of the devices used to generate the spot include a spark discharged near the surface, a puff of fluid ejected through a small orifice in the surface, an expandable diaphragm on the surface, and a pin momentarily inserted into the flow from the surface. In any case, Elder, who examined the effects of puff generators, Wygnanski et al. [1976], who examined the effects of spark generators, and Chambers and Thomas, who examined the effects of pin, puff, and diaphragm generators, found that their devices had little effect on the development of the spot except in the immediate neighborhood of the generator, providing that the amplitude of the disturbance was "reasonable" and that a turbulent spot was initially formed.

Gaster and Grant [1975], Amini and Lespinard, and Breuer and Haritonidis [1990] experimentally examined the situation when turbulent spots were not initially formed by these generators. In particular, they investigated the evolution of a very small-amplitude, three-dimensional disturbance through its linear, nonlinear, and final breakdown into a turbulent spot in an otherwise undisturbed laminar boundary layer. In this case, the linear and weak-nonlinear development compares well with the analytical "wave-packet" models proposed by Gaster [1968, 1975], Craik [1981], Kachanov [1987], and Herbert [1988]. The breakdown stage begins with the growth of high and low frequency oscillations resulting from their sums and differences within the disturbance, and ends with the rapid development of the broadband spectrum of oscillations typically associated with turbulent spots.[1] As experimentally determined by Cohen et al. [1991] and Breuer et al. [1997], the "gestation" distance for a turbulent spot, that is, the distance from the disturbance to where the spot finally forms, is rela-

---

[1] While highly recommended to those pursuing a more thorough understanding of spot formation from isolated disturbances and its early development, much of this work is beyond the scope of this book.

tively large when the amplitude of the disturbance is small. In their case, it was about $Re_{\Delta x} = 10^6$, which, as we will see in the next section, is about twice the distance it takes an immature spot to fully develop. However, if we only consider the distance it takes for the nonlinear oscillations to breakdown and become a "recognizable" turbulent spot, the gestation distance Reynolds number reduces to $1.3(10)^5$, or about one-quarter of the distance for a spot to fully develop. As might be expected, the initial gestation stages of a spot depend on the disturbance amplitude but, as pointed out by Morkovin [1969], they may be completely bypassed when the amplitude is sufficiently high.

In this section we will examine the physical attributes of turbulent spots, their growth and propagation rates, and various velocity and Reynolds stress components within them.

5.1.1 Some general features of a turbulent spot.

Spot shape. Propagation speeds. Spot development. The becalmed zone.

As shown in the high-speed photos of Fig. 5.2, a turbulent spot is aptly named. These photos (presented here as photographic negatives of the spot) were obtained by illuminating a layer of dye swept up into the spot with a sheet of light oriented in the spot's vertical plane of symmetry to photograph the elevation view, and in a plane parallel to and just slightly above the wall to photograph the plan view. Shown nearly to scale, the spot is clearly a *very* flat patch of turbulence.

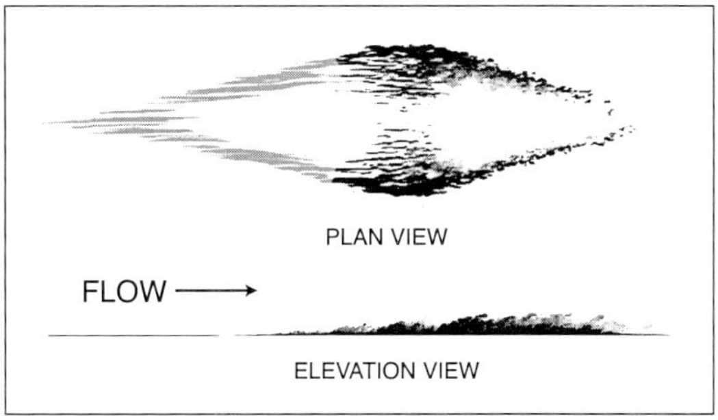

Figure 5.2.
Snapshots of a turbulent spot with $dp/dx = 0$. After Gad-el-Hak et al. [1981].

In the plan view, the typical "arrowhead" shape of the spot and its triangular shaped laminar "recovery tail" are clearly outlined. As explained by Gad-el-Hak et al. [1981], who obtained the original photos, only the outlines are visible in this view because the spot entrains the dyed fluid near the wall and carries it into regions above the illuminated plane. Cantwell et al. [1978] graphically describe this action as "the spot picks up dye from the wall like a vacuum cleaner." Gad-el-Hak et al. attribute the streaks in the recovery tail, seen here only at its edge, to counter-rotating, streamwise oriented vortices. Streaks were also found by Elder, Cantwell et al., and Chambers and Thomas in their flow visualization experiments. In their experiments, Chambers and Thomas show that streaks of this sort are produced when two-dimensional Tollmien-Schlichting waves in the surrounding laminar boundary layer turn streamwise within the spot's recovery tail. Glezer et al. [1989], who investigated "wave packets" emanating from the outboard tips of the spot's trailing edge, suggest that the streaks are crests of these waves that normally turn streamwise as they propagate inboard. This author suggests that they are simply the "stretched residue" of spot turbulence. Schubauer and Klebanoff, who first examined this region, and later Wygnanski et al. [1979], who first investigated the oblique wave structure, didn't detect any streaks within this region. Nonetheless, Schubauer and Klebanoff did discover that the recovering laminar boundary layer behind the spot is highly stable and that another spot cannot be generated within it. They also found that previously transitioned flow entered by a spot becomes laminar again within its tail. This phenomenon they described as a "calming" effect.

In the elevation view of Fig. 5.2, we see eddies as large as the height of the spot with their "tops" being "blown" downstream. The small eddies seen near the wall in the plan view are buried beneath these. The eddy structure in a spot was studied by Wygnanski et al. [1982]. Contours of the velocity perturbations, obtained by averaging low-pass-filtered hot-wire measurements in the plane of symmetry of a typical spot, are shown in Fig. 5.3. In this figure, the height is scaled by a factor of about five. The different size eddies in the upper and lower portions of the spot suggest that a turbulent spot may be divided into three primary regions; namely, a *base*, a *cap*, and a *recovery tail*. As we will soon see, the base and recovery tail are about as thick as the surrounding laminar boundary layer, while the cap protrudes into the free stream. Hence, we can expect that the two regions near the wall are viscously dominated, while the cap is inertially dominated.

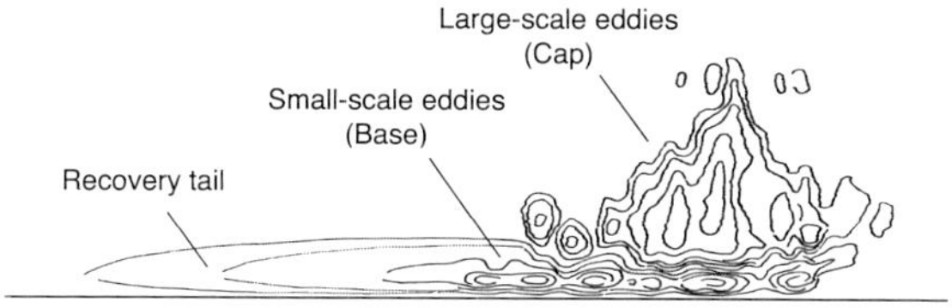

Figure 5.3.
Mean-velocity perturbations contours in the midspan plane of a turbulent spot. After Wygnanski et al. [1982].

As a spot passes over any given position in the boundary layer, its leading- and trailing-edge laminar-turbulent interfaces can be detected by the jumps in a hot-wire signal similar to those seen in Fig. 5.1. By averaging the interval of time it takes these interfaces to pass for a large number of spots, known as *ensemble-averaging*, the average *duration* of the spot at any position can be determined. And by positioning a second probe slightly downstream from the first, the average *speed* of these interfaces can be determined.[1] As we will learn in §5.3, the temporal description of a spot can be transformed into a spatial description to obtain an instantaneous snapshot of the spot at any time. Likewise, using the averaging techniques described in §7.2, all the features within the spot can also be determined. Since a spot is three-dimensional, turbulent, and varies in size and shape with distance from its point of origin, obtaining any information about it requires multitudinous measurements and an enormous amount of data processing, even when its origin is fixed. Nevertheless, Schubauer and Klebanoff obtained the first such measurements and provided us with an average spot shape and propagation speed. Much later, Wygnanski et al. [1976] obtained a more detailed set of measurements on the shape, propagation speeds, and mean-velocity field of a turbulent spot, while Antonia et al. [1981] measured the Reynolds stresses. Furthermore, Makita and Nishizawa obtained a comprehensive set of measurements for spots without a pressure gradient, while Chong and Zhong [2003, 2005] obtained data for spots both with and without pressure gradients.

In general, it is found that a spot develops according to the free-stream pressure gradient and the displacement thickness Reynolds number at its inception. Without a pressure gradient, a turbulent spot expands approximately linearly across the surface as it propagates downstream. This behavior is sketched in Fig. 5.4 where a plan and elevation view of the ensemble-averaged interface of a spot are shown at a particular instant of time. The dashed outlines in the figure represent the surface "footprint" and midspan profile of the spot. Measured as shown in the figure, h is the height of the spot, b is its breadth, and l is its *midspan* length along the surface which is about two-thirds its overall length.

For a typical spot, the leading edge is swept back about 75° such that the included half angle of the leading edge θ is about 15°. In addition, the leading- and trailing-edge propagation velocities u_{le} and

[1] The intricacies associated with this process are thoroughly discussed by Wygnanski et al. [1976].

u_{te} are roughly 0.90 and 0.55 times the free-stream velocity. From these values, the average speed of a spot is about 0.72 times that in the free stream and its average expansion rate fore and aft is roughly equal to 0.18 times the free-stream velocity. If the shape of the spot remains similar as it propagates downstream, we calculate that its leading edge propagates into the laminar boundary layer (perpendicular to itself) with a velocity roughly equal to $0.72 \sin(15°) = 0.19$ times the free-stream velocity, which is the same as the spot's streamwise expansion rate. Consequently, the spot appears to expand across the surface at about the same rate in all directions as it propagates downstream, and that this rate is directly proportional to the free-stream velocity. Carrying this simplistic analysis a little further, we can also calculate that the expansion half-angle is $\alpha \approx \tan^{-1}(0.19) \approx 11°$, which is about that measured. The effects of the displacement-thickness Reynolds number and free-stream pressure gradient on these numbers will be examined in §5.3.2 and §5.3.3.

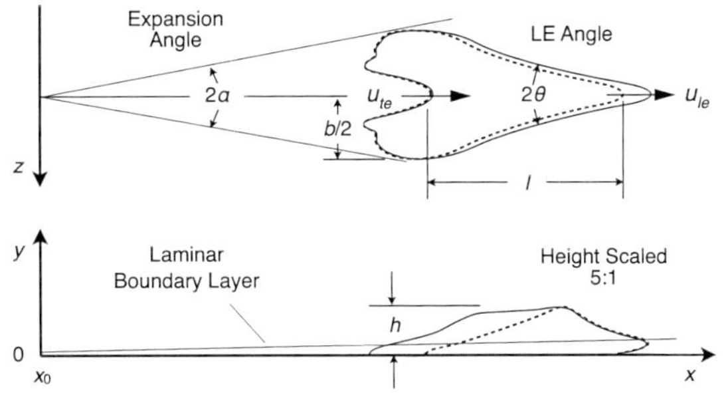

Figure 5.4.
The ensemble-averaged shape, midspan profile, and surface footprint of a typical turbulent spot with $dp/dx = 0$. Data from Wygnanski et al. [1976].

First observed by Schubauer and Klebanoff, the height of a spot grows approximately as a turbulent boundary layer with an initial thickness equal to that of the laminar boundary layer at the spot's inception.[1] Wygnanski et al. [1982] report that it grows as 0.8 times the turbulent-boundary-layer thickness, but also note that the spot's height is difficult to determine. Nevertheless, the growth of a spot as it propagates downstream is shown in Fig. 5.5a together with the growth of the measured laminar and calculated turbulent boundary layers. In this case, the growth of the turbulent boundary layer was calculated using the results in §3.2.3 with the thickness at inception set equal to that for the laminar boundary layer there. As noted, the growth of the spot is nearly the same as a two-dimensional turbulent boundary layer. Furthermore, we see that the height of the leading-edge bulge of the spot is nearly independent of distance, and that the shape of the early spots are noticeably different from those farther downstream. In particular, it appears that the early spots are still developing from the outer portion of the laminar boundary layer where they initially appear to form. On the other hand, the downstream spots appear quite similar to one another, which is more consistent with that expected of a fully-developed or mature spot.

If we plot the ratio of the leading-edge-tip height to the peak height of the spot against the spot's distance from its source, as measured for the spot shown in Fig. 5.5a, we obtain the filled circles shown in part (b) of the figure. The results for spots created farther downstream at larger source Reynolds numbers $Re_{\delta 1,s}$ are also shown.[2] In addition to the data shown, a point with $Re_{\delta 1,s} = 1500$ lies off to

[1] We note that while the word "grows" presently refers to the spot's change in height, it sometimes also refers to the spot's overall increase in size. In contrast, the word "expands" always refers to its increase in size over the surface.

[2] The subscript "s" is being used to denote the source of the turbulent spot. Since these spots were generated by a spark, the subscript refers to the conditions at the spark, which do not necessarily correspond to the conditions that would exist at the birth of an unforced spot propagating, growing, and expanding downstream in a similar fashion.

the right of the figure at $Re_{\Delta x} = 1.3(10)^6$ and $h_{tip}/h_{peak} \approx 0.24$. Although no trend with inception Reynolds number is immediately obvious, a trend with distance Reynolds number clearly exists. An equation for this trend is given by

$$h_{tip}/h_{peak} = 0.21 + 0.85\exp\left[-5(10)^{-6}\left(Re_x - Re_{x,s}\right)\right]$$

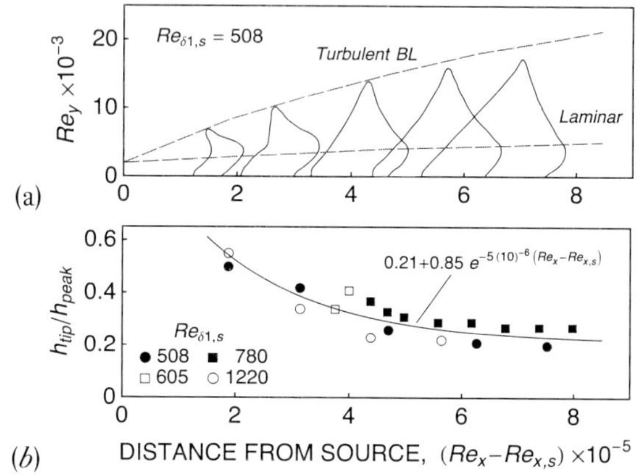

Figure 5.5a,b.
Snapshots of the midspan profile of a spot taken at increasing distances downstream from inception, and the variation of the leading-edge-tip to spot-peak height ratio with distance. Data from Wygnanski et al. [1976, 1982] with $dp/dx = 0$.

If we associate a spot developing to maturity with the asymptotic decrease of h_{tip}/h_{peak} to the value of about one-fifth, we might then consider a spot to be fully-developed when the Reynolds number based on the distance from its source is greater than about $5(10)^5$. This is particularly enlightening because the length of transition with a turbulent free stream is generally less than this. In fact, transition with a free-stream turbulence level of one to two percent is nearly complete by this point. In other words, turbulent spots in flows with an elevated turbulence level are never fully developed before they merge, unless spots developing in a turbulent free stream behave differently. Surprisingly, after a half century of research on turbulent spots, this still remains to be investigated.

The shape of a fully-developed spot as measured by Wygnanski et al. [1976] is shown in Fig. 5.6. These data were obtained for a spot propagating in a laminar boundary layer with low free-stream turbulence and zero pressure gradient. The thicker lines in parts (*a*) and (*b*) of the figure, are the same midspan profile and footprint shown by the dashed lines in Fig. 5.4. The thicker line in part (c) is a cross section close to where the spot has its maximum breadth. From this section forward an "overhang" extends all along the leading edge of the spot to its leading-edge tip just forward of the cross section at $(x - x_{te})/l = 1.0$. This overhang is also seen as a leading-edge bulge in the profiles and the contours extending forward beyond the footprint in parts (a) and (b) of the figure.

The overhang is a direct result of the larger eddies in the cap being convected over the slower moving fluid in the boundary layer. In the boundary layer below the overhang, Gad-el-Hak et al. observed a systematic breakdown of the flow into turbulent eddies. As Cantwell et al. and Wygnanski et al. [1979], they also observed large-amplitude oscillations occurring in the laminar boundary layer immediately ahead of the spot (see Plate 1, Fig. 4 of their paper). These oscillations, which had a wavelength of about one-and-a-half times the boundary-layer thickness and a propagation speed roughly half that of the spot, rapidly amplified as the spot approached. Thus, they concluded that a spot expands by destabilizing and breaking down the already susceptible laminar boundary layer ahead of it, not by entraining non-turbulent fluid as in a turbulent boundary layer.[1] The instability of the flow near and downstream of the outboard tip of a spot, and its effect on the spot's expansion is discussed by Wygnanski et al. [1979], Glezer et al. [1989], and Seifert and Wygnanski [1995].

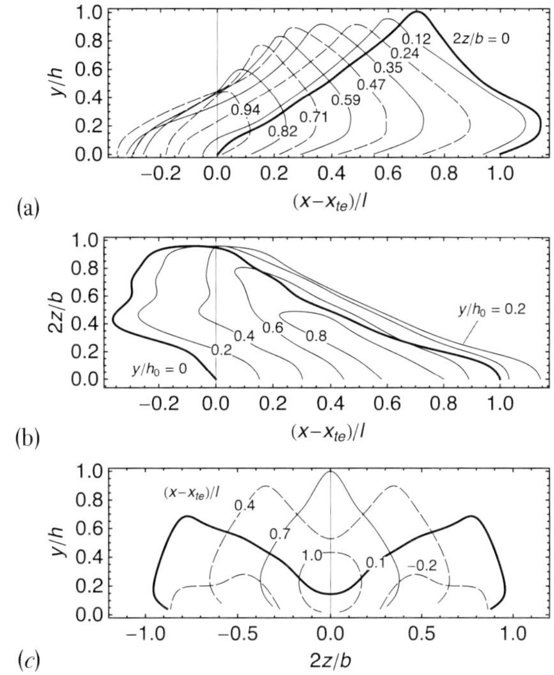

Figure 5.6a,b,c.
Spot cross-sections at (*a*) various distances from the spot's mid-plane, (*b*) various elevations above the wall, and (*c*) various distances from the mid-plane trailing edge for a fully-developed turbulent spot. Data from Wygnanski et al. [1976] with $dp/dx = 0$, $b/2x_s = 0.18$, $h/x_s = 0.025$, $l = x_{le} - x_{te}$, $u_{le}/u_{te} = 1.8$, and $u_\infty = 9.4$m/s.

Schubauer and Klebanoff also showed that there is a significant difference in the development of a spot depending on whether the spot is produced before or after the critical Reynolds number for laminar instability. If the spot is produced *after* the critical Reynolds number is attained, the spot expands linearly over the surface as it propagates downstream. On the other hand, if the spot is produced *before* the critical Reynolds number, it expands over the surface with an initial rate significantly less than that farther downstream (see Fig. 5.35a). Since a spot produced before the critical Reynolds number is forced to exist within an initially stable boundary layer, this behavior is easily understood when we consider the combined effect of the "locally" unstable flow beneath the leading-edge bulge of the spot, which is caused by the spot itself, and the instability of the surrounding boundary layer. Consequently, it appears that the base of a spot expands by a rapid transition from laminar to turbulent flow along its leading edge, while the turbulent cap of the spot expands as the turbulent eddies in the base are swept upward into the free-stream fluid above.

Presuming this is the case, it then follows from stability considerations that the included half angle θ of a spot, which is roughly proportional to the rate of expansion, should be greater in a boundary layer with an adverse pressure gradient than the included half angle of a spot in a boundary layer without a

[1] If we assume that spots expand by turbulent entrainment at the same rate as a turbulent boundary layer grows, then from Eqs. (3.39) and (3.41) we have $(db/dt)_{entr} = u_\infty d\delta_1/dx \approx 0.06 u_\infty Re_x^{-1/5}$, which is about $0.004u_\infty$ for a Reynolds number corresponding to a fully-developed spot. Recalling that the measured expansion rate is $db/dt \approx 0.2u_\infty$, we see that turbulent entrainment is about fifty times too small to account for its expansion.

pressure gradient, and smaller in one with a favorable gradient. As we will see in §5.3.3, this is indeed the behavior observed by Katz et al. [1990], Seifert and Wygnanski, Gostelow et al. [1996], and D'Ovido et al. [2001] among others. In addition, if the average propagation speed of these spots are about the same, we should then expect that the expansion angle will be greater for spots in an adverse pressure gradient than in a favorable gradient. We will see that this is also true.

5.1.2 The mean-flow field associated with a turbulent spot.

Sample ensemble-averaged velocity profiles from fore to aft along the midspan of a spot are shown in Fig. 5.7. The locations for these profiles, except the very first and last, are shown in the elevation view of the spot. The exact locations for the first and last profiles were not reported. Both of these profiles are laminar, with the profile in the recovery tail being somewhat fuller than that ahead of the spot. The other profiles appear more turbulent, that is, fuller than the laminar profiles near the wall with a velocity deficit farther away. As a consequence, the mean velocity first increases then decreases near the wall as the spot passes over, while farther away it first decreases and then increases. The variation of the boundary-layer thickness is plotted in the upper portion of the figure. It increases up to the peak of the spot, whereafter it decreases aft to a thickness *less* than that ahead of the spot. Thus, upon passing, the spot leaves the laminar boundary layer in an accelerated, more stable state; which explains why Schubauer and Klebanoff's attempts to generate spots within it failed.

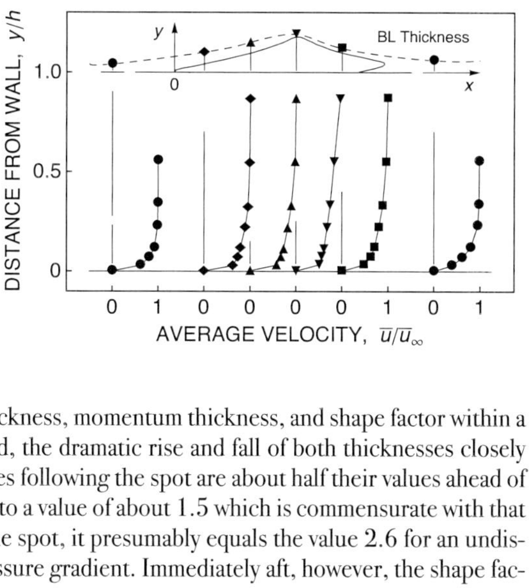

Figure 5.7.
Midspan mean-velocity profiles before, after, and within the spot. Data from Wygnanski et al. [1976] with $dp/dx = 0$, $Re_{\delta 1,s} = 508$, $h = 2$mm, $x_m - x_s = 1000$mm, and $u_\infty = 9.4$m/s.

The midspan distributions of displacement thickness, momentum thickness, and shape factor within a spot are plotted in Fig. 5.8. As to be expected, the dramatic rise and fall of both thicknesses closely follow the thickness of the spot, and their values following the spot are about half their values ahead of it. Within the spot, the shape factor decreases to a value of about 1.5 which is commensurate with that for a turbulent boundary layer. Far ahead of the spot, it presumably equals the value 2.6 for an undisturbed laminar boundary layer with a zero pressure gradient. Immediately aft, however, the shape factor is about 2.3 which corresponds to that for an accelerated boundary layer in stagnation flow. These results are similar to those later measured by Mautner and Van Atta [1986] in a spot, and by Blair [1992] in the turbulent segment of a transitional boundary layer.

The streamwise variation of the skin-friction coefficient along the spot's midspan, as measured by Mautner and Van Atta, is shown in Fig. 5.9. In this figure, the abscissa is the distance from the spot source scaled by the distance traveled by a particle in the free stream during the elapsed time from spot formation. The wall shear stress was obtained from a calibrated hot-film sensor placed flush with the plate's surface, and not from the slope of a velocity profile near the wall which they found was always larger. This experiment was conducted with a mild favorable pressure gradient, which produced a laminar profile at x_m with a Falkner-Skan parameter $\beta = 0.1$, where $\beta = 2m/(1+m)$ and $m = 0.053$

(see Appendix D.2 and "wedge flows" in §4.2.1). The displacement-thickness Reynolds number and shape factor of the undisturbed flow at this location were 1,520 and 2.47.

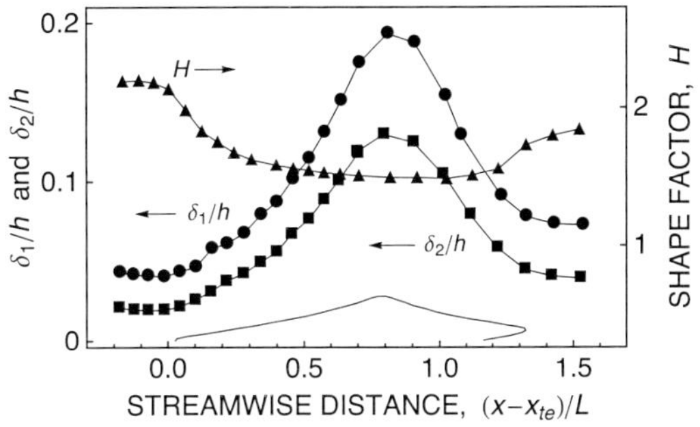

Figure 5.8.
Midspan boundary-layer parameters in a turbulent spot. Data from Wygnanski et al. [1976] with $dp/dx = 0$, $Re_{\delta 1,s} = 508$, $h = 2$mm, $x_m - x_s = 1000$mm, and $u_\infty = 9.4$m/s.

(a)

(b)

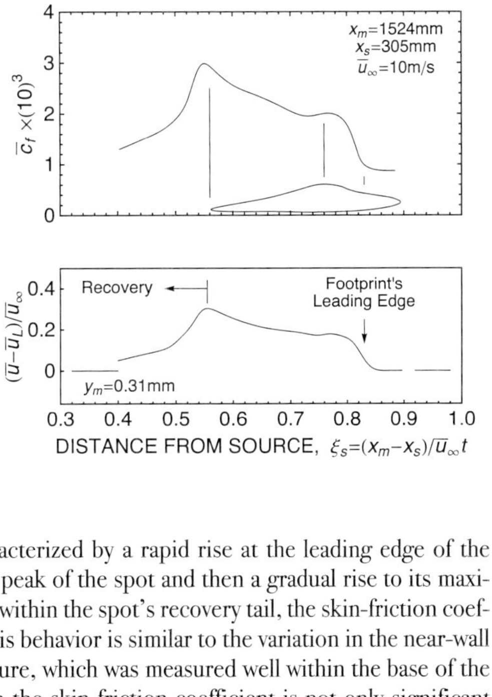

Figure 5.9a,b.
Streamwise variation of the skin-friction coefficient and near-wall velocity excess on the midspan plane of a turbulent spot. Data from Mautner and Van Atta [1986] for a free-stream flow with $\beta = 0.1$.

The variation in skin-friction coefficient is characterized by a rapid rise at the leading edge of the spot's footprint, followed by a leveling off at the peak of the spot and then a gradual rise to its maximum value at the spot's trailing edge. Afterward, within the spot's recovery tail, the skin-friction coefficient slowly relaxes to its undisturbed value. This behavior is similar to the variation in the near-wall velocity perturbation shown in part (b) of the figure, which was measured well within the base of the spot at $y = 0.031$mm. The effect of the spot on the skin-friction coefficient is not only significant within the spot, but well beyond within its recovery tail; which, as shown in Figs. 5.2 and 5.3, is about the same length as the spot itself. Hence, while the average skin-friction coefficient through the spot is about $2.5(10)^{-3}$, its average through the recovery tail is about $2.25(10)^{-3}$. In other words, the spot increases the skin-friction coefficient by at least a factor of two over a distance equal to approximately twice its length. This immediately raises a question regarding the area disturbed by the spot. Usually, as we will see in §7.1, an increase in wall-shear stress is assumed to occur only beneath the spot. Clearly, it increases beneath the recovery tail as well.

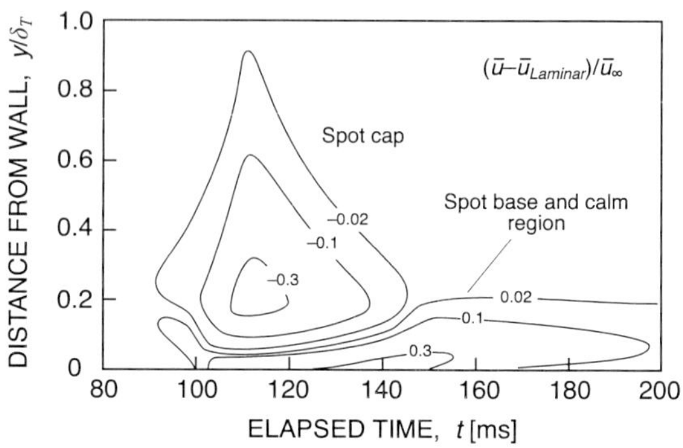

Figure 5.10.
Midspan contours of the mean streamwise velocity deficits and excesses in a turbulent spot. Data from Chong and Zhong [2005] with $dp/dx = 0$, $Re_{\delta 1,s} = 803$, $x_s = 275$mm, $x_m - x_s = 780$mm, and $u_\infty = 11$m/s.

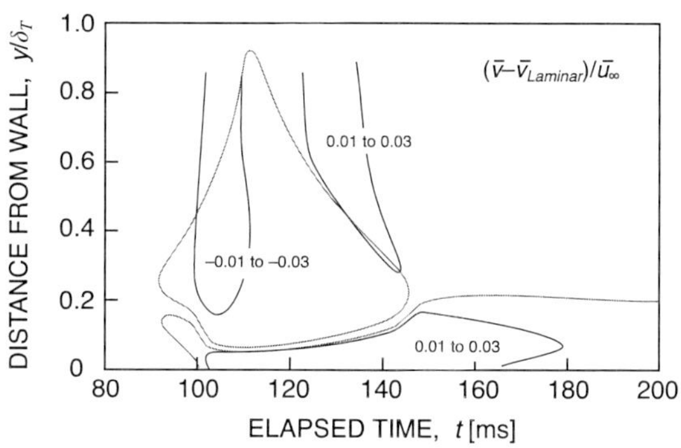

Figure 5.11.
Midspan contours of the mean normal velocity deficits and excesses in a turbulent spot. Data from Chong and Zhong [2005] with $dp/dx = 0$, $Re_{\delta 1,s} = 803$, $x_s = 275$mm, $x_m - x_s = 780$mm, and $u_\infty = 11$m/s.

Contours of the streamwise velocity perturbations on the midspan of a spot are shown in Fig. 5.10. The abscissa in this figure is the time measured since the spot was initiated. Therefore, the flow moves from right to left, and the spot appears to be moving "backwards" when compared with the spots drawn in all the previous figures. This coordinate system corresponds to that used in the laboratory when recording the temporal variation of a quantity as a spot passes over a probe placed at the distance x_m. To obtain the data in this and the following figures, a rake of hot wires placed 780mm downstream from the spark initiating the spot simultaneously recorded data at different distances from the wall. The contours in this figure are curves of constant velocity excess or deficit relative to the mean velocity of the *undisturbed* laminar boundary layer. As pointed out earlier, the mean velocity near the wall first increases then decreases as the spot passes over any given streamwise position, while farther away it decreases and then increases. That is, the velocity perturbations within the base and tail of the spot are positive as shown in the figure, while those in its cap are negative.

The corresponding contours of the normal velocity perturbations are shown in Fig. 5.11. For reference, the light gray lines are the ±0.02 contours of streamwise perturbations. This data indicates a large outward (positive) velocity near the wall which is the "vacuum cleaner" effect noted by Cantwell et al. For comparison, the maximum of these near-wall normal perturbations is an order of magnitude larger than the normal velocity at the edge of the undisturbed boundary layer. A large outward flow is also found trailing the cap while a large inward flow of about the same magnitude occurs within its forward portion. Thus, the general mean motion within the cap is outward behind the peak and inward before. Likened by many to the motion of a large counterclockwise rotating vortex, the elevation view in Fig. 5.2 and the results in Fig. 5.3 show that the motion in the cap, which would be rotating clockwise in these figures, really consists of a series of large-scale eddies with a combined average rotating motion. We also see in Fig. 5.3 that the region of large outward motion near the wall in Fig. 5.11 is associated with the small-scaled eddies within the base.

5.1.3 The Reynolds-stress field associated with a turbulent spot.

Midspan contours of the root-mean-square of the streamwise velocity fluctuations within a spot are shown in Fig. 5.12. Since its two percent contour nearly coincides with the ensemble-averaged *laminar-turbulent interface* of the spot (not shown), this contour can be used to define the spot. In the past, the two-percent contours of the mean velocity perturbation were arbitrarily used to define the spot except near the trailing edge where the contour from the cap was usually extrapolated to the wall. This is the method that was used to define the spot shape shown in Figs. 5.6 and 5.9a. As we see in Fig. 5.12, the shape of a spot defined by using these contours (shown gray) does provide a reasonable outline to the two percent root-mean-square contour and, therefore, to the ensemble-averaged interface of the spot except along its trailing edge.

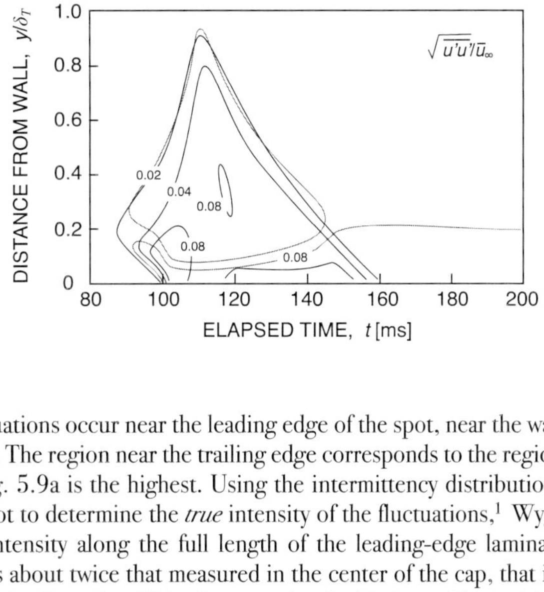

Figure 5.12.
Midspan contours of the root-mean-square streamwise velocity fluctuations in a turbulent spot. Data from Chong and Zhong [2005] with $dp/dx = 0$, $Re_{\delta 1,s} = 803$, $x_s = 275$mm, $x_m - x_s = 780$mm, and $u_\infty = 11$m/s.

In any case, the highest streamwise fluctuations occur near the leading edge of the spot, near the wall aft in its base, and in the center of its cap. The region near the trailing edge corresponds to the region where the skin-friction coefficient in Fig. 5.9a is the highest. Using the intermittency distribution associated with ensemble-averaging a spot to determine the *true* intensity of the fluctuations,[1] Wygnanski et al. [1979] showed that the intensity along the full length of the leading-edge laminar-turbulent interface *under* the overhang is about twice that measured in the center of the cap, that is, about twice that shown in the figure at the leading edge. This, they associated with the rapid transition occurring there. Although not shown herein, the intensity of the normal-velocity fluctuations are also the highest in this region and have a maximum root-mean-square value equal to about five percent of the free-stream velocity.[2]

The Reynolds shear-stress contours on the midspan of a spot are shown in Fig. 5.13. The region of highest stress coincides with that of the highest intensity near the wall at the leading edge of the spot, which again emphasizes that this is one of the most crucial regions in the spot. This is even more evi-

---

[1] We will discuss this method in §7.2.

[2] See Fig. 6e in Chong and Zhong [2005] for a corresponding plot of this stress component.

dent in the measurements obtained by Makita and Nishizawa. Directly behind this, however, the shear-stress is relatively low, presumably because much of the viscously damped fluid near the wall is "lifted" into the spot. Since some investigators use the change in Reynolds-shear stress to detect the laminar-turbulent interface of a spot, rather than the change in intensity, a comparison of the smallest-valued contours in Figs. 5.12 and 5.13 indicate that significantly different spot shapes will be obtained if proper attention isn't paid to the "threshold" of turbulence marking the change.[1]

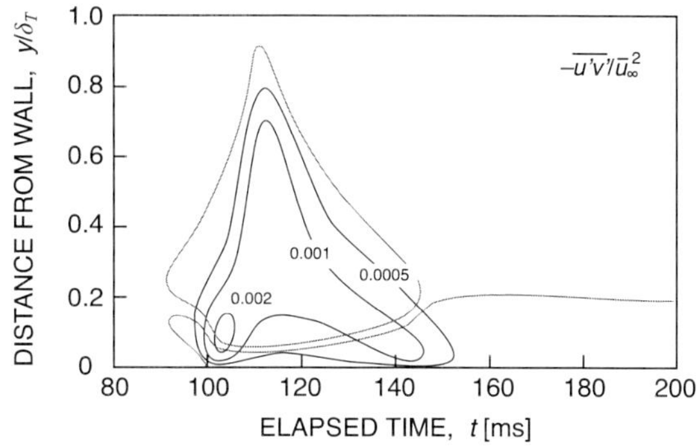

Figure 5.13.
Midspan contours of the Reynolds shear stress in a turbulent spot. Data from Chong and Zhong [2005] with $dp/dx = 0$, $Re_{\delta 1,s} = 803$, $x_s = 275$mm, $x_m - x_s = 780$mm, and $u_\infty = 11$m/s.

5.1.4 The calm behind the storm.

As previously noted, the laminar boundary layer in the recovery tail is thinner and has a smaller shape factor than that to either side in the undisturbed boundary layer. Consequently, the flow within the recovery tail is more stable, which, as discovered by Schubauer and Klebanoff [1955], both impedes turbulent spot formation and "calms" intruding turbulent flow. However, in contrast to the number of experiments conducted on turbulent spots, virtually no experiment has been conducted to determine the general shape and size of what is now commonly referred to as the "becalmed zone." Furthermore, the effects of free-stream turbulence level, pressure gradients, and spot (breakdown) Reynolds number on the geometry and kinematics of the laminar-becalmed zone are virtually unknown. In fact, since the physics associated with laminar breakdowns still remains uncertain, there is no hope to understanding the "calming" effect of the flow following a turbulent spot on spot formation even if we know everything about the flow in the region. This is particularly unsettling when we know in some cases, such as in flows with favorable pressure gradients, that spots are formed over extended streamwise distances.[2] In any event, the best we can do at this time is to estimate when the flow behind a spot again becomes susceptible to spot formation and turbulent interference.

In this regard, the measurements and observations made behind a spot by Schubauer and Klebanoff, Wygnanski et al. [1979], Gad-el-Hak et al. [1981], and Makita and Nishizawa [1998, 1999] for a zero pressure gradient, Glezer et al. [1989] and Chong and Zhong [2003, 2005] for zero and favorable pressure gradients, and Clark et al. [1994], Seifert and Wygnanski [1995], and Gostelow et al. [1997] for adverse pressure gradients are useful.

A sketch of the triangular-shaped recovery tail, defined by the trailing-edge loci of the laminar streaks observed by Gad-el-Hak et al. and the trailing edge of the spot, is shown in Fig. 5.14a. In this figure,

[1] This will be discussed further in §5.2.5.

[2] It is also important to understand the calming effect when transition occurs through a variety of modes known as *multi-mode transition*. We will investigate this mode of transition in §8.2.

the streak pattern shown in Fig. 5.2 has been scaled to the spot footprint shown in Fig. 5.6b using the spot expansion half-angle of 10°. Based on this footprint, the length of the recovery tail is $l_t \approx 1.3\, l$, which is roughly the length Schubauer and Klebanoff found was needed for the mean velocity along the mid-plane of the tail to decay to its undisturbed value. Hence its area is about $0.65\, bl$. The trailing edge half-angle θ_t in this case is about 13°. Further evidence that the flow in this region is stable to outside disturbances is the fact that the oblique wave system measured by Wygnanski et al. and Glezer et al., and shown by the wave crests in the figure, decay within it. That is, the inward extent of these waves coincide with the trailing boundaries of the recovery tail.

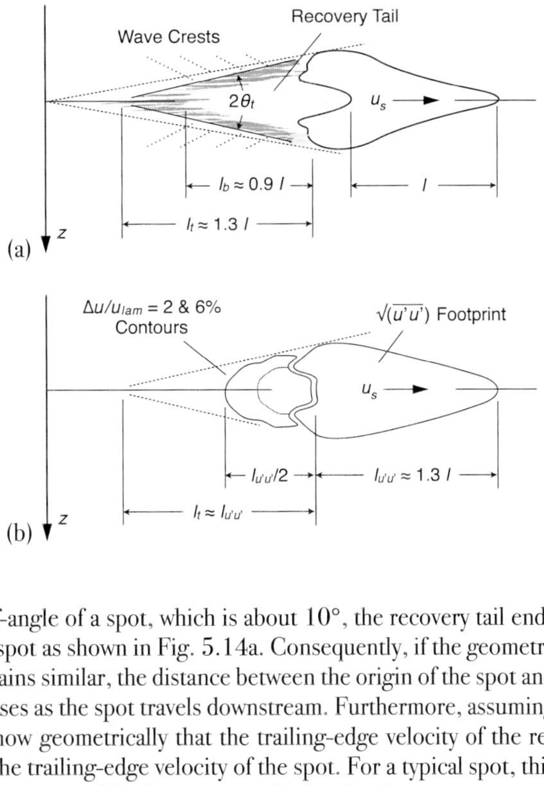

Figure 5.14a,b.
The near-wall (a) recovery tail, and (b) contours of velocity excess following a turbulent spot for $dp/dx \approx 0$. Data from (a) Wygnanski et al. [1976], Wygnanski et al. [1979], Gad-el-Hak et al. [1981], Glezer et al. [1989], and (b) Makita and Nishizawa [1998].

Since θ_t is greater than the expansion half-angle of a spot, which is about 10°, the recovery tail ends downstream from the virtual origin of the spot as shown in Fig. 5.14a. Consequently, if the geometry of the spot and recovery tail ensemble remains similar, the distance between the origin of the spot and the trailing edge of the recovery tail increases as the spot travels downstream. Furthermore, assuming similarity, it's rather straightforward to show geometrically that the trailing-edge velocity of the recovery tail is about $(1-\tan\alpha/\tan\theta_t)$ times the trailing-edge velocity of the spot. For a typical spot, this yields a velocity $u_{t,te}$ equal to about thirteen percent of the free-stream velocity; that is,

$$u_{t,te} \approx 0.13\bar{u}_\infty \; ; \quad (dp/dx = 0)\,^{1} \tag{5.1}$$

If the instabilities in the laminar boundary layer are two-dimensional Tollmien-Schlichting waves, they travel downstream at a speed that depends on their frequency, the boundary-layer Reynolds number and pressure gradient. For zero pressure gradient, instabilities at the critical frequency travel at about forty percent of the free-stream velocity (see §4.2.1), while lower frequency instabilities travel slower. Since the trailing edge of the spot travels faster than this, waves *downstream* of the spot's origin will be "scrubbed clean" by the spot, while waves *upstream* of the origin will never catch up. Consequently, a region completely devoid of waves will develop directly behind the spot that is at

[1] This and the following estimates are based on the spot trailing- and leading-edge velocities equal to fifty-five and ninety percent that of the free stream.

least $(u_{s,te}-c_{r,crit})l/(u_{le}-u_{te})_s$, or roughly $0.45l$ long (presuming that the celerity of the unstable flow doesn't change). Furthermore, since the wave speed is greater than the trailing-edge velocity of the recovery tail $u_{te,t}$, the waves upstream of the origin must decay completely as they propagate into the tail by the time they reach this position. This behavior can be detected in the upper traces of Fig. 9 in Schubauer and Klebanoff [1955].

In addition, since the leading edge of a spot travels at almost twice the speed of its trailing edge, a spot will always catch up with the one before it. Knowing the propagation speeds and expansion rate (see §5.3), it is also possible to predict when they will merge. As previously mentioned, Schubauer and Klebanoff found that a spanwise turbulent "strip" entering the recovery tail from behind "propagates" only so far into the tail. In particular, they found that the highly stable flow within the tail prevents turbulence from being generated directly in front of the strip, thus reducing its downstream rate of growth. From a few, rather clever set of tests on an unstable laminar boundary layer with zero pressure gradient, they also determined that the propagation speed of a turbulent *front* into the recovery tail is roughly one quarter that of the free-stream velocity. Because all of their measurements were obtained along the mid plane of the spot and no other measurements of this kind have since been reported, we can only presume that this result applies across the full span of the recovery tail at that location. If this is the case, then the length of the region completely devoid of turbulence directly behind the spot is roughly equal to $0.9\,l$ as shown in part (a) of Fig. 5.14.

Accordingly, it appears that the length of the *becalmed zone*, defined as a region following a turbulent spot that is devoid of both turbulence and laminar instabilities, is about 0.45 to 0.9 times the length of the spot for a zero pressure gradient flow, and that the speed of its trailing edge is between 25 and 40 percent of the free-stream velocity; that is,

$$0.45 \le l_b/l \le 0.9 \quad \text{and} \quad 0.25 \le u_{b,te}/u_\infty \le 0.4 \; ; \qquad (dp/dx = 0) \tag{5.2}$$

The ensemble-averaged mean-velocity-perturbation contours measured by Makita and Nishizawa [1998] *following* a spot are shown in part (b) of Fig. 5.14.[1] The spot footprint in this figure corresponds to their two-percent root-mean-square velocity-fluctuation contour near the surface scaled linearly to the footprint shown in part (a). These two footprints differ from one another not only because Wygnanski et al. used the two-percent mean-velocity perturbation contours to determine its shape, but also because they extrapolated data from the cap of the spot to the surface to obtain the trailing-edge contour of their footprint (see §5.1.3 and Fig. 5.12). In any case, the two and six percent perturbation contours shown following the spot in part (b) are seen to lie well within the recovery tail or maximum extent of the becalmed zone. If the spot/recovery-tail ensemble expands linearly as it propagates downstream, then the celerity of the trailing edge for these contours are about 0.32 and 0.38 times the free-stream velocity respectively, which in turn lies within the range just cited for the trailing-edge celerity of the becalmed zone.

With no other information, nor any justification other than its trailing-edge velocity should be close to the celerity of a turbulent front entering the zone, our best guess for the size and shape of the becalmed zone roughly corresponds to the two-percent contour shown in part (b) of Fig. 5.14. If this is the case, its length and breadth is about equal to about eighty percent of the breadth of the spot b, and

[1] The author obtained the footprints in this figure by averaging the data given in a zt-plane for two spots, assuming that the footprints expand linearly in xzt-space, and transforming them accordingly into the physical xz-plane. The transformation is briefly discussed in §5.1.2 and more fully in §5.3. The actual zt-data for the two spots are shown individually in Fig. 5.49b by the dashed and dot-dashed lines. The spot footprint shown in Fig. 5.14b was obtained by taking symmetrical averages of the zt-data for the two-percent velocity-fluctuation contours associated with the two spots. In other words, the footprint is an average of four velocity-fluctuation contours. The mean-velocity-perturbation contours in the tail were similarly obtained by taking the symmetrical averages of zt-data for the two- and six- percent velocity-perturbation contours associated with the two spots and their tails, and "subtracting" the intersections of them with the velocity-fluctuation contour of the "average" spot.

the included angle between its sides is $2\theta_t \approx 2\alpha$. Accordingly, the area of the becalmed zone is then about $A_b \approx 0.64\, b^2(1 - \tan\alpha)$, which is roughly $5b\,(1 - \tan\alpha)/4l$ times the area of the spot. That is, the size of the becalmed zone's footprint in a zero pressure-gradient flow (based on the two percent velocity-excess contour) is approximately given by

$$l_b \approx b_{b,le} \approx 0.8b,\ \ b_{b,te} \approx 0.8b\left(1-2\tan\alpha\right),\ \text{ and }\ A_b \approx 0.64b^2\left(1-\tan\alpha\right) \tag{5.3}$$

where the nomenclature should be evident. For a typical spot with $\alpha \approx 10°$, and an included leading-edge half-angle of $\theta \approx 15°$, the becalmed/spot area ratio is roughly one quarter.

Recovery-tail flow theory. As it turns out, a very simple theory can be developed for modeling the flow in the recovery tail. Denote the excess in mean velocity at the trailing edge of the spot by $\Delta\bar{u}_0$. Since the flow within the tail is viscously dominated, suppose that the excess velocity simply diffuses and decays with time after the passing of the spot's trailing edge according to the diffusion equation,

$$\frac{\partial \Delta\bar{u}}{\partial t} = \nu \frac{\partial^2 \Delta\bar{u}}{\partial y^2}$$

At $t = 0$, the excess-velocity profile equals the difference between the undisturbed and spot trailing-edge mean-velocity profile. The boundary conditions are that the excess velocity at the surface and its derivative in the free stream must equal zero. The solution to this equation, using the aft velocity profile measured by Wygnanski et al. [1976] and the undisturbed laminar profile for the same location, is shown in Fig. 5.15. For reference, the mean-velocity profile aft of the spot is also plotted. In this case, the maximum value of $\Delta\bar{u}_0$ at $t = 0$ is about one-half the free-stream velocity and it occurs slightly below the distance $y/h = 0.1$ from the wall. As before, h and l denote the midspan height and length of the spot. Compared with the results of Chong and Zhong's in Fig. 5.10, even though their experimental conditions are somewhat different, the calculated velocity-excess contours are remarkably similar.[1] In particular, the streamwise distance between the trailing edges of the 0.1 and 0.3 contours are roughly the same.

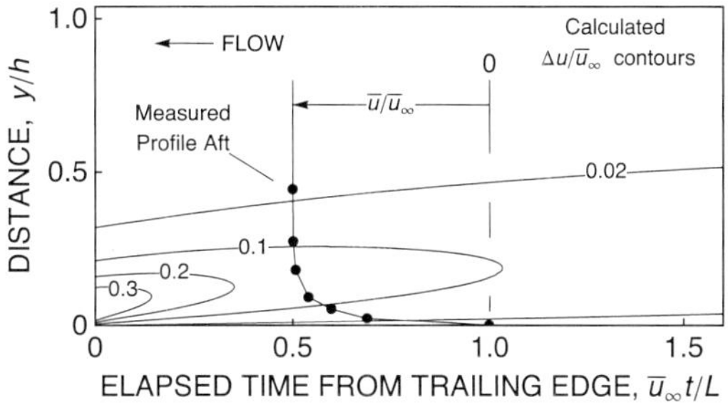

Figure 5.15.
Calculated contours of the mean streamwise velocity excess and measured velocity profile in the recovery tail following a turbulent spot. Data from Wygnanski et al. [1976].

5.2 Emmons' Turbulent Spot Theory.

A stochastic theory of transition. A two- and three-dimensional turbulent spot theory.

In his paper on turbulent spots, Emmons [1951] also presented a stochastic spot theory. The beauty of this theory is that he managed without too much fuss to reduce a difficult fluid mechanics problem to one of geometry and probability. The mechanics of transition he "tucked" into his spot *production*

[1] The difference can be attributed in part to the relatively strong spanwise diffusion of the mean flow within the recovery tail that is not considered in the simple model (see Chong and Zhong's Fig. 6e–h).

and *propagation* parameters, which characterize the formation and development of a *single* spot.[1] As a result, if we embrace his theory, we only need to examine (either theoretically or experimentally) the behavior of a single turbulent spot. This unprecedented and powerful reduction of the transitional flow problem is known as *Emmons' turbulent-spot theory*. We will now examine his theory and its consequence for various types of spot production distributions.

5.2.1 A stochastic theory of transition.

Propagation and dependence cones. Emmons' production function. Emmons' intermittency factor. Spot probability density function and passing frequency.

Consider transition on a flat plate aligned parallel to the incident flow. As before, consider a rectangular coordinate system such that the plate lies in the $y = 0$ plane with x measured downstream along the plate from its leading edge and z measured along its leading edge. Mindful of the spot's three-dimensionality, we will initially follow Emmons' analysis by considering only the surface area covered by the spot. That is, we will consider the flow turbulent at any point on the surface within the spot's footprint, otherwise it is laminar. By comparing the contours in Fig. 5.6b, we see the cross-sectional area of the spot is nearly the same as that at the surface up to the height of the bulge. Therefore, as found by many experimentalists, this theory will be valid up to the edge of the laminar boundary layer.[2]

Following Emmons, consider only one of the many spots within the transition region and suppose that it expands and propagates independently of the others, even when it merges with (that is, overlaps) one of them.[3] Thus, the basic physical element in his theory is the turbulent spot, and the basic physical process is its evolution.

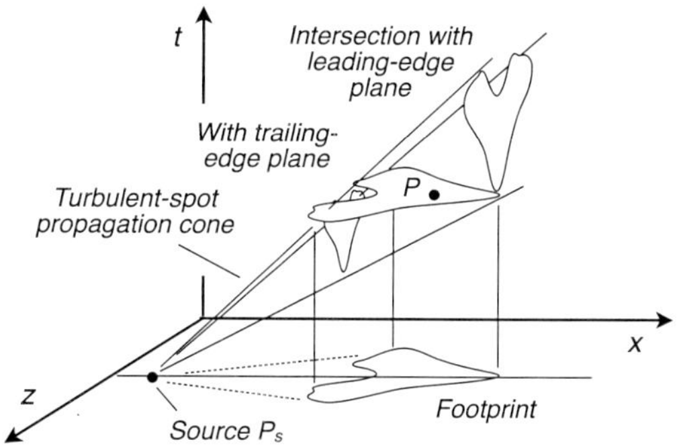

Figure 5.16.
The propagation cone for a turbulent spot produced at the point $P_s(x, z, 0)$.

Its evolution, or more precisely, the evolution of its footprint in xzt-space is shown in Fig. 5.16. In this space the footprint sweeps out a conical volume with a vertex at its point of origin P_s. Lines traced from this point to the circumference of the footprint generate the cone's surface. The cone and its vertex are respectively known as the spot's *propagation cone* and *source*. Clearly, the shape of the cone depends on the rates at which the footprint expands and propagates along the surface. If the

---

[1] The catch of course, as we will see, is that the fluid mechanics of the spot must eventually be addressed.

[2] Note that Emmons' turbulent spot theory from a three-dimensional viewpoint is a theory for a turbulent "right-cylinder" with the same cross-section as the spot's footprint.

[3] As pointed out in the introduction of §5.1, this is only true before separate spots become close enough to interact and merge. Therefore, we should expect that this theory will be most appropriate near the beginning of transition where most of the spots are separated and least appropriate near the "end" of transition where they merge.

footprint expands linearly with distance and propagates at a constant velocity, its generators will be straight lines as shown in the figure, and all parallel sections of the cone will be similar.

From geometry, if P lies within a propagation cone emanating from the upstream point P_s, then P will *sometime* be turbulent. The volume containing all such points *upstream* of P is the image of a propagation cone emanating from P projected upstream through a focal point at P.[1] Shown in Fig. 5.17, this cone is known as the spot's *dependence cone*. If only one spot is produced and it is produced at a point P' within this cone, then the flow at point P will be turbulent at least for the duration it takes the spot to pass over it – otherwise the flow at P will be laminar. Therefore, the probability that the flow at P is sometime turbulent due to a spot produced at P' is

$$\phi(P;P') = \begin{cases} 0; & P' \notin R \\ 1; & P' \in R \end{cases} \tag{5.4}$$

where R is the volume of the dependence cone.

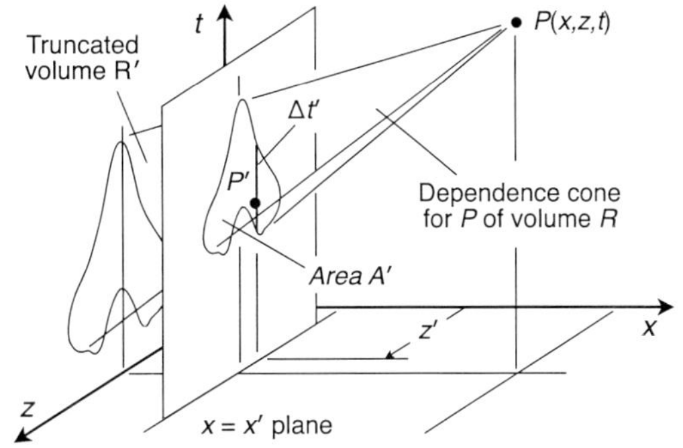

Figure 5.17.
The dependence cone for a point P downstream of sources in the $x = x'$ plane.

Let's suppose that the number of spots produced at P' per unit surface area per unit time, called the "spot production rate," is $g(x', z', t')$ and ask what is the probability that the flow somewhere downstream at the point $P(x, z, t)$ is turbulent? If $\Delta V' = \Delta x'\Delta z'\Delta t'$ is the element of volume surrounding the point P', then the probability of a spot being produced at P' is $g\Delta V'$. This follows by realizing that if there are n spots (items) randomly produced (placed) in $N > n$ volume elements (boxes), then the probability of a spot produced (an item found) in any element (box) is n/N. This in turn equals $(n/V)\Delta V = g\Delta V$ where V is the sum of all elemental volumes ΔV. Hence the probability that the flow at P is turbulent becomes

$$\Phi(P;P') = \phi(P;P')g(P')\Delta V' \tag{5.5}$$

We cannot determine the probability that the flow is turbulent at P for any number of sources by simply adding up all their contributions, however, since this might include simultaneous contributions (overlapping spots). In other words, the events leading to the flow being turbulent at P are *not* mutually exclusive. But given two events that are not mutually exclusive, the probability that at least one event occurs can be obtained by adding the probability that each occurs, and then subtracting the probability that both occur.[2] That is, the probability or fraction of time the flow is turbulent at P as a result of turbulent spots being produced only at P_1 and P_2, say, is

---

[1] This is true whether the spot expands linearly or not.

[2] The development from this point on follows Steketee's [1955] probability analysis.

$$\gamma(P)=\Phi(P;P_1)+\Phi(P;P_2)-\Phi(P;P_1)\Phi(P;P_2)=1-\prod_{i=1}^{2}\left[1-\Phi(P;P_i)\right] \tag{5.6}$$

where, following convention, γ is called the "intermittency factor" or simply the "intermittency." Defined as the fraction of time the flow at any position $P(x, z, t)$ is turbulent, it is the same quantity previously introduced in §3.2.2 regarding the intermittent behavior in the outer region of a turbulent boundary layer. As we will see in §7.2, it can be determined from an instantaneous hot-wire signal such as shown in Fig. 5.1. Nevertheless, varying between zero and one, the flow at P is *always* laminar (non-turbulent) when the intermittency equals zero and *always* turbulent when it equals one.

The above expression for two spots (events) is easily extended to include turbulent spots being produced at any number of points, say N, by simply increasing the upper limit of the product in Eq. (5.6). Then the probability for turbulent flow at point P caused by spots being produced at P_i is

$$\gamma(P)=1-\prod_{i=1}^{N}\left(1-\Phi_i\right) \tag{5.7}$$

where the less cumbersome notation $\Phi_i = \Phi(P; P_i)$ has been introduced. Since $\Phi_i < 1$, the product may be rewritten as

$$\prod_{i=1}^{N}\left(1-\Phi_i\right)=\exp\left\{\sum_{i=1}^{N}\ln\left(1-\Phi_i\right)\right\}=\exp\left\{-\sum_{i=1}^{N}\sum_{m=1}^{\infty}\Phi_i^m/m\right\} \tag{5.8}$$

where the logarithm has been expanded in a power series. Substituting this into Eq. (5.7) and eliminating Φ_i using Eq. (5.5) yields

$$\gamma(P)=1-\exp\left\{-\sum_{i=1}^{N}\sum_{m=1}^{\infty}\left[\phi(P;P_i)g(P_i)\Delta V_i\right]^m/m\right\}$$

This in turn may be evaluated by taking the limit as ΔV_i becomes infinitesimal such that the number of elements increases indefinitely according to $N = V/\Delta V$, where V is the volume in xzt-space. Then, since all but the first term in the series for m may be neglected, and since the summation in i becomes a sum of all infinitesimal elements within the volume V, we obtain

$$\gamma(P)=1-\exp\left\{-\int_V\phi(P;P')g(P')dV'\right\}$$

for an arbitrary distribution of sources at P' in xzt-space. Finally, since ϕ equals zero unless P' is in volume R of the dependence cone, where according to Eq. (5.4) it then equals one, the above expression reduces to

$$\gamma(P)=1-\exp\left\{-\int_R g(P')dV'\right\} \tag{5.9}$$

which is identical to the expression first obtained by Emmons.

As should be expected, the probability lies between the values of zero and one inclusively, and increases with increasing turbulent spot production. Note that most of the fluid mechanics remains buried in the production function g and the dependence volume R, which describe the formation and subsequent motion of a single turbulent spot. We should also note that Eq. (5.9) is valid for any number of space-time dimensions even though it was developed for a "two-dimensional" spot. For a three-dimensional spot, g becomes the number of spots produced per unit volume per unit time and $dV' = dx'dy'dz'dt'$. Of course, the dependence cone in this case cannot be visualized (except perhaps by some "very interesting" beings).

The *probability density function* is defined as the probability the flow at P is turbulent because of sources within the elemental volume dV_m centered about the point P_m. Denote this probability function by $\psi(P, P_m)$. Then

$$\psi(P;P_m)=\frac{\Phi_{m+1}-\Phi_m}{\Delta V_{m+1}}=\frac{\Phi_{m+1}}{\Delta V_{m+1}}\prod_{i=1}^{m}\left(1-\Phi_i\right)$$

where i includes all events occurring up to and including m. Substituting Eqs. (5.4) and (5.8) into this expression, we obtain

$$\psi(P;P_{m+1})=\phi(P;P_{m+1})g(P_{m+1})\exp\left\{-\sum_{i=1}^{m}\sum_{n=1}^{\infty}\left[\phi(P;P_i)g(P_i)\Delta V_i\right]^n/n\right\}$$

which upon taking the limit for indefinitely small ΔV provides

$$\psi(P;P')=\phi(P;P')g(P')\exp\left\{-\int_{R'}g(P')dV'\right\} \tag{5.10}$$

where the volume R' is the portion of the dependence volume upstream of the point P'. With this, the frequency and average duration of the turbulent spots passing over the point P can be determined.

While measurements taken near the wall at any point within transition would show intermittent segments of turbulent and laminar flow, spots passing over that point may not be separated by laminar segments because they have merged with one another (overlapped). If we consider a sampling period equal to T, then point P will be turbulent for a time equal to $\psi(P;P')TdV'$ because of the spots produced in the volume dV'. As shown in Fig. 5.17, each spot from P' will cause P to be turbulent for an interval of time equal to $\Delta t\,(P;P')$, where Δt is the temporal chord of the dependence cone corresponding to the position (x', z'). It will also be seen to be the duration of a spot passing over P that was produced at P' when considering the propagation cone. Then the number of spots passing P from spots produced in dV' is $\psi(P;P')TdV'/\Delta t(P;P')$, and the number of spots passing per unit time becomes $\psi(P;P')\,dV'/\Delta t\,(P;P')$. Consequently, the total number of spots per unit time passing over the point P, or the spot *passing frequency* is

$$\omega_s(P)=\int_V\frac{\psi(P;P')}{\Delta t(P;P')}dV'=\int_R\frac{g(P')e^{-\int_{R'}g(P'')dV''}}{\Delta t(P;P')}dV' \tag{5.11}$$

Since γ is the fraction of time the flow is turbulent at P, the *average duration* of a turbulent spot at P is given by $\overline{\Delta t_s}=\gamma(P)/\omega_s(P)$.

5.2.2 The classical two-dimensional solution.

A line source of spots. Narasimha's production function. Emmons' propagation parameter.

The classical theory considers a two-dimensional boundary layer with a constant time-averaged, free-stream velocity, and assumes that the production of turbulent spots is both *stationary* in time and *homogeneous* in the transverse direction across the surface. The latter simply guarantees a stationary, two-dimensional transition. Supposing that the turbulent spots propagate in the x direction along the $y=0$ surface, the dependence cone R appears as drawn in Fig. 5.17, and Eq. (5.9) reduces to

$$\gamma(x)=1-\exp\left\{-\int_R g(P')dx'dz'dt'\right\} \tag{5.12}$$

A line source of spots. Spurred by his experimental observations of spots being produced over much of the laminar portion of a surface, Emmons [1951] assumed g was constant (see §5.4.3). Subsequent experiments by Narasimha [1957], however, indicated that spots are produced in a relatively narrow band stretching across the surface. This and the fact that the intermittency distributions for different production distributions are all nearly similar prompted him to consider g to be a *line source* of spots. In reality, the truth is somewhere between Emmons' and Narasimha's observations with the width of the band depending on the free-stream pressure gradient, turbulence level, and other factors.

Nonetheless, in this section, we will first consider that all spots are produced at one streamwise position, say at $x = x_s$, which is Narasimha's model. There are several advantages for doing this; namely, the solution for the intermittency is easy to obtain, analytical, and can be used to construct the solutions for any other two-dimensional source distribution. Furthermore, since all spots are produced at one position, we don't have to consider the effect of the becalmed zones on their *production*.[1]

The production rate for such a source may be expressed as

$$g(x')=\dot{n}'\delta(x'-x_s) \tag{5.13}$$

where $\dot{n}'$ is the number of turbulent spots produced per unit time per unit span and $\delta(z)$ is the Dirac delta function.[2] Then Eq. (5.12) becomes

$$\gamma(x)=1-\exp\left\{-\dot{n}'\int_{x'=0}^{x}\delta(x'-x_s)\,dx'\int_{A'}dz'\,dt'\right\} \tag{5.14}$$

where A' is the cross-sectional area of the dependence cone in the x' equal-constant plane (see Fig. 5.17).

If we assume, as initially observed by Emmons, that spots in a flow without a pressure gradient expand linearly in both the streamwise and spanwise directions, and travel downstream at a constant rate, then area A' will be directly proportional to $(x-x')^2/\bar{u}_\infty$ such that

$$A'(x)=\int_{A'}dz'\,dt'=\frac{\sigma}{\bar{u}_\infty}\left(x-x'\right)^2 \tag{5.15}$$

where σ, known as *Emmons' propagation parameter,* is a dimensionless constant of proportionality that depends on the shape and velocity of the spot. Narasimha [1985], using the measurements of Schubauer and Klebanoff [1955], determined that σ has a value of about 0.27. We will investigate this further in §5.5.1. Nevertheless, substituting the above expression into Eq. (5.14), assuming σ is constant and integrating yields

$$\gamma_0(x)=1-\exp\left[-\frac{\dot{n}'\sigma}{\bar{u}_\infty}\left(x-x_t\right)^2\right]; \qquad (x>x_t,\ \text{otherwise } \gamma_0=0) \tag{5.16}$$

where the subscript zero has been applied to the intermittency as a reminder that it corresponds to the "near-wall" solution, and where x_s has been replaced by x_t which, in this case, is the beginning of transition. Clearly, the intermittency increases monotonically from zero to unity through transition as the distance from the onset of transition increases. This expression was first obtained by Narasimha [1957] and has since become the basic equation for the intermittency distribution in a two-dimensional transitional boundary layer.

Introducing a dimensionless *spot production parameter*, defined by

$$\hat{n}=\dot{n}'\nu^2/\bar{u}_\infty^3 \tag{5.17}$$

and the distance Reynolds number, one obtains

---

[1] The effect of becalmed zones interfering with turbulent spot production will be examined in §5.4, where spot and becalmed flow interaction and the appropriate modification to Emmons' theory will be considered.

[2] The Dirac delta function is defined as $\delta(z-z_0) = 0$ for $|z-z_0| > \varepsilon$, undefined at $z = z_0$, satisfies $\int_{z_0-\varepsilon}^{z_0+\varepsilon}\delta\left(z-z_0\right)dz=1$ as $\varepsilon \to 0$, and has the property that $\int_{-\infty}^{+\infty} f(z+z_0)\delta(z)\,dz = f(z_0)$ where f is any continuous function at $z = z_0$.

$$\gamma_0(x) = 1 - \exp\left[-\hat{n}\sigma\left(Re_x - Re_{xt}\right)^2\right]; \quad (Re_x \geq Re_{xt}) \tag{5.18}$$

where Re_{xt} refers to the Reynolds number at the onset of transition, and γ_0 equals zero before onset. For $dp/dx = 0$, the transition Reynolds number can vary from about $4(10)^4$ to $4(10)^6$ and $\hat{n}$ is of the order $(10)^{-11}$. If we suppose that transition "ends" when $\gamma_0 = 0.99$, then the length of transition for flows without a pressure gradient is given by

$$Re_{LT} \approx 2.15/\sqrt{\hat{n}\sigma} \tag{5.19}$$

where the quantity $\hat{n}\sigma$ will henceforth be referred to as the "turbulent spot parameter." Clearly, as either the spot propagation or production rate increases, the length of transition decreases.

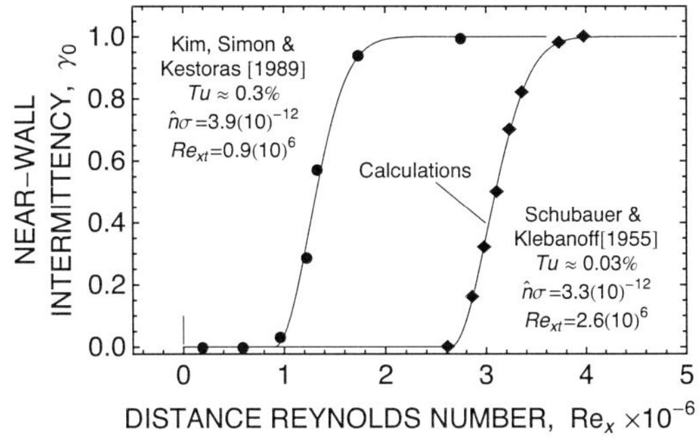

Figure 5.18.
Variation of the near-wall intermittency with distance Reynolds number for $dp/dx = 0$ and low free-stream turbulence.

A comparison of Eq. (5.18) with the near-wall data of Schubauer and Klebanoff [1955] and Kim et al. [1989] is shown in Fig. 5.18. The intermittency in both cases was determined from hot-wire measurements. The values of $\hat{n}\sigma$ and Re_{xt} reported in the figure were obtained by fitting Eq. (5.18) to the data as will be described in §6.1.2. The agreement is excellent.

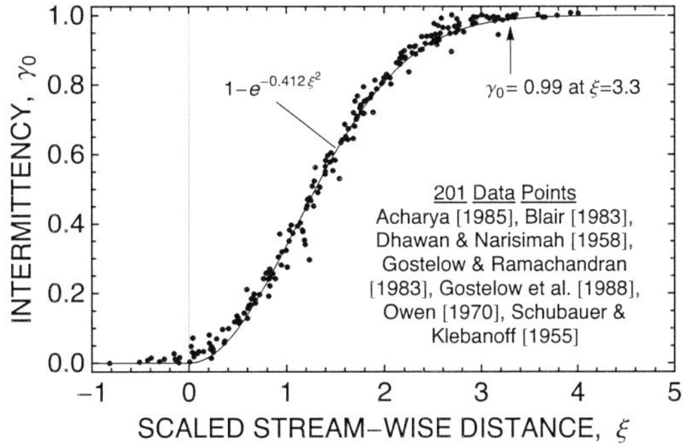

Figure 5.19.
Plot of the near-wall intermittency using Narasimha's similarity variable. Data are for tests with $dp/dx = 0$ and various free-stream turbulence levels.

A more comprehensive comparison with data is provided in Fig. 5.19. In this figure, the abscissa is Narasimha's similarity variable

$$\xi = \frac{x - x_t}{x_{0.75} - x_{0.25}} \tag{5.20}$$

where $x_{0.25}$ and $x_{0.75}$ refer to the distances where γ_0 equals 0.25 and 0.75 respectively. Thus, $\xi_{0.75} - \xi_{0.25} = 1$. Introducing ξ and evaluating the quantity $(x_{0.75} - x_{0.25})$, Eq. (5.18) transforms to

$$\gamma_0(\xi) = 1 - \exp\left(-0.412\xi^2\right) \tag{5.21}$$

Again, if we suppose that transition ends when $\gamma_0 = 0.99$, then the length of transition is given by $\xi_{LT} = 3.34$, consequently Re_{x75}–$Re_{x25} = 0.30 Re_{LT}$. Nevertheless, plotted in this format, all data appear to collapse onto Narasimha's curve within the region $0.2 \le \gamma_0 \le 0.9$. For $\gamma_0 \le 0.2$, the data is probably higher because of errors in measuring the small values of intermittency and spots forming before $\xi = 0$. For $\gamma_0 \ge 0.9$, the data is probably higher because of errors in measuring the high values and incorrectly modeling the interactions between merging turbulent spots.

Spot passing frequency and duration. The probability density function, obtained from Eqs. (5.10), (5.13) and (5.14), is given by

$$\psi(P;P') = \phi(P;P')\dot{n}'\delta(x'-x_t)\exp\left\{-\frac{\dot{n}'\sigma}{\bar{u}_\infty}\left[x^2 - \left(x-x'\right)^2\right]\right\} \tag{5.22}$$

where, according to Eq. (5.4), ϕ equals one if P' is in R and equals zero if P' is not. Since the Dirac delta function is nonzero only when $x' = x_t$, ψ is clearly nonzero only then, and furthermore only when P' is in the dependence cone.

Substituting this expression into Eq. (5.11), the spot passing frequency is given by

$$\omega_s(x) = \dot{n}'\int_R \delta(x'-x_t)\exp\left\{-\frac{\dot{n}'\sigma}{\bar{u}_\infty}\left[x^2 - \left(x-x'\right)^2\right]\right\}\frac{1}{\Delta t}dx'\,dz'\,dt'$$

The integral over z' and t' may be evaluated using Eq. (5.15) but the integral with respect to x' depends on Δt. In §5.5.1,[1] we will show that the time during which point P is turbulent due to a spot produced at P' is given by

$$\Delta t(P;P') = \frac{\sigma}{\tan\alpha}\left(\frac{x-x'}{\bar{u}_\infty}\right)$$

where α is one-half the spot's propagation angle (see Fig. 5.4). Consequently, the spot passing frequency is

$$\omega_s(x) = \dot{n}'\tan\alpha\left(x-x_t\right)\exp\left\{-\frac{\dot{n}'\sigma}{\bar{u}_\infty}x_t\left(2x-x_t\right)\right\} \tag{5.23}$$

Expressed as a dimensionless *spot-passing frequency parameter* Ω_s, this result may be written as

$$\Omega_s = \frac{\omega_s \nu}{\bar{u}_\infty^2} = \hat{n}\tan\alpha\left(Re_x - Re_{xt}\right)\exp\left\{-\hat{n}\sigma Re_{xt}\left(2Re_x - Re_{xt}\right)\right\} \tag{5.24}$$

after introducing various other dimensionless quantities. Hence, the dimensionless average duration of the spot is

$$\overline{\Delta\tau_s} = \bar{u}_\infty^2\overline{\Delta t_s}/\nu = \gamma_0/\Omega_s$$

These expressions are plotted in Fig. 5.20 for $Re_{xt} = 100{,}000$, $\hat{n} = 5(10)^{-10}$, $\sigma = 0.2$, and $\alpha = 10°$ (values which roughly correspond to transition with a free-stream turbulence level of three percent) together with the corresponding intermittency distribution. In general, as the intermittency in-

[1] See Eqs. (5.103) and (5.104) with $f_s = 1$.

creases, the spot passing frequency first increases and then decreases while the average spot duration continues to increase. The maximum spot passing frequency occurs where

$$Re_x - Re_{xt} = \left(2\hat{n}\sigma Re_{xt}\right)^{-1}$$

at which position its value is

$$\Omega_{s,\max} = \frac{\tan\alpha}{2\sigma Re_{xt}}\exp\left[-\left(1+\hat{n}\sigma Re_{xt}^2\right)\right] \tag{5.25}$$

In this case, the maximum spot passing frequency occurs at $Re_x – Re_{xt} = 50{,}000$, or almost one-quarter of the distance through transition, and has the value of about $6(10)^{-7}$. Looking back at Table 4.2, this is about three orders of magnitude smaller than the frequency of the most-effective turbulent eddy causing transition. On the other hand, $\Omega_{s,\max} Re_{\delta,t}^2 = 1.6$ which is a moderate-frequency boundary-layer occurrence.

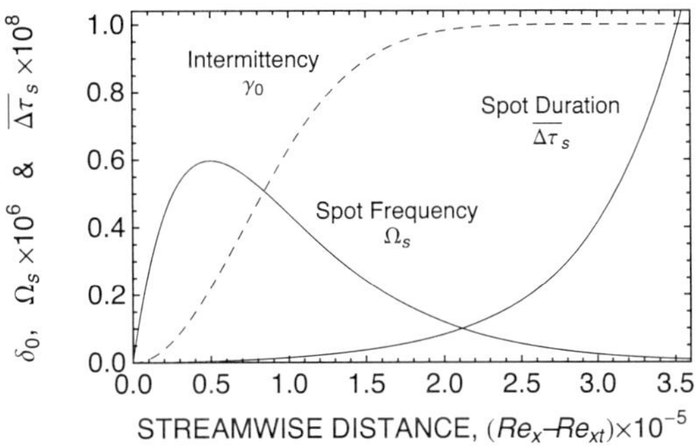

Figure 5.20.
Distributions of the intermittency factor and spot passing frequency plotted as functions of the distance from transition.

5.2.3 Distributed line sources of spots.

An arbitrary distribution of sources. A Gaussian distribution.

Let's now consider a distribution of line sources over the surface. In this case, since spots can be produced downstream from other spots, the effect of becalmed zones on their production must be taken into account, which in turn requires us to modify Emmons' theory because this flow now contains three, instead of two, separate regions (turbulent, laminar becalmed, and laminar non-becalmed). However, as we will see in §5.4, where a transition theory including the effect of becalmed flow is developed, the solution to the distributed source problem *without* considering the becalmed flow is not only a good first approximation to the solution *with* becalmed flow, particularly near the beginning of transition, but is also required to solve the problem.

Therefore, in this section, we will develop a general expression for the intermittency with an arbitrary distribution of line sources without considering the becalmed zones and then as an example provide its solution for a Gaussian distribution of these sources.

An arbitrary distribution of sources. The intermittency for any distribution of line sources stationary in time and homogeneous in z can be obtained directly from Eq. (5.16) by first replacing x_t with x_s. Then an equivalent way of writing it is

$$\gamma_0(x) = 1 - \exp\left[-\frac{\dot{n}'\sigma}{\bar{u}_\infty}\left(x - x_s\right)^2 \mathrm{U}(x - x_s, x_s)\right] \tag{5.26}$$

where $\mathrm{U}(x_1, x_2, \ldots, x_i)$ is the multidimensional unit step function which equals one only when *none* of its arguments are negative; otherwise it equals zero.

Consequently, the probability that the flow near the wall is *laminar* when a line source of strength $\dot{n}'_i$ exists at $x_{s,i}$ is

$$1-\gamma_{0,i}=\exp\left[-\frac{\dot{n}'_i\sigma}{\bar{u}_\infty}\left(x-x_{s,i}\right)^2 \mathrm{U}(x-x_{s,i},x_{s,i})\right]$$

presuming that Emmons' propagation parameter remains independent of the source's position. Then, according to Eq. (5.7), the probability that the flow is laminar for any number of discrete line sources is

$$1-\gamma_0=\prod_{i=1}^{N}\left(1-\gamma_0\right)_i=\exp\left\{-\frac{\sigma}{\bar{u}_\infty}\sum_{i=1}^{N}\dot{n}'_i\left(x-x_{s,i}\right)^2 \mathrm{U}(x-x_{s,i},x_{s,i})\right\}$$

and the probability that the flow is turbulent due to these sources is

$$\gamma_0=1-\exp\left\{-\frac{\sigma}{\bar{u}_\infty}\sum_{i=1}^{N}\dot{n}'_i\left(x-x_{s,i}\right)^2 \mathrm{U}(x-x_{s,i},x_{s,i})\right\}$$

Hence, the intermittency for a *continuous* distribution of line sources on the surface from $\xi = 0$ to x_1 is

$$\gamma_0(x;x_1)=1-\exp\left\{-\frac{\sigma}{\bar{u}_\infty}\int_0^{x_1}\dot{n}''(\xi)\left(x-\xi\right)^2 \mathrm{U}(x-\xi,\xi)\,d\xi\right\};\quad (\sigma \text{ constant}) \tag{5.27}$$

where $\dot{n}''(\xi)$ is the source strength per unit time per unit surface area. Here, the dummy variable ξ for x_s should not be confused with Narasimha's similarity variable.

A Gaussian distribution. As an example, let's consider a Gaussian distribution of sources centered at x_t. Then

$$\dot{n}''(\xi)=\frac{\dot{n}'}{\sqrt{2\pi}\chi}\exp\left[-\frac{1}{2}\left(\frac{\xi-x_t}{\chi}\right)^2\right] \tag{5.28}$$

where $\dot{n}'$ is the number of spots produced per unit time per unit span, and χ is the distribution's standard deviation. If we define the onset of transition to be the position where turbulent spots are most often produced, the onset of transition is then x_t even though spots are produced, that is, laminar breakdowns occur, upstream of it. We should note here that in the limit $\chi \to 0$, this function behaves as $\dot{n}'$ times a Dirac delta function positioned at x_t, which is exactly the production function we used for a line source of spots given by Eq. (5.13).

For the sake of simplicity we will assume that x_t is much greater than the standard deviation. Then the lower limit of the integral in Eq. (5.27) may be replaced by $-\infty$ with virtually no loss in accuracy, and the integral reduces to

$$\frac{\dot{n}'}{\sqrt{2\pi}}\chi\left(x-x_t\right)\exp\left[-\frac{1}{2}\left(\frac{x-x_t}{\chi}\right)^2\right]+\frac{1}{2}\left[\chi^2+\left(x-x_t\right)^2\right]\left[1+\mathrm{erf}\left(\frac{x-x_t}{\sqrt{2}\chi}\right)\right]$$

Substituting this into Eq. (5.27) and introducing the various Reynolds numbers together with the dimensionless spot production parameter

$$\hat{n}=\dot{n}'\nu^2/\bar{u}_\infty^3$$

we obtain

$$\gamma_0(x)=1-\exp\left\{-\hat{n}\sigma\left[\frac{1}{\sqrt{2\pi}}Re_\chi\left(Re_x-Re_{xt}\right)\exp\left(-\frac{1}{2}\frac{(Re_x-Re_{xt})^2}{Re_\chi^2}\right)\right.\right.$$

$$\left.\left.+\frac{1}{2}\left(Re_\chi^2+\left(Re_x-Re_{xt}\right)^2\right)\left(1+\operatorname{erf}\left(\frac{Re_x-Re_{xt}}{\sqrt{2}Re_\chi}\right)\right)\right]\right\};\quad (\sigma \text{ constant}) \tag{5.29}$$

where Re_χ is the Reynolds number based on the standard deviation. By expanding the argument of the exponential function on the right-hand side of this expression in a Taylor's series for small values of Re_χ/Re_{xt}, it can be shown that Eq. (5.18) for a line source of spots is recovered when $Re_\chi \rightarrow 0$.

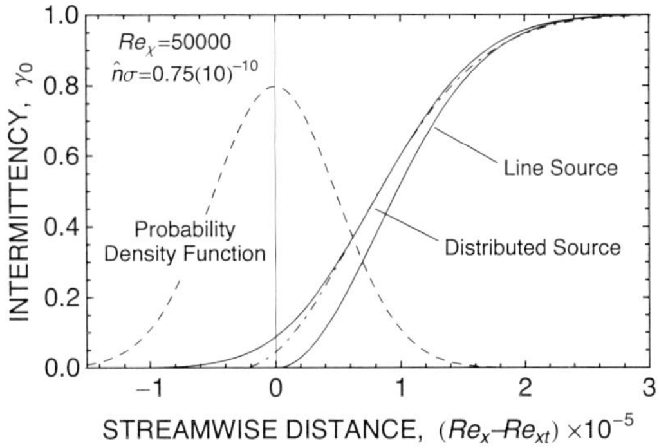

Figure 5.21.
The variation of near-wall intermittency for a Gaussian distribution and line source of spots with the same total production rate.

The variation of the near-wall intermittency given by Eq. (5.29) is shown in Fig. 5.21 for $\hat{n}\sigma$ = $0.75(10)^{-10}$ and Re_χ = 50,000, which is about one-fifth the length of transition. In addition, the probability density function for the sources, that is the right-hand side of Eq. (5.28) divided by $\hat{n}'$, and the intermittency distribution for a line source of the same total strength are shown. Clearly, distributing sources over a broad area affects the intermittency distribution as might be expected. In particular, γ_0 increases before the position of maximum spot production and increases slower than that for an equivalent line source. For continually smaller standard deviations, the intermittency curve for the distributed source simply morphs into that for the line source.

The intermittency distribution given by Eq. (5.29) may be expressed as a one-parameter family of curves in $\hat{n}\sigma Re_\chi^2$ when written using $(Re_x-Re_{xt})/Re_\chi$ as the independent variable. Again, if we consider transition is completed when $\gamma_0 = 0.99$, then the transition length Reynolds number Re_{LT} may be calculated and other parameters related to it may be determined. Some of these results are provided in the first four rows of Table 5.1 where the information in the third and fourth rows is provided in terms of Narasimha's similarity variable with ξ_χ and ξ_{LT} defined by

$$\xi_\chi = Re_\chi/\left(Re_{x,0.75}-Re_{x,0.25}\right) \quad \text{and} \quad \xi_{LT} = Re_{LT}/\left(Re_{x,0.75}-Re_{x,0.25}\right)$$

For the example plotted in Fig. 5.21, where $\hat{n}\sigma Re_\chi^2 = 0.19$ we obtain from Table 5.1 that $Re_\chi/Re_{LT} \approx 0.2$. Furthermore, we obtain from the table that

$$Re_{LT} = 2.10/\sqrt{\hat{n}\sigma}$$

which should be compared with Eq. (5.19) for a line source. Here we note that the values in Table 5.1 for $\hat{n}\sigma Re_\chi^2 = 0$ correspond to those for a line source. They are obtained by letting Re_χ approach zero for a finite value of the turbulent spot parameter $\hat{n}\sigma$.

Table 5.1
Parameters for matching the intermittencies given by a Gaussian distribution of sources by a line source.

$n\sigma Re_\chi^2$	0	0.05	0.1	0.2	0.4	0.6	0.8	1.0
Re_χ/Re_{LT}	0	0.1	0.15	0.21	0.31	0.39	0.46	0.53
$n\sigma Re_{LT}^2$	4.61	4.56	4.51	4.41	4.21	4.01	3.81	3.62
ξ_χ	0	0.33	0.45	0.60	0.75	0.84	0.90	0.95
ξ_{LT}	3.34	3.20	3.05	2.79	2.43	2.17	1.97	1.82
$Re_{xt,shift}/Re_\chi$	0	0.31	0.46	0.63	0.85	1.00	1.10	1.17
n_r	1	0.91	0.84	0.72	0.58	0.48	0.42	0.38

A reasonable match to Eq.(5.29) using the line source solution can be obtained from

$$\gamma_0 = 1 - \exp\left[-n_r\hat{n}\sigma\left(Re_x - Re_{xt} + Re_{xt,shift}\right)^2\right]$$

where $Re_{xt,shift}$ is the shift in transition Reynolds number and $n_r = (\hat{n}\sigma)_{line}/\hat{n}\sigma$ is the spot parameter ratio. Their values can be obtained by fitting the above expression to Eq. (5.29) within the range $0.25 < \gamma_0 < 0.75$ for different values of $\hat{n}\sigma Re_\chi^2$. The results of this exercise are shown in the last two rows of Table 5.1. Reasonably good approximations of these results are given by

$$Re_{xt,shift} \approx 1.56\sqrt{\hat{n}\sigma}Re_\chi\left(1 - 0.153\sqrt{\hat{n}\sigma}Re_\chi\right) \quad \text{and} \quad n_r \approx \left(1 + 1.78\hat{n}\sigma Re_\chi^2\right)^{-1} \; ; \quad (\hat{n}\sigma Re_\chi^2 \leq 1)$$

Again, for the example plotted in Fig. 5.21 where $\hat{n}\sigma Re_\chi^2 = 0.19$, we obtain from the table that $Re_{xt,shift}/Re_{LT} \approx 0.13$ or $Re_{xt,shift} \approx 32{,}000$. The approximating line-source match in this case is shown by the dot-dashed line in the figure where the largest discrepancy between it and the distributed-source solution occurs near the onset of transition.

The best fit of the distributed-source solution with data and its comparison with the line-source solution, are shown in Fig. 5.22. Plotted using Narasimha's similarity variable ξ, the data is identical to that shown in Fig. 5.19. In this figure, however, the abscissa near onset has been expanded to emphasize the major difference between the two solutions. As it turns out, the best fit is obtained from Eq. (5.29) by using $\hat{n}\sigma Re_\chi^2 = 0.125$, which, by interpolating the results in Table 5.1, corresponds to a standard deviation of $\xi_\chi = 0.5$ and a shift in the onset of transition, that is, in the mean of the source-probability-density function (shown using an arbitrary vertical scale by the dot-dashed line in the figure) of $\xi_{xt} = 0.26$. This standard deviation and shift equal about one-sixth and one-twelve of the transition length respectively.

If we consider this fit an accurate description of the measurements on average, and the measurements an accurate description of reality, then the breakdown of laminar flow into turbulent spots is a random occurrence that not only begins before the onset of transition as defined by Narasimha's line-source solution, but can also be described quite well by a Gaussian distribution.[1] Furthermore, as we will see

[1] Steelant and Dick [1996] proposed another spot production model for distributed laminar breakdowns. See §7.3.2.

in §5.4.3, this observation is unaffected by the effect of becalmed zones on spot production.

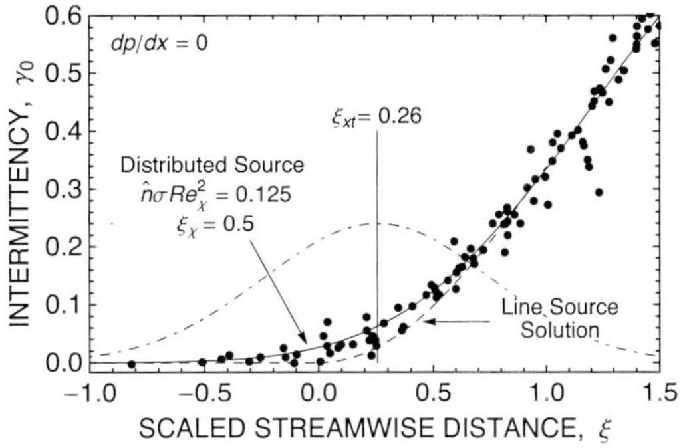

Figure 5.22.
Comparison of the near-wall intermittency for a distributed source with data and Narasimha's solution. See Fig. 5.19 for data source.

5.2.4 A point source of spots.

Spot transit function. General solution for point source distributions.

If we consider spots are produced only at the point (x_s, z_s) on the surface, the production rate is given by

$$g(P') = \dot{n}\delta(x' - x_s)\delta(z' - z_s) \tag{5.30}$$

where $\dot{n}$ is the number of turbulent spots produced per unit time. Then Eq. (5.12) becomes

$$\gamma_0(x,z) = 1 - \exp\left\{ -\dot{n} \int_{x'=0}^{x} \int_{z'=-\infty}^{\infty} \delta(x' - x_s)\delta(z' - z_s)\, dx'\, dz' \int_{\Delta t'} dt' \right\}$$

where $\Delta t'$ is the transit time of a spot from a source at P' passing over the point P as shown in Fig. 5.17. In this case, $\Delta t'$ is proportional to $F_0(z; z')\ (x-x')/u_\infty$, where F_0, which we will call the "near-wall spot-transit function," is the distance from the leading to trailing edge in the x direction for a spot footprint of *unit length*. Substituting this expression into the above equation and integrating yields

$$\gamma_0(x,z) = 1 - \exp\left[-\frac{\dot{n}\sigma'}{\bar{u}_\infty} F_0(z; z_s)\left(x - x_s\right) \right] \tag{5.31}$$

where the proportionality factor σ' is a *one-dimensional* propagation parameter.

The transit function F_0 can be obtained by determining the length of the spot's footprint parallel to the x axis at various spanwise positions. Given the three-dimensional shape of the spot, this length can also be determined at various heights above the wall to obtain what might be called a "generalized spot-transit function," the simplest functional form of which would be $F(y, z; z_s)$. Then $F_0(z; z_s)$ is given by $F(0, z; z_s)$.

The generalized transit function for a fully-developed spot (in particular, that shown in Fig. 5.6) is plotted in Fig. 5.23, where b and h are the *breadth* and *midspan height* of the spot. Near the wall, within the laminar boundary layer, this function changes little. Above the leading-edge bulge of the spot, however, it decreases with height.

A rather good analytical fit to the transit function near the wall is

$$F_0(\zeta) = \left(1 - 0.85\zeta^2 - 0.15\zeta^8\right)\left[\mathrm{U}(\zeta + 1) - \mathrm{U}(\zeta - 1)\right] \tag{5.32}$$

where $\zeta = 2(z-z_s)/b$ and $b = 2(x-x_s)\tan\alpha$. This may be substituted into Eq. (5.31) to obtain the near-wall intermittency for a point source located at (x_s, z_s). The results are plotted in Fig. 5.24 for a dimensionless source strength $\dot{n}\nu\sigma'/u_\infty^2 = 4(10)^{-7}$ and $\alpha = 10°$. Upstream of the $\gamma_0 = 0$ contour, which corresponds to loci of the spot's spanwise tips, the intermittency equals zero.

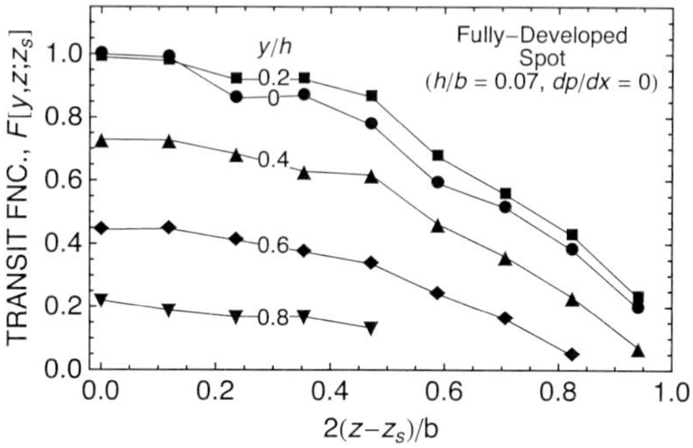

Figure 5.23.
The generalized transit function for a fully-developed spot normalized by a footprint of unit length. Calculated from the measurements by Wygnanski et al. [1976].

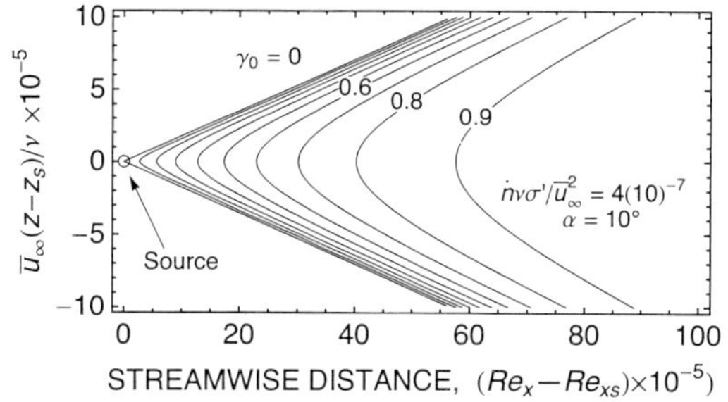

Figure 5.24.
The near-wall intermittency distribution downstream of a point source.

The solution given by Eq. (5.31) can be used to obtain the intermittency for other source distributions. As an example, we can use it to construct the solution given in Eq. (5.16) for a line source. Since the probability that the flow is turbulent from any number of sources, say N, is not a mutually exclusive event, we use the equivalent of Eq. (5.6) to obtain

$$\gamma_0 = 1 - \prod_{k=1}^{N}\left(1-\gamma_{0,k}\right) \tag{5.33}$$

Upon substituting Eq. (5.31), we obtain

$$\gamma_0(x,z) = 1 - \exp\left[-\frac{\sigma'}{\bar{u}_\infty}\sum_{k=1}^{N}\dot{n}_k(x_{s,k},z_{s,k})F_0(z;z_{s,k})\left(x-x_{s,k}\right)\right]$$

which is valid for any number of sources located at the points $(x_{s,k}, z_{s,k})$. For sources distributed along the line $x = x_s$ at $z_{s,k} = \zeta_k$, we obtain

$$\gamma_0(x,z) = 1 - \exp\left[-\frac{\sigma'}{\bar{u}_\infty}\left(x-x_s\right)\sum_{k=1}^{N}\dot{n}_k(\zeta_k)F_0(z;\zeta_k)\right]$$

and if we let the number of sources increase indefinitely while maintaining a finite source strength it follows that

$$\gamma_0(x,z)=1-\exp\left[-\frac{\sigma'}{\bar{u}_\infty}\left(x-x_s\right)\int_{-\infty}^{\infty}\dot{n}'(\zeta)F_0(z;\zeta)d\zeta\right] \tag{5.34}$$

where $\dot{n}'$ is the strength per unit time per unit length of the line. Note that this is a general formula for the intermittency resulting from a line source with an arbitrary strength distribution. For a uniform strength distribution, we obtain

$$\gamma_0(x)=1-\exp\left[-\frac{\dot{n}'\sigma'}{\bar{u}_\infty}\left(x-x_s\right)\int_{-\infty}^{\infty}F_0(z;\zeta)d\zeta\right]$$

Substituting Eq. (5.32) and integrating yields

$$\gamma_0(x)=1-\exp\left[-\sqrt{2}\frac{\dot{n}'\sigma'}{\bar{u}_\infty}\left(x-x_s\right)^2\right]$$

which is functionally identical to that which we derived for a line source of spots. The numerical value $\sqrt{2}$ in the exponential argument depends on the transit function F_0, that is, the shape of the footprint. For a uniform value across the spot, the number would be two. If it decreased linearly with z from its maximum value of one at the midspan to zero at $z = b/2$, the number would be one. For the typical turbulent spot in a zero-pressure-gradient flow, it just happens to be a number very close to $\sqrt{2}$. Nevertheless, comparing the above expression with Eq. (5.16), we see that the dimensionless point and line source propagation parameters are related by

$$\sigma=\sqrt{2}\sigma'$$

Another example of constructing solutions from the elemental point source solution is given in Appendix G.

5.2.5 A three-dimensional spot theory.

A four-dimensional dependence cone. A three-dimensional production function. A two-dimensional intermittency factor.

Consider now a line source of three-dimensional spots and determine the intermittency distribution within the flow downstream. In this case, dV' is an elemental volume in four-dimensional space and Eq. (5.9) becomes

$$\gamma(P)=1-\exp\left\{-\int_R g(P')dx'dy'dz'dt'\right\}$$

where g is the number of spots produced per unit volume per unit time. If the production of turbulent spots is both stationary in time and homogeneous in the spanwise direction, the production rate for a line source at $x = x_t$ and $y = 0$, can be expressed as

$$g\left(x',y'\right)=\dot{n}'\delta\left(x'-x_t\right)\delta\left(y'\right) \tag{5.35}$$

where as before $\dot{n}'$ is the number of turbulent spots produced per unit time per unit span. Then

$$\gamma(x,y)=1-\exp\left\{-\dot{n}'\int_{x'=0}^{x}\int_{y'=0}^{y}\delta(x'-x_t)\delta(y')dx'dy'\int_{A'}dz'dt'\right\}$$

where A' is the cross-sectional area of the dependence cone in an x' = constant plane at the distance y' above the surface. For $y' = 0$, this is the area of the spot footprint shown in Fig. 5.17. Similar to the classical analysis carried out for spot footprints, if we assume that the *footprint* expands linearly with respect to x across the surface, then A' will be proportional to $G(x,y)(x-x')^2/\bar{u}_\infty$, where G, which we will simply call the "spot-transit function," is the area of the spot in planes parallel to the wall for a spot footprint of *unit area*. Consequently, the intermittency distribution for a three-dimensional spot becomes

$$\gamma(x,y)=1-\exp\left[-\frac{\dot{n}'\sigma}{\bar{u}_\infty}G(x,y)\left(x-x_t\right)^2\right];\quad (x\geq x_t) \tag{5.36}$$

where, since $G(x, 0) = 1$, σ equals Emmons' propagation parameter because the above expression must reduce to Eq. (5.16) when $y = 0$. Expanding this expression for small values of $(x-x_t)$ shows that the intermittency initially varies according to $G(x_t, y)(x-x_t)^2$. That is, the initial distribution of the transit function $G(x_t, y)$ may be determined directly from the intermittency profiles near the beginning of transition.

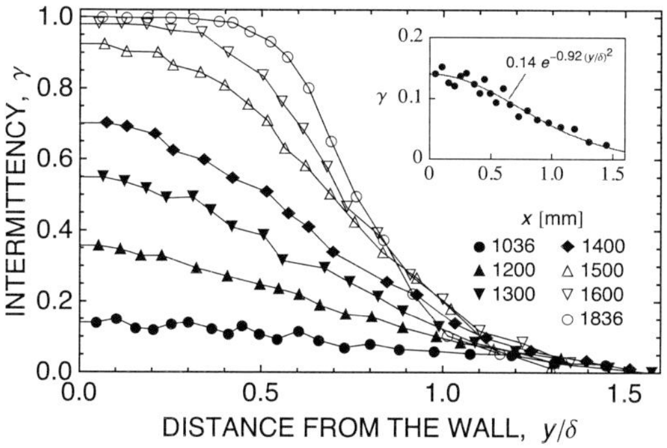

Figure 5.25.
Intermittency profiles through transition. Data from Acharya's test case TRA with $Tu \approx 0.7$ percent, $u_\infty/\nu = 1.26(10)^6$ m^{-1} (filled circles), and $u_\infty/\nu = 1.31(10)^6$ m^{-1} (open circles).

Intermittency profiles have been measured by a number of investigators,[1] a typical set of which is shown in Fig. 5.25.[2] Here, it must be remembered that δ in the abscissa of this figure equals the laminar boundary-layer thickness near the beginning of transition, the turbulent boundary-layer thickness near its end, and their time or intermittency-weighted average in between. In the initial stage of transition, the first few intermittency profiles are Gaussian-like with a maximum near the wall. A fit for Acharya's first profile using a Gaussian distribution is given in the insert of the figure, namely,

$$\gamma(x_1,y)/\gamma_0(x_1)=\exp\left\{-0.92\left[y/\delta(x_1)\right]^2\right\}$$

where x_1 is the position at the first profile.

The transit function can be determined from the above expression once the ratio of the boundary-layer thickness at the first intermittency profile δ and the height of the spot at the onset of transition is known. As we will momentarily discover, this ratio can be expressed as a ratio between the local boundary-layer thickness and the turbulent boundary-layer thickness at onset. For Acharya's first profile, this ratio is equal to 1.74. Therefore, the transit function near onset is approximately given by

---

[1] As previously noted in §5.1.3 and further discussed in Appendix H, their results depend on the scheme they use to discriminate between the laminar and turbulent portions of the flow. In general, these may be classified as discriminating schemes that are "more sensitive" and "less sensitive". In the following, we will use the results obtained by those using the less sensitive schemes.

[2] In this figure, Acharya's [1985] data has been corrected to account for differences in the definition of boundary-layer thickness as explained in Appendix I.

$$G_1(x,y) = \exp\left[-2.78\left(y/\delta_T\right)^2\right] \tag{5.37}$$

where δ_T is the turbulent boundary-layer thickness. This function is plotted using a solid line in Fig. 5.26. Whether or not fortuitous, this result indicates that the immature-spot height is approximately equal to δ_t, which also agrees with that found for turbulent spots in general (see Fig. 5.5).

The transit function for a fully-developed spot is different. If the $F(y, z; z_s)$ curves in Fig. 5.23 are integrated with respect to z, the filled circles in Fig 5.26 are obtained. In this case, the distance y is normalized by the spot height h, which in turn is proportional to the turbulent boundary layer thickness. Assuming that $h = \delta_T$, as reported by Schubauer and Klebanoff [1955], a reasonable approximation for this distribution is

$$G_2(x,y) = \exp\left[-6.5\left(y/\delta_T\right)^3\right]; \quad (y \geq 0) \tag{5.38}$$

which is plotted using a dashed line in the figure.

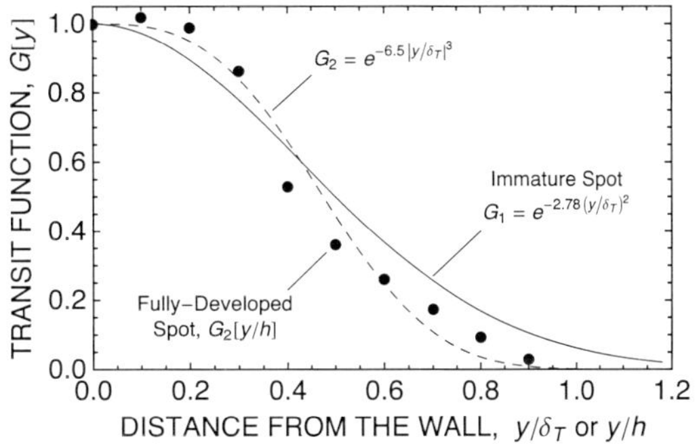

Figure 5.26.
Variation of the spot transit function G with distance from the wall for an immature and fully-developed spot with $dp/dx = 0$.

Given these two limits for the transit function and the fact that they aren't significantly different, the variation of $G(x, y)$ between them can easily be approximated using

$$G(x,y) = \left(1 - f_{tr}\right)G_1(x,y) + f_{tr}G_2(x,y) \tag{5.39}$$

where f_{tr} is an interpolation function that varies from zero to one as the spot develops from onset to maturity. Considering the results plotted in Fig. 5.5b and regarding the rate at which spots become fully-developed, a good guess for this function would be

$$f_{tr} = 1 - \exp\left[-5(10)^{-6}\left(Re_x - Re_{xt}\right)\right] \tag{5.40}$$

Substituting Eqs. (5.37)–(5.39) into Eq. (5.36), and introducing the spot production parameter given by Eq. (5.17), we obtain

$$\gamma(x,y) = 1 - \exp\left\{-\hat{n}\sigma\left(Re_x - Re_{xt}\right)^2\left[\left(1 - f_{tr}\right)e^{-2.78\left(y/\delta_T\right)^2} + f_{tr}e^{-6.5\left(y/\delta_T\right)^3}\right]\right\}; \ (Re_x \geq Re_{xt}) \tag{5.41}$$

which satisfies the condition that $\gamma(x, 0)$ equals the near-wall intermittency distribution $\gamma_0(x)$ given by Eq. (5.18), and reduces to Gaussian profiles near the beginning of transition.

To compare the above expression with data, we need $\delta_T(x)$ for each set of data and, since most intermittency profiles are plotted as in Fig. 5.25, we also need the local boundary-layer thickness $\delta(x)$. Anticipating the results given in §7.1, the boundary-layer thickness at any position within transition is approximately given by

$$\delta(x)=\left[1-\gamma_0(x)\right]\delta_L(x)+\gamma_0(x)\delta_T(x) \tag{5.42}$$

which is the intermittency-weighted average of the local laminar and turbulent boundary-layer thicknesses. If we consider flows without a pressure gradient, these thicknesses can be approximately determined from the results given in §2.1 and §3.2.3 once the initial conditions are known.

According to Schubauer and Klebanoff [1955], the height of a spot not only increases with distance as a fully-developed turbulent boundary layer, but with an initial value $h_t \approx \delta_{T,t}$ at onset equal to the thickness of the laminar boundary layer there $\delta_{L,t}$. On the other hand, the skin friction measurements of Dhawan and Narasimha [1958] indicate that the initial height is nearly zero at onset. Consequently, if we consider that the height of the spot at onset is given by $h_t \approx \delta_{T,t} = r\delta_{L,t}$, where r is either less than one or thereabouts, then Eqs. (2.12), (3.39) and (3.41) yield

$$\delta_L \approx 5\left(\nu/\bar{u}_\infty\right)^{1/2} x^{1/2}, \quad \delta_T \approx 0.37\left(\nu/\bar{u}_\infty\right)^{1/5}\left(x-x_0\right)^{4/5}$$

and (5.43)

$$\delta_{T,t} = r\delta_{L,t} \approx 5r\left(\nu/\bar{u}_\infty\right)^{1/2} x_t^{1/2}$$

where the laminar and turbulent boundary layers are assumed to "begin" at $x = 0$ and x_0, respectively. Setting $x = x_t$ and eliminating x_0, we obtain

$$\delta_T = 0.37\left(\nu/\bar{u}_\infty\right)\left[Re_x - Re_{xt} + 25.9r^{5/4}Re_{xt}^{5/8}\right]^{4/5} \tag{5.44}$$

which, by setting $\delta_T = 0$ when $x = x_0$, yields

$$Re_{x0} = Re_{xt}\left(1-25.9r^{5/4}Re_{xt}^{-3/8}\right) \tag{5.45}$$

Given the transition Reynolds number and turbulent spot parameter, the intermittency distribution according to Eq. (5.41) may be evaluated for different values of r and compared to data to obtained its best value. Using the profile data of Owen [1970], Acharya [1985], Kim, et al. [1989], and Blair [1992], the best fit to the data nearest onset is obtained when r is about equal to one-half, that is, a value halfway between that suggested by Schubauer and Klebanoff's and Narasimha's measurements.[1]

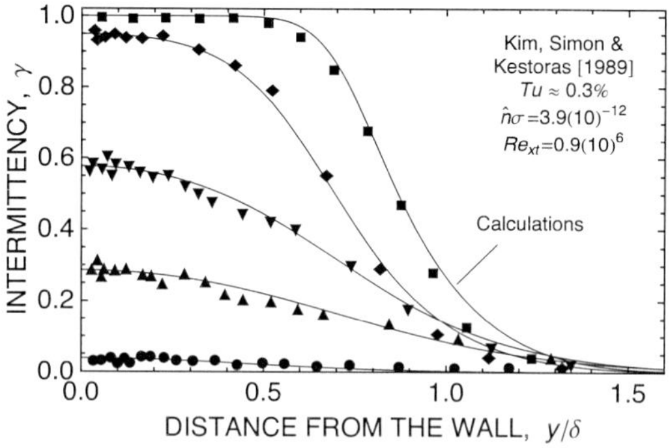

Figure 5.27.
A comparison between calculated and measured intermittency profiles through transition.

A comparison of Eq. (5.41) with the data from Kim et al. [1989] is shown in Fig. 5.27. The values of

[1] This is the value used to obtain the factor of 1.74 when determining Eq. (5.37).

Re_{xt} and $\hat{n}\sigma$ are those obtained by fitting their near-wall intermittency data as shown in Fig. 5.18.[1] The local boundary-layer thicknesses used for the calculations were obtained from Eqs. (5.42)–(5.45) with $r = 1/2$. The agreement between the calculations and data is good. As it turns out, the agreement between similarly calculated results and the data plotted in Fig. 5.25 is equally as good.

A dimensionless form of Eq. (5.41) is plotted in Figs. 5.28 and 5.29 compared with data for low free-stream turbulence levels. All but Blair's data are for transition with $dp/dx = 0$. His data are for transition in a favorable pressure gradient with an acceleration parameter K equal to $0.75(10)^{-6}$. The data shown in the first figure are for the near-wall values $0.27 \le \gamma_0 \le 0.35$, while those in the second figure are for $0.93 \le \gamma_0 \le 1.0$. In other words, the figures show data taken near the beginning and end of transition, respectively. The calculations are for the near-wall values of 0.3 and 0.97 using the values of $\hat{n}\sigma = 3(10)^{-12}$ and $Re_{xt} = 1.1(10)^6$, which roughly correspond to those for transition with a free-stream turbulence level of 0.4 percent. In addition, the ordinate in these figures is the intermittency normalized by its near-wall value, while the abscissa is the distance y normalized by the profile's half-width.

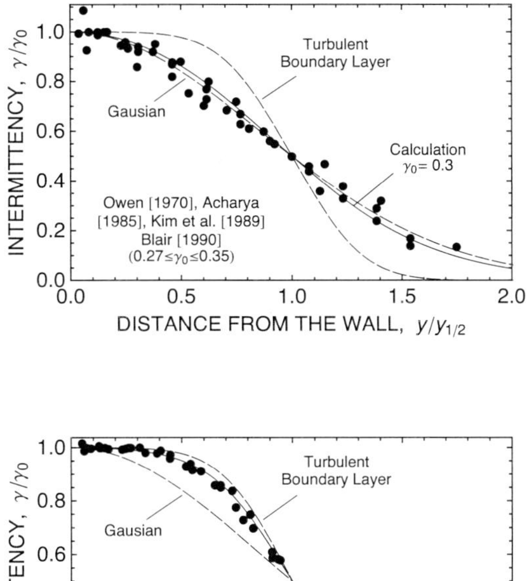

Figure 5.28.
Intermittency profiles near the beginning of transition. Data and calculations for transition with a free-stream turbulence level of 0.4 to 1.0 percent.

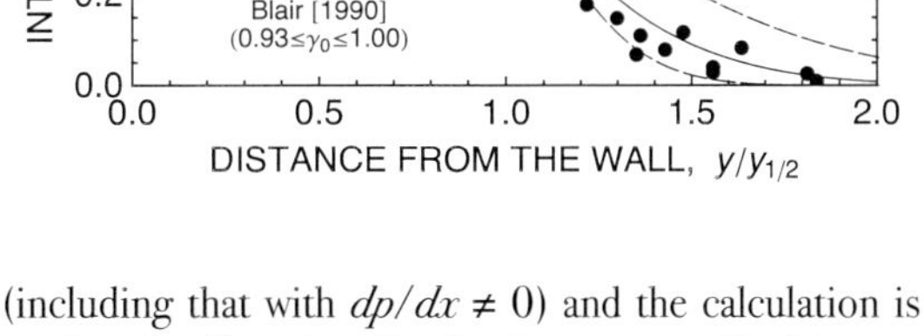

Figure 5.29.
Intermittency profiles near the end of transition. Data and calculations for transition with a free-stream turbulence level of 0.4 to 1.0 percent.

In both cases, the agreement between data (including that with $dp/dx \neq 0$) and the calculation is good. In Fig. 5.28, the calculation closely matches the Gaussian distribution as prescribed by our choice for the transit function $G_1(x, y)$, while in Fig. 5.29 the theory nearly matches the intermittency

[1] To match the measured value of γ_0 using Eq. (5.18), the distance Reynolds number was shifted slightly for the second and third lowest profiles in Fig. 5.27. In both cases, the shift was less than two percent.

profile for a fully-developed turbulent boundary layer given by Eq. (3.38).

A point to note here, however, is that unlike the intermittency profile for a turbulent boundary layer, the profile calculated using Eq. (5.41) continues to steepen with increasing distance downstream. Realizing that the interaction between the individual spots becomes less important near the end of transition than the large-scale intermittent motion between the developing turbulent layer and free stream, this behavior can be corrected by introducing a function that forcibly blends the two intermittency profiles near the end of transition. Without going into any details, this function is easily conceived and would presumably become important as the near-wall value of the intermittency given by Eq. (5.41) approaches one. As a consequence, the spanwise and time averaged intermittency profile obtained from this three-dimensional spot theory would blend smoothly with the time averaged intermittency profile for a fully-developed turbulent boundary layer near the end of transition.

5.3 Spot Kinematics.

Some preliminaries. Spots in dp/dx = 0 flow. Spots in dp/dx ≠ 0 flow.

Emmons' turbulent spot theory might more appropriately be called an "intermittency theory" because, as mentioned in the introduction to the previous section, he managed to tuck all the fluid mechanics associated with the spots into his propagation parameter. As a consequence, he left us the more difficult problem of determining how the spots grow and propagate.

Information regarding the general shape and velocity of a "fully-developed" spot was presented in §5.1. As we noted there, it appears that a spot developing in a turbulence-free zero pressure-gradient flow becomes fully developed when the Reynolds number based on the distance from inception is greater than about $5(10)^5$. As we also noted, the length and width of a spot quickly becomes orders of magnitude larger than its height, which in turn scales approximately as a turbulent boundary layer developing from where the spot was formed.

In this section, we will more thoroughly investigate the propagation rates of spots developing in flows with and without pressure gradients. But first we will consider some basic relations concerning the geometry and kinematics of a propagating spot.

5.3.1 Some preliminaries.

Spot geometry. Spot kinematics.

There are two basic methods used to describe a spot and its motion. One, as shown in Figs. 5.2–5.9, provides a snapshot of the spot at any instant and the other, as shown in Figs. 5.10–5.13, provides the history of the spot at any position. If we are only concerned with the footprint of a spot, the first describes the footprint at different instants in an xz-plane, while the second describes the footprint in a zt-plane at different streamwise positions. The projection of the footprint onto these planes are shown in Fig. 5.16 where the geometry of the footprint in an xzt-space is described.

Spot geometry. Since it is natural to describe the physical size and shape of the spot's footprint in an xz-plane, as shown in Fig. 5.4, we will define its length at any instant by

$$l(t) = x_{le}(t) - x_{te}(t) \tag{5.46}$$

where x_{le} and x_{te} are the leading- and trailing-edge positions of the spot along its midspan. Furthermore, if z_{tp} is the position of the spot's outboard tip, which may or may not be at the same streamwise distance as the spot's trailing edge, the breadth of the spot at any instant is

$$b(t) = 2z_{tp}(t)$$

Following Emmons and many investigators after, consider using an isosceles triangle as a "skeleton" for the footprint.[1] Then the breadth and length can be related by

$$b(t) = 2l(t)\tan\theta(t) \tag{5.47}$$

where θ is the skeleton's leading-edge half-angle. As a result, we obtain

$$b(t) = 2\left(x_{le} - x_{te}\right)\tan\theta$$

If we now introduce f_s as the ratio of the true-to-skeletal footprint area of the spot, the *true* area covered by the spot at any instant may be written as

$$A_{s,xz}(t) = f_s\, bl/2 = f_s\left(x_{le} - x_{te}\right)^2 \tan\theta$$

This corresponds to the cross-sectional area in any xz-plane of either the propagation or dependence cone in Figs. 5.16 and 5.17. When a spot changes shape relative to its skeleton as it propagates downstream, the ratio f_s also changes with time. This most likely occurs as the spot initially develops and in flows with a nonzero pressure gradient. Data on this effect are minimal. For the fully-developed spot drawn in Fig. 5.6b, however, the value of f_s lies roughly between 1.5 and 1.8 depending on which height *within* the surrounding laminar boundary layer the footprint is measured.

Since it is natural to describe the duration of the spot's footprint in the zt-plane of our xzt-space, we define the duration of its midspan at any streamwise position by

$$\Delta t(x) = t_{te}(x) - t_{le}(x) \tag{5.48}$$

where t_{le} and t_{te} are the times corresponding to the passing of the spot's leading and trailing edge over the position x. Together with this, the breadth of a spot at any position is given by

$$b(x) = 2z_{tp}(x) \tag{5.49}$$

and the "area" of the spot in any zt-plane of the propagation or dependence cone becomes

$$A_{s,zt}(x) = f_s\, b\Delta t/2 = f_s z_{tp}\left(t_{te} - t_{le}\right) \tag{5.50}$$

which has the dimension of [length×time]. Although the ratio of the areas $A_{s,zt}$ and $A_{s,xz}$ appears to be $\Delta t/l$, it must be remembered that Δt is a function of x, while l is a function of t. We will return to this point shortly.

Spot kinematics. If we consider the most general motion of the spot's footprint, its rate of elongation is

$$\frac{dl}{dt} = u_{le} - u_{te}$$

where

$$u_{le}(t) = \frac{dx_{le}}{dt} \quad \text{and} \quad u_{te}(t) = \frac{dx_{te}}{dt} \tag{5.51}$$

are the leading- and trailing-edge velocities of the spot at any instant.

In addition, the spot's rate of expansion is given by

$$\frac{db}{dt} = 2\left(u_{le} - u_{te}\right)\tan\theta + 2l\sec^2\theta\frac{d\theta}{dt} \tag{5.52}$$

The first term on the right-hand-side of this expression is the contribution from the spot's elongation, while the second term is the contribution from the rotation of the spot's leading edge about its nose.

---

[1] For some examples, see McCormack [1968], Chen and Thyson [1971], and Solomon et al. [1996].

In addition, since the outward velocity of the spot's outboard tip is $dz_{tp}/dt = \frac{1}{2}\, db/dt$, the *instantaneous* angle α of the tip's trajectory along the surface is given by

$$\tan\alpha = \frac{u_{le} - u_{te}}{u_{te}} \tan\theta + \frac{l}{u_{te}} \sec^2\theta \frac{d\theta}{dt} \tag{5.53}$$

where the tip, being on the trailing edge of the skeletal spot, is presumed to be moving in the streamwise direction with the velocity u_{te}.

Using this to eliminate the contribution caused by the leading-edge rotation in Eq. (5.52), we obtain

$$\frac{db}{dt} = 2u_{te} \tan\alpha$$

In similar fashion, considering the motion of the footprint in the zt-plane, the change of its duration with respect to x is given by

$$\frac{d\Delta t}{dx} = \frac{1}{u_{te}(x)} - \frac{1}{u_{le}(x)} \tag{5.54}$$

where

$$u_{le}(x) = \frac{1}{dt_{le}(x)/dx} \quad \text{and} \quad u_{te}(x) = \frac{1}{dt_{te}(x)/dx} \tag{5.55}$$

are the leading- and trailing-edge velocities of the spot at any position. Furthermore, we obtain

$$\frac{db}{dx} = 2\frac{dz_{tp}}{dx} = 2\tan\alpha \tag{5.56}$$

All of the above equations are universally valid for an approximately triangular spot. Under certain circumstances, they can be simplified and algebraic relations between the length and duration of a spot can be developed.

5.3.2 Spots in dp/dx = 0 flow.

Streamwise development. Spanwise development. Self-similarity. Approach to self-similarity.

In this section we will examine the streamwise and spanwise development of a turbulent spot when the free-stream velocity is constant, examine the effects of Reynolds number on the duration, length, propagation velocities, and breadth of a spot, and consider spot similarity and the approach to it.

Streamwise development. The variation of a spot's leading- and trailing-edge positions with time are shown in Fig. 5.30 for two different "source" Reynolds numbers. These Reynolds numbers are based on the displacement thickness of the undisturbed boundary layer where the spot was produced. The ordinate is the distance from the spark that created it (located at x_s), and the abscissa is a distance based on the time from ignition. The streamwise Reynolds numbers, based on the distance from the spark, range from about $5.3(10)^5$ to $(10)^6$ for the $Re_{\delta 1,s} = 780$ data, and almost twice as large for the $Re_{\delta 1,s} = 1060$ data. Therefore, the data in both cases have been acquired for a fully-developed spot (see Fig. 5.5). Furthermore, since the critical Reynolds number for a $dp/dx = 0$ flow is 520, the spots in both cases were produced in an unstable boundary layer.

As remarked before, the leading- and trailing-edge positions of the spot and, hence, its length increase linearly with respect to time. It then follows directly from the previous set of equations that the spot's elongation rate together with its leading- and trailing-edge velocities are constant. The propagation velocities, which differ slightly from Wygnanski's et al. reported values, are given in the figure. The elongation rates dl/dt are about $0.32\bar{u}_\infty$ and $0.40\bar{u}_\infty$ for the low and high Reynolds numbers respectively. Extrapolating the lines back in time to ignition, which is significantly beyond the range of data in both cases, yields leading- and trailing-edge intercepts that lie in the relatively small region indicated by the thick vertical line centered about the origin. If the best straight-line fit of the data includes the origin, the propagation velocities would be about $0.86\overline{u}_\infty$ and $0.57\overline{u}_\infty$ for $Re_{\delta 1,s} = 780$, and $0.91\overline{u}_\infty$ and $0.52\overline{u}_\infty$ for $Re_{\delta 1,s} = 1060$, which are nearly the same as those given in the figure.

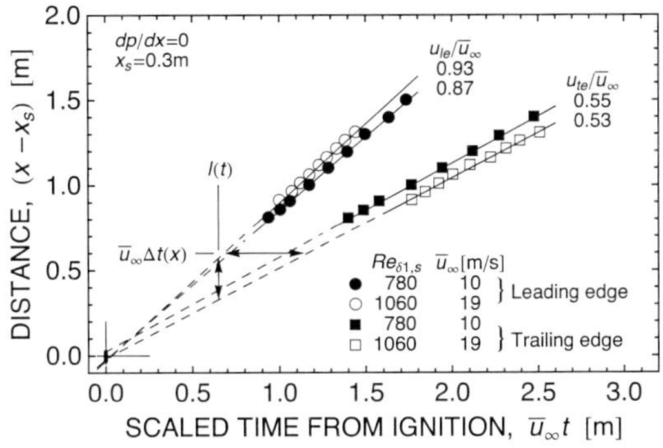

Figure 5.30.
Propagation of a fully developed turbulent spot produced at two different Reynolds numbers. Data from Wygnanski et al. [1982].

When the leading- and trailing-edge velocities of a spot are constant, it follows from Eq. (5.51) that

$$x_{le}(t) = x_{le,0} + u_{le}t \quad \text{and} \quad x_{te}(t) = x_{te,0} + u_{te}t \tag{5.57}$$

where the "0" subscripted quantities are the x intercepts. Consequently, the length of a spot is given by

$$l = \left(u_{le} - u_{te}\right)t + \left(x_{le,0} - x_{te,0}\right); \quad (u_{le} \text{ and } u_{te} \text{ constant}) \tag{5.58}$$

When $x_{le,0} < x_{te,0}$, the virtual origin of the spot occurs after ignition, otherwise it occurs before. By definition, we also obtain from Eq. (5.57) that

$$t_{le}(x) = \frac{x - x_{le,0}}{u_{le}} \quad \text{and} \quad t_{te}(x) = \frac{x - x_{te,0}}{u_{te}}; \qquad (u_{le} \text{ and } u_{te} \text{ constant})$$

It then follows from Eq. (5.48) that the duration of a spot moving with constant leading- and trailing-edge velocities is given by

$$\Delta t = \frac{u_{le} - u_{te}}{u_{le}u_{te}}x - \frac{u_{le}x_{te,0} - u_{te}x_{le,0}}{u_{le}u_{te}} \tag{5.59}$$

When the intercepts coincide, $x_{le,0} = x_{te,0} = x_0$, and a variety of expressions can be developed relating the variables. For example, from Eqs. (5.58) and (5.59) we immediately obtain

$$l = \left(u_{le} - u_{te}\right)t \quad \text{and} \quad \Delta t = \frac{u_{le} - u_{te}}{u_{le}u_{te}}\left(x - x_0\right); \quad (u_{le} \text{ and } u_{te} \text{ constant}, x_{le,0} = x_{te,0} = x_0) \tag{5.60}$$

These are the length-time relations upon which most of the intermittency analyses in §5.2 are based. However, it is noteworthy that these relations apply only to the leading and trailing edge of the spot and not necessarily to points between unless the propagation velocity of these points vary linearly from the leading to trailing edge.

It is best at this time to distinguish between the various spot lengths reported in the literature. As will become immediately obvious, there are *two* values of $l(t)$ for each value of $\Delta t(x)$. This is seen by noting that two right-triangles can be formed in Fig. 5.30 with Δt as the adjacent side; one with the right-angle vertex on the leading-edge trajectory as shown in the figure, the other with the right-angle vertex on the trailing-edge trajectory (not drawn). These correspond to the length of a spot *at the instant* its leading and trailing edges are over the position x. Hence, in the following, they will be denoted by $l(x; t_{le})$ and $l(x; t_{te})$. Obviously, for an elongating spot, the former is always smaller than the latter.

In addition, although not usually reported, there are two values of $\Delta t(x)$ for each value of $l(t)$. They will be denoted by $\Delta t(t; x_{le})$ and $\Delta t(t; x_{te})$. These correspond respectively to the duration of a spot whose leading edge is presently at x_{le} and has yet to pass, and the duration of a spot whose trailing edge is presently at x_{te} and has already passed. In this case, the former is always larger than the latter, that is, $\Delta t(t; x_{le}) > \Delta t(t; x_{te})$ for an elongating spot.

When the leading- and trailing-edge velocities of a spot are constant and the leading- and trailing-edge intercepts coincide, several relations between these quantities can immediately be obtained from geometry. Using the four different right triangles that can be drawn with l and $u_\infty \Delta t$ as the opposite and adjacent sides of the triangle, and recalling that the slopes of the trajectories are either u_{le} or u_{te} divided by the free-stream velocity, we obtain

$$\Delta t(t;x_{le}) = l(t)/u_{te},\ \Delta t(t;x_{te}) = l(t)/u_{le} \text{ and } l(x;t_{le}) = u_{te}\Delta t(x),\ l(x;t_{te}) = u_{le}\Delta t(x) \tag{5.61}$$

which relate the duration of a spot in the zt-plane to its length in the xz-plane and vice versa when considering either the propagation or dependence cone in xzt-space. It then follows from Eq. (5.60) that

$$\Delta t(t;x_{le}) = \frac{u_{le}-u_{te}}{u_{te}}t,\quad \Delta t(t;x_{te}) = \frac{u_{le}-u_{te}}{u_{le}}t,$$

and (5.62)

$$l(x;t_{le}) = \frac{u_{le}-u_{te}}{u_{le}}\left(x-x_0\right),\quad l(x;t_{te}) = \frac{u_{le}-u_{te}}{u_{te}}\left(x-x_0\right)$$

when u_{le} and u_{te} are constant, and $x_{le,0} = x_{te,0} = x_0$.

Now that the different durations and lengths are properly defined, a universally valid set of equations for them can be obtained by taking the derivatives of Eqs. (5.46) and (5.48), and using Eqs. (5.51) and (5.55). The results are

$$\frac{d\Delta t(t;x_{le})}{dt} = \frac{u_{le}-u_{te}}{u_{te}},\quad \frac{d\Delta t(t;x_{te})}{dt} = \frac{u_{le}-u_{te}}{u_{le}}$$

and (5.63)

$$\frac{dl(x;t_{le})}{dx} = \frac{u_{le}-u_{te}}{u_{le}},\quad \frac{dl(x;t_{te})}{dx} = \frac{u_{le}-u_{te}}{u_{te}}$$

from which the previous expressions immediately follow.

The unknown quantities in all of these and the above equations are the propagation velocities. The variation of the leading- and trailing-edge velocities for spots produced at different displacement thickness Reynolds numbers are shown in Fig. 5.31a. The open symbols are the values determined by fitting a straight line through Wygnanski's et al. data *and* the origin (see Fig. 5.30). In general, the leading-edge velocity appears to be independent of the Reynolds number and equal to 0.89 times the free-stream velocity, while the trailing-edge velocity decreases with source Reynolds number. This decrease can be evaluated reasonably well, except for the point nearest the critical Reynolds number, by using

$$u_{te}/\overline{u}_{\infty} = 3Re_{\delta 1,s}^{-0.25} \tag{5.64}$$

which is also plotted in the figure. For $Re_{\delta 1,s} = 1000$, we obtain $u_{te}/\overline{u}_{\infty} = 0.53$ and $u_{le}/\overline{u}_{\infty} = 0.89$. Here it is worth noting that this function is virtual identical to 1.63 times the wave speed of a neutrally stable disturbance on the *lower* branch of the stability curve (see Fig. 4.2 and Appendix D). The difference between this trailing-edge velocity and wave speed accounts for the becalmed zone following the spot as discussed earlier in §5.1.4.

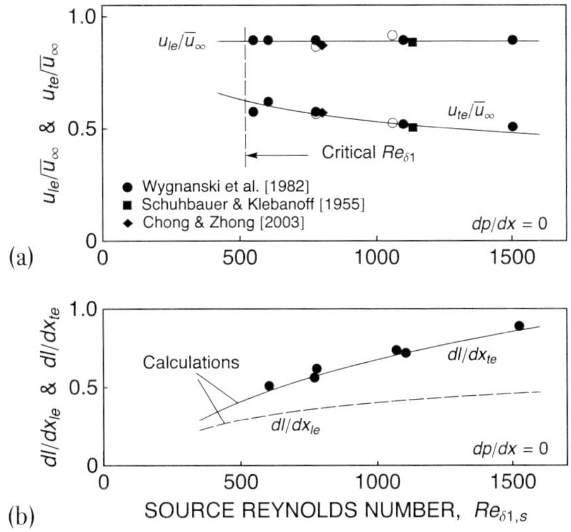

Figure 5.31a,b.
Variation of a spot's leading- and trailing-edge velocity with the displacement thickness Reynolds number at the source.

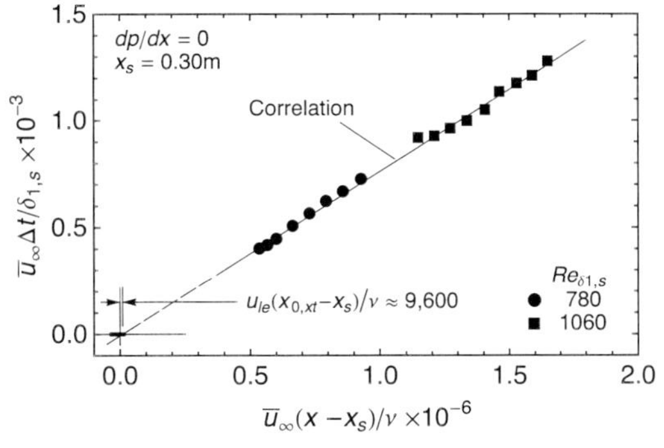

Figure 5.32.
A correlation for the duration of a turbulent spot with distance for two source Reynolds numbers. Data from Wygnanski et al. [1982].

Wygnanski et al. also measured the rate of change of the spot length $l(x; t_{te})$ relative to the distance along the wall. Expressions for it and its companion for $l(x; t_{le})$ are given in Eq. (5.63). These were evaluated using Eq. (5.64) and are plotted in Fig. 5.31b compared with Wygnanski's et al. data. Although the agreement is excellent, it is important to realize that this result is extremely sensitive to the expression used for the trailing-edge velocity. In this regard, no other fit than that given in Eq. (5.64), except a reasonable approximation to it, will provide the agreement seen here.

A correlation for the duration of a spot is provided in Fig. 5.32. Here, the ordinate is the duration scaled by the velocity and displacement thickness at the source, while the abscissa is the Reynolds number based on the free-stream velocity and the distance from the source of the spot. The x intercepts of the straight-line fits shown in Fig. 5.30 lie within the thick band shown near the origin. In these tests the virtual origin is approximately given by

$$\bar{u}_\infty (x_{0,xt} - x_s)/\nu \approx 9{,}600 \tag{5.65}$$

which is relatively small. Since these measurements were obtained at a height roughly equal to the displacement thickness of the undisturbed boundary layer, this result agrees with those presented later in Fig. 5.38 where the duration measured at different heights is shown. There, the virtual origin also coincides with the source for the data taken closest to the wall.

The data in Fig. 5.32 is correlated by

$$\frac{\bar{u}_\infty \Delta t}{x - x_{0,xt}} = 7.7(10)^{-4} Re_{\delta 1,s} \tag{5.66}$$

The variation of spot length with distance follows directly from Eq. (5.61). Depending on whether the leading or trailing edge is positioned at x, we obtain

$$\frac{l(x;t_{le})}{x - x_{0,xt}} = 7.7(10)^{-4} Re_{\delta 1,s} \left(u_{te}/\bar{u}_\infty \right) \quad \text{and} \quad \frac{l(x;t_{te})}{x - x_{0,xt}} = 7.7(10)^{-4} Re_{\delta 1,s} \left(u_{le}/\bar{u}_\infty \right) \tag{5.67}$$

For $Re_{\delta 1,s} = 1000$ with $u_{te}/\bar{u}_\infty = 0.53$, and $u_{le}/\bar{u}_\infty = 0.89$, these expressions yield 0.41 and 0.69, respectively.

Spanwise development. The path of a spot's outboard tip is plotted in Fig. 5.33. The measurements were made in an xz-plane slightly above the wall for spots formed at different source Reynolds numbers and free-stream velocities in a zero-pressure-gradient flow. The ordinate of this figure is the position of the tip measured from the spot's mid-plane, and the abscissa is the position of the tip measured from the virtual origin for the streamwise development of the spot in the xt-plane, $x_{0,xt}$, which corresponds to x_0 in Eq. (5.60).

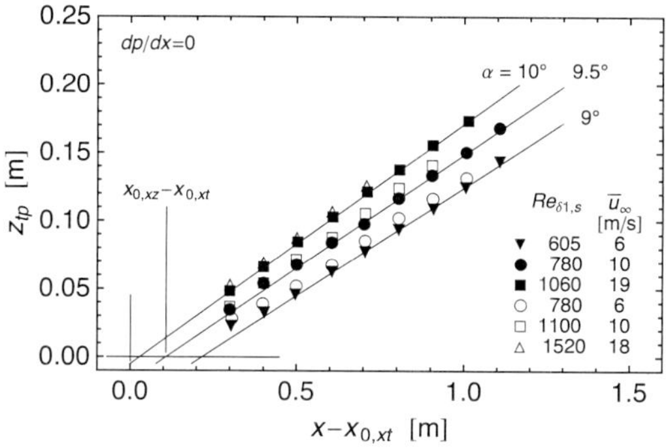

Figure 5.33.
Trajectories of the outboard tip of a spot for a variety of source Reynolds numbers and free-stream velocities. Data from Wygnanski et al. [1982]. Filled symbols, $x_s = 0.3$m; open symbols, $x_s = 0.6$m.

In general, the spots expand in a linear fashion making an angle α of about ten degrees with the x-axis. The exceptions are the spots with the lowest free-stream velocity which initially expand at a smaller angle near the origin. In any event, extrapolating the straight-line fit of the downstream data upstream yields a different virtual origin for the spanwise than for the streamwise development. That is, denoting the virtual origin for the spanwise development of the spot in the xz-plane by $x_{0,xz}$, it and $x_{0,xt}$ are different. As shown in the figure, the difference decreases markedly with increasing velocity for a given source Reynolds number, but changes little with Reynolds number for a given velocity.

A correlation of this data is provided in Fig. 5.34. Here, the half-width of the spot is divided by the displacement thickness, and the Reynolds number based on the distance from the *source* is divided by the displacement-thickness Reynolds number raised to the 0.6 power. The straight-line fit in this figure is given by

$$\frac{z_{tp}}{\delta_{1,s}} = 11.7\left[\frac{u_\infty(x - x_s)}{\nu} Re_{\delta 1,s}^{-0.6}(10)^{-3} - 2.2\right] \tag{5.68}$$

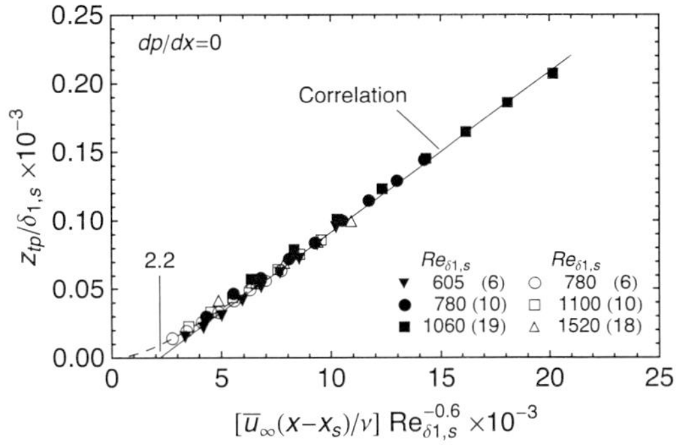

Figure 5.34.
A correlation for the half-width of a spot with distance for a variety of source Reynolds numbers and free-stream velocities. Data from Wygnanski et al. [1982].

The virtual origin $x_{0,xz}$ of the spot is given by

$$\overline{u}_\infty(x_{0,xz} - x_s)/\nu \approx 2200 Re_{\delta 1,s}^{0.6} \tag{5.69}$$

which when evaluated for $Re_{\delta 1,s} = 1000$ is about $1.4(10)^5$. If we estimate that it takes a distance of about twice this before the breadth of the spot increases linearly with distance, the required distance is roughly one-half of that for a spot to become fully developed. Combining Eqs. (5.68) and (5.69) to eliminate x_s yields

$$\frac{z_{tp}}{(x - x_{0,xz})} = \frac{b}{2(x - x_{0,xz})} = 11.7(10)^{-3} Re_{\delta 1,s}^{0.4} \tag{5.70}$$

which equals 0.19 when $Re_{\delta 1,s} = 1000$. The *asymptotic* angle of the tip's trajectory along the wall, obtained by using Eq. (5.56), is then

$$\tan\alpha = 11.7(10)^{-3} Re_{\delta 1,s}^{0.4} \tag{5.71}$$

which when $Re_{\delta 1,s} = 1000$ yields $\alpha = 10.5°$.

A comparison of Eq. (5.71) with data is provided in Fig. 5.35a. The agreement is marginal and not any better than using $\alpha = 10°$ (shown by the dashed line). The figure also includes Schubauer and Klebanoff's data for spots produced at inception Reynolds numbers lower than the critical Reynolds number. These data are shown by the small filled dots connected by a line. The lower of these points correspond to the best estimate for the initial value of α, while the higher points correspond to the angle farther downstream where it becomes constant. In both cases, the *virtual origin* of the spanwise expansion corresponds to a source Reynolds number of about 480. This suggests that a more meaningful plot would result by graphing the angle α against the displacement-thickness Reynolds number at the virtual origin $x_{0,xz}$, not at the source. The other Schubauer and Klebanoff "small-dot" data point with a Reynolds number of about 1100 shows the difference between taking the initial size of the disturbance into account. As reported by them, this difference is about 1.3° which is roughly half the measured variation of α with respect to Reynolds number.

The leading-edge angle can be obtained from Eq. (5.47) and the correlations given by Eqs. (5.67) and (5.68). To use Eq. (5.47), however, the length and breadth must be evaluated at a coinciding time. Since the breadth is measured at the time t_{te}, we must use the expression for $l(x; t_{te})$ in Eq. (5.67).

Therefore, the correct expression for θ is

$$\tan\theta = b(x)/2l(x;t_{te}) \tag{5.72}$$

Substituting for the length and breadth yields

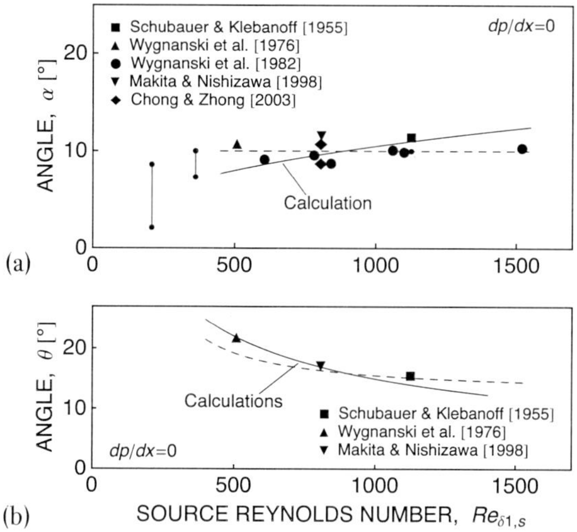

Figure 5.35a,b.
The variation of the expansion and leading-edge angles with source Reynolds number.

$$\tan\theta = 15 Re_{\delta 1,s}^{-0.6} \frac{u_\infty}{u_{le}} \frac{(x - x_{0,xz})}{(x - x_{0,xt})}$$

which is only valid when x is greater than both $x_{0,xz}$ and $x_{0,xt}$. Even so, when the virtual origins are different, the leading-edge angle becomes constant only asymptotically. Its asymptotic value is easily obtained by considering a distance far enough downstream from the virtual origins that their difference can be neglected. In this limit, the above equation then reduces to

$$\tan\theta \approx 15 Re_{\delta 1,s}^{-0.6} \left(u_\infty / u_{le}\right) \quad \text{(large } x\text{)} \tag{5.73}$$

where, of course, the leading- and trailing-edge velocities are equal to their values at large streamwise distances. This relation is plotted as the solid line in Fig. 5.35b, where 0.89 was used as the leading-edge to free-stream velocity ratio. As a result, the above equation reduces to

$$\tan\theta \approx 16.8 Re_{\delta 1,s}^{-0.6}$$

which for $Re_{\delta 1,s} = 1000$ yields $\theta \approx 15°$. The data for this figure is surprisingly limited. That for Schubauer and Klebanoff is found in their paper. That for Wygnanski et al., and Makita and Nishizawa are the author's best estimates based on the measurements and information provided in their papers. Of these two data points, the former is the least certain. Nevertheless, it appears that the asymptotic leading edge angle decreases with an increasing source Reynolds number.

Self similarity. A fully-developed spot is by definition a spot whose shape remains unchanged along its path when appropriately scaled by distance; that is, the spot is "self similar." For $dp/dx = 0$, we found that the height of a fully-developed spot scales by the turbulent boundary-layer thickness or the four-fifths power of distance, while both its length and breadth scale linearly with distance. The self-similarity of a fully-developed spot in the yt-plane on the spot's midspan is shown in Fig. 5.36 where positions of the two-percent deficit in mean velocity are plotted at nine streamwise locations.[1] In this figure, the abscissa is the elapsed time from ignition normalized by the time of flight to the measurement location x_m, while the ordinate is the distance from the wall divided by the turbulent boundary-layer thickness at x_m.

The self-similarity of a spot's footprint, that is, the similarity in the xz-plane near $y = 0$, requires that the leading-edge half-angle θ of the spot must remain constant as it propagates downstream. When this is the case, it follows from Eqs. (5.47) and (5.53) that

$$b(t)/l(t) = \left(u_{le}/u_{te} - 1\right)^{-1} \tan\alpha = \tan\theta = \text{constant} \tag{5.74}$$

for self-similar turbulent spots; independent of how the length and breadth of the spot change with distance and however the velocity ratio may vary. When this expression is used to evaluate the leading-edge angle for $dp/dx = 0$ and a fully-developed spot, we obtain the dashed line shown in Fig. 5.35b. For this calculation, the trailing-edge velocity was obtained from Eq. (5.64) while the leading-edge velocity was assumed equal to 0.89 times the free-stream velocity.

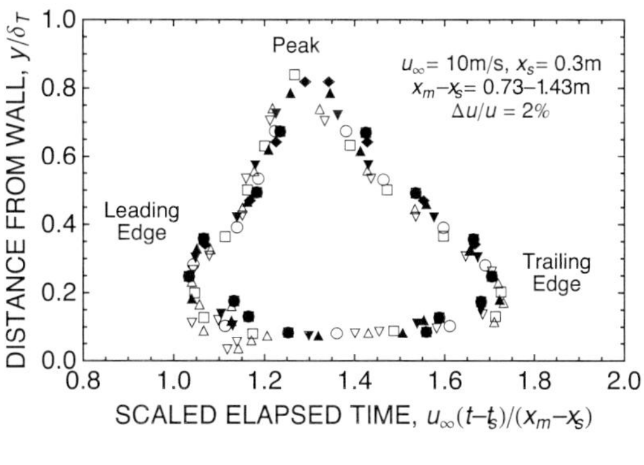

Figure 5.36.
Self similarity in the midspan plane of a fully-developed turbulent spot. Data from Wygnanski et al. [1976].

Approach to self-similarity. Since the virtual origins for the streamwise and spanwise development of a spot are different, at least when the free-stream turbulence level is low, its approach to similarity in the two directions will be different. For high turbulence levels, we might expect this difference to be much smaller even though, as far as the author knows, no data exists.

Presently, the approach to self similarity can only be estimated. We can accomplish this by plotting the straight-line correlations for l and b as functions of x_{te}, and drawing plausible curves that fair into them from the source of the spot. This fairing for b is shown by the dashed line near the origin in Fig. 5.34. For our example, consider the development of a spot with a source Reynolds number $Re_{\delta 1,s} = 1000$. Consequently, its asymptotic leading- and trailing-edge velocities are $u_{le}/\bar{u}_\infty = 0.89$ and $u_{te}/\bar{u}_\infty = 0.53$, where the latter is obtained from Eq. (5.64). The *asymptotic* variation of the spot's length and breadth are easily determined from Eqs. (5.67) and (5.70). They are

[1] For another investigation into the self similarity of turbulent spots, see Cantwell et al. [1978].

$$l(x;t_{te})=0.69\left(x-x_{0,xt}\right) \text{ and } b/2=0.19\left(x-x_{0,xz}\right); \quad (Re_{\delta 1,s}=1000)$$

where upon using Eqs. (5.65) and (5.69), we obtain

$$x_{0,xt}-x_s \approx 0.07\left(x_{0,xz}-x_s\right); \quad (Re_{\delta 1,s}=1000)$$

In addition, evaluating Eqs. (5.71) and (5.73), we obtain the asymptotic angles α_{asym} = 10.5° and θ_{asym} = 15°.

The results of this exercise are shown in Fig. 5.37. In this case, a cubic polynomial passing through the origin and matching the correlation and its first and second derivatives at a match point was used for the fairing. The match points for these polynomials correspond to $x_{l,m}=x_s+1.23(x_{0,xz}-x_s)$ and $x_{b,m}=x_s+3(x_{0,xz}-x_s)$. Their positions are indicated by the short vertical lines in each part of the figure.

The estimated variation in the length, breadth, and skeletal-area of the spot are shown in part (*a*). They are plotted against the distance from the source scaled by the distance to the virtual origin $x_{0,xz}$ from the source. As noted in the figure, this virtual origin has been designated by x_0 while the distance x_0-x_s has been arbitrarily scaled to equal one. The portions of the curves for b and l to the left of the vertical hash marks correspond to the faired polynomials. The dashed line, which is proportional to the square of the distance from the source, is barely distinguishable from the calculated variation for the area of the spot. Since the dashed line represents the variation used by most of the intermittency analyses in §5.2, a quadratic distribution is indeed a respectable substitute for the actual variation in area despite the initial nonlinear variations in spot expansion and associated spot non-similarity.

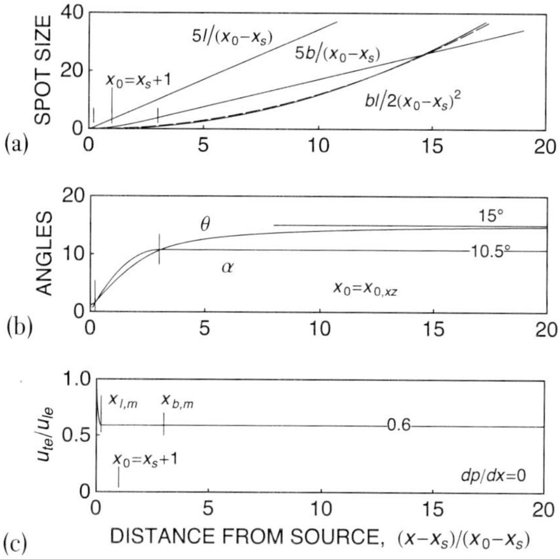

Figure 5.37a,b,c.
Estimated variations of the size, angles and propagation velocities of a developing turbulent spot for $Re_{\delta 1,s}$ = 1000.

The estimated variation in the half-angles α and θ are shown in part (b) of the figure. These were determined from Eqs. (5.56) and (5.72). Clearly, both angles change significantly as the spot develops, and although α attains its asymptotic value within the distance it takes the spot to develop in the spanwise direction (vertical line on the right), the variation in the leading-edge angle θ indicates that takes almost two to three times longer to attain "spot similarity."

Finally, the estimated variation of the trailing-to-leading edge velocity ratio is shown in part (c). This was determined from Eq. (5.63). As should be expected, it becomes constant once the length of the spot begins to increase linearly with distance (vertical line on the left). If $dl/dx = 0$ at inception, as in this example, the initial value of u_{te}/u_{le} equals one. Measurements for both favorable and unfavorable pressure gradients show that the trailing-edge to free-stream velocity ratio remains nearly independent of distance as the spot develops.[1] Presuming that this is also true for zero pressure gradient, the calculated decrease shown in Fig. 5.37c is mostly caused by the increase of u_{le}.

5.3.3 Spots in dp/dx ≠ 0 flow.

Favorable pressure gradients. Adverse pressure gradients.

The major differences between a spot propagating in non-turbulent flows with and without a pressure gradient are the rates by which it grows and spreads across the surface. Although it again grows as a turbulent boundary layer, this growth depends on the pressure gradient. Likewise, the rates by which it elongates and expands, which depend on the stability of the flow both ahead and beneath the overhang of the spot, also depend on the pressure gradient. In this section, we will examine the effect of pressure gradients on the duration and lateral expansion of a spot and their subsequent effect on the intermittency distribution.

Favorable pressure gradients. The effects of favorable pressure gradients on turbulent spots have been investigated by Cantwell et al. [1978], Wygnanski [1981], Glezer et al. [1989], Katz et al. [1990], Clark et al. [1994], and Chong and Zhong [2003, 2005]. Cantwell et al. examined the similarity of spots in a very mild pressure gradient using flow visualization techniques.[2] Wygnanski, Glezer et al., and Katz et al. examined spots propagating in Falkner-Skan flows, Clark et al. examined spots in constant-pressure-gradient flows with elevated Mach numbers,[3] and Chong and Zhong investigated the detail structure within a spot in flows with a constant acceleration parameter. In the following, we will concentrate primarily on the work reported by Wygnanski and his coworkers.

The advantages to investigating spots propagating in a Falkner-Skan flow is that the laminar velocity profiles within the boundary layer remain similar and the laminar-stability results are well documented. As pointed out in §4.2.1, the free-stream velocity for these flows vary according to x^m. Furthermore, as mentioned, the pressure-gradient parameter $\lambda_{\delta 2}$ is independent of distance and proportional to the Falkner-Skan parameter $\beta = 2m/(1+m)$. Wygnanski, Glezer et al., and Katz et al. examined flows with $\beta = 0.12$, 0.2 and 1.0 respectively, which correspond to $m = 0.064$, 0.11 and 1.0.[4] These may be characterized as "very mild," "mild," and "strong" favorable pressure gradients with the latter corresponding to that for stagnation flow. Since the results from Katz' et al. strong pressure-gradient test are the most complete of the set, we will concentrate on them.

In this case, the free-stream velocity increases linearly with distance while the boundary-layer thickness remains constant. Furthermore, the critical Reynolds number is about 13,000, which is about twenty times larger than the Reynolds numbers at their spot source and about fifteen times the largest Reynolds number in the boundary layers they investigated. Therefore, since all of their results are for spots propagating in a *stable* boundary layer, spot growth depends *only* on the instability within the boundary layer caused by the spot itself.

The duration measurements for this flow are shown in Fig. 5.38 (normalized as in Fig. 5.32). The

---

[1] See Figs. 5.40 and 5.43.

[2] These tests were conducted with a small favorable pressure gradient corresponding to a Falkner-Skan flow with $m = 0.143$ ($\beta = 0.25$).

[3] Their constant pressure-gradient flow at low Mach numbers corresponds to a Falkner-Skan flow with $m = 0.5$ ($\beta = 0.67$).

[4] The corresponding velocity profiles can be found in Fig. 4.9. The corresponding values of the pressure-gradient parameter can be found in Appendix D, Table D.3.

velocity used in both coordinates is the *local* free-stream velocity given by

$$\overline{u}_\infty / \overline{u}_{\infty,s} = (x - x_0)/(x_s - x_0) \tag{5.75}$$

where $\overline{u}_{\infty,s}$ is the free-stream velocity at the source and x_0 is the position where the upstream-extrapolated free-stream velocity equals zero. For these tests, x_0 is roughly four meters *upstream* of the plate's leading edge.

The data in this figure were obtained at four different heights above the surface; one well within the undisturbed laminar boundary layer, one at its edge, and two above with the farthest away near the peak of the spot. Clearly, the duration at every height varies linearly with distance.[1] The "virtual" origin, however, changes with height, which makes sense after giving some thought regarding the mid-span profile of a spot. For the data measured closest to the surface, the virtual origin coincides with the source. For that measured at the height of the undisturbed boundary layer, the virtual origin shifts *upstream* as shown in the figure by an amount roughly equal to $Re_{\Delta x 0}$ = 68,000. This shift corresponds to the distance between the leading edge of the spot's footprint and the leading-edge nose of the spot, whereas the largest *downstream* shift corresponds to the distance between the footprint's leading edge and the peak of the spot.

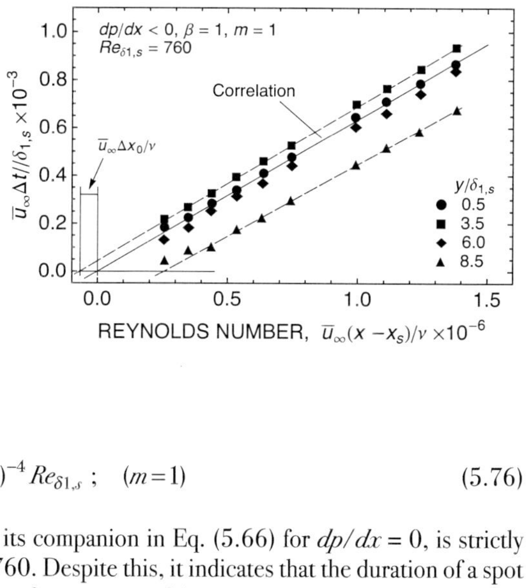

Figure 5.38.
The measured duration of a turbulent spot at several heights in its midspan plane. Data from Katz et al. [1990].

The data nearest the wall is correlated by

$$\frac{\overline{u}_{\infty,s}\Delta t}{x - x_s} = 6.34(10)^{-4}\, Re_{\delta 1,s}\,; \quad (m=1) \tag{5.76}$$

This result, which is functionally the same as its companion in Eq. (5.66) for $dp/dx = 0$, is strictly valid only for the source Reynolds number of 760. Despite this, it indicates that the duration of a spot in an accelerating flow with $m = 1$ is about eighty-four percent of that in a non-accelerating flow.

A correlation of their measured spot length for several Reynolds numbers is shown in Fig. 5.39. The length in this case is based on the time when the *leading edge* of the spot is at the position x, and the coordinates are normalized by using the *local* free-stream velocity as suggested in Eq. (5.67). The equation for a straight-line fit is

$$\frac{l(x;t_{le})}{\delta_{1,s}} = 5.5(10)^{-4}\frac{u_{le}}{\overline{u}_\infty}\left[\overline{u}_\infty(x - x_s)/\nu - \Delta Re_{x0}\right] \tag{5.77}$$

where the displacement of the virtual origin from the source is given approximately by

[1] Clark's et al. data taken in a "mild" favorable pressure gradient displays a similar linear variation.

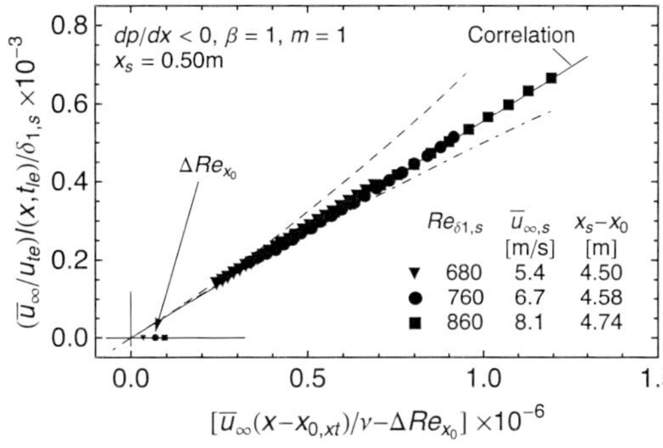

Figure 5.39.
A correlation for the length of a turbulent spot with distance in a flow with a strong favorable pressure gradient. Data from Katz et al. [1990].

$$\Delta Re_{x0} = (\overline{u}_\infty)_{0,xt}\left(x_{0,xt} - x_s\right)/\nu = 1.87(10)^5\left(1-0.0018\,Re_{\delta 1,s}\right)$$

Since the ratio of the trailing-edge to free-stream velocity is nearly constant as shown in Fig. 5.40, and since the displacement thickness is constant for this flow, the length $l(x;\, t_{le})$ varies essentially as $(x-x_0)(x-x_s)/(x_s-x_0)$. The values of (x_s-x_0) for these data are listed in Fig. 5.39, while the maximum value of $(x-x_s)$ is about 1.5m. Therefore, it is quickly ascertained that the variation of $l(x;\, t_{le})$ with respect to x is mildly nonlinear with respect to distance.

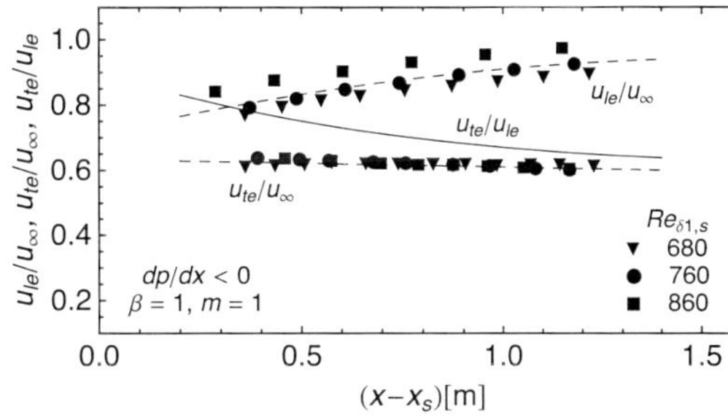

Figure 5.40.
Leading- and trailing-edge propagation velocities of a turbulent spot as a function of the streamwise distance in a flow with a strong favorable pressure gradient. Data from Katz et al. [1990].

Given $l(x;\, t_{le})$ and the propagation velocities, the length $l(x;\, t_{te})$ can be determined from Eq. (5.63). The mean of this length for the three Reynolds numbers (obtained by using the mean of the velocity ratio shown by the solid line in Fig. 5.40) is plotted as the dashed line in Fig. 5.39. Its nonlinear behavior with respect to x is already obvious before taking the variation of the free-stream velocity into account.

Finally, Katz' et al. correlation for the mean Reynolds numbers is plotted as the dot-dashed line in the figure. They assumed a linear expansion and obtained the correlation

$$l(x;t_{le}) = 4.63(10)^{-4}\,Re_{\delta 1,s}\left(x - x_s\right)$$

which when plotted in the present format becomes nonlinear because of the variation in the free-stream velocity. It is relatively easy to show from Eqs. (5.54) and (5.63) that either the length or duration must vary nonlinearly with distance when one of them varies linearly and the propagation velocities are not constant. In this case, the variation of the propagation velocities affects the length and duration by an amount that can easily be buried within the correlation and/or experimental error.

The half-width of the spot correlated with distance is shown in Fig. 5.41. In this case, the correlation is given by

$$\frac{z_{tp}}{\delta_{1,s}} = 14.9\left[\frac{\overline{u}_{\infty,s}(x - x_s)}{\nu}(10)^{-5} - 0.93\right] \tag{5.78}$$

The virtual origin $x_{0,xz}$ of the spot in the xz-plane is given by

$$\overline{u}_{\infty,s}(x_{0,xz} - x_s)/\nu \approx 93{,}000$$

which when combined with Eq. (5.78) to eliminate x_s yields

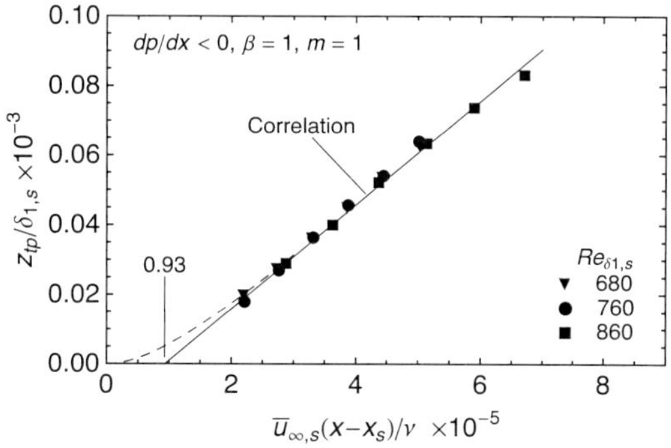

Figure 5.41.
A correlation for the half-width of a turbulent spot with distance in a flow with a strong favorable pressure gradient. Data from Katz et al. [1990].

$$\frac{z_{tp}}{x - x_{0,xz}} = 14.9(10)^{-5}\,Re_{\delta 1,s}\,; \quad (m=1)$$

Furthermore, by taking the derivative of the above and using Eq. (5.56), we obtain an expression for the *asymptotic* expansion angle which is given by

$$\tan\alpha = 14.9(10)^{-5}\,Re_{\delta 1,s}\,; \quad (m=1) \tag{5.79}$$

For $Re_{\delta 1,s} = 1000$ this yields $\alpha = 8.5°$, which when compared to the angle of 10.5° for $dp/dx = 0$ (see Fig. 5.37b), indicates that a spot in an accelerating flow expands less than a spot in a non-accelerating flow. Furthermore, since the spot expands linearly with respect to distance but elongates quadratically, at least for stagnation flow with $\beta = m = 1$, its leading-edge angle continually decreases with distance such that the spot never attains self-similarity.

Adverse pressure gradients. The effects of an adverse pressure gradient on turbulent spots have been investigated by Seifert and Wygnanski [1995], Gostelow et al. [1996], and D'Ovidio et al. [2001]. Most of these investigations were conducted in Falkner-Skan flows. The data obtained by Seifert and Wygnanski are for $\beta = -0.1$, whereas Gostelow et al. and D'Ovidio et al. investigated spots propagating in nearly separating flows with β close to -0.19. These are respectively moderate and strong adverse pressure gradients. Since the latter is complicated by interacting separation bubbles, which will be dealt with in §8.3, we will examine the moderate adverse pressure gradient results obtained by Seifert and Wygnanski in this section.

In their case, with $\beta = -0.1$, the critical Reynolds number is about 200 or roughly half of that for zero pressure gradient.[1] Since the source Reynolds numbers are all about four times larger than this, their data are for spots propagating in a highly *unstable* boundary layer. Therefore, their expansion de-

[1] Refer to Appendix D, Table D.3.

pends on both the instability within the boundary layer and the local instability caused by the spot. Since $m = -0.048$ when $\beta = -0.1$, the free-stream velocity varies according to

$$\overline{u}_\infty / \overline{u}_{\infty,s} = \left[(x - x_0) / (x_s - x_0) \right]^{-0.048}$$

where again x_0 is the virtual origin of the flow, which in this case corresponds to where the free-stream velocity equals infinity. For these tests, x_0 is 0.23m *downstream* of the plate's leading edge. Finally, from Eq. (4.20), it can be determined that the boundary-layer-thickness Reynolds number increases with distance raised to the 0.48 power, which is only slightly less than that for flow without a pressure gradient.

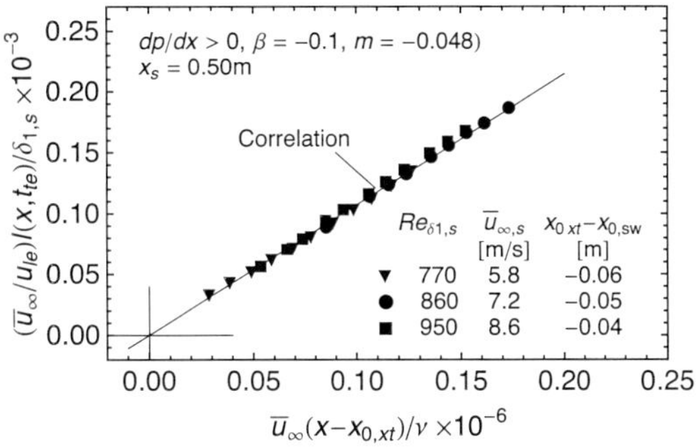

Figure 5.42.
A correlation for the length of a turbulent spot with distance in a flow with a moderate adverse pressure gradient. Data from Seifert and Wygnanski [1995].

A correlation of Seifert and Wygnanski's measured spot lengths with distance is shown in Fig. 5.42 for several source Reynolds numbers. Since their lengths are based on the time when the *trailing edge* of the spot is positioned at x, their data are plotted using the leading edge instead of the trailing edge velocity ratio in the ordinate. This is the form suggested by Eq. (5.67) and the fact that the duration of a spot, which increases linearly with distance for flows with a zero and favorable pressure gradient, might increase linearly irrespective of the pressure gradient.[1] Here, however, it should be noted that the abscissa in this figure is an order a magnitude shorter than the abscissa in either Fig. 5.32 or Fig. 5.38 for a zero and favorable pressure gradient. This is easily understood once it is remembered that spots generated in an undisturbed flow with an adverse pressure gradient are propagating in a highly unstable boundary layer which will naturally and quickly transition. In their experiments, Seifert and Wygnanski report that natural transition occurred at a Reynolds number based on the distance from the source of roughly $3(10)^5$. This also explains the relatively small range of source Reynolds numbers associated with these data compared to those for the other pressure gradients.

In any case, the correlation shown by the straight line in this figure is given by

$$\frac{l(x;t_{te})}{\delta_{1,s}} = 1.07(10)^{-3} \frac{u_{le}}{\overline{u}_\infty} \left[\overline{u}_\infty (x - x_{0,xt}) / \nu \right] \tag{5.80}$$

where the difference between $x_{0,xt}$ and the Seifert and Wygnanski's virtual origin $x_{0,sw}$ is given in the figure. Since the difference between $x_{0,xt}$ and x_s were not reported, a correlation using the distance from the source cannot be obtained.

The variations of the spot's leading- and trailing-edge velocities with distance are shown in Fig. 5.43.

[1] As we will see in §7.3.1, it is best to measure the spot's duration and breadth rather than its length and breadth. This follows directly from the fact that Emmons' turbulent spot theory depends on the spot's "area" measured in the zt-plane $A_{s,zt}$, and not on its physical area $A_{s,xz}$.

The leading-edge velocity increases with distance as does that for a spot in a favorable pressure gradient. More than likely, the leading-edge velocity for the zero pressure gradient case also increases with distance as the spot develops even though no data exists to corroborate it.[1] As indicated by the lowest dashed line in the figure, Seifert and Wygnanski report that the trailing-edge velocity for the adverse pressure gradient is independent of distance. This is similar to what was found for the favorable pressure gradient case, and probably what will be found for the case without a pressure gradient. The level shown in Fig. 5.43 for this velocity is the mean of some very scattered data for several source Reynolds numbers. This average is about 0.48 which, when compared to the mean values of 0.55 and 0.61 for the zero and favorable pressure gradient cases respectively, indicates that the trailing-edge propagation velocity increases with an increasing Falkner-Skan parameter β or pressure-gradient parameter $\lambda_{\delta 2}$.

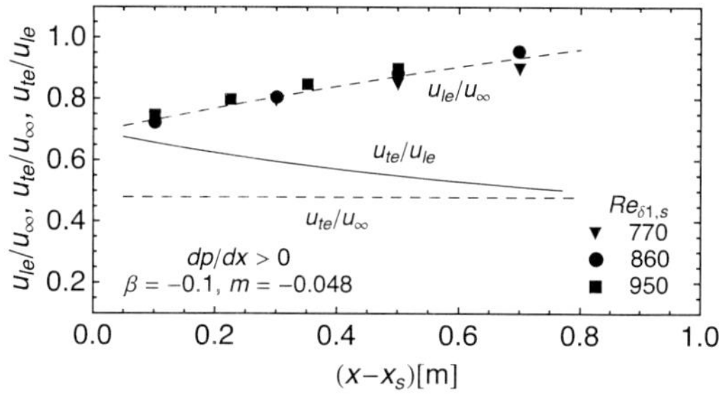

Figure 5.43.
Leading- and trailing-edge propagation velocities of a turbulent spot as a function of the streamwise distance in a flow with a moderate adverse pressure gradient. Data from Seifert and Wygnanski [1995].

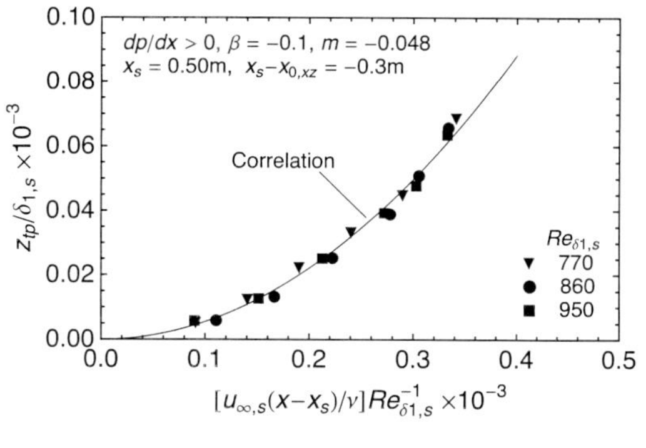

Figure 5.44.
A correlation for the half-width of a turbulent spot with distance in a flow with a moderate adverse pressure gradient. Data from Seifert and Wygnanski [1995].

The spanwise expansion of the spot is shown in Fig. 5.44. This is dramatically different from the asymptotically linear expansion displayed by both the zero and favorable pressure-gradient results. Although the difference between $x_{0,xz}$ and x_s was not reported by the investigators, the author's best guess for it is given in the figure. In any case, the correlation shown in the figure for the spot-tip trajectory is approximately given by

$$\frac{z_{tp}}{\delta_{1,s}} = 5.5 Re_{\delta 1,s}^{-2} (10)^{-4} \left[\bar{u}_{\infty,s} (x - x_s) / \nu \right]^2 \; ; \quad (m = -0.048) \tag{5.81}$$

[1] Spot propagation measurements for the zero pressure gradient case are only available for fully-developed spots.

In other words, the spots never reach an asymptotic state of linear expansion. By taking the derivative of the above expression and using Eq. (5.56), an expression for the *local* expansion angle is obtained, namely,

$$\tan\alpha = 1.1(10)^{-3}\, Re_{\delta 1,s}^{-1}\left[\overline{u}_{\infty,s}(x - x_{0,xz})/\nu\right]; \quad (m = -0.048) \tag{5.82}$$

which for $Re_{\delta 1,s}$ = 1000 yields an angle that varies from α = 6.3° to 18° as the distance Reynolds number from the source increases from about 100,000 to 300,000. This result is significantly different from Seifert and Wygnanski's, who correlated the data assuming a linear asymptotic expansion and obtained an asymptotic angle of twenty-one degrees. Nevertheless, we see that a spot in a decelerating flow expands faster than a spot in a non-accelerating flow.

A comparison between the spanwise expansion of a spot propagating in the three different pressure gradients studied by Wygnanski and associates is shown in Fig. 5.45. The curves were obtained from Eqs. (5.68), (5.78) and (5.81) for $Re_{\delta 1,s}$ = 1000. The length of each curve corresponds roughly to the distance over which they obtained data. As previously mentioned, the curve for the adverse pressure gradient ends just before natural transition occurs in the surrounding boundary layer. The dashed line beyond this position is an extrapolated calculation. A comparison between the intermittency distributions for these three cases will be presented in §7.3.1 where we will investigate various turbulent spot models.

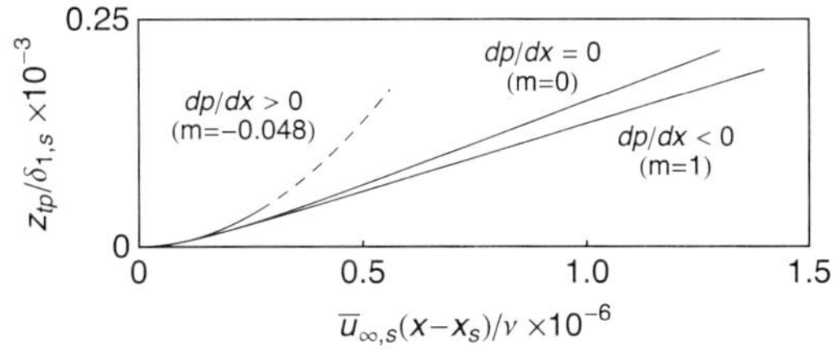

Figure 5.45.
The effect of pressure gradients on the spanwise expansion of a turbulent spot in a non-turbulent flow with $Re_{\delta 1,s}$ = 1000.

5.4 Spots and Becalmed Matters.

Stochastic theory of transition for multiple regions of flow. Spot, becalmed, and non-becalmed flow interaction. Effect of becalmed flow on spot production.

Because Emmons' turbulent spot theory *appears* to predict the intermittency through transition so well, a general theory of transition including becalmed zones and other regions of the flow has yet to be developed. Moreover, as mentioned in §5.1.4, virtually no experiments have been conducted to both define and investigate becalmed zones despite knowing that they can affect transition by; 1) their interaction with turbulent spots, and 2) their influence on spot production.

In this section, we will first develop a stochastic theory of transition for multiple regions of flow, including becalmed zones, examine the interactions that occur between these different regions, and then estimate the effect of becalmed zones on turbulent spot production.

5.4.1 Stochastic theory of transition for multiple regions of flow.

The three regions of transitional flow. A stochastic becalmed-zone theory. The becalmed-zone propagation parameter.

If we consider a portion of the laminar flow within transition as being becalmed, then transition consists of three primary regions of flow. As illustrated in Fig. 5.46, these are the turbulent, laminar becalmed, and laminar non-becalmed regions.[1] The spot and becalmed footprints shown in this figure correspond to the two-percent $\sqrt{(\overline{u'u'})}/u_\infty$ and $\Delta u/u_{lam}$ contours drawn in Fig. 5.14b.

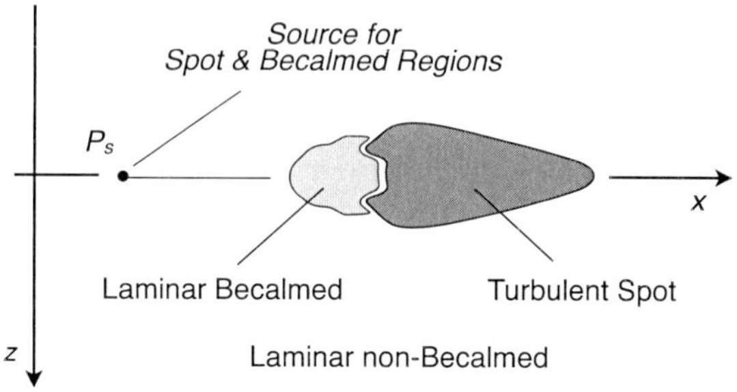

Figure 5.46.
The three regions of laminar-turbulent transitional flow. Footprints based on data from Makita and Nishizawa [1998].

Following Emmons analysis, if we consider a point $P(x, z, t)$ in xzt-space and denote the probability that the flow there is turbulent, laminar becalmed, and laminar non-becalmed by γ_t, γ_b, and γ_{nb}, respectively, then according to probability theory, since these regions are by definition *mutually exclusive*, we obtain

$$\gamma_t(P)+\gamma_b(P)+\gamma_{nb}(P)=1\,; \quad (0\le\gamma_t,\gamma_b,\gamma_{nb}\le 1) \tag{5.83}$$

The problem is now to determine γ_t, γ_b, and γ_{nb} – and this begins by reinterpreting and applying Emmons' solution when three regions of transitional flow are considered instead of two.

First define $\gamma_{t,E}$ as Emmons' solution for the probability of turbulent flow when the flow is either *laminar* or *turbulent*, that is, composed of only two regions. At this point in the development, the quantity $\gamma_{t,E}$ should not be called the "intermittency," since, according to Eq. (5.83), this term is strictly reserved for the quantity γ_t. Nevertheless, as defined, $\gamma_{t,E}$ is given by Eq. (5.9), namely,

$$\gamma_{t,E}(P)=1-\exp\left\{-\int_R g(P')\,dx'\,dz'\,dt'\right\}; \quad (0\le\gamma_{t,E}\le 1) \tag{5.84}$$

where $g(P')$ is the production rate of turbulent spots at P', and R is the volume of the spot's dependence cone in xzt-space (shown in Fig. 5.17). In this case, $1-\gamma_{t,E}(P)$ is the probability the flow at P is non-turbulent, be it "laminar becalmed" or "laminar non-becalmed."

A similar solution can be obtained for the probability that the flow is becalmed when the flow is either *becalmed* or "something else" such as combination of *turbulent* and *non-becalmed* flows. In this case, realizing that 1) becalmed zones only exist when spots are produced, and 2) the number of spots and becalmed zones produced at any point on the surface must be the same, it follows from either the development of Eq. (5.9) or by direct inspection of Eq. (5.84) that the probability of a becalmed flow at P when only a becalmed and some other region exists is

$$\gamma_{b,E}(P)=1-\exp\left\{-\int_{R_b} g(P')\,dx'\,dz'\,dt'\right\}; \quad (0\le\gamma_{b,E}\le 1) \tag{5.85}$$

where R_b is the volume of the becalmed dependence cone in xzt-space as shown in Fig. 5.47.[2] Similar to its definition for a turbulent spot, the dependence cone for a becalmed zone is the volume containing all sources of becalmed zones upstream of the point P that will cause the flow at P to be becalmed

[1] Actually, four regions may be distinguished if we consider dividing the region of laminar non-becalmed into regions of "laminar stable" and "laminar unstable" flow, while considering "laminar stable" flow different than "laminar becalmed."

[2] The shapes of the spot/becalmed ensemble in Figs. 5.46 and 5.47 are different because they are shown in different planes of xzt-space.

at least for the time it takes the becalmed zone to pass over it. It also follows, that $1-\gamma_{b,E}(P)$ is the probability the flow at P is something other than laminar becalmed, be it "turbulent" or "laminar non-becalmed."

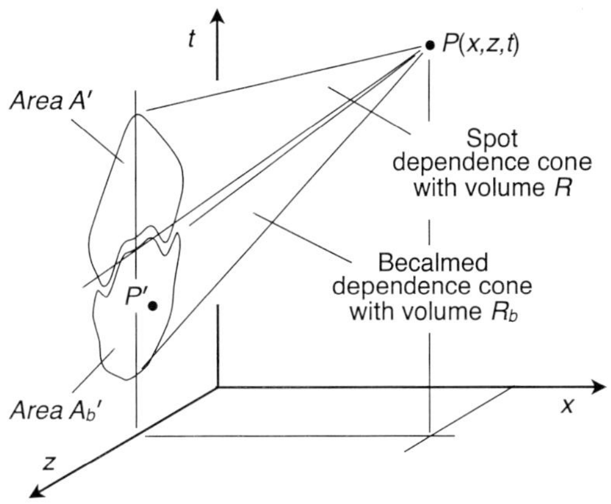

Figure 5.47.
The dependence cone associated with the footprints of a turbulent spot and its becalmed zone for the point $P(x,z,t)$. Based on data from Makita and Nishizawa [1998].

For transition downstream of a line source of turbulent spots at $x = x_t$ when the spot/becalmed ensemble expands linearly across the surface, Eqs. (5.84) and (5.85) after referring to the development of Eq. (5.16) may be written as

$$\gamma_{t,E}(x;x_t)=1-\exp\left[-\frac{\dot{n}'\sigma}{\overline{u}_\infty}\left(x-x_t\right)^2 \mathrm{U}(x-x_t)\right] \tag{5.86}$$

and

$$\gamma_{b,E}(x;x_t)=1-\exp\left[-\frac{\dot{n}'\sigma_b}{\overline{u}_\infty}\left(x-x_t\right)^2 \mathrm{U}(x-x_t)\right] \tag{5.87}$$

where $\mathrm{U}(z)$ is the unit step function of z, and σ_b is the propagation parameter for the becalmed zone. According to the expression given for σ in Eq. (5.15), the latter is defined by

$$\sigma_b=\frac{\overline{u}_\infty}{(x-x')^2}\int_{A_b'}dz'\,dt'=\left[\int_{A_b'}dz'\,dt' \Big/ \int_{A'}dz'\,dt'\right]\sigma \tag{5.88}$$

where A' and A_b' are the "areas" of the spot and becalmed footprints as shown in Fig. 5.47.

Once the extent of the becalmed zone is defined, the value of σ_b, or more relevant, the ratio σ_b/σ, can be determined. Evidently, it is most easily obtained from measurements of the spot/becalmed "footprints" in the zt-plane. In particular, when the spot/becalmed areas A' and A_b' corresponding to the two percent velocity fluctuation and perturbation contours in Fig. 5.14b are used, a direct integration of the measurements taken by Makita and Nishizawa [1998] yields $\sigma_b/\sigma \approx 1.0$.[1] On the other hand, when the flow within the recovery tail corresponding to the triangular region in Fig. 5.14a is considered as becalmed, $\sigma_b/\sigma \approx 4.0$. The numerical values of σ and σ_b as determined by the author are provided in Table 5.2. For further information about the becalmed zone, see the discussion in §5.1.4.

Whether $\gamma_t = \gamma_{t,E}$ and $\gamma_b = \gamma_{b,E}$, or not, depends upon the interactions occurring between the various regions of flow. Presently, these interactions must be inferred from experiments.

[1] The data used for this calculation are shown in part (b) of Fig. 5.49.

Table 5.2
Propagation parameter of a turbulent spot, becalmed zone, and recovery tail for $Re_{\delta 1,s} = 805$. Based on the data of Makita and Nishizawa [1998].

Region	σ	σ_b	σ_b/σ
Turbulent Spot (two percent $\overline{u'u'}$ contour)	0.17	–	–
Becalmed Zone (two percent $\Delta\overline{u}$ contour)	–	0.18	1.04
Becalmed Zone (triangular recovery tail)	–	0.69	4.03

5.4.2 Spot, becalmed, and non-becalmed flow interactions.

Experimental results. Interaction rules. Some calculations.

Since the leading-edge velocity of a spot is greater than its trailing-edge velocity, a spot produced later than another will eventually catch-up and interact with its nearby predecessor. Three of the four primary interactions, namely, spot/spot, spot/becalmed, and becalmed/becalmed, are illustrated in Fig. 5.48, where the footprints of individual spots and becalmed zones are sketched superimposed on each other. Here, the interaction represents the results as measured at $x = x_m$ in xzt-space for spots produced by two sources located at P_1 and P_2 with $x_1 = x_2 = x_s$, $z_2 = z_1 + \Delta z_s$, and $t_2 = t_1 + \Delta t$. Note that the shape and size of each ensemble is the same in this figure because the sources are the same distance upstream of the measurement plane, that is, $x_m - x_s$ is the same for each. The fourth interaction, which occurs between the laminar becalmed and non-becalmed unstable regions, will be discussed momentarily.

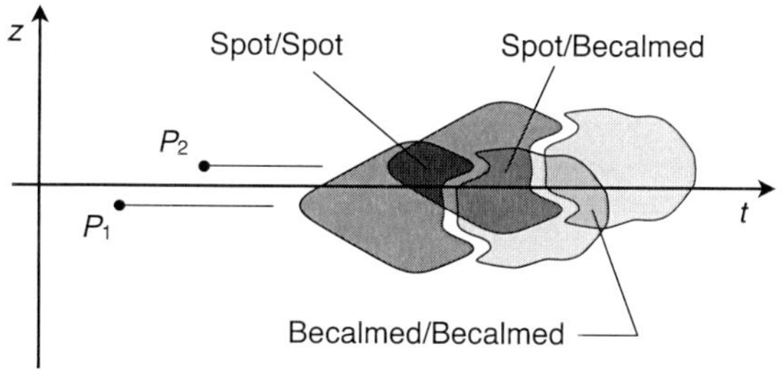

Figure 5.48.
The intersections of two turbulent spots and their appended becalmed regions in a zt-plane. Based on data from Makita and Nishizawa [1998].

Makita and Nishizawa [1998, 1999] provide the only helpful information regarding spot, becalmed, and non-becalmed flow interactions known to the author. Their experiments were conducted on a flat plate placed parallel to the incident flow with $dp/dx = 0$, and consisted of temporal measurements taken throughout the yz-plane 600mm downstream of two sources placed 40mm apart, 694mm downstream from the plate's leading edge, that is, at x_m = 1294mm. The incident velocity was 5m/s yielding a displacement-thickness Reynolds number $Re_{\delta 1,s}$ equal to 805 at the sources.

Their experimental results near the surface are shown in Fig. 5.49, where the difference between parts (a) and (b) of the figure is the delay in producing the second spot. In part (a), the delay is $t_2 - t_1$ = 60ms which produces a "fore/quarter" interaction. The "beam/beam" interaction shown in part (b) of the figure occurs when the spots are produced simultaneously. The merged results in both parts of the figure are drawn using solid lines, while the results for the first and second spot/becalmed ensembles without the other are drawn using dashed and dot-dashed lines. To the author's knowledge, no data exists for spots following one another, that is, for a "fore/aft" interaction.

Upon comparing the contours of the merged spot and becalmed zone in part (a) of this figure with the overlapped areas of the previous figure, it becomes evident that interactions between similar regions of flow, that is, a spot/spot or becalmed/becalmed interaction, do not change the flow to something else, whereas a spot/becalmed interaction changes most of the flow to turbulent; the exception here being the inboard tails of the spots which change to laminar-becalmed as seen most readily in part (b) of the figure.

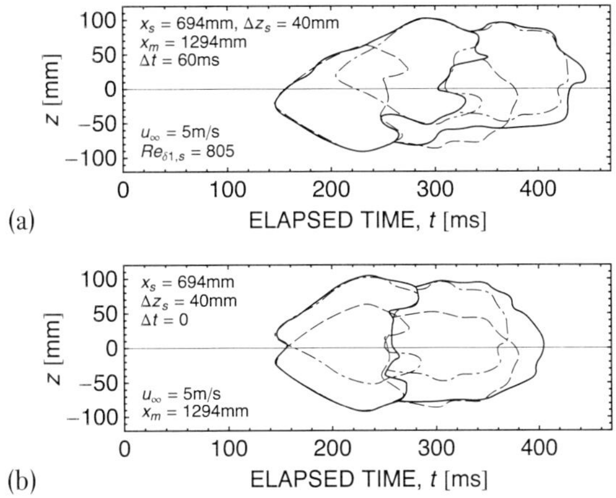

Figure 5.49a,b.
Overlays of two turbulent spots and their appended becalmed regions (dashed and dot-dashed lines) compared to their merged regions (solid lines) for (a) spots produced with a 60ms delay, and (b) spots produced without a delay. Data for $dp/dx = 0$ from Makita and Nishizawa [1998, 1999].

The combined-overlapping areas of the first-and-second spots and first-and-second becalmed zones in parts (a) and (b) of Fig. 5.49 and the actual merged areas (solid contours) for the two types of interactions are provided in Table 5.3. The first column in the table is the area of the combined overlapping spots A_t, the second is that of the combined overlapping becalmed zones A_b, while the third and fourth are the areas of the actual merged regions; all given in [ms]. The ratios of the actual-to-combined areas are provided in the fifth and sixth columns.

Table 5.3
Areas associated with the intersections of two turbulent spots and their becalmed regions in [ms]. Based on the data presented in Fig. 5.49.

Interaction	A_t	A_b	$A_{t,actual}$	$A_{b,actual}$	$A_{t,actual}/A_t$	$A_{b,actual}/A_b$
Fore/Quarter	0.0217	0.0157	0.0218	0.0167	1.00	1.06
Beam/Beam	0.0177	0.0161	0.0172	0.0204	0.97	1.27

A ratio of unity in the last two columns implies that the merged region may be calculated by properly superimposing the spot/becalmed ensembles and using three interaction rules. Stated somewhat differently than above, these rules are: 1) all interactions with a turbulent spot become turbulent, 2) all interactions with a becalmed zone except those with a turbulent spot become becalmed, and 3) all interactions with the non-becalmed flow can be neglected. Clearly, according to Table 5.3 and Fig. 5.49b, these rules apply for all but the beam/beam interaction which doesn't satisfy rule "3." In this case, the interaction between the becalmed and unstable non-becalmed flow increases the area of becalmed flow by about twenty-seven percent. Fortunately, however, as we will see in the following section, this effect on transition is small enough such that, to the first approximation, the above rules can be applied to all interactions (at least for transitional flow with $dp/dx = 0$).

Mathematically, the above rules may be stated as follows. Let $A_{s,i}$ be the set of all points within the spot footprint in the zt-plane associated with a source located at P_i.[1] Likewise let $A_{b,i}$ and $A_{nb,i}$ be the corresponding sets for the becalmed footprint and non-becalmed region of flow. Then the above rules for any number of spots n within an area A_{zt} of the zt-plane at a distance x are given by

$$\begin{aligned} A_t(x) &= A_{s,1} \cup A_{s,2} \cup \ldots A_{s,i} = \bigcup_{i=1}^{n} A_{s,i} \\ A_b(x) &= \bigcup_{i=1}^{n} A_{b,i} \cap \left(\bigcup_{i=1}^{n} A_{s,i} \cap \bigcup_{i=1}^{n} A_{b,i} \right)' \\ A_{nb}(x) &= A \cap \left(A_t \cup A_b \right)' \end{aligned} \tag{5.89}$$

where A_t is the set of all points of turbulent flow within the area A_{zt}, whether it is composed of merged turbulent spots or not, A_b and A_{nb} are the corresponding sets of points of becalmed and non-becalmed flow, and $A = A_t \cup A_b \cup A_{nb}$ is the set of all points within the area A_{zt}.[2] Furthermore, the notation Q' denotes the complement of any set Q, that is, the set of points that do not belong to Q. For example, $A_t' \cup A_t = A$ and $A_t' \cap A_t = \varnothing$, where $\varnothing$ is the empty set of points. Physically, the second term of the second expression in Eq. (5.89) represents that portion of the flow which, following rule "1," changes from becalmed to turbulent as a result of the interaction. Likewise, the second term of the third expression represents that portion of the flow that is neither turbulent nor becalmed. To account for an interaction between the becalmed and unstable non-becalmed flow, the second expression in Eq. (5.89) must be modified accordingly.

Presuming all three rules of interacting flow apply, the first implies that the probability of the near-wall flow being turbulent at any point through transition is given by

$$\gamma_t = \gamma_{t,E} \tag{5.90}$$

which was originally observed by Emmons [1951] and incorporated into his theory.

Since $(1 - \gamma_{t,E})$ is the probability the flow is something other than turbulent, and since $(1 - \gamma_{b,E})$ is the probability the flow is something other than becalmed, the third rule implies that the probability of the flow being neither turbulent nor laminar-becalmed, in other words, laminar non-becalmed, is

$$\gamma_{nb} = \left(1 - \gamma_{b,E}\right)\left(1 - \gamma_{t,E}\right) \tag{5.91}$$

As a result, from Eq. (5.83) we obtain

$$\gamma_b = \gamma_{b,E}\left(1 - \gamma_{t,E}\right) \tag{5.92}$$

Hence, the distribution of flow throughout transition can be expressed in terms of Emmons' solutions for both the spots and becalmed zones as defined by Eqs. (5.84) and (5.85).

Before proceeding, it is worthwhile to define the fraction of time the *laminar* flow is becalmed. Since $\gamma_t = \gamma_{t,E}$, it is given by

---

[1] Rather than changing notation, the symbol A is being used to denote both the set of points within an area and the area. That is, the integral with respect to the elemental area $dA = dzdt$ over all the points within $A_{s,i}$ equals the area of the spot footprint in the zt-plane, namely, $A_{s,zt}(x_m)$. Similarly for $A_{b,i}$ and $A_{nb,i}$.

[2] Thus, the integral with respect to the elemental area $dA = dzdt$ over all the points within the sets A_t and A_b equals the areas denoted by A_t and A_b in Table 5.3.

$$f_b = \gamma_b / (1 - \gamma_t) = \gamma_{b,E} \tag{5.93}$$

where Eq. (5.92) has been used. In other words, Emmons' solution for the becalmed zones is the fraction of time the laminar flow is becalmed.[1]

Some calculations. As an example, consider calculating the probability distributions for transitional flow downstream of a line source of turbulent spots on a flat plate. Since all spots are produced at one position, we don't have to consider the effect of the becalmed zones on their production. Presuming the interaction rules given in Eq. (5.89) apply, Eqs. (5.90)–(5.92) may be used to determine the probabilities γ_t, γ_b, and γ_{nb}. In this case, Emmons' solutions $\gamma_{t,E}$, and $\gamma_{b,E}$ are given by Eqs. (5.86) and (5.87).

The results of these calculations are plotted in Fig. 5.50 for $\hat{n}\sigma = 7.5(10)^{-11}$ and two propagation parameter ratios σ_b/σ. Since $\gamma_t = \gamma_{t,E}$ and $\gamma_b + \gamma_{nb} = (1 - \gamma_{t,})$, the only difference between parts (a) and (b) of the figure are the probability distributions of the becalmed and non-becalmed flow. In other words, only the partition between the two different regions of laminar flow is affected by the propagation ratio.

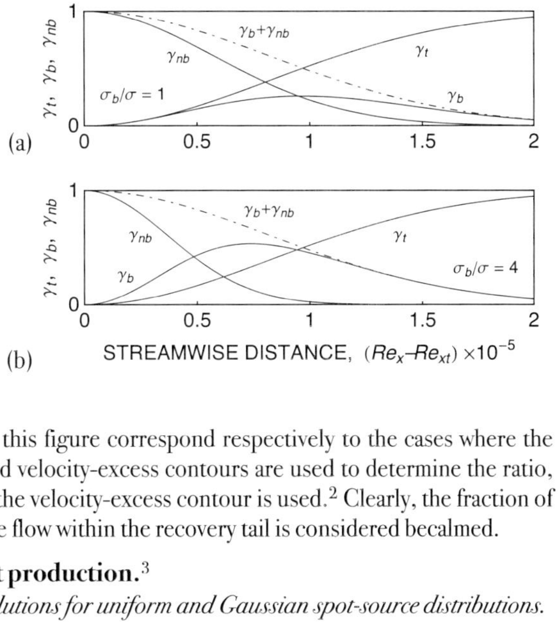

Figure 5.50a,b.
Probability distributions for transitional flow when the becalmed-to-spot propagation ratio $\sigma_b/\sigma = 1$ and 4. Calculated for $\hat{n}\sigma = 7.5(10)^{-11}$.

The curves for $\sigma_b/\sigma = 1$ and $\sigma_b/\sigma = 4$ in this figure correspond respectively to the cases where the measured areas of the two percent spot and velocity-excess contours are used to determine the ratio, and the area of the recovery tail instead of the velocity-excess contour is used.[2] Clearly, the fraction of becalmed laminar flow is larger when all the flow within the recovery tail is considered becalmed.

5.4.3 Effect of becalmed flow on spot production.[3]

Integral equations for transitional flow. Solutions for uniform and Gaussian spot-source distributions.

The stabilizing effect of the becalmed flow on transition is easier to describe when we consider uniform flow over a flat plate aligned parallel to it, and a rectangular coordinate system such that the plate lies in the $y = 0$ plane with x measured downstream along the plate from its leading edge, and z measured along its leading edge.

Since becalmed flow only affects the intermittency, that is, γ_t, when spots are *produced* over an area of the surface, let's consider the effect of becalmed flow on the production of turbulent spots from a dis-

---

[1] Indeed, in the first edition of this book, the quantity $\gamma_{b,E}$, which was simply denoted by γ_b, was defined to be the fraction of laminar flow that is becalmed.

[2] See Table 5.2 and Fig. 5.14.

[3] This analysis is essentially that proposed by Mayle [1993], but developed in more detail.

tribution of line sources.[1] The solution to this problem without considering the becalmed flow has previously been given by Eq. (5.27), namely,

$$\gamma_t^{(0)}(x)=1-\exp\left\{-\frac{\sigma}{\bar{u}_\infty}\int_0^\infty \dot{n}''(\xi)\left(x-\xi\right)^2 \mathrm{U}(x-\xi,\xi)\,d\xi\right\} \tag{5.94}$$

where the superscripted index "0" refers to when the becalmed flow does *not* affect production, $\dot{n}''(\xi)$ is the source strength per unit time per unit surface area, σ is the spot's propagation parameter, $\mathrm{U}(z_1, z_2, \ldots, z_i)$ is the multidimensional unit step function which equals one when *none* of its arguments are negative and is otherwise zero, and where the integration is taken over the whole surface of the plate.

To account for the becalmed flow, recall that γ_{nb} is the probability or fraction of time the flow is laminar non-becalmed at any point on the surface and therefore susceptible to the formation of turbulent spots. Hence, the quantity $\gamma_{nb}\dot{n}''(x)$ is the source strength per unit time per unit surface area when production is affected by the becalmed flow. Consequently, the probability of turbulent flow accounting for this reduced production is

$$\gamma_{t,E}(x)=1-\exp\left\{-\frac{\sigma}{\bar{u}_\infty}\int_0^\infty \gamma_{nb}(\xi)\dot{n}''(\xi)\left(x-\xi\right)^2 \mathrm{U}(x-\xi,\xi)\,d\xi\right\}$$

where the subscript "E" reminds us that the solution applies only when the flow is either turbulent or non-turbulent. Furthermore, by inspecting the solutions given by Eqs. (5.86) and (5.87) for a line source of turbulent spots, the corresponding expression for $\gamma_{b,E}$ is

$$\gamma_{b,E}(x)=1-\exp\left\{-\frac{\sigma_b}{\bar{u}_\infty}\int_0^\infty \gamma_{nb}(\xi)\dot{n}''(\xi)\left(x-\xi\right)^2 \mathrm{U}(x-\xi,\xi)\,d\xi\right\}$$

Presuming the interaction rules given in Eq. (5.89) apply, Eqs. (5.90)–(5.92) yield

$$\gamma_{nb}=\left(1-f_b\right)\left(1-\gamma_t\right)$$

$$\begin{aligned}\gamma_t(x)&=1-\exp\left\{-\frac{\sigma}{\bar{u}_\infty}\int_0^\infty\left(1-f_b(\xi)\right)\left(1-\gamma_t(\xi)\right)\dot{n}''(\xi)\left(x-\xi\right)^2 \mathrm{U}(x-\xi,\xi)\,d\xi\right\}\\ f_b(x)&=1-\exp\left\{-\frac{\sigma_b}{\bar{u}_\infty}\int_0^\infty\left(1-f_b(\xi)\right)\left(1-\gamma_t(\xi)\right)\dot{n}''(\xi)\left(x-\xi\right)^2 \mathrm{U}(x-\xi,\xi)\,d\xi\right\}\end{aligned} \tag{5.95}$$

The last two expressions provide a pair of integral equations for γ_t and f_b which are best solved by iteration. In this case, using the notation introduced in Eq. (5.94), we rewrite them as

$$\begin{aligned}\gamma_t^{(i)}(x)&=1-\exp\left\{-\frac{\sigma}{\bar{u}_\infty}\int_0^\infty\left(1-f_b^{(i-1)}(\xi)\right)\left(1-\gamma_t^{(i-1)}(\xi)\right)\dot{n}''(\xi)\left(x-\xi\right)^2 \mathrm{U}(x-\xi,\xi)\,d\xi\right\}\\ f_b^{(i)}(x)&=1-\exp\left\{-\frac{\sigma_b}{\bar{u}_\infty}\int_0^\infty\left(1-f_b^{(i-1)}(\xi)\right)\left(1-\gamma_t^{(i-1)}(\xi)\right)\dot{n}''(\xi)\left(x-\xi\right)^2 \mathrm{U}(x-\xi,\xi)\,d\xi\right\}\end{aligned} \tag{5.96}$$

where $i = 1, 2, 3, \ldots, m$, and the functions $\gamma_t^{(0)}$ and $f_b^{(0)}$ are the initial guesses which, in turn, are most conveniently set equal to $\gamma_{t,E}$ and $\gamma_{b,E}$ respectively. To obtain $\gamma_t(x)$ and $f_b(x)$, we evaluate $\gamma_t^{(i)}$ and $f_b^{(i)}$ for successive values of i until the solutions for two consecutive solutions agree within the desired accu-

[1] The solution for an arbitrary distribution of point sources, although not necessarily easy to obtain, is treated similarly.

racy. As it turns out, the iteration converges quickly and usually only one or two steps are required to obtain a reasonably accurate result.

At this point it is worthwhile to differentiate between a "prescribed" and "probable" or "realized" source distribution. Mathematically, the prescribed distribution is $\dot{n}''(\xi)$, that is, where and how often turbulent spots *could* form, while the probable distribution, given by $\gamma_{nb}\dot{n}''(x)$, represents where and how often they *most likely* form. From an experimental viewpoint, the former corresponds to spot generators operating at different frequencies distributed along the surface, while the latter corresponds to the fraction of these generators at each point that most likely produce spots.

Let's now consider the solutions to the above equations for two *prescribed* spot-source distributions: 1) a uniform distribution of line sources and 2) a Gaussian distribution.

Uniform distribution. In this case, suppose that turbulent spots after the onset of transition could form at a uniform rate along the surface of a plate according to

$$\dot{n}''(\xi) = a\,\mathrm{U}(\xi - x_t) \tag{5.97}$$

where a is a constant having the units $1/\mathrm{m}^2\mathrm{s}$. Substituting this into Eq. (5.94) and integrating yields

$$\gamma_t^{(0)} = 1 - \exp\left[-\left(a\sigma/3\bar{u}_\infty\right)\left(x - x_t\right)^3 \mathrm{U}(x - x_t)\right]$$

This solution was first obtained by Emmons, who considered a uniform spot production rate beginning at the leading edge of the plate. Comparing this with the solution given by Eq. (5.86) for a line source of turbulent spots reveals that the probability of turbulent flow for the uniform source eventually increases faster than that for the line source since spots are continuously produced.

The corresponding solution for the fraction of time the laminar flow is becalmed is obtained by simply replacing σ with σ_b in the above equation. Therefore,

$$f_b^{(0)} = 1 - \exp\left[-\left(a\sigma_b/3\bar{u}_\infty\right)\left(x - x_t\right)^3 \mathrm{U}(x - x_t)\right]$$

which together with the previous expression become the initial guesses to the solutions for $\gamma_t(x)$ and $f_b(x)$ in Eq. (5.96). Furthermore, from Eq. (5.95), we obtain

$$\gamma_{nb}^{(0)} = \left(1 - f_b^{(0)}\right)\left(1 - \gamma_t^{(0)}\right) = \exp\left[-\frac{1}{3}\frac{a(\sigma + \sigma_b)}{\bar{u}_\infty}\left(x - x_t\right)^3 \mathrm{U}(x - x_t)\right] \tag{5.98}$$

In this case, the solution converges rapidly and the result from the first iteration is sufficiently independent of the propagation parameter ratio to be used as the general solution for the problem. Furthermore, the solution can be expressed analytically by

$$\begin{aligned}\gamma_t^{(1)}(Re_x) = 1 - \exp\Bigg\{-\hat{n}\sigma\Bigg[&\frac{1 - \gamma_{nb}^{(0)}(Re_x)}{\hat{n}(\sigma + \sigma_b)} + (Re_x - Re_{xt})^2 \frac{\Gamma(1/3) - \Gamma[1/3, \hat{n}(\sigma + \sigma_b)(Re_x - Re_{xt})^3/3]}{3^{2/3}[\hat{n}(\sigma + \sigma_b)]^{1/3}} \\ &- 2(Re_x - Re_{xt})\frac{\Gamma(2/3) - \Gamma[2/3, \hat{n}(\sigma + \sigma_b)(Re_x - Re_{xt})^3/3]}{3^{1/3}[\hat{n}(\sigma + \sigma_b)]^{2/3}}\Bigg]\Bigg\}\end{aligned} \tag{5.99}$$

where $\hat{n} = av^3/\bar{u}_\infty^4$ is the dimensionless production rate for a uniform distribution, $\Gamma(z)$ is Euler's gamma function, $\Gamma(a, z)$ is the incomplete gamma function,[1] and where

[1] These functions are defined as $\Gamma(z) = \int_0^\infty t^{z-1} e^{-t}\,dt$ and $\Gamma(a,z) = \int_z^\infty t^{a-1} e^{-t}\,dt$.

$$\gamma_{nb}^{(0)}(Re_x) = \exp\left[-\frac{1}{3}\hat{n}(\sigma+\sigma_b)\left(Re_x - Re_{xt}\right)^3 \mathrm{U}(Re_x - Re_{xt})\right]$$

which is the dimensionless version of Eq. (5.98).

These expressions are plotted in Fig. 5.51 for $\hat{n}\sigma = 7.5(10)^{-11}$ and two propagation parameter ratios. The intermittency for $\sigma_b/\sigma = 1$ and the resulting spot-source distribution are plotted in part (a) of the figure. For comparison, the results without the stabilizing effect of the becalmed flow on spot production are shown by the dashed lines. Clearly, the effect of the becalmed flow is to reduce the downstream portion of the turbulent-spot production rate. Despite this somewhat significant effect on production, however, the effect on intermittency appears surprisingly minimal.

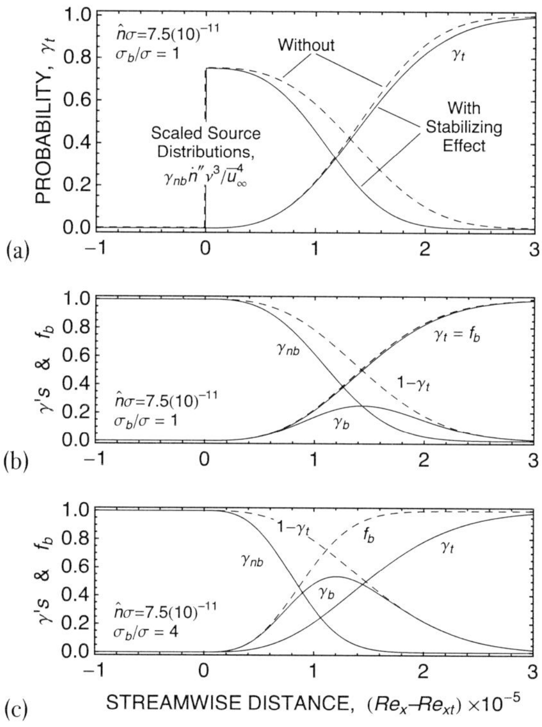

Figure 5.51a,b,c.
The stabilizing effect of becalmed flow on a uniform distribution of line-sources and the various probability distributions through transition when (b) $\sigma_b/\sigma = 1$, and (c) $\sigma_b/\sigma = 4$.

The probability of finding a particular type of flow at any location when $\sigma_b/\sigma = 1$ is shown in part (b) of the figure. According to Eq. (5.83), the sum of γ_t, γ_b, and γ_{nb} (solid lines) must equal unity. Since f_b is the fraction of laminar flow that is becalmed, it varies from zero before any spots and becalmed zones are produced, to unity where all the remaining laminar flow is becalmed. In this case, since $\sigma_b = \sigma$, the distributions f_b and γ_t are identical. This is not true of course when σ_b and σ differ, as shown in part (c) of the figure where the probability distributions for $\sigma_b/\sigma = 4$ are plotted. In this case, it is noteworthy that most of the laminar flow is becalmed, that is, f_b nearly equals one, before half of the flow is turbulent. A comparison of the distributions plotted in parts (b) and (c) of the figure quickly reveals that the intermittency is minimally affected by the stabilizing effect of the becalmed flow following the spots, even though the partition between the becalmed and non-becalmed regions of laminar flow appreciably differs. In other words, the solution for either σ_b/σ provides a good estimate for the stabilizing effect of the becalmed flow on transition when we are only interested in the turbulent portion of the flow.

In this example, the "prescribed" source is the uniform distribution given by Eq. (5.97), while the dimensionless form of the "probable" distribution is given by

$$\gamma_{nb}^{(0)}\dot{n}''\nu^3/\overline{u}_\infty^4 = \hat{n}\exp\left[-\frac{1}{3}\hat{n}(\sigma+\sigma_b)\left(Re_x - Re_{xt}\right)^3 \mathrm{U}(Re_x - Re_{xt})\right]\mathrm{U}(Re_x - Re_{xt}) \tag{5.100}$$

This expression is plotted in Fig. 5.52 for the ratios $\sigma_b/\sigma = 0, 1, 4$, which correspond respectively to Emmons' solution, the ratio determined from Makita and Nishizawa's measurements, and the ratio obtained by presuming all spots produced within the recovery tail are suppressed. In this figure, the prescribed source distribution is plotted by the dashed line. As noted above, the effect of changing σ_b/σ from zero to four is minimal on the intermittency even though the effect on the spot production rates are obviously significant.

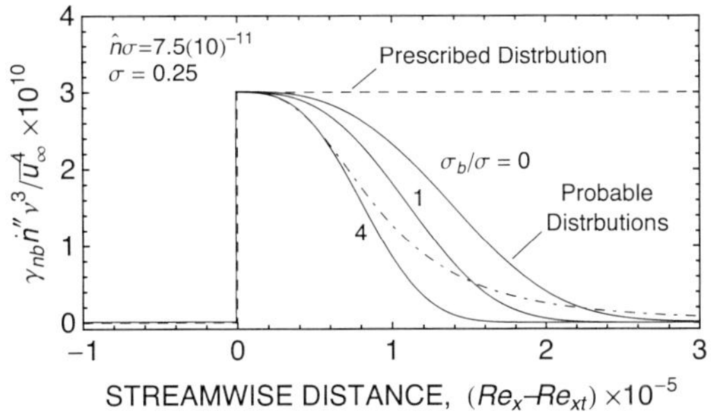

Figure 5.52.
The stabilizing effect of becalmed flow on a uniform distribution of line-sources for various propagation-parameter ratios σ_b/σ. Also Ramesh and Hodson's [1999] relation (dot-dashed curve).

The dot-dashed curve in Fig. 5.52 corresponds to Ramesh and Hodson's [1999] relation for the stabilizing effect of the becalmed flow, namely,

$$\frac{\dot{n}''\nu^3}{\overline{u}_\infty^4} = \frac{\hat{n}(\sigma+\sigma_b)}{1+\hat{n}(\sigma+\sigma_b)(Re_x - Re_{xt})^3/3}$$

where they suggest using $\sigma_b/\sigma = 4.44$.[1] This relation, which cuts diagonally across the curves given by Eq. (5.100), can also be shown to yield an intermittency distribution which is not much different than that shown in Fig. 5.51a. However, the corresponding solutions for γ_b and γ_{nb} are substantially different as can be deduced from the results shown in Figs. 5.51b and 5.51c.

Gaussian distribution. In this case, consider calculating the probability distributions for a prescribed Gaussian distribution of line sources. In particular, again consider the distribution given by Eq. (5.28), namely,

$$\dot{n}''(\xi) = \frac{\dot{n}'}{\sqrt{2\pi}\chi}\exp\left[-\frac{1}{2}\left(\frac{\xi - x_t}{\chi}\right)^2\right]$$

where $\dot{n}'$ is the total number of spots produced over the surface per unit time per unit span, and χ is

[1] Their approach uses this expression for the spot production rate $\dot{n}''(\xi)$ in Eq. (5.85) but double book keeps the effect of spots on the non-turbulent portion of the flow. Since Emmons' theory already accounts for the mutual exclusivity of spots, there is no need to take it into account again. See the discussion leading to Eq. (5.6) in §5.2.1.

the distribution's standard deviation.

The solution when the production of spots is unaffected by the becalmed flow is given analytically by Eq. (5.29). It corresponds to $\gamma_t^{(0)}$ in the new notation and is plotted in Fig. 5.21 for $\hat{n}\sigma = 7.5(10)^{-11}$ and $Re_\chi = 50{,}000$, where Re_χ is the Reynolds number based on the standard deviation. The solution for $f_b^{(0)}$ is also given by Eq. (5.29) after replacing σ with σ_b. Therefore, the initial guesses to the solutions for $\gamma_t(x)$ and $f_b(x)$ are again analytical.

However, in this case, the solutions to Eq. (5.96) for the first and following iterations must be obtained numerically. The results of the first iteration, which again are found sufficiently accurate, are plotted in parts (a) and (b) of Fig. 5.53 for $\sigma_b/\sigma = 1$. The results without the stabilizing effect of the becalmed flow are shown by the dashed lines in part (a). Here it is seen that the stabilizing flow in the becalmed region following a spot reduces the downstream production of turbulent spots, which subsequently reduces the rate of transition to turbulent flow. The various probability distributions for $\sigma_b/\sigma = 1$ are plotted in part (b) of the figure, while those for $\sigma_b/\sigma = 4$ are plotted in part (c). Again, the major effect of suppressing turbulent-spot production is to change the partitioning between the two regions of laminar flow.

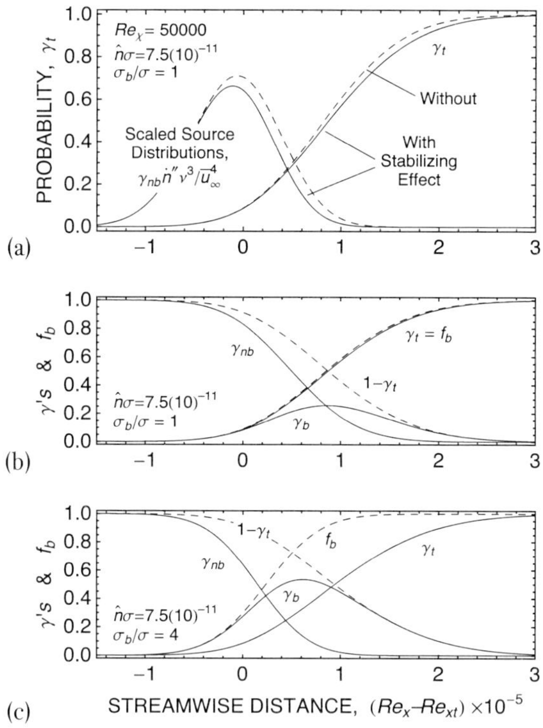

Figure 5.53a,b,c.

The stabilizing effect of becalmed flow on a Gaussian distribution of line-sources and the various probability distributions through transition when (b) $\sigma_b/\sigma = 1$, and (c) $\sigma_b/\sigma = 4$.

The suppression of spot production is plotted in Fig. 5.54 for various becalmed-to-spot propagation-parameter ratios. Note that all the source distributions are "Gaussian-like," which implies that all the corresponding intermittency distributions will be indistinguishable when plotted using Narasimha's similarity variable ξ (see §5.2.2). In other words, the effect of the becalmed zones cannot be detected from intermittency measurements alone. We will return to this point in Chapter 6.

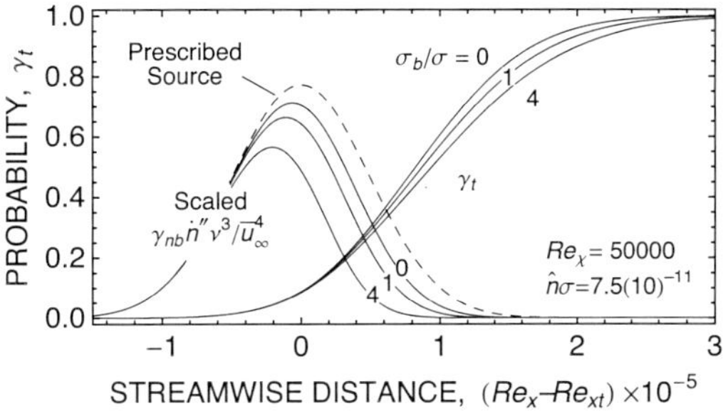

Figure 5.54.
The stabilizing effect of becalmed flow on a Gaussian distribution of line-sources and the transition to turbulence for various propagation-parameter ratios σ_b/σ.

5.5 Propagation Parameters.

Emmons' propagation parameter. Becalmed propagation parameter.

There are four quantities that must be determined prior to calculating the near-wall intermittency distribution. Three of them appear explicitly in Emmons-Narasimha intermittency solution given by Eq. (5.16). They are the spot production rate $\dot{n}'$, Emmons' spot propagation parameter σ, and the beginning of transition x_t. The fourth, introduced in §5.4.1, is the becalmed propagation parameter σ_b.

In this section we will investigate the propagation parameters σ and σ_b. The onset of transition and spot production rate will be examined in §6.2.1 & 2, and §6.2.3 & 4 respectively.

5.5.1 The spot propagation parameter σ.

Linear expansion. Nonlinear expansion.

This is perhaps the least understood concept introduced by Emmons. For a point $P(x, t, z)$ at the apex of the dependence cone in Fig. 5.17, the propagation parameter σ as defined by Eq. (5.15) is

$$\sigma(x;x') = \frac{\bar{u}_\infty}{(x-x')^2}\int_{A'} dz'\,dt'$$

where A' is the area of the *dependence* cone in the x' = constant plane.

Since the integral is the same as the area of the *propagation* cone $A_{s,zt}$ in an x = constant plane with the source at x_s, the spot propagation parameter using the notation we introduced in §5.3 is given by

$$\sigma(x;x_s) = \frac{\bar{u}_\infty A_{s,zt}}{(x-x_s)^2} \tag{5.101}$$

In other words, Emmons' propagation parameter is a dimensionless substitute for the area $A_{s,zt}$ which equals a constant only when the area increases *quadratically* with distance from the spot's source, that is, when the spot expands linearly with distance in both the spanwise and streamwise directions, and the free-stream velocity remains constant.

When spot data in the zt-plane are available, such as that obtained by Makita and Nishizawa [1998, 1999] and Chong and Zhong [2005], it is relatively easy to determine the spot propagation parameter at any point along the surface by using either Eq. (5.15) or (5.101); as was done to obtain the value of σ in Table 5.2. Obtaining the streamwise distribution σ entails taking temporal measurements across the span of the spot at various streamwise distances and integrating over each area of the spot. To the author's knowledge, a streamwise distribution of neither σ nor $A_{s,zt}$ has been measured.

On the the hand, when data in the xz-plane are available, such as that reported by Wygnanski et al. [1976] and shown in Fig. 5.6, a transformation relating the spot area in the physical plane $A_{s,xz}$, which is a function of time, to the area $A_{s,zt}$ must be made before the spot propagation parameter can be determined. As might be imagined, this transformation depends not only on the free-stream velocity distribution, but also on the shape of the surface. In the Appendix of his paper, Emmons developed a transformation for spots on a flat surface expanding linearly in both the streamwise and spanwise directions as they propagate downstream from their source. A slightly different version of his analysis is provided in Appendix J. Subsequently, Emmons and Bryson [1952] developed a transformation for spots expanding on the surface of a cone, while Narasimha [1985] developed another for spots expanding on the surface of a circular cylinder with a hemispherical leading edge. In this section, we will consider spots expanding along a flat surface and, following the method initially presented by Chen and Thyson [1971], develop some of the more commonly used expressions for the propagation parameter.

For simplicity, suppose the footprint of a spot can be described reasonably well by its triangular skeleton. As a consequence, the area $A_{s,zt}$ as given by Eq. (5.50) is

$$A_{s,zt}(x)=\tfrac{1}{2}f_s b(x)\Delta t(x) \tag{5.102}$$

where f_s is the ratio of the true-to-skeletal footprint area of the spot. It then follows that a general expression for Emmons' propagation parameter is given by

$$\sigma(x)=\frac{1}{2}f_s\frac{\bar{u}_\infty b(x)\Delta t(x)}{(x-x_s)^2} \tag{5.103}$$

Let's now consider this expression when the spot expands both linearly and nonlinearly across the surface.

Linear expansion. When the instantaneous angle α of the tip's trajectory, and leading- and trailing-edge velocities u_{le} and u_{te} are constant, it follows from Eqs. (5.56), (5.60) and (5.63) that the breadth, duration, and length of a spot increase *linearly* with distance from the source. In particular, the dimensions of the spot are given by

$$\begin{gathered} b(x)=2\left(x-x_s\right)\tan\alpha,\quad \Delta t(x)=\frac{u_{le}-u_{te}}{u_{le}u_{te}}\left(x-x_s\right) \\ l(x;t_{le})=\frac{u_{le}-u_{te}}{u_{le}}\left(x-x_s\right),\quad \text{and}\quad l(x;t_{te})=\frac{u_{le}-u_{te}}{u_{te}}\left(x-x_s\right) \end{gathered} \tag{5.104}$$

Then the area $A_{s,zt}(x)$, which is readily determined from Eq. (5.102), is

$$A_{s,zt}(x)=f_s\frac{u_{le}-u_{te}}{u_{le}u_{te}}\left(x-x_s\right)^2\tan\alpha \tag{5.105}$$

and the propagation parameter, according to Eq. (5.103), is given by

$$\sigma=f_s\frac{\bar{u}_\infty}{u_{le}}\frac{u_{le}-u_{te}}{u_{te}}\tan\alpha \tag{5.106}$$

A comparison between the latter and the limited amount of data is plotted in Fig. 5.55 against the source Reynolds number. Here it appears that a value slightly less than f_s = 1.8 provides the best result. For these calculations, the leading-edge velocity was assumed to be 0.89 times the free-stream velocity, and Eqs. (5.64) and (5.71) were used to determine the trailing-edge velocity and expansion angle. The propagation parameter for Schubauer and Klebanoff's data is the average of that evaluated by Narasimha [1985], while those for Wygnanski's et al. and Makita and Nishizawa's [1998] data have been evaluated by the author using the footprints shown in Figs. 5.6b and 5.49 respectively. For

$Re_{\delta 1,s} = 1000$ with $u_{te}/\bar{u}_\infty = 0.53$, $u_{le}/\bar{u}_\infty = 0.89$, and $\alpha = 10.5°$, Eq. (5.106) yields $\sigma = 0.26$.

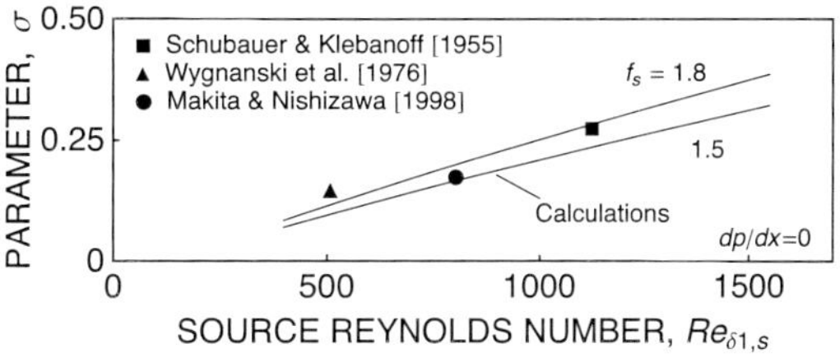

Figure 5.55.
The calculated and measured variation of the spot's propagation parameter with source Reynolds number for a fully-developed turbulent spot.

Clearly, the propagation parameter is aptly named. As written in Eq. (5.106), σ is seen to depend only on the spot's propagation velocities and the angle of its outboard-tip's trajectory. But α is really a combined *result* of the spot's motion and expansion, and not a fundamental attribute of the spot itself. However, since a linearly expanding spot is also self similar, this shortcoming can be rectified by using Eq. (5.74) to eliminate α in favor of the spot's leading-edge angle θ. As a result, we obtain

$$\sigma = f_s \frac{\bar{u}_\infty}{u_{le}} \frac{(u_{le} - u_{te})^2}{u_{te}^2} \tan\theta \tag{5.107}$$

which relates the propagation parameter to attributes of the spot only.

Nonlinear expansion. Initially developed by Chen and Thyson for flow with a variable free-stream velocity, the quickest way of dealing with a nonlinear expanding spot is to obtain the breadth and duration of the spot by integrating Eqs. (5.54) and (5.56), If both the duration and breadth of the spot equals zero at its source, this yields

$$\Delta t(x) = \int_{x_s}^{x} \frac{u_{le} - u_{te}}{u_{le} u_{te}} dx' \quad \text{and} \quad b(x) = 2\int_{x_s}^{x} \tan\alpha \, dx' \tag{5.108}$$

Then from Eq. (5.102), the area $A_{s,zt}$ is given by

$$A_{s,zt} = f_s \int_{x_s}^{x} \tan\alpha \, dx' \int_{x_s}^{x} \frac{u_{le} - u_{te}}{u_{le} u_{te}} dx' \tag{5.109}$$

which reduces to Eq. (5.105) when the angle α and propagation velocities are constant. Substituting the above into Eq. (5.101) yields

$$\sigma(x;x_s) = \frac{\bar{u}_\infty f_s}{(x - x_s)^2} \int_{x_s}^{x} \tan\alpha \, dx' \int_{x_s}^{x} \frac{u_{le} - u_{te}}{u_{le} u_{te}} dx' \tag{5.110}$$

Consequently, from Eq. (5.16), after realizing that x_s and x_t are the same, we obtain

$$\gamma_0(x) = 1 - \exp\left[-\dot{n}' f_s \int_{x_s}^{x} \tan\alpha \, dx' \int_{x_s}^{x} \frac{u_{le} - u_{te}}{u_{le} u_{te}} dx' \right]; \quad (x \geq x_t) \tag{5.111}$$

which is a generally valid expression for the intermittency either with or without a pressure gradient. We will investigate this expression further in §7.3.1.

For now, however, let's estimate the variation of the propagation parameter with distance by considering the development of a spot in a flow with $dp/dx = 0$ and an inception Reynolds number of $Re_{\delta 1,s}$ = 1000. The corresponding variation of the angles and propagation velocities are then as provided in

Fig. 5.37. For this case, the asymptotic value of the propagation parameter using $f_s = 1.8$ is $\sigma_{asym} = 0.261$. The calculation using Eq. (5.110) is shown in Fig. 5.56a. Clearly, the asymptotic value is only attained very far downstream from the source. As it turns out, the variation of both integrals in Eq. (5.110) are significantly different from linear, which is the result when their *integrands* are constant.

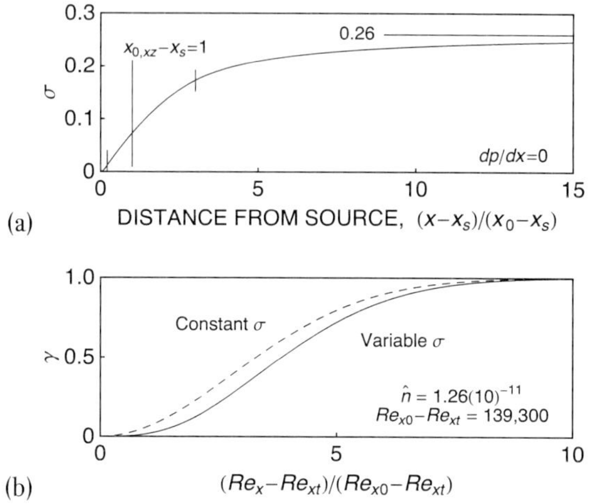

Figure 5.56a,b.
Calculated propagation parameter of a developing turbulent spot for an inception Reynolds number of $Re_{\delta 1,s} = 1000$.

The resulting near-wall intermittency distribution, evaluated using Eq. (5.111) with $\hat{n} = 3.3(10)^{-12}/\sigma_{asymp}$ is shown by the solid line in part (b) of the figure. Here it is important to stress that the same result can more readily be obtained by disregarding Emmons' propagation parameter altogether and simply using

$$\gamma(x)=1-\exp\left\{-\frac{1}{2}\hat{n}' f_s b(x;x_t)\Delta t(x;x_t)\right\};\quad (x\ge x_t) \tag{5.112}$$

which is obtained directly from Eqs. (5.16), (5.101) and(5.102) providing the faired curves for $\Delta t(x; x_t)$ and $b(x; x_t)$ shown in Figs. 5.32 and 5.34 are used.

The dashed line in Fig. 5.56b corresponds to an intermittency calculated from Eq. (5.18) with $\hat{n}\sigma = 3.3(10)^{-12}$ a constant. Differing mostly in their streamwise position, the variation of both are virtual identical. In fact, the two curves are barely distinguishable when the result using σ constant is shifted downstream by a distance of about $Re_x = 58{,}000$, an amount in this case equal to about five percent of the transition length. Thus, the major effect of a spot expanding nonlinearly across the surface is to shift the intermittency distribution downstream. That is, as we found with distributed sources of turbulent spots either with or without becalmed zones, the shape of the intermittency distribution is similar to that for a line source of spots, but the "real" onset of transition is different. Consequently, if we determine the onset of transition by fitting Narasimha's line-source solution to the data, the resulting value will most certainly not represent reality! and the conclusions we draw from it, such as the propagation parameter being constant, are misleading. We will return to this issue in the Chapter 6.

5.5.2 The becalmed propagation parameter σ_b.

Until now, this quantity has received little attention. From Mayle's [1993] analysis, the simplest expression for σ_b is given by

$$\sigma_b=\left(u_s l_t / u_t l\right)\sigma \tag{5.113}$$

where u_s and u_t are the average spot and recovery-tail velocities, and l and l_t are their lengths. This assumes that all the flow in the recovery tail is becalmed and that the tail as shown in Fig. 5.14a is triangular in shape.

A general expression for the becalmed propagation parameter can be obtained by starting with its definition given in Eq. (5.88). After realizing that A_b' and $A_{b,zt}$ refer to the same area, we obtain

$$\sigma_b(x;x')=\frac{\bar{u}_\infty}{(x-x')^2}\int_{A_b'}dz'\,dt'=\frac{\bar{u}_\infty A_{b,zt}}{(x-x')^2}=\frac{A_{b,zt}}{A_{s,zt}}\sigma$$

which implies that the ratio of the propagation parameters σ_b/σ is a constant when the shape of the spot/becalmed ensemble remains similar as it propagates downstream.

Once the becalmed zone is defined, the methods used to determine σ_b are the same as those used to determine the spot propagation parameter. That is, when data on the spot/becalmed ensemble in the zt-plane are available, σ_b can be obtained directly from the above equation since any of the expressions on its right-hand side can be evaluated. This was the method used to obtain the value of σ_b in Table 5.2. On the other hand, when data in the xz-plane are available, the area transformations discussed in the previous section on spots, or an analysis similar to the following, must be used.

Even though the shape and size of a becalmed zone is presently unknown, it is easy to determine an upper limit of the becalmed propagation parameter by assuming that all the flow in the triangular-shaped recovery tail is becalmed. In this case, the relations given by Eqs. (5.102) and (5.103) for a triangular-shaped spot may be applied. Rewritten in terms of quantities related to a becalmed zone and arranged as in the above equation, we obtain

$$\sigma_b(x;x_s)=\frac{1}{2}f_b\frac{\bar{u}_\infty b(x)\Delta t_b(x)}{(x-x_s)^2}=\frac{f_b}{f_s}\frac{\Delta t_b(x)}{\Delta t_s(x)}\sigma \tag{5.114}$$

where f_b is the ratio of the actual-to-triangular area of the becalmed zone,[1] $b(x)$ is the breadth of the spot and presumably the becalmed zone, and $\Delta t_b(x)$ is the duration of the becalmed zone along its midspan.

From Eq. (5.54) and the fact that the leading-edge velocity of the becalmed zone equals the trailing-edge velocity of the spot, we obtain

$$\frac{d\Delta t_b}{dx}=\frac{1}{u_{b,te}(x)}-\frac{1}{u_{b,le}(x)}=\frac{1}{u_{b,te}(x)}-\frac{1}{u_{s,te}(x)}$$

Integrating this with respect to x and substituting the result into Eq. (5.114), we obtain a general expression for the ratio of propagation parameters in terms of the propagation velocities and area ratios f_s and f_b, namely,

$$\frac{\sigma_b}{\sigma}=f_b\int_{x_s}^{x}\frac{u_{s,te}-u_{b,te}}{u_{s,te}u_{b,te}}dx'\Bigg/f_s\int_{x_s}^{x}\frac{u_{s,le}-u_{s,te}}{u_{s,le}u_{s,te}}dx' \tag{5.115}$$

If the propagation velocities are constant, then the ratio of propagation parameters is given by

$$\frac{\sigma_b}{\sigma}=\frac{f_b}{f_s}\left[\frac{(u_{s,te}/u_{b,te})-1}{1-(u_{s,te}/u_{s,le})}\right] \tag{5.116}$$

which when evaluated using data for the typical triangular shaped recovery tail and turbulent spot yields $\sigma_b/\sigma\approx 3.9$. This is virtually the same value given in Table 5.2.

[1] Which we note has nothing to do with the fraction of time the laminar flow is becalmed as defined in Eq. (5.93).

References

Acharya, M., 1985, "Pressure-Gradient and Free-Stream Turbulence Effects on Boundary-Layer Transition," Forschungsbericht KLR 85-127 C, Brown Boveri Forschungszentrum, Baden, Switzerland; also Feiereisen, W.J., and Acharya, M., 1986, "Modeling of Transition and Surface Roughness Effects in Boundary-Layer Flows," AIAA Journal, 24, pp. 1642–1649.

Amini, J., and Lespinard, G., 1982, "Experimental Study of an Incipient Spot in a Transitional Boundary Layer," Phys. Fluids, 25, pp. 1743–1750.

Antonia, R.A., Chambers, A.J., Sokolov, M., and Van Atta, C.W., 1981, "Simultaneous Temperature and Velocity Measurements in the Plane of Symmetry of a Transitional Turbulent Spot," J. Fluid Mech., 108, pp. 317–343.

Blair, M.F., 1983, "Influence of Free-Stream Turbulence on Turbulent Boundary Layer Heat Transfer and Mean Profile Development, Part I–Experimental Data," J. Heat Transfer, 105, pp. 33–40.

Blair, M.F., 1992, "Boundary Layer Transition in Accelerating Flows with Intense Free-Stream Turbulence: Part 1–Disturbances Upstream of Transition Onset, and Part 2–The Zone of Intermittent Turbulence," J. Fluids Engng., 114, pp. 313–332.

Breuer, K.S., and Haritonidis, J.H., 1990, "The Evolution of a Localized Disturbance in a Laminar Boundary Layer. Part 1. Weak Disturbances," J. Fluid Mech., 220, pp. 569–594.

Breuer, K.S., Cohen, J., and Haritonidis, J.H., 1997, "The Late Stages of Transition Induced by a Low-Amplitude Wavepacket in a Laminar Boundary Layer," J. Fluid Mech., 340, pp, 395–411.

Burgers, J.M., 1924, "The Motion of a Fluid in the Boundary Layer along a Plane Smooth Surface," Proc. 1st Intern. Congress for Applied Mech., Delft, p. 113.

Cantwell, B., Coles, D., and Dimotakis, P., 1978, "Structure and Entrainment in the Plane of Symmetry of a Turbulent Spot," J. Fluid Mech., 87, pp. 641–672.

Chambers, F.W., and Thomas, A.S.W., 1983, "Turbulent Spots, Wave Packets, and Growth," Phys. Fluids, 26, pp. 1160–1162.

Chen, K.K., and Thyson, N.A., 1971, "Extension of Emmons' Spot Theory to Flows on Blunt Bodies," AIAA Journal, 9, pp. 821–825.

Chong, T.P., and Zhong, S., 2003, "On the Three-Dimensional Structure of Turbulent Spots," ASME Paper GT2003-38435.

Chong, T.P., and Zhong, S., 2005, "On the Momentum and Thermal Structures of Turbulent Spots in a Favorable Pressure Gradient," ASME Paper GT2005-69046.

Clark, J.P., Jones, T.V. and Lagraff, J.E., 1994, "On the Propagation of Naturally-Occurring Turbulent Spots," Jour. Engng. Math., 28, pp. 1–19.

Cohen, J., Breuer, K.S., and Haritonidis, J.H., 1991, "On the Evolution of a Wave Packet in a Laminar Boundary Layer," J. Fluid Mech., 225, pp. 575–606.

Craik, A.D.D., 1981, "The Development of Wavepackets in Unstable Flows," Proc. Roy. Soc. London A, 373, pp. 457–476.

Dhawan, S., and Narasimha, R., 1958, "Some Properties of Boundary Layer Flow During Transition from Laminar to Turbulent Motion," J. Fluid Mech., 3, pp. 418–436.

D'Ovidio, A., Harkins, J.A., and Gostelow, J.P., 2001, "Turbulent Spots in Strong Adverse Pressure Gradients, Part 1–Spot Behavior, and Part 2–Spot Propagation and Spreading Rates," ASME Papers 2001-GT-0194 and 0406.

Dryden, H.L., 1934, "Boundary Layer Flow Near Flat Plates," Proc. 4th Intern. Congress for Appl. Mech., 4, p. 175.

Dryden, H.L., 1936, "Airflow in the Boundary Layer Near a Plate," NACA Rept. 562.

Dryden, H.L., 1939, "Turbulence and the Boundary Layer," JAS, 6, pp. 85–100 and 101–105.

Elder, J.W., 1960, "An Experimental Investigation of Turbulent Spots and Breakdown to Turbulence," J. Fluid Mech., 9, pp. 235–246.

Emmons, H.W., 1951, "The Laminar-Turbulent Transition in a Boundary Layer, Part I," J. Aero. Sci., 18, pp. 490–498.

Emmons, H.W., and Bryson, A.E., 1952, "The Laminar-Turbulent Transition in a Boundary Layer, Part II," Proc. 1st U.S. Natl. Cong. Theor. Appl. Mech., pp. 859–868.

Gad-el-Hak, M., Blackwelder, R.F., Riley, J.J., 1981, "On the Growth of Turbulent Regions in Laminar Boundary Layers," J. Fluid Mech., 198, pp. 72–100.

Gaster, M., 1968, "The Development of Three-Dimensional Wave Packets in a Boundary Layer," J. Fluid Mech., 32, pp. 173–184.

Gaster, M., 1975, "A Theoretical Model for the Development of a Wave Packet in a Laminar Boundary Layer," Proc. Roy. Soc. London A, 347, pp. 271–289.

Gaster, M., and Grant, I., 1975, "An Experimental Investigation of the Formation and Development of a Wave Packet in a Laminar Boundary Layer," Proc. Roy. Soc. London A, 347, pp. 253–269.

Glezer, A., Katz, Y., and Wygnanski, I., 1989, "On the Breakdown of the Wave Packet Trailing a Turbulent Spot in a Laminar Boundary Layer," J. Fluid Mech., 198, pp. 1–26.

Gostelow, J.P. and Ramachandran, R.M., 1983, "Some Effects of Free Stream Turbulence on Boundary Layer Transition," Proc. 8th Australasian Fluid Mech. Conference.

Gostelow, J.P., Blunden, A.R., and Blunden, W.R., 1988, "Measurements and Stochastic Analysis of Boundary Layer Transition for a Range of Free-Stream Turbulence Levels," Proc. 2nd Int'l Symp. on Transport Phenomena, Dynamics and Design of Rotating Machinery, Honolulu, pp. 199–211.

Gostelow, J.P., Blunden, A.R., and Walker, G.J., 1994, "Effects of Free-Stream Turbulence and Adverse Pressure Gradients on Boundary Layer Transition," J. Turbomachinery, 116, pp. 392–404.

Gostelow, J.P., Melwani, N., and Walker, G.J., 1996, "Effects of Streamwise Pressure Gradient on Turbulent Spot Development," J. Turbomachinery, 118, pp. 737–743.

Gostelow, J.P., Walker, G.J., Solomon, W.J., Hong, G., and Melwani, N., 1997, "Investigation of the Calmed Region Behind a Turbulent Spot," J. Turbomachinery, 119, pp. 802–809.

Herbert, T., 1988, "Secondary Instability of Boundary Layers," Ann. Rev. Fluid Mech., 20, pp. 487–526.

Hegge Zijnen, B.J. van der, 1924, "Measurements of the Velocity Distribution in the Boundary Layer along a Plane Surface," Thesis, Delft.

Kachanov, Y.S., 1987, "On the Resonant Nature of the Breakdown of a Laminar Boundary Layer," J. Fluid Mech., .184, pp. 43–74.

Katz, Y., Seifert, A., and Wygnanski, I., 1990, "On the Evolution of a Turbulent Spot in a Laminar Boundary Layer in a Favorable Pressure Gradient," J. Fluid Mech., 221, pp. 1–22.

Kim, J., Simon, T.W., and Kestoras, M., 1989, "Fluid Mechanics and Heat Transfer Measurements in Transitional Boundary Layers Conditionally Sampled on Intermittency," Heat Transfer in Convective Flows, 1989 Nat'l Heat Transfer Conf., ASME Publication HTD-107, pp. 69–81.

Makita, H., and Nishizawa, A., 1998, "Interaction Between Two Horizontally Displaced Turbulent Spots," JSME Transaction (B), 64, pp. 3682–3689 (in Japanese).

Makita, H., and Nishizawa, A., 1999, "Effects of Mutual Interaction Between Turbulent Spots on Their Streamwise Growth," JSME Transaction (B), 65, pp. 573–580 (in Japanese).

Mautner, T.S., and Van Atta, C.W., 1986, "Wall Shear Stress Measurements in the Plane of Symmetry of a Turbulent Spot," Experiments in Fluids, 4, pp. 153–162.

Mayle, R.E., 1991, "The Role of Laminar-Turbulent Transition in Gas Turbine Engines," J. Turbomachinery, 113, pp. 509–537.

Mayle, R.E., 1993, "Unsteady, Multimode Transition in Gas Turbine Engines," Heat Transfer and Cooling in Gas Turbines, AGARD CP 527.

McCormack, M.E., 1968, "An Analysis of the Formation of Turbulent Patches in the Transitional Boundary Layer," J. Applied Mech., 35, pp. 216–219.

Morkovin, M.V., 1969, "On the Many Faces of Transition," Viscous Drag Reduction (ed., C.S. Wells), Plenum Press, New York, pp. 1–31.

Narasimha, R., 1957, "On the Distribution of Intermittency in the Transition Region of a Boundary Layer," J. Aero. Sci., 24, pp. 711–712.

Narasimha, R., 1984, "Subtransitions in the Transition Zone," IUTAM Symposium on Laminar-Turbulent Transition (ed., V.V. Kozlov), Springer-Verlag, Berlin, pp. 140–151.

Narasimha, R., 1985, "The Laminar-Turbulent Transition Zone in the Boundary Layer," Progress Aerospace Sci., 22, pp. 29–80.

Owen, F.K., 1970, "Transition Experiments On a Flat Plate at Subsonic and Supersonic Speeds," AIAA Journal, 8, pp. 518–523.

Prandtl, L., 1927, "Über den Reibungswiderstand strömender Luft," Ergebnisse AVA Göttingen, 3rd Series; also 1932, "Zur turbulenten Strömung in Rohren und längs Platte," Ergebnisse AVA Göttingen, 4th Series.

Prandtl, L., and Schlichting, H., 1934, "Das Widerstandsgesetz rauher Platten," Werft, Reederei, Hafen, 15, pp. 1–4.

Ramesh, O.N., and Hodson, H.P., 1999, "A New Intermittency Model Incorporating the Calming Effect," Third European Conference on Turbomachinery.

Rotta, J.C., 1972, Turbulente Strömungen, B.G. Teubner, Stuttgart (in German).

Schlichting, H., 1979, Boundary-Layer Theory, McGraw-Hill, New York.

Schubauer, G.B., and Klebanoff, P.S., 1955, "Contribution to the Mechanism of Boundary-Layer Transition," NACA TN 3489.

Seifert, A., and Wygnanski, I.J., 1995, "On Turbulent Spots in a Laminar Boundary Layer Subjected to a Self-Similar Adverse Pressure Gradient," J. Fluid Mech., 296, pp. 185–209.

Solomon, W.J., Walker, G.J., and Gostelow, J.P., 1996, "Transition Length Prediction for Flows with Rapidly Changing Pressure Gradients," J. Turbomachinery, 118, pp. 744–751.

Steelant, J., and Dick, E., 1996, "Modeling of Bypass Transition with Conditioned Navier-Stokes Equations Coupled to an Intermittency Transport Equation," Int'l Jour. Numerical Methods in Fluids, 23, pp. 193–220.

Steketee, J.A., 1955, "A Note on a Formula of Emmons," J. Aero. Sci., 22, pp. 578–580.

Wygnanski, I., 1981, "The Effects of Reynolds Number and Pressure Gradients on the Transitional Spot in a Laminar Boundary Layer," Lecture Notes in Physics, 136, pp. 304–332.

Wygnanski, I., Sokolov and Friedman, 1976, "On a Turbulent 'Spot' in a Laminar Boundary Layer," J. Fluid Mech., 78, pp. 785–819.

Wygnanski, I., Haritonidis, J.H., and Kaplan, R.E., 1979, "On a Tollmien-Schlichting Wave Packet Produced by a Turbulent Spot," J. Fluid Mech., 92, pp. 505–528.

Wygnanski, I., Zilberman, M., and Haritonidis, J.H., 1982, "On the Spreading of a Turbulent Spot in the Absence of a Pressure Gradient," J. Fluid Mech., 123, pp. 69–90.

Zilberman, M., Wygnanski, I., and Kaplan, R.E., 1977, "Transitional Boundary Layer Spot in a Fully Turbulent Environment," Phys. Fluids Supplement, 20, pp. 258–271.

6 Transition Correlations

Beginning and end of transition. Zero pressure-gradient flows. Nonzero pressure-gradient flows. Effect of surface curvature. Effect of surface roughness.

Anything that affects the mean-velocity profile in a pre-transitional laminar boundary layer, or its unsteadiness, affects the onset and length of transition. To name a few, these include pressure gradients, surface curvature, compressibility, and heat transfer, which mostly affect the profile, and free-stream turbulence, surface roughness, noise, and surface vibrations, which mostly affect the unsteadiness. To correlate the effects of all of these on transition is clearly a daunting task – a task which even today remains unfinished despite all the data acquired over the past seven decades or so. What is even more difficult to accept, however, is that we are still struggling to define the relevant parameters associated with the various modes of transition, and that, as we will discover in this chapter, not enough data has been acquired to corroborate any of the new ideas related to the onset and length of transition.

Strictly speaking, transition begins when pre-transitional fluctuations first *break down* into small-scale turbulent eddies and terminates when the last *remnants* of laminar flow within the boundary layer disappear. From a practical standpoint, however, the "onset" and "end" of transition are where transition *appears* to begin and end, which in turn depends upon some accepted useful definitions.

In this chapter, we will first consider several definitions for the "apparent" beginning and end of transition and the methods used to obtain them. We will then present some of the widely accepted correlations for the effects of free-stream turbulence, pressure gradient, surface curvature, and surface roughness on these quantities. In most cases, new correlations are also proposed. For the effect of free-stream turbulence on transitional flows *without* a pressure gradient, new correlations based on the effective frequency of turbulence are examined. For the effect of free-stream turbulence on transitional flows *with* a pressure gradient, a completely different viewpoint is taken of the data presently available and a set of new correlations for the onset and rate of transition proposed.

Some of the important results in this chapter are:

1. The errors associated with most correlations are generally *unacceptable* for accurately calculating transitional flow.
2. Correlations for the beginning and end of transition obtained by fitting an intermittency solution to transition data are inherently subject to irreconcilable errors.
3. Transition correlations should be based on the displacement-thickness Reynolds number at transition rather than the momentum-thickness Reynolds number.

Beside the references at the end of this chapter, other relevant articles may be found in the Bibliography.

6.1 Beginning and End of Transition.

Wall shear-stress based methods. Intermittency based methods. Translating between methods.

Early definitions for the onset of transition were based on the position where some mean-flow quantity such as boundary-layer thickness, wall shear stress, etc., "begins" to deviate from its expected laminar boundary-layer behavior. Although practical, one difficulty with these definitions is that the position can easily differ from one investigator to another. Another difficulty is that the laminar flow before transition may not be what is "expected." That is, disturbances such as free-stream turbulence, roughness, etc., affect the mean flow in a previously unknown way.[1]

The first experimental investigations of laminar-to-turbulent boundary-layer transition on a flat plate

[1] See §4.3.3.

were carried out by Burgers [1924] and van der Hegge Zijnen [1924] who measured the mean velocity profiles through transition. In 1928, Hansen determined the now well-known rule of thumb that transition occurs on a flat plate when $Re_x = 350{,}000$. This result was obtained by plotting the measured boundary-layer thickness along the surface divided by the square root of its distance from the leading edge (which should be a constant according to Blasius' solution) and noting where the data turned sharply upwards from a horizontal straight line. The distance Reynolds number at which this break occurred was *considered* the onset of transition. After using Blasius' solution to convert this result to an equivalent transition momentum-thickness Reynolds number, which yields $Re_{\delta 2,t} = 400$, his result was then used for flows with a variable free-stream velocity. However, we now know that transition on a flat surface without a pressure gradient can begin at almost any distance depending on a large number of conditions, and that Hansen's experiments were probably conducted in a wind tunnel with a background turbulence of about one percent.[1]

6.1.1 Wall shear-stress based methods.

The earliest experiments concerning transition and the effects of free-stream turbulence were conducted by Dryden [1936], Hall and Hislop [1938], Hislop [1940], and Schubauer and Skramstad [1948], most of whom defined the beginning and end of transition as the positions where the wall shear stress are respectively a minimum and maximum. Whereas Dryden, and Schubauer and Skramstad used hot-wire anemometry to measure the mean-velocity profiles at various distances along their test surface to determine these positions, Hall and Hislop, and Hislop "dragged" a small Pitot tube downstream along the surface until they detected a minimum and maximum in the dynamic pressure near the surface which, in turn, indicate the minimum and maximum in the wall shear stress. As can easily be determined from a sketch of the wall shear stress versus distance through transition, or as seen in Fig. A of the introduction to this book, the minimum in shear stress always occurs somewhat downstream of where the wall shear stress begins to deviate from its pre-transitional value, that is, downstream of the onset of transition, while the maximum always occurs upstream of where the wall shear stress conjoins its post-transitional value. Here it is also worthwhile to note that both Dryden and Hislop measured *momentum* thicknesses at the position of minimum wall stress that was larger than that determined from Blasius' solution, while Hall and Hislop, and Schubauer and Skramstad measured *displacement* thicknesses at this position that were in close agreement with Blasius' solution. These results, which suggest that the displacement thickness might be a better correlation parameter for the onset of transition, were also confirmed by Roach and Brierley [2000].

Abu-Ghannam and Shaw [1980], who studied the effect of free-stream pressure gradients and turbulence on transition, used a slightly different technique to determine the beginning and end of transition. They mounted a surface Pitot tube at a fixed distance from the leading edge of their plate, varied the free-stream velocity, and noted the velocity where the ratio of the measured "surface" to free-stream velocity was a minimum and maximum. Once these velocities were determined, the boundary-layer profiles and turbulence levels for these velocities were measured at the Pitot tube's location and the various "onset" boundary-layer parameters determined.

Most investigators now measure boundary-layer velocity profiles at various distances along their test surface to determine the onset and length of transition, and profiles of several other quantities as well. In this case, the skin-friction coefficient is determined from the near-wall velocity profile and its extremes obtained by fairing a curve through the streamwise distribution of the so determined skin-friction coefficients. While this appears to be a relatively straightforward process, accurate values for the onset and length of transition can only be determined when profile data at enough streamwise positions, particularly near the extremes, are obtained. As a good rule of thumb, this requires data at

[1] This can be seen by looking up the turbulence level in Fig. 4.1 for transition Reynolds number of 350,000.

streamwise intervals that are at least one-tenth of the distance between the extremes.[1]

6.1.2 Intermittency based methods.

Initially attempted by Emmons and Bryson [1952], but mainly developed and promoted by Narasimha [1957] and Dhawan and Narasimha [1958], another way to obtain the onset of transition is to compare near-wall intermittency measurements with theory and determine where the intermittency equals zero. After measuring the intermittency distributions for a large variety of transitional flows without a pressure gradient, Narasimha and his coworkers noted that the data are correlated reasonably well by the solution for a line source of turbulent spots providing we're not too critical about the fit near the beginning and end of transition.[2] Consequently, they suggested that the onset and rate of transition should be determined by fitting this solution to the data.

Briefly described, their technique is to fit Eq. (5.18), namely,

$$\gamma_0(x) = 1 - \exp\left[-\hat{n}\sigma\left(Re_x - Re_{xt}\right)^2\right]; \quad (\sigma \text{ constant}, Re_x \geq Re_{xt}) \tag{6.1}$$

to the intermittency data by adjusting the values of Re_{xt} and $\hat{n}\sigma$, the latter of which we have referred to as the spot parameter.[3] This is best accomplished by plotting the square root of $-\ln(1-\gamma_0)$ against Re_x as shown in Fig. 6.1 and fitting a straight line to the data.

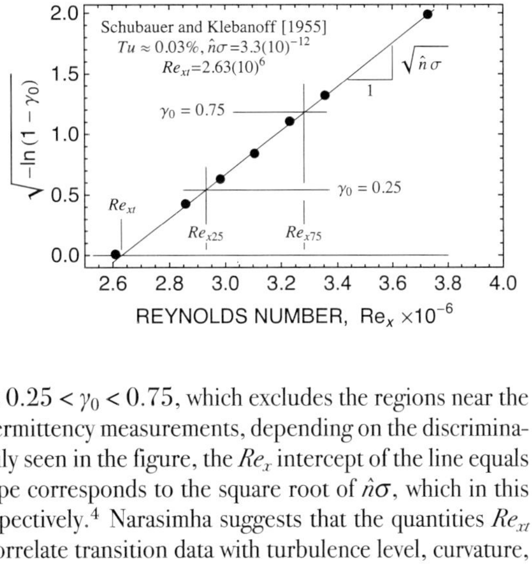

Figure 6.1.
Narasimha's method for determining the onset of transition from intermittency measurements. (See Fig. 5.18 for corresponding intermittency plot.)

Narasimha suggests fitting it within the range $0.25 < \gamma_0 < 0.75$, which excludes the regions near the beginning and end of transition where the intermittency measurements, depending on the discrimination scheme, can be quite inaccurate. As readily seen in the figure, the Re_x intercept of the line equals the onset Reynolds number Re_{xt} while its slope corresponds to the square root of $\hat{n}\sigma$, which in this example equal $2.63(10)^6$ and $3.3(10)^{-12}$ respectively.[4] Narasimha suggests that the quantities Re_{xt} and $\hat{n}\sigma$ so determined may now be used to correlate transition data with turbulence level, curvature, roughness, etc., as done in the following sections. Furthermore, as long as the propagation parameter

---

[1] Since heat transfer measurements have also been used to obtain the onset and length of transition, it is important to note that the positions of the minimum and maximum Stanton number are generally different from those for the skin-friction coefficient, especially for transitional flows developing with a pressure gradient. This is easy to understand when it is realized that pressure gradients affect the momentum and thermal development of the flow differently, depending on the Prandtl number. The reader interested in this effect is referred to the papers by Blair [1982], Sharma [1987], Mayle [1991], and Zhong et al. [2002],

[2] See Fig. 5.19.

[3] Here it should be noted that as long as the propagation parameter σ is constant, the rate of transition always depends on the product of $\hat{n}$ and σ whether we consider transition with distributed sources, becalmed zones, or otherwise. See the intermittency solutions given by Eqs. (5.18), (5.29) and (5.99).

[4] These are the same values used to calculate the intermittency curve in Fig. 5.18 for Schubauer and Klebanoff's data.

σ is constant, the length of transition can also be determined from Eq. (5.19), that is,

$$Re_{LT} \approx 2.15 / \sqrt{\hat{n}\sigma} \tag{6.2}$$

An obvious difficulty arises with fitting the intermittency data using Narasimha's method. To fit the data between his recommended limits and then extrapolate it a full range beyond itself is clearly a source for error. In fact, it is easy to show that the error incurred in Re_{xt} is about three times the error in fitting the slope of the data within the limits, while the error in the spot parameter is about twice the error in slope. Of course any suitable range may be used to fit the data depending on one's confidence in the measurements. For the example shown in Fig. 6.1, the data outside the range of recommended limits certainly helps to determine the fit, but this isn't always the case.[1] Again, as in the case of using the wall-shear-stress methods, the errors associated with using Narasimha's method requires taking measurements at smaller streamwise intervals through transition, particularly near its beginning and end. As we will discuss in §7.2.1, intermittency measurements near zero and one are difficult to obtain.[2]

An example of using Narasimha's method for transition with a high free-stream turbulence level is shown in Fig. 6.2. The solid lines in the figure are the best fit using the distributed source solution given by Eq. (5.29) for a Gaussian probability distribution with a standard deviation $Re_\chi = 40{,}000$, while the dashed lines are the best fit using the line source solution. The transition Reynolds numbers and spot parameters so obtained are 160,000 and 136,000, and $1.15(10)^{-10}$ and $8.8(10)^{-11}$, respectively, which agree well with the interpolated results found by using Table 5.1. As it turns out, the standard deviation is about one-fifth the length of transition, which also agrees with most of the intermittency results for elevated turbulence levels as shown in Fig. 5.22.

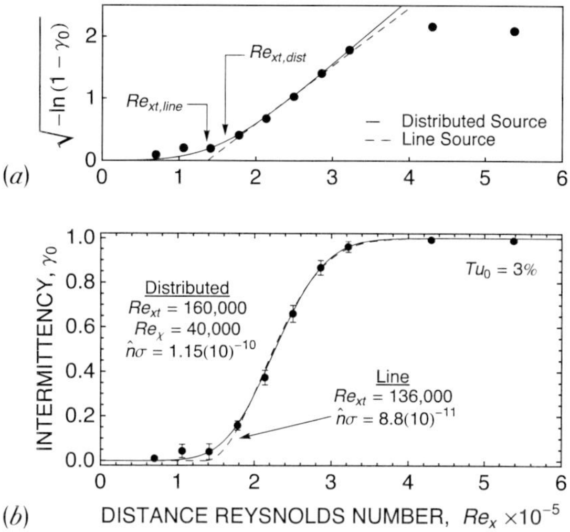

Figure 6.2a,b.
Narasimha's method for determining the onset of transition for a distributed and line source intermittency solution. Data from Roach and Brierley's [1990] T3A test case with a nominal three percent turbulence level.

If we consider transition begins somewhere near the *first* laminar breakdowns, say two standard deviations before where they *most likely* breakdown, that is, $Re_x = 160{,}000$, then transition begins near $Re_{xt} = 80{,}000$. On the other hand, if we assume that transition begins when the intermittency equals

[1] As an exercise, the author suggests using Narasimha's technique to fit the data by Kim et al. plotted in Fig. 5.18.
[2] Errors associated with measuring intermittency are also discussed in §5.2.5 and Appendix H.

one percent, as many of the early investigators including Gostelow and his coworkers[1] did, then Re_{xt} = 114,000. Thus the definition for the onset of transition as determined from fitting intermittency data, which in this case varies from 80,000 to 136,000, clearly makes a difference that cannot be ignored. In particular, as a quick calculation will show, this difference is equivalent to roughly a ±13 percent difference in the transition Reynolds number based on the average of either integral boundary-layer thickness.

This raises a question about the validity of using Narasimha's intermittency-based method. Clearly, as pointed out in §5.2.3 and §5.5.1, the values of Re_{xt} and $\hat{n}\sigma$ obtained by fitting the line-source solution to intermittency data are "real" only when the assumptions used to obtain the solution are satisfied; that is, only when all turbulent spots 1) form at one streamwise location; 2) spread linearly with distance as they propagate downstream from that location; and 3) propagate without a becalmed zone. Otherwise Narasimha's method provides a pair of "pseudo values" for Re_{xt} and $\hat{n}\sigma$ which when substituted into Eq. (6.1) simply reproduces that measured intermittency distribution within the range $0.25 < \gamma_0 < 0.75$. At first glance it might appear that this is not a problem, because the major portion of the intermittency distribution is reproduced even though the assumptions are not satisfied (see Fig. 6.2b for example). The problem arises, however, when correlating a multitude of results with respect to a parameter, say turbulence level, that not only negates one or more of the assumptions but also affects Re_{xt} and $\hat{n}\sigma$ in some unknown way. In this case, a correlation of the pseudo values for Re_{xt} and $\hat{n}\sigma$ with turbulence level becomes an attempt at correlating not only the effect of turbulence level, but also the discrepancies it causes in the assumptions. Since this aspect of Narasimha's method remains to be investigated, we must keep in mind that any correlations based on his method can be misleading and inherently subject to errors that can not be evaluated.

Despite these uncertainties about its usefulness, however, his method can also be used to correlate intermittency distributions for flows with a *constant* pressure-gradient or Falkner-Skan parameter, that is, for flows where the pre-transitional velocity profiles are similar at every streamwise location. As demonstrated by Narasimha et al. [1984], Narasimha [1985], Gostelow and his coworkers, and more recently by Roberts and Yaras [2003], Eq. (6.1) also fits the near-wall intermittency distribution for these flows as long as we limit the fit to the range $0.25 < \gamma_0 < 0.75$.

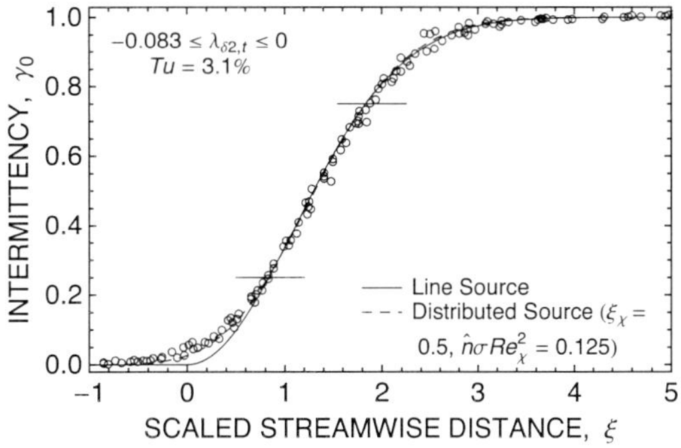

Figure 6.3.
Plot of near-wall intermittency measurements for flows with various adverse pressure gradients and a comparison with the solutions for a line source and a Gaussian distribution of line sources. Data from Gostelow et al. [1994].

A good example of this is shown in Fig. 6.3, where intermittency distributions for attached flow over essentially the whole range of adverse pressure gradients are plotted against Narasimha's similarity variable.[2] The calculated curves are for the line-source solution (solid line) and the solution for a

[1] Gostelow and Blunden [1988], Gostelow [1989], and Gostelow and Walker [1990].

[2] See Eq. (5.20) for its definition.

Gaussian distribution of sources (dashed line passing directly through the data) having the same parametric values as those used in Fig. 5.22 for $0.03 < Tu < 8$ percent and $dp/dx = 0$. As a consequence, values of Re_{xt} and $\hat{n}\sigma$ for these flows can also be determined, but as just discussed, since transition is obviously caused by distributed sources of spots, the so determined results will certainly include the unknown effects of turbulence on pre-transitional flow, becalmed flow, etc.

Finally, as shown by Narasimha [1984, 1985], Gostelow and his coworkers, Blair [1992], and Roberts and Yaras [2003], it is important to note that the intermittency distributions for flows with a *variable* pressure gradient cannot be correlated using Narasimha's method. In this case, as we will learn in §7.3.1, the intermittency distribution can vary dramatically from the classical solution given by Eq. (6.1).

6.1.3 Converting between the two definitions.

To convert between the wall-shear and intermittency based definitions, Narasimha [1985] empirically determined that

$$\begin{aligned} Re_{xt} &= Re_{cf,\min} - 0.26\left(Re_{cf,\max} - Re_{cf,\min}\right) \\ Re_{LT} &= 1.34\left(Re_{cf,\max} - Re_{cf,\min}\right) \end{aligned} \tag{6.3}$$

where $Re_{cf,min}$ and $Re_{cf,max}$ correspond to the distance Reynolds numbers where the skin-friction coefficient is a minimum and maximum, and Re_{xt} and Re_{LT} are the transition and transition-length Reynolds number obtained by fitting the line-source solution, Eq. (6.1), to the intermittency measurements together with using Eq. (6.2). It is the author's experience, however, that the coefficients in both of these expressions are not constant and that the first is too large.

To examine this, we can equate the skin-friction coefficient through transition to an intermittency-weighted average of the local laminar and turbulent skin-friction coefficients (see §7.1) and use Eq. (6.1). It is then relatively easy to demonstrate that the coefficients in the above expressions depend on both Re_{xt} and $\hat{n}\sigma$ once the laminar and turbulent skin-friction-coefficient distributions are provided. In particular, for flow on a flat plate with $dp/dx = 0$ where the skin-friction distributions are approximately given by Eqs. (2.10) and (3.40), both coefficients decrease as Re_{xt} and $\hat{n}\sigma$ increase. The results of these calculations are provided in Appendix K, where it is found that the coefficient in the above expression for Re_{xt} is about two-to-three times the highest expected value (see the values of f_1 in the gray rows of Table K.1), while the coefficient for Re_{LT} is about the same as the highest expected value (see the f_2 values). However, since the coefficient in the expression for Re_{xt} varies by a factor of roughly twenty-five over the whole range of transition Reynolds numbers, using the conversions similar to that given in Eq. (6.3) for the onset of transition are highly questionable.

Nevertheless, if we must convert between the two, the relations given by Eq. (K.2) in Appendix K for flows with $dp/dx = 0$ provide a better alternative than those presented in Eq. (6.3), namely,

$$\left.\begin{aligned} Re_{xt} &= Re_{cf,\min} - 15Re_{xt}^{-0.5}\left(Re_{cf,\max} - Re_{cf,\min}\right) \\ Re_{LT} &= 2Re_{xt}^{-0.04}\left(Re_{cf,\max} - Re_{cf,\min}\right) \end{aligned}\right. \quad ; \quad (dp/dx = 0) \tag{6.4}$$

For the example shown in Fig. 6.2, the minimum and maximum of the best fit to the skin-friction-coefficient data are $Re_x = 135{,}000$ and $330{,}000$ respectively. Substituting these into the relations given directly above yields $Re_{xt} = 127{,}000$ and $Re_{LT} = 244{,}000$, while Eq. (6.3) provides 84,000 and 262,000. Compared with the result in Fig. 6.2b, where we found $Re_{xt} = 136{,}000$ using the line-source curve fit, the onset of transition is better determined from the shear-stress measurements by using Eq. (6.4). In this case, for $Re_{xt} = 127{,}000$, the coefficients of the terms in parentheses become 0.042 and 1.25, compared with the coefficients 0.26 and 1.34 in Eq. (6.3).

If the propagation parameter is indeed constant, then Eq. (6.2) may be used to relate the spot parameter to the "min-max" locations of the skin-friction coefficient. Eliminating Re_{LT} between Eqs. (6.2) and (6.4) yields

$$\hat{n}\sigma = 1.15 Re_{xt}^{0.08}\left(Re_{cf,\max} - Re_{cf,\min}\right)^{-2} \;;\quad (dp/dx = 0,\ \sigma \text{ constant}) \tag{6.5}$$

while using Eq. (6.3) yields a coefficient that is constant and equal to 2.57. For the example just considered where we found $\hat{n}\sigma = 8.8(10)^{-11}$, the above expression provides a value of $7.7(10)^{-11}$ while that with a constant coefficient of 2.57 provides an even smaller value of $6.6(10)^{-11}$. In this case, with $Re_{xt} = 127{,}000$, the coefficient in the above equation becomes 2.94.

6.2 Zero Pressure-Gradient Flows.

Onset of transition – standard and new correlations. Rate of transition – standard correlations and production rate models

The first serious experimental investigations into the effect of free-stream turbulence on transition were conducted in the late 1930s. Covering most of the range for bypass transition, these investigations measured the effect for turbulence levels ranging from about one to seven percent. Between the early 1940s and late 1970s, when wind tunnels with free-stream turbulence levels less than about one percent were consciously developed and confirmation of Tollmien-Schlichting's theory was very much in vogue, and when transition on both low and high speed aircraft was very much of interest, most transition investigations were conducted at very low free-stream turbulence levels. Beginning in the late 1970s, however, spurred mainly by the concern of transition in gas-turbine engines, the number of investigations at elevated turbulence levels increased with more investigators measuring intermittency distributions as well.

Although all investigators measured the effect of turbulence level on the onset of transition, only a few measured its effect on both the beginning and end of transition. In some investigations, the onset of transition was based on the distance Reynolds number, while in others on either the displacement- or momentum-thickness Reynolds number. Since the boundary-layer thickness before transition depends on the leading-edge shape, diameter, and the free-stream turbulence, converting between these different transition Reynolds numbers is always questionable. Furthermore, since most investigators measured the free-stream turbulence (many using the streamwise fluctuations) at only one streamwise position, some at the leading edge of the plate, others at the onset of transition, and still others halfway between, nothing can be determined about the isotropy, length scales, and decay of turbulence, nor their effect on pre-transition and transition itself. As a result, little can be said about the actual turbulence level and the accuracy of the correlations based on this data, such as those presented in this and the following sections, except that they provide only estimates for the effect of turbulence on the onset and rate of transition.

Clearly, this situation can be greatly improved by taking more comprehensive boundary-layer and free-stream turbulence measurements than in the past. Regarding the turbulence measurements, it takes more to describe turbulence than just the turbulence level, even when its isotropic.[1] For isotropic turbulence, at least two quantities such as the intensity and turbulent Reynolds number are required, and from the variation of these with distance all other turbulent quantities, including its dissipation, can be determined. Consequently, while measuring the streamwise variation of the streamwise intensity would be an improvement over the documentation reported by most investigators, measuring the variation of all turbulent components would not only provide the necessary information on turbulent intensity, length scale and decay, but also provide the much needed boundary conditions for

---

[1] See §3.1 and §3.4. Also see Eqs. (3.75) and (3.81) for definitions of the free-stream turbulence level.

modeling and calculating pre-transitional laminar flow.[1]

In this section, we will consider the effect of free-stream turbulence on transition for flows without a pressure gradient. First we will present the standard correlations for the onset of transition and then some newer correlations. We will then consider the standard correlations for the rate of transition and finally two models for the production rate of turbulent spots.

6.2.1 Onset of transition – standard correlations.

Numerous correlations for the effect of free-stream turbulence on the onset of transition have been developed. The early correlations are based on either the distance to transition as shown in Fig. 4.1, or the displacement thickness at transition. Now, however, most correlations for the onset of transition are based on the momentum thickness at transition.[2]

Two of the most popular correlations for the momentum-thickness Reynolds number at transition are

$$Re_{\delta 2,t} = 163 + \exp\left(6.91 - 100Tu\right)\,; \quad \left(dp/dx = 0\right) \tag{6.6}$$

and

$$Re_{\delta 2,t} = 22.5\,Tu^{-5/8}\,; \quad \left(dp/dx = 0,\, Tu > 0.005\right) \tag{6.7}$$

The first is that developed by Abu-Ghannam and Shaw [1980], who correlated transition results based on wall-shear-stress measurements, while the second is that developed by the author [1991], who correlated the results based on intermittency and corrected wall-shear measurements.[3]

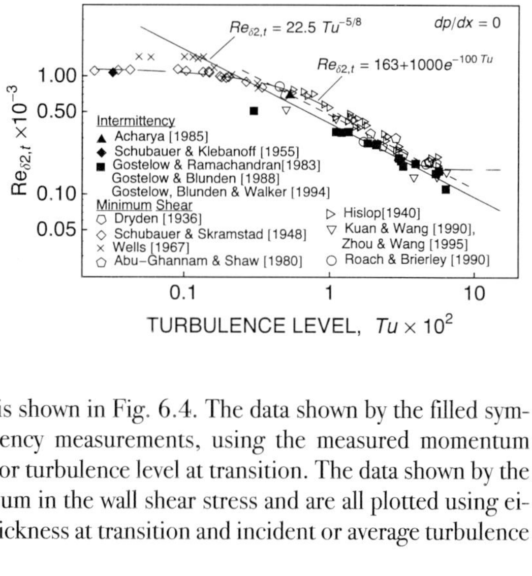

Figure 6.4.
Momentum-thickness Reynolds number at the onset of transition plotted against the free-stream turbulence level for zero pressure gradients flows.

A comparison of these correlations with data is shown in Fig. 6.4. The data shown by the filled symbols were obtained from near-wall intermittency measurements, using the measured momentum thickness at transition and either the incident or turbulence level at transition. The data shown by the empty symbols were obtained from the minimum in the wall shear stress and are all plotted using either the measured or calculated momentum thickness at transition and incident or average turbulence

[1] Here it must be kept in mind that all free-stream turbulence measurements must be obtained at a position outside the kinematic boundary layer of turbulence which roughly equals the size of the large, energy-containing eddies. See discussion in §3.5.1 and results plotted in Figs. 3.14 and 4.24.

[2] There is now some evidence that the displacement thickness is better choice. See §6.2.2 and §8.3.3.

[3] The author's original correlation was developed using Tu in percent, in which case the coefficient in Eq. (6.7) becomes 400.

level to transition.[1] Schubauer and Klebanoff, who used the same tunnel as Schubauer and Skramstad, acknowledge that their result is affected by wind-tunnel acoustics. Wells [1967], Acharya [1985], and Gostelow et al. [1994] also acknowledge that their results for turbulence levels less than about one-half of a percent are facility dependent.

For turbulence levels higher than about a half of a percent, Abu-Ghannam and Shaw's correlation snakes through the wide swath of data while the author's lies along its lower edge. In each case, the correlations approximately fit the data they were intended to represent. It is worthwhile to point out, however, that Eq. (6.6) was forced to fit Schubauer and Skramstad's data for very low turbulence levels (as shown in the figure) and to level out for high turbulence levels at the stability limit calculated by Tollmien of $Re_{\delta 2} = 163$. Since transition for $Tu < 0.005$ is strongly affected by disturbances other than turbulence, and transition at the higher turbulence levels occurs in a bypass mode completely independent of Tollmien-Schlichting instabilities, there is no need to do either. Thus, Abu-Ghannam and Shaw's correlation at each of these extremes is misleading. The data from Fig. 6.4 and a few other investigations are plotted in Fig. 6.5 using the more conventional linear format. In this figure, the data obtained from intermittency measurements are plotted using the filled circles, while those from shear and heat transfer measurements are presented using empty circles.

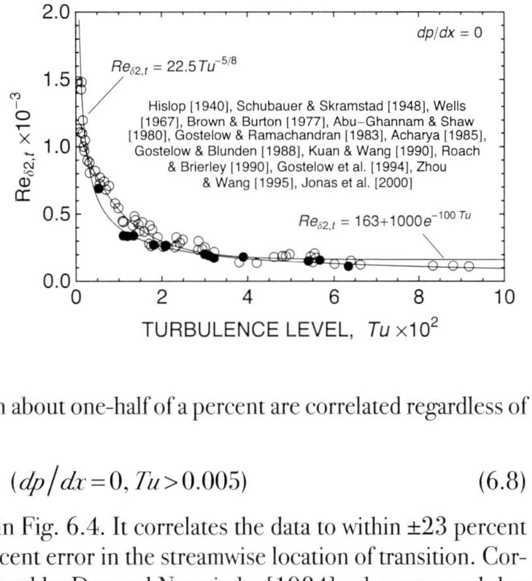

Figure 6.5.
Comparison of the standard correlations for the onset of transition using the conventional format. Data are from both wall-shear-stress measurements (empty circles) and intermittency measurements (filled circles).

If all the data for turbulence levels greater than about one-half of a percent are correlated regardless of the method used to obtain it, we obtain

$$Re_{\delta 2,t} = 26.5\,Tu^{-5/8}\;;\quad (dp/dx = 0,\ Tu > 0.005) \qquad (6.8)$$

This correlation is shown by the dashed line in Fig. 6.4. It correlates the data to within ±23 percent which translates to roughly a forty-to-fifty percent error in the streamwise location of transition. Correlations similar to this were previously obtained by Dey and Narasimha [1984], who proposed that $Re_{\delta 2,t} = 26.5\,Tu^{-0.6}$ for a turbulence level between 0.1 and 4 percent, and Hourmouziadis [1989] who proposed $Re_{\delta 2,t} = 23\,Tu^{-0.65}$. When plotted, the former lies close to Eq. (6.7) while the latter is nearly the same as Eq. (6.8).

In any case, the errors associated with these correlations are substantial and generally unacceptable when transition occurs over a significant portion of the surface. The shifts associated with measuring the turbulence level at different locations easily span the swath of data shown in either Fig. 6.4 or

[1] Data plotted using the incident turbulence level, Tu_0: GB, GBW, GR, H, HH, SK, KW, SS, SW, and ZW. Data plotted using the onset turbulence level, Tu_t: A and RB. Data using $(Tu_0 + Tu_t)/2$: AGS. The turbulence level of KW, SK, SS, SW, and ZW changes little from the leading edge of the test plate to transition. Data using calculated momentum-thickness Reynolds number: H, HH, SK, SS, and SW.

6.5.[1] As shown in Fig. 6.6, the mesh size of the turbulence-producing screen M (think free-stream turbulent length scale) also spans the swath. Thus, besides the different methods used to define the onset of transition, the position where the free-stream turbulence is measured and the effect of turbulent length scale must be considered before these correlations can be improved. As it turns out, the effect of length scale is automatically taken into account by the two newer correlations we will now examine.

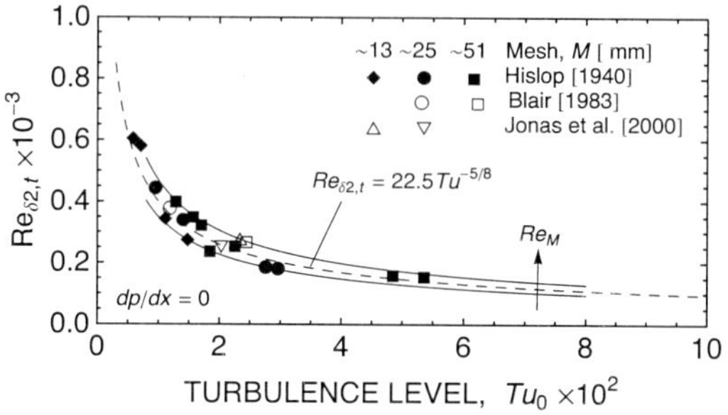

Figure 6.6.
Effect of screen mesh size on the transition Reynolds number.

6.2.2 New correlations.

These correlations, which pertain to bypass transition only, are the result of considering the pressure fluctuations impressed on the boundary layer by the free-stream turbulence. There are two primary effects to consider: 1) the effect of the resulting time-averaged pressure gradient on the mean flow; and 2) the effect of the work done by the fluctuating pressure forces on the Reynolds stress. Since the latter is discussed at some length in §4.3, we will examine the correlation based on this effect first.

A new correlation based on the effective turbulence. When considering the effect of the fluctuating pressure forces on the Reynolds stress, the primary parameters associated with turbulence are those related to the "effective" scales of turbulence. One result of including these scales in a dimensional analysis when considering transition is the new transition criterion introduced in §4.3.4. Given by Eq. (4.53) and repeated here, this criterion is

$$Re_{\delta 1,t} = 1.25 \big/ \Omega_{eff,t}; \quad (dp/dx = 0) \tag{6.9}$$

where Ω_{eff} is the frequency parameter associated with the most effective eddies. Defined in Eq. (4.30), it is given by

$$\Omega_{eff} = \omega_{eff} \nu \big/ \bar{u}_\infty^2 = 0.3 (\nu \varepsilon_\infty)^{1/4} \big/ \bar{u}_\infty \tag{6.10}$$

where ε_∞ is the dissipation of turbulence in the free stream, which is the same as either $\hat{\varepsilon}_\infty$ or $\tilde{\varepsilon}_\infty$ for homogeneous, isotropic turbulence. Furthermore, for isotropic turbulence, this parameter can be evaluated from free-stream turbulence measurements by using Eq. (4.51), which yields

$$\Omega_{eff} = 0.3 \left| \frac{3}{2} \frac{dTu^2}{dRe_x} \right|^{1/4}$$

For anisotropic turbulence, Eqs. (3.76)–(3.79) must be solved to obtain ε_∞.

In addition, upon introducing the free-stream turbulent Reynolds number based on Taylor's microscale of turbulence using Eqs. (3.6) and (3.7), namely,

[1] See Fig. 6.9b for some typical shifts.

$$Re_\lambda = \lambda_\infty (\overline{u'u'})_\infty^{1/2} \big/ \nu \quad \text{where} \quad \lambda = \sqrt{15\nu(\overline{u'u'})\big/\varepsilon} \tag{6.11}$$

it is also possible to show that

$$\Omega_{eff} = 0.59\, Tu \big/ \sqrt{Re_\lambda} \tag{6.12}$$

Then the transition criterion given in Eq. (6.9) may be reformulated as

$$Re_{\delta 1,t} = 2.12\sqrt{Re_{\lambda,t}} \big/ Tu_t \tag{6.13}$$

where all the values are evaluated as indicated at the onset of transition. When sufficient free-stream turbulence data is available, every quantity in these expressions can be evaluated. Unfortunately, very few experiments have been conducted where both this and all the other necessary information about transition is given.

The data from four of these experiments are plotted in Fig. 6.7. The new criterion, Eq. (6.9), is shown in part (a) of the figure.[1] Here, the nominal turbulence levels for each test is given in parentheses and only the data that can be determined with some confidence is plotted. Clearly, Eq. (6.9) correlates this limited amount of data exceptionally well. The criterion when plotted in the format given by Eq. (6.13) against the turbulence level Tu_t is shown in part (b) of the figure with the values of $Re_{\lambda,t}$ given in parentheses. As explained in §4.3.3 and §4.3.4, the values for Tu_t and $Re_{\lambda,t}$ were determined after fitting the measured free-stream turbulence distributions to obtain the variation of k_∞ and ε_∞ with distance. The range of $Re_{\lambda,t}$ for the calculations covers the typical range for grid-generated turbulence. In this case, most of the data was gathered near the lower limit of this range. If the data and assumptions are consistent, the data and calculations should agree as they clearly do. The upshot here, however, is that the turbulent Reynolds number has a significant effect on the onset of transition and that it may be correlated by using either Eq. (6.9) or (6.13). Of course, more data is required to confirm these relations.

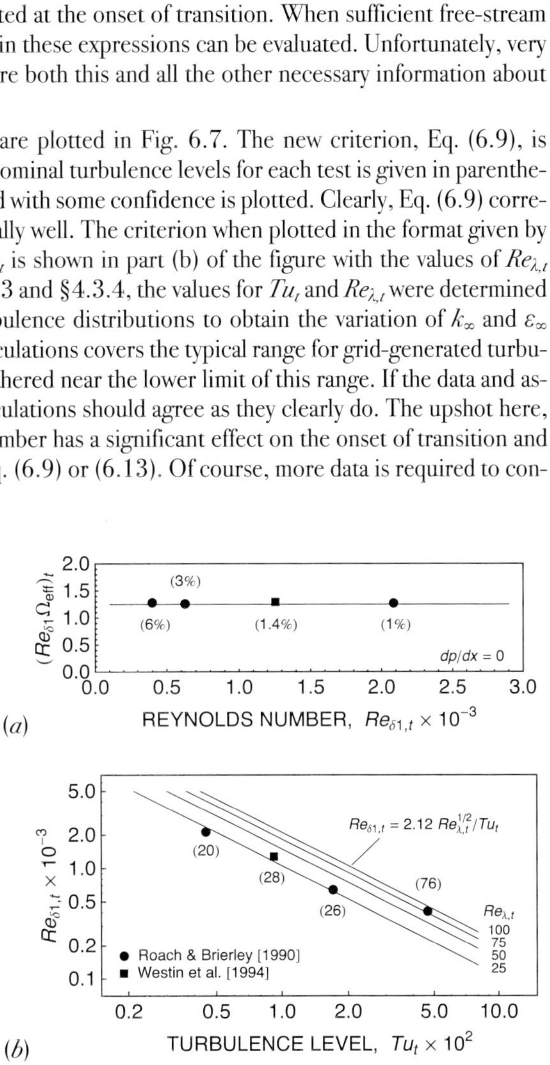

Figure 6.7a,b.
A new correlation for the onset of transition based on the effective turbulence, and its predicted effect of turbulent Reynolds number on transition.

[1] This comparison was originally plotted in Fig. 4.32 using the effective wave number of turbulence and Reynolds number at the onset of transition.

Writing Eq. (6.13) in terms of the momentum-thickness Reynolds number yields

$$Re_{\delta 2,t} = 2.12\sqrt{Re_{\lambda,t}} \,/\, H_t Tu_t \tag{6.14}$$

where H_t is the shape factor at transition. This relation is plotted in Fig. 6.8 compared with the same data shown in Fig. 6.4. In this figure the lines for various turbulent Reynolds numbers have been calculated for a transition shape factor equal to 2.2, which roughly corresponds to its value for elevated turbulence levels.Clearly, the data appears to drift higher with turbulence level across each contour of increasing $Re_{\lambda,t}$, similar to the result plotted in Fig. 6.6 for increasing mesh Reynolds number. Since the turbulent Reynolds number increases with both increasing turbulence level and length scale, this drift agrees with physical expectation. Thus, if we accept the new criterion as being correct, then all the standard correlations previously discussed represent an attempt to correlate the combined effects of turbulence level and length scale by a correlation using only the turbulence level.

For lower turbulence levels, the shape factor is closer to that for a laminar boundary layer, that is, $H =$ 2.6. Plotting Eq. (6.14) using $H_t = 2.6$ and $Re_{\lambda,t} = 20$, which correspond to the conditions for Roach and Brierley's test with the lowest turbulence level, yields the dashed line in Fig. 6.8. As it should, this line passes directly through their data point (given by the empty circle). Furthermore, the results obtained from the equation by changing either H_t to 2.2 or $Re_{\lambda,t}$ to 25, yields a line roughly halfway between the dashed and the solid line for $Re_{\lambda,t} = 25$ in the figure. Consequently, correlating the transition momentum-thickness Reynolds number incurs errors caused by differences in both the turbulent Reynolds number and shape factor. This strongly suggests, following the standard practice used in laminar stability analyses (see §4.2.1), that we should correlate the onset of transition using the transition displacement-thickness Reynolds number $Re_{\delta 1,t}$ instead of the momentum-thickness Reynolds number $Re_{\delta 2,t}$.

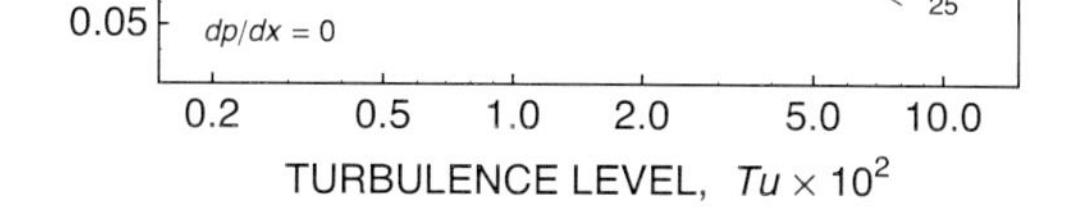

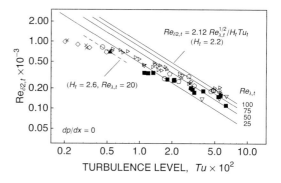

Figure 6.8.
The new correlation compared with data using the momentum-thickness Reynolds number at onset. (See Fig. 6.4 for legend.)

A problem with using the correlations given by Eqs. (6.9) and (6.13), as well as those presented in the following section, however, is that the local turbulence level and either the turbulent Reynolds number or dissipation of turbulence must be known. In many practical situations, these are usually difficult to evaluate or accurately measure. Consequently, correlations like those given by Eqs. (6.6), (6.7) and (6.8) with their notorious errors are still favored.

A new correlation based on Taylor's pressure gradient. The effect of turbulence on the mean flow was examined theoretically by Taylor [1936]. In particular, he determined that turbulence increased the mean pressure gradient on the boundary layer by an amount equal to $2\sqrt{2}\rho\overline{u'u'}/\lambda$, where λ is the micro-scale of turbulence defined in Eq. (6.11). Defining the mean pressure-gradient parameter using the displacement- rather than the momentum-thickness and its increase caused by turbulence by $\Delta\lambda_{\delta 1}$, we obtain

$$\Delta\lambda_{\delta 1} = \frac{\delta_1^2}{\mu \bar{u}_\infty}\sqrt{\overline{\frac{dp'}{dx}\frac{dp'}{dx}}} = 2\sqrt{2}\frac{\delta_1^2}{\nu \bar{u}_\infty}\frac{\overline{u'u'}}{\lambda}$$

Rearranging the quantities on the right into familiar dimensionless parameters yields

$$\Delta\lambda_{\delta 1} = 2\sqrt{2}\, Re_{\delta 1}^2 Tu^3 / Re_\lambda \tag{6.15}$$

In 2000, Roach and Brierley presented a correlation for the transition Reynolds number Re_{xt} as a function of $\Delta\lambda_{\delta 1}$ for flows with $dp/dx = 0$. Although their correlation includes the effect of plate thickness, its effect can be neglected for most of the data shown in the previous figures since they were obtained on plates with relatively sharp leading edges. Furthermore, it is generally acknowledged that the effect of the plate's leading edge shape and diameter on the correlation is minimized by correlating either the displacement- or momentum-thickness Reynolds number at onset instead of the distance Reynolds number, unless the large-scale eddies are distorted enough about the leading edge to change the character of the flow within the boundary layer.[1] Thus using their "thin-plate" correlation, we obtain

$$Re_{xt} = 9(10)^4 \Delta\lambda_{\delta 1}^{-2}$$

where Tu and Re_λ in *their* definition of $\Delta\lambda_{\delta 1}$ are evaluated at the *leading-edge* of the plate.[2]

Eliminating Re_{xt} in favor of $Re_{\delta 1,t}$ using Blasius solution (recalling that several investigators including Roach and Brierley have shown this is generally valid for pre-transitional flow with free-stream turbulence),[3] and substituting Eq. (6.15) for $\Delta\lambda_{\delta 1}$ yields

$$Re_{\delta 1,t} = 5.68 Re_{\lambda,0}^{1/3} / Tu_0; \quad (dp/dx = 0) \tag{6.16}$$

Although similar to Eq. (6.13), the variation with turbulent Reynolds number is weaker, and the incident turbulence level and turbulent Reynolds number are used. While it can be shown that $Re_{\lambda,0} \approx Re_{\lambda,t}$ when $Re_\lambda > 10$, the turbulence levels at the leading edge and transition can differ significantly.[4]

This correlation and its modified version for the momentum-thickness transition Reynolds number are compared with data in Fig. 6.9. Except for Roach and Brierley's data in part (b), only data obtained using the measured incident turbulence levels Tu_0 are plotted in this figure. As noted before, the shift between using the incident and onset turbulence level for this data is significant, while the change in the turbulent Reynolds number is negligible (compare the numbers in parentheses in Figs. 6.7b and 6.9a).

Nevertheless, the agreement shown in part (a) between the correlation given by Eq. (6.16) and data using the incident turbulence level is about as good as that shown in Fig. 6.7b using the correlation given in Eq. (6.13) and the onset turbulence level. On the other hand, as shown in part (b) of the figure, the comparison with data for $Re_{\delta 2,t}$ is not as good, even after taking proper account of the shape factor. In particular, most of the data lies below the correlation's range of turbulent Reynolds num-

[1] This only occurs when the large-scale, energy-containing eddies are about the same size as the plate's leading-edge diameter.

[2] To include Roach and Brierley's correlation for the effect of plate thickness, multiply the right-hand side of this equation by $\exp[-0.4(t/\lambda)]$, where t is the plate's thickness.

[3] This is not valid however when eliminating Re_{xt} in favor of $Re_{\delta 2,t}$, since the momentum thickness and skin-friction coefficient distributions in pre-transitional flow with free-stream turbulence are usually different from that given by the laminar solution.

[4] When $Re_\lambda > 10$, which corresponds roughly to $Re_T > 10$, the length scales of turbulence vary roughly as the square root of the distance from the grid while the intensity varies roughly as the inverse square root of the distance (see §3.4.1). As a consequence, Re_λ is almost independent of distance, while Tu decreases.

bers, and this remains true even when a shape factor $H_t = 2.6$ is used to calculate the case for $Re_{\lambda,0} = 25$ as shown by the dashed line in the figure, which should pass through the data point for their lowest turbulence level. Clearly, more data is required to evaluate either of these "newer" correlations.

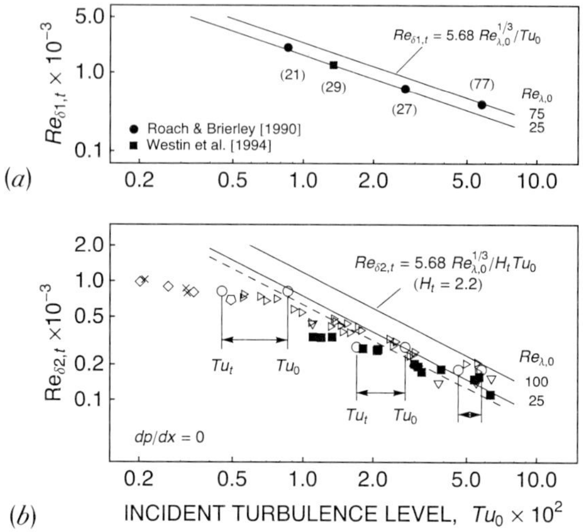

Figure 6.9a,b.
A new correlation for the onset of transition based on Taylor's parameter, and its comparison with data using the momentum-thickness Reynolds number at onset. (See Fig. 6.4 for legend.)

6.2.3 Rate of transition – standard correlations.

The effect of turbulence level on the spot parameter $\hat{n}\sigma$ is shown in Fig. 6.10a. The data in this and Fig. 6.4 provide pairs of results that when used together in Eq. (5.18) will reproduce either the measured intermittency distribution or positions of the minimum and maximum in the shear-stress distribution when a reasonable transition model is used. As in Fig. 6.4, the data shown by the open symbols represent those obtained from transition length measurements using Eq. (6.5), while those shown by the filled symbols are from near-wall intermittency measurements. Again, most of the data suffers from a lack of information about the turbulence. Nevertheless, as might be expected, the production of turbulent spots or more correctly the spot parameter increases with increasing turbulence level. A correlation of these results, which was also provided previously by the author [1991], yields

$$\hat{n}\sigma = 4.74(10)^{-8}\, Tu^{7/4}\ ; \quad (dp/dx = 0,\ Tu > 0.005) \tag{6.17}$$

when the effect of disturbances other than the turbulence level are ignored. In general, it correlates the data to within ±50 percent, which represents an error of roughly ±25 percent in the length of transition.

A relation between $\hat{n}\sigma$ and $Re_{\delta 2,t}$ can be obtained by eliminating the turbulence level between Eq. (6.7) and the above. This provides

$$\hat{n}\sigma = 2.9(10)^{-4}\, Re_{\delta 2,t}^{-14/5}\ ; \quad (dp/dx = 0,\ Tu > 0.005) \tag{6.18}$$

suggesting that the boundary-layer thickness and not the turbulence level is the controlling variable here. A comparison between this relation and data is shown in part (b) of Fig. 6.10. Since the transition length according to Eq. (6.2) is proportional to the inverse square root of $\hat{n}\sigma$, the above correlation provides

$$Re_{LT} = 120\, Re_{\delta 2,t}^{7/5} \tag{6.19}$$

This result is similar to correlations proposed by other investigators. From their data, Dhawan and Narasimha [1958] found $Re_{LT} \approx 38(Re_{\delta 2,t})^{1.6}$, which was corroborated by Abu-Ghannam and Shaw [1980]. This was later revised by Narasimha [1985] who suggested using $Re_{LT} \approx 68(Re_{\delta 2,t})^{1.5}$. Investigating transition in compressible flows at low turbulence levels, Chen and Thyson [1971] found that $Re_{LT,\gamma=0.95} \approx 104(Re_{\delta 2,t})^{1.34}$ for low Mach numbers, which when translated to our definition for the end of transition yields $Re_{LT} \approx 169(Re_{\delta 2,t})^{1.34}$. In addition, based on their data and Walker's [1987] analysis, Walker and Gostelow [1990] suggest using $Re_{LT} \approx 60(Re_{\delta 2,t})^{1.5}$, which is about twelve percent lower than Narasimha's version.

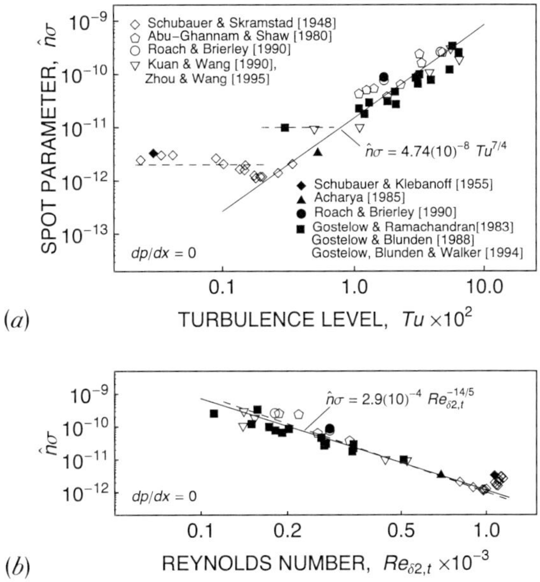

Figure 6.10a,b.
Spot parameter as a function of free-stream turbulence level and momentum-thickness Reynolds number at onset for zero pressure gradient flows.

Nevertheless, using Eq. (6.2) to convert Chen and Thyson's low free-stream turbulence result for transition length into an expression for the spot parameter, we obtain

$$\hat{n}\sigma = 1.62(10)^{-4}\, Re_{\delta 2,t}^{-2.68}\ ; \quad (dp/dx = 0,\ Tu < 0.005) \tag{6.20}$$

which is virtually the same as that given by Eq. (6.18) when plotted in Fig. 6.10b. In similar fashion we find that Narasimha's result provides $\hat{n}\sigma = 10^{-3}(Re_{\delta 2,t})^{-3}$, which as shown by the dashed line in Fig. 6.10b is again about the same, while Walker and Gostelow's yields $\hat{n}\sigma = 1.28(10)^{-3}(Re_{\delta 2,t})^{-3}$, which falls twenty-eight percent higher. Both of these yield $N = \hat{n}\sigma(Re_{\delta 2,t})^3$ = constant; a condition first proposed by Narasimha [1985]. In any case, all of these expressions lie well within the band of data except at the far right where transition is affected by acoustics and vibrations. More recently, Roberts and Yaras [2004] correlated Narasimha's spot parameter N for flows with free-stream turbulence and a pressure gradient using the shape factor at transition H_t. For transition without a pressure gradient, their correlation yields $N = 3.38(10)^{-3}$ for very low turbulence levels, which is significantly higher than any of the other correlations.

6.2.4 Production rate models.

Natural transition. Bypass transition.

Several relations have been developed for the spot production rate using a phenomenological approach. Of those proposed by Chen and Thyson [1971], Walker [1989], Walker and Gostelow [1990], and Mayle [1999], the first three apply to natural transition,[1] while the last applies to bypass transition. All, however, begin with the same following premise.

For a line source of spots, the spot production rate $\dot{n}'$ is defined as the number of spots produced per unit time per unit distance across the surface in the spanwise direction. If we define ω as the characteristic frequency of the unsteadiness in the boundary layer and l as the characteristic spanwise length of the disturbance causing the spots, and presume that these are the only relevant quantities governing the production of turbulent spots, then the dimensionless production rate is given by

$$\hat{n} = \frac{\dot{n}'\nu^2}{\bar{u}_\infty^3} \propto \frac{\omega\nu}{\bar{u}_\infty^2}\frac{\nu}{\bar{u}_\infty l} = \frac{\Omega}{Re_l} \tag{6.21}$$

where the definitions for $\hat{n}$ and the frequency parameter Ω are given by Eqs. (5.17) and (2.24). This is where the first three analyses begin to differ from the last.

Natural transition. Consider the production of turbulent spots in an unstable boundary layer. In this case, it is reasonable to expect that the characteristic frequency ω is related to the frequency of the Tollmien-Schlichting wave causing the instability.[2] It is also reasonable to expect, since spots are produced at the peaks of the streamwise vortices as described by Klebanoff et al. [1962], that the characteristic length l is related to the transverse wavelength of these vortices.[3]

Chen and Thyson used the frequency associated with the upper branch of the stability curve for $dp/dx = 0$ in their analysis, while Walker, and Walker and Gostelow used the frequency associated with the maximum amplification rate for all pressure gradients as determined by Ombrewski et al. [1969]. Curve fits to these variations in frequencies with Reynolds number are given respectively by

$$\Omega_{TS2} = 0.38 Re_{\delta 2}^{-1.34}\ ; \quad (dp/dx = 0, Re_{\delta 2,t} > 280) \tag{6.22}$$

and

$$\Omega_{TS\max} = 0.76 Re_{\delta 2}^{-3/2} \tag{6.23}$$

The first correlation, obtained by fitting the upper branch of the stability curve given in Table D.1, Appendix D, is slightly different from Chen and Thyson's who used Tollmien's result. If we use Eq. (6.7), the restriction on Reynolds number in Eq. (6.22) requires that the turbulence level must be *less* than roughly two percent, which is above that we normally associate with natural transition.

Since data on the transverse wavelength of the secondary vortices are limited, the characteristic length l is normally assumed proportional to the distance between spots when they first merge, that is, their breadth at the end of transition. Presuming that this is the case, we obtain

$$Re_l \propto Re_{LT} \tan\alpha$$

As a result, Eq. (6.21) yields

$$\hat{n} \propto \frac{\Omega_{TS}}{Re_{LT} \tan\alpha}$$

where Ω_{TS} stands for either Ω_{TS2} or Ω_{TSmax}, and any of the correlations for Re_{LT} may be used.

Using Chen and Thyson's relation for Re_{LT} and Eq. (6.22) for Ω_{TS}, the spot production rate is given

[1] Walker's analysis applies to natural transition in separation bubbles (see §8.3).

[2] See Reshotko [1969]. Also refer to §4.2.2 regarding initial amplification rates and their dependence on frequency.

[3] See §4.2.2, "Three-dimensional instability" and "Secondary-instability and laminar breakdown" regarding the secondary flow vortices and spot formation.

by

$$\hat{n} = C_1 \frac{1}{Re_{\delta 2,t}^{2.68} \tan\alpha} \tag{6.24}$$

where C_1 is a constant of proportionality. If Eq. (5.71) is used to evaluate the tangent of α, then the production rate varies inversely with the transition Reynolds number raised to the 3.08 power. On the other hand, if α is assumed constant as shown by the dashed line in Fig. 5.35a, it then varies inversely with the Reynolds number raised to the 2.68 power. Since no data on $\hat{n}$ itself is available, neither of these variations nor any other model for the production rate parameter can be verified.

However, if we assume σ constant, a comparison of Eqs. (6.20) and (6.24) yields

$$C_1 = 1.62(10)^{-4}\left(\sigma/\tan\alpha\right)^{-1}$$

Using the values $\alpha = 10.5°$ and $\sigma = 0.26$, as determined in §5.3.2 and §5.5.1 for a source Reynolds number of $Re_{\delta 1,s} = 1000$, we estimate that $C_1 = 1.15(10)^{-4}$.

Bypass transition. For bypass transition, the author proposed that spots are produced at the same frequency affecting the pre-transitional boundary layer, namely, the effective frequency of turbulence. Consequently, the characteristic frequency parameter in Eq. (6.21) is Ω_{eff}. Although it was originally thought that the characteristic Reynolds number in the equation would be associated with one of the turbulent length scales, all attempts to correlate the limited amount of data using them failed. Subsequently, since the large-scale eddies are stretched within the boundary layer and hence intensified only there, it was reasoned that the characteristic length scale in Eq. (6.21) associated with their upwellings must be related to the boundary-layer thickness. Choosing the displacement thickness as the characteristic length rather than the momentum thickness, the characteristic Reynolds number in Eq. (6.21) becomes $Re_{\delta 1}$. As a result we obtain

$$\hat{n} \propto \Omega_{eff,t} / Re_{\delta 1,t}$$

Using Eqs. (6.9) and (6.13), to rewrite the right-hand side of this expression in terms of the free-stream turbulence level and Reynolds number, we obtain

$$\hat{n} \propto Tu_t^2 / Re_{\lambda,t}$$

which, similar to the result obtained from the new correlation in Eq. (6.13) for the onset of transition, directly relates the rate of transition to both the turbulence level and turbulent Reynolds number.

When σ is assumed constant, these expressions may be written as

$$\hat{n}\sigma = C_2 \Omega_{eff,t} / Re_{\delta 1,t} \quad \text{and} \quad \hat{n}\sigma = 0.28 C_2 Tu_t^2 / Re_{\lambda,t} \tag{6.25}$$

where C_2 is a constant of proportionality. A curve fit of the first expression with Roach and Brierley's [1990] data yields $C_2 = 2.6(10)^{-5}$, which in turn provides a coefficient equal to $7.3(10)^{-6}$ in the second.

Graphs of each expression in Eq. (6.25) are provided in Fig. 6.11 compared with data. The numbers given in the parentheses in part (a) of the figure correspond respectively to the turbulence level and Reynolds number. Although only two data points are available for the plot, the first expression in Eq. (6.25) provides a reasonable fit especially after considering the fits using Eq. (6.21) with the other possible options for Re_l. Of course, Eq. (6.21) represents the simplest result of the analysis by considering the fewest relevant dimensional quantities. The second expression in Eq. (6.25) is compared with data in part (b) of the figure. Here, the calculated band for $25 \leq Re_\lambda \leq 100$ roughly spans the measurements. Since most of this data is plotted using the incident turbulence level, we can expect it to shift leftward into the band when plotted against the turbulence level at transition. Furthermore, similar to the effect of turbulent Reynolds number on the onset of transition (see §6.2.2), a drift to lower transition rates for higher turbulence levels across the band is to be expected. Consequently, the correlation given in Eq. (6.25), although far from being verified, appears in part to explain the large scatter associated with the data in Fig. 6.10a.

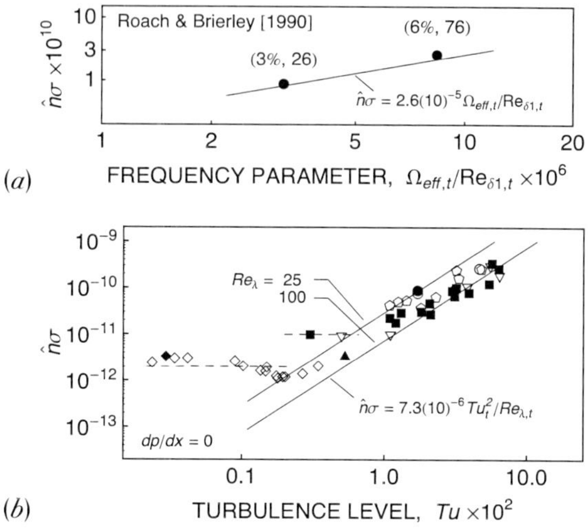

Figure 6.11a,b.
Spot parameter as a function of the frequency parameter at onset and free-stream turbulence level for zero pressure gradient flows.

If we eliminate the frequency parameter in the first expression of Eq. (6.25) by using Eq. (6.9), we obtain

$$\hat{n}\sigma = 1.25 C_2 Re_{\delta 1,t}^{-2} \tag{6.26}$$

where the coefficient with $C_2 = 2.6(10)^{-5}$ equals $3.25(10)^{-5}$. Although a graph of this relation passes through Roach and Brierley's data in Fig. 6.10b, using $Re_{\delta 1,t} = H_t Re_{\delta 2,t}$ with $H_t = 2.2$, it lies substantially above the average bypass-transition data. When examined, it was found that the frequency parameters for Roach and Brierley's experiments were larger than the frequency parameters estimated by the author for any of the other investigations. This discrepancy remains to be investigated.

6.3 Nonzero Pressure-Gradient Flows.

Onset of transition. Length of transition.

Information regarding the effect of pressure gradients on the onset of transition hasn't changed much since the author's last review of the situation about the twenty-five years ago. Similar to the investigations conducted for transition in flows without a pressure gradient, detailed measurements of the developing flows are still lacking, turbulence levels are reported only at one location, and the onset and length of transition are determined by various methods. More disturbing, however, little effort has been expended in measuring the distributions of either the spot production rate or the propagation parameter, which are critical to understanding transition in flows with a pressure gradient.

Consequently, rather than presenting the older correlations and variations thereof,[1] we will consider the data from a slightly different perspective and provide some new correlations for the effect of pressure gradient and turbulence level on the transition Reynolds number $Re_{\delta 2,t}$ and spot parameter $\hat{n}\sigma$.

[1] Those interested in the older correlations should see Mayle [1991], Gostelow et al. [1994], and Solomon et al. [1996].

6.3.1 Onset of transition.

The most useful data regarding boundary-layer transition with a free-stream pressure gradient have been obtained by Feindt [1956], Blair [1983], and Wang and Keller [1999] for favorable pressure gradients, Sharma et al. [1982] and Gostelow and his coworkers [1988, 1989, 1990, 1994] for adverse gradients, and Abu-Ghannam and Shaw [1980] and Roberts and Yaras [2003, 2004] for both favorable and adverse gradients. Gostelow's et al. [1994] data presumably supersede their earlier results and will therefore be used herein. As shown by Abu-Ghannam and Shaw among others, not only does the pressure gradient affect the onset of transition in these experiments, but also its variation before transition which is usually referred to as its "pressure-gradient history." In this regard, Gostelow et al. conducted their tests with the free-stream velocity distribution adjusted to produce a pressure-gradient parameter $\lambda_{\delta 2}$ independent of distance. As discussed in §4.2.1 regarding Fig. 4.9, the shape of the pre-transitional velocity profiles for these flows are similar when the free-stream turbulence level is zero. Blair on the other hand maintained a constant acceleration parameter K which results in a pressure gradient parameter that increases with distance, while both Feindt and Abu-Ghannam and Shaw held the pressure gradient dp/dx nearly constant which results in a continually decreasing pressure-gradient parameter for an adverse gradient and a continually increasing parameter for a favorable gradient. Thus, except for the flows with a constant $\lambda_{\delta 2}$, the shape of the velocity profiles and, therefore, the stability of the flow, before transition continually changes with distance regardless of the turbulence level.

A summary of the data is shown in Fig. 6.12 where the transition Reynolds number is plotted against the pressure-gradient parameter for various incident free-stream turbulence levels.[1] The data for turbulence levels of approximately one percent are shown in part (a) of the figure, where the data of Feindt and Abu-Ghannam and Shaw for $\lambda_{\delta 2} < 0$ are seen to differ significantly from Gostelow's et al., and differ from one another for $\lambda_{\delta 2} > 0$. The result obtained from the correlation given by Eq. (6.7) for $dp/dx = 0$ is shown for comparison by the filled circle.

The results for higher turbulence levels are shown in part (b) of the figure. Most of the data for adverse pressure gradients were obtained by Gostelow et al., with some of the data obtained by Roberts and Yaras (vertically struck data points). The data for favorable pressure gradients were obtained by Abu-Ghannam and Shaw (obliquely struck points), Blair (horizontally struck points), and Roberts and Yaras. Roberts and Yaras observed that the intermittency distribution in many of their experiments differed significantly from Narasimha's solution, and hence determined the onset of transition from a visual inspection of where turbulent spots first appeared in their velocity traces.

The dashed lines in this figure represent the boundaries between separated and attached flows, defined according to Thwaites [1949] by

$$\lambda_{\delta 2,sep} = -0.082 \qquad (6.27)$$

between unstable and stable flows, obtained from columns five and seven in Table D.3 of Appendix D, and between flows that can and cannot transition. In the region to the right of the latter boundary, the favorable pressure gradient is strong enough to prevent any turbulent spots from forming.[2] We refer to it here as "non-transitional flow." It is also the region where all fluctuations or turbulence within a boundary layer decay and where even a turbulent boundary layer eventually becomes laminar. Discussed briefly in §8.4.2, this process is called either "laminarization" or "reverse transition." According to Jones and Launder [1972], reverse transition occurs when the acceleration parameter $K \geq 3(10)^{-6}$, or in terms of the pressure-gradient parameter when

[1] Gostelow's earlier data are plotted against the acceleration parameter by Mayle [1991]. Referring to the discussion regarding velocity profiles in §5.3.3 with $dp/dx \neq 0$, it is best to correlate this data using the pressure-gradient parameter.

[2] See Schraub and Kline [1965],

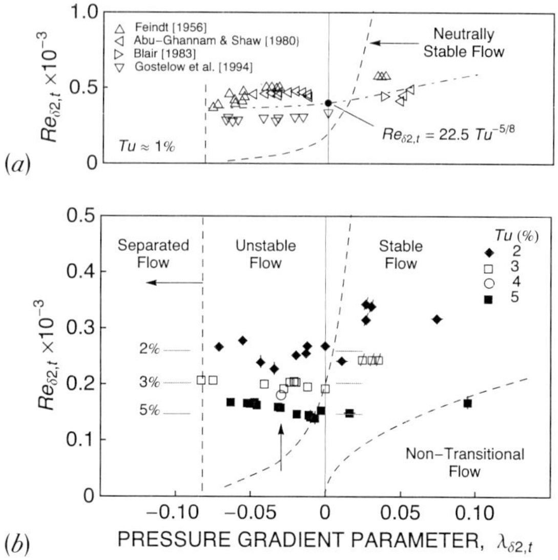

Figure 6.12a,b.
Momentum thickness Reynolds number at the onset of transition as a function of the free-stream pressure-gradient parameter for different turbulence levels. Data from Feindt [1956], Abu-Ghannam and Shaw [1980], Blair [1983], Gostelow et al. [1994] and Roberts and Yaras [2003].

$$\lambda_{\delta 2} \geq 3(10)^{-6} Re_{\delta 2}^2 \qquad (6.28)$$

This equation is used to plot the boundary between transitional and non-transitional flow in the figure. Consequently, as we cross these dashed boundaries from right to left, we pass through regions of laminar flow that is non-transitional, stable, unstable, and separated. From a slightly different perspective, these boundaries separate regions where reverse, bypass, natural, and separated-flow transition may occur.

In general, we see that the transition Reynolds number increases with either an increase in the pressure-gradient or a decrease in the free-stream turbulence level. This behavior is found by all investigators. For high turbulence levels, the onset of transition appears to be unaffected by the pressure gradient. The horizontal lines marked "2%", "3%", and "5%" in the figure correspond to the correlation given by Eq. (6.7) for $dp/dx = 0$. The slight rise in the transition Reynolds number for these turbulence levels near separation, which was also observed by Abu-Ghannam and Shaw, is probably caused in part by the decay of turbulence from its initial value, and in part by an increase in the mean velocity near the wall caused by the turbulence which in turn delays transition. In this regard, we note that the free-stream velocity distribution for the empty-circled data, pointed to by the arrow in the figure, nearly corresponds to that for a wedge flow with an exponent $m = -0.12$. That is, without a turbulent free stream this flow would separate (see Fig. 4.10).

6.3.2 Rate of transition.

It is usual to display the effects of pressure gradient on the spot parameter by plotting $\hat{n}\sigma$ against $\lambda_{\delta 2}$, or K, with the free-stream turbulence level as a parameter. Because these plots are more confusing than enlightening, particularly regarding the effect of free-stream-turbulence level, neither will be presented here.[1] In general, however, they show that the spot parameter decreases as the pressure gradient becomes more favorable and increases as it becomes more adverse.

Instead we will take a different approach. In particular, if we presume that turbulence mainly affects

[1] See Gostelow [1989] and Mayle [1991] for these plots.

the transition Reynolds number, and that this effect is indirectly taken into account by correlating the spot parameter with the Reynolds number as in Fig. 6.10b, then another option for displaying the effect of pressure gradient is to plot $\hat{n}\sigma$ against $Re_{\delta 2,t}$ with Tu and $\lambda_{\delta 2}$ as parameters. Blair's [1983] and Gostelow's et al. [1994] data are plotted this way in Fig. 6.13. The $\lambda_{\delta 2} = 0$ curve in this figure is given by Eq. (6.18). The results using Gostelow's et al. data were obtained by selecting the data points in Fig. 13 of their paper nearest the values of $\lambda_{\delta 2}$ listed in the legend Fig. 6.13 regardless of the turbulence levels. In this case, the cross plotting significantly reduces the number of data points to only about two to five per each value of the pressure-gradient parameter. Nevertheless, with a stretch of the imagination and some physical considerations, there does appear to be a trend.

As previously noted, the effect of pressure gradient depends on whether the gradient is favorable or adverse. Remarkably, Blair's data for favorable gradients, which is for two different acceleration parameters, three different turbulence levels, and four different pressure-gradient parameters, all lie along the same line. The equation for this line is given by

$$\hat{n}\sigma = 1.2(10)^{-6} Re_{\delta 2,t}^{-2} \,; \quad (\lambda_{\delta 2,t} \geq 0) \tag{6.29}$$

If we disregard the rise in the transition Reynolds number for the high turbulence levels near separation, the 2, 3 and 5 percent lines of constant turbulence plot as shown by the short-dashed, vertically oriented curves in the figure. As indicated, the turbulence level increases from right to left for these curves following the trend shown in Fig. 6.12. Nearly vertical in the adverse pressure-gradient region, they cross the $\lambda_{\delta 2} = 0$ line to somehow fair into the lower line given by the above equation. Hence, labeled "FPG Locus" in the figure, it appears that Eq. (6.29) is the *locus* for all the constant turbulence-level contours with a favorable-pressure-gradient when plotted in this format.

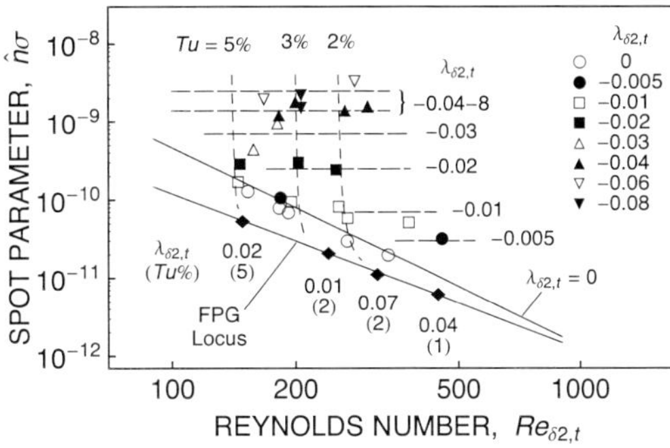

Figure 6.13.
Spot parameter as a function of the transition Reynolds number for different free-stream pressure-gradient parameters and turbulence levels. Data from Blair [1983] and Gostelow et al. [1994].

For adverse pressure gradients, the data in the upper portion of the figure above the $\lambda_{\delta 2} = 0$ curve appear to be independent of $Re_{\delta 2,t}$, whereas data in the lower portion fair into the curve at the lower Reynolds numbers. The data for $\lambda_{\delta 2,t} \approx -0.01$ illustrates this behavior particularly well. Hence, contrary to the propositions of previous investigators (author included), we will consider that the spot parameter somewhat above the $\lambda_{\delta 2} = 0$ curve, shown by any of the long-dashed horizontal lines, is independent of both the turbulence level and Reynolds number. It's variation with pressure gradient, as we will momentarily see, is given quite well by

$$\hat{n}\sigma = 1.2(10)^{-9}\left[1-\cos(12\pi\lambda_{\delta 2,t})\right]; \quad (-0.082 < \lambda_{\delta 2,t} < 0, \hat{n}\sigma > 5.6(10)^{-5} Re_{\delta 2,t}^{-2.54}) \tag{6.30}$$

where the conditions in parentheses restrict its use to adverse pressure gradients in the upper portion of the figure without separation. Furthermore, we will see that this relation is the *locus* for all the constant turbulence-level contours when plotted in the standard $\hat{n}\sigma$ versus $\lambda_{\delta 2}$ format. Therefore, Eq.

(6.30) may be considered an adverse-pressure-gradient (APG) locus. The maximum of $\hat{n}\sigma$ according to this expression, which occurs when $\lambda_{\delta 2,t} \approx -0.082$, is $2.4(10)^{-9}$. Using Eq. (6.2), this would imply that the transition length Reynolds number for a flow with a constant pressure gradient of $\lambda_{\delta 2} = -0.082$ and a Falkner-Skan separation velocity profile (see Fig. 4.9, with $m = -0.091$) is about $4.4(10)^4$.

6.3.3 Topology of the pressure-gradient effect.

By considering the trends in Figs. 6.12 and 6.13, and the behavior of the constant turbulence-level contours as they approach their loci in these figures, a new topology emerges for the effects of turbulence level on transitional flows with a constant pressure-gradient parameter.

This topology is sketched in Figs. 6.14 and 6.15. In the first figure, the best guess for the variations of $Re_{\delta 2,t}$ with $\lambda_{\delta 2}$ are drawn for the turbulence levels of 1, 2, 3 and 5 percent. For turbulence levels higher than about five percent the transition Reynolds number should be independent of the pressure gradient and given by Eq. (6.7) or its equivalent. For turbulence levels lower than five percent, the transition Reynolds number is very nearly independent of the pressure gradient when $\lambda_{\delta 2,t} < 0$, and increases when $\lambda_{\delta 2,t} > 0$, particularly for the lowest turbulence levels. For comparison, the curve for $Tu = 0.01$ is also plotted in Fig. 6.12a as the dot-dashed line.

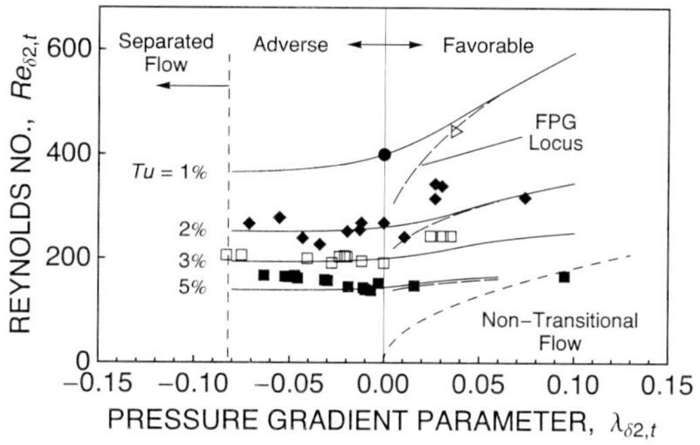

Figure 6.14.
Faired correlations for the transition Reynolds number as a function of the free-stream pressure-gradient parameter for different turbulence levels. Symbols are the same as in Fig. 6.12.

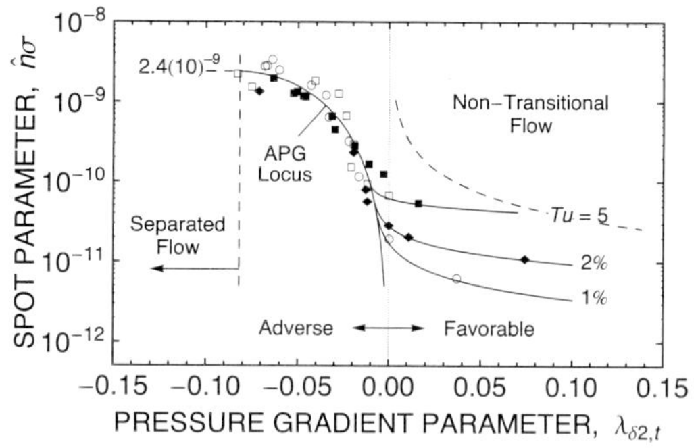

Figure 6.15.
Faired correlations for the spot parameter as a function of the free-stream pressure-gradient parameter for different turbulence levels. Symbols are the same as in Fig. 6.12.

In the second figure, the best guess for the variations of $\hat{n}\sigma$ with $\lambda_{\delta 2}$ are drawn for the turbulence levels of 1, 2 and 5 percent. For adverse pressure gradients, the spot parameter increases dramatically as the gradient becomes more adverse, leveling out as separation is approached. In this range, the spot parameter is very nearly independent of the turbulence level and varies according to Eq. (6.30), that is, according to the APG locus. The scatter of the data in this region, while significant, is not obviously related to the free-stream turbulence level. The spot parameter for favorable pressure gradients decreases as the gradient becomes more favorable, leveling out in this case as the non-transitional region is approached. It of course increases with turbulence level. The boundary between transitional and non-transitional flow in this figure is obtained by using Eqs. (6.28) and (6.29) to eliminate $Re_{\delta 2}$, which yields

$$\hat{n}\sigma \geq 3.6(10)^{-12} / \lambda_{\delta 2,t} \; ; \quad (\lambda_{\delta 2,t} \geq 0)$$

In both figures, the curves of constant turbulence levels for large favorable pressure gradients were obtained by fairing curves through the meager amount of data using the FPG locus given by Eq. (6.29) as a constraint. That is, once the curve fit is determined in Fig. 6.15, Eq. (6.29) was used to eliminate the spot parameter and draw the FPG curves in Fig. 6.14. In the best situation, they would be empirically determined. Nevertheless, the remaining portions of these constant-turbulence-level curves were drawn by fairing them into their corresponding horizontal lines in Fig. 6.12 and the APG curve in Fig. 6.15 for $\lambda_{\delta 2,t} < 0$. Whereas all of this is expected to change as more data becomes available, the fairing of the two percent curve between its FPG and horizontal line for $\lambda_{\delta 2,t} < 0$ in Fig. 6.14 is approximately given by

$$Re_{\delta 2,t} = 258 + 1.2(10)^4 \left(0.01 + \lambda_{\delta 2,t}\right)^{1.9} \; ; \quad (-0.03 < \lambda_{\delta 2,t} < 0.05, \; Tu = 0.02)$$

The curve for two percent data in Fig. 6.15 is given by

$$\hat{n}\sigma = 3.5(10)^{-12} \left(0.01 + \lambda_{\delta 2,t}\right)^{-0.45} \; ; \quad (\lambda_{\delta 2,t} > -0.03, \; Tu = 0.02)$$

These expressions were used to plot the two percent curve between the FPG and $\lambda_{\delta 2,t} = 0$ lines in Fig. 6.13, which fairs the vertical two percent line into the FPG locus. The curves for the one and five percent turbulence levels in these figures were drawn by fitting the limited amount of data with similar expressions to provide a similar behavior. The three percent curves in Figs. 6.13 and 6.14 were simply interpolated from the two and five percent curves using its data for $\lambda_{\delta 2,t} < 0$.

The process for determining the combined effect of pressure gradients and turbulence levels on transition can now be summarized as follows: 1) for $dp/dx = 0$, use Eqs. (6.7) and (6.18); and 2) for $dp/dx \neq 0$, determine $Re_{\delta 2,t}$ from Fig. 6.14 by interpolation, or use Eq. (6.7) when $Tu > 0.03$. Then determine $\hat{n}\sigma$ from Fig. 6.15 by interpolation, or use Eq. (6.30) when $\lambda_{\delta 2,t} < 0$ and Eq. (6.29) together with the interpolated result of Fig. 6.14 when $\lambda_{\delta 2,t} \geq 0$. As we see, much of this is speculative and requires empirical confirmation. What is apparent, however, is that an extraordinary amount of attention has been paid to transitional flow in adverse pressure gradients, which occurs over short streamwise distances, rather than to transitional flow in favorable gradients where transition occurs over long distances, and the effect of free-stream turbulence is obviously significant.

6.4 Effect of Surface Curvature.

Curvature effects on transition were first examined by Görtler [1940], who considered the theoretical aspects of stability, and Liepmann [1943], who conducted transition experiments on curved surfaces. Görtler determined that a laminar boundary layer on a concave surface can be unstable to three-dimensional disturbances because of the imposed centrifugal force, and that the amplified instability leads directly to a stable spanwise array of streamwise vortices within the layer, known as *Görtler vortices*.[1] Liepmann demonstrated that the onset of transition on a *convex* surface occurs downstream of that on a flat surface, as did Wang and Simon [1987], and upstream on a *concave* surface. Although

[1] See Schlichting [1979].

all of this work was done at relatively low turbulence levels, we can deduce from the results for different pressure gradients that the onset of transition on a convex surface at high turbulence levels will be about the same as that on a flat surface.

For $Tu = 0.06$ percent, Liepmann found that transition occurs on a concave surface when

$$G\ddot{o} = Re_{\delta 2,t}\left(\delta_{2,t}/r\right)^{1/2} \geq 7$$

where $G\ddot{o}$ is the Görtler number and r is the radius of curvature of the surface. More recent experiments on concave surfaces at higher turbulence levels were conducted by Riley et al. [1989]. Their results together with Liepmann's are plotted in Fig. 6.16 where the transition Reynolds number is plotted against the square root of the surface radius-to-momentum thickness ratio at transition. In this format, all straight lines passing through the origin correspond to a constant Görtler number. As we see, Liepmann's data lie close to the $G\ddot{o} = 7$ line.

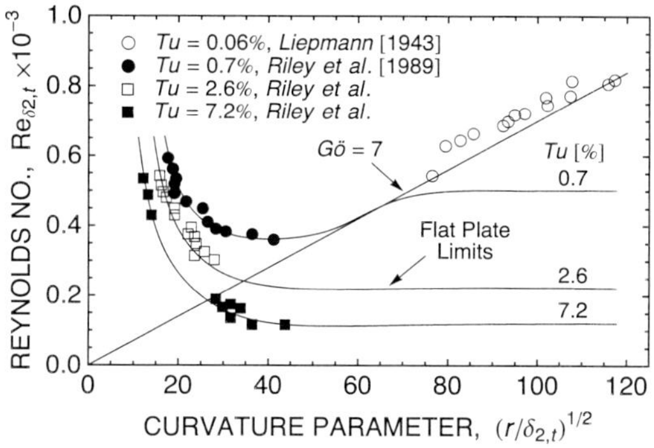

Figure 6.16.
Transition Reynolds number for flow on a concave surface with different turbulence levels.

The rise in transition Reynolds number above this line as found by Riley et al. is caused by the Görtler vortices increasing the mean velocity gradients near the wall thereby delaying transition. For highly curved surfaces, this effect dominates that of turbulence. Since curvature effects diminish towards the right in this figure, data for elevated turbulence levels must asymptotically approach the flat plate results in this region. Using this fact, estimates of the effect of curvature for these turbulence levels with $(r/\delta_{2,t})^{1/2} \geq 50$ have been drawn in the figure by fairing curves between the data and the horizontal lines given by Eq. (6.7) to the right. From this result, it is seen that concave curvature can either decrease, as found by Liepmann, or increase the transition Reynolds number depending on the strength of curvature and turbulence level.

There is virtually no reliable data from which one may determine the spot parameter $\hat{n}\sigma$ for transition on a concave surface. For flow on a convex surface, the heat transfer data of Wang and Simon were examined by the author. Here, it was found that $\hat{n}\sigma$ is relatively unaffected up to a Görtler number of four and then decreases to about 80 percent of its flat plate value at $G\ddot{o} = 7$. The latter situation corresponds to an increase of roughly ten percent in the transition length. Since this is based on only three data points, its validity remains to be confirmed.

6.5 Effect of Surface Roughness.

Although much work has been done on the effects of roughness on laminar and turbulent flows, little has been done on its effect on boundary layer transition except for single roughness elements. The earliest experiments investigating the effect of distributed roughness on transition was conducted by Feindt [1956], who demonstrated that the transition Reynolds number Re_{xt} decreases with an increasing "roughness" Reynolds number Re_k based on the free-stream velocity and equivalent "sand-grain" roughness k_s. An approximate conversion between the equivalent sand-grain roughness and half the average height of a roughness element in a packed array is to multiply the latter by six, but this depends on the geometry of the roughness element.[1]

Roughness investigations in the 1990s were mainly conducted on gas turbine airfoils to study its effect on heat transfer and aerodynamic loss. The effect of roughness on a flat plate in a zero pressure-gradient flow were measured by Gibbings and Al-Shukri [1997], who were interested in its effect on pre-transitional flow as well as on transition, and by Pinson and Wang [1997, 2000], and Wang and Rice [2003], who were primarily interested in the effect of its distribution on heat transfer in transitional flows. Other investigations include those of Roberts and Yaras [2005, 2006], who measured the effect of uniformly distributed roughness on a flat plate at elevated turbulence levels in an accelerated/decelerated flow with separation.

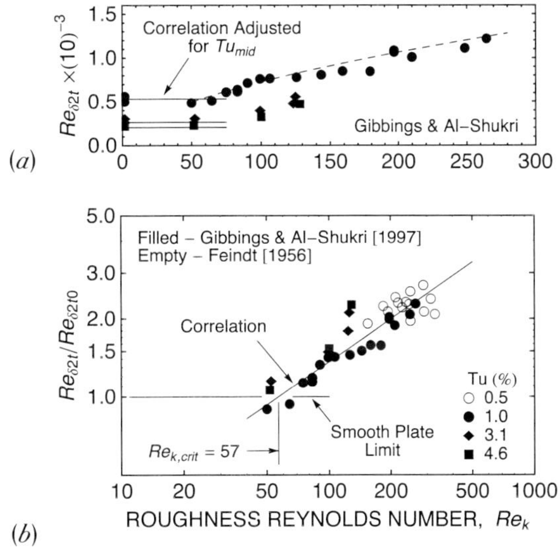

Figure 6.17a,b.
Transition Reynolds number for flow on a rough surface with different turbulence levels. Data from Feindt [1956] and Gibbings and Al-Shukri [1997].

The transition results of Feindt, and Gibbings and Al-Shukri are plotted in Fig. 6.17. This data is for uniform roughness and zero pressure gradient flow. The abscissa is the roughness Reynolds number Re_k. The effect of roughness for various turbulence levels is shown in part (a) of the figure, while a preliminary correlation of the results is shown in part (b).[2] The correlation shown in this figure is given by

$$\frac{Re_{\delta 2t}}{Re_{\delta 2t,0}} = \begin{cases} 1 & ; \; Re_k < Re_{k,crit} \\ \left(Re_k / Re_{k,crit}\right)^{0.55} & ; \; Re_k \geq Re_{k,crit} \end{cases} \qquad (dp/dx = 0) \tag{6.31}$$

where $Re_{k,crit} = 57$ is the critical roughness Reynolds number, and $Re_{\delta 2t,0}$ is the transition Reynolds

---

[1] See Schlichting [1979]. Also, see Mick [1987] for a discussion about modeling roughness. Not surprising, it requires modeling the tip of the profilometer used to measure the roughness as well as the other parameters of the problem.

[2] Because Gibbings and Al-Shukri's data for a turbulence level of 1.7 percent behaves significantly different from the rest of their data, it is not included in this figure.

number for a smooth surface at a given turbulence level. Its value in this case was obtained by adjusting the constant in Eq. (6.8) to match the data for $Re_k = 0$, as plotted in part (a) of the figure, to correct for measuring the turbulence level midway between the leading edge of the plate and the onset of transition. To include Feindt's data in the correlation, the momentum-thickness Reynolds number at transition was calculated from the distance Reynolds number using Gibbings and Al-Shukri's correlation for the effect of roughness on the growth of the laminar boundary layer, namely,

$$Re_{\delta 2} = 0.664\sqrt{Re_x}\,\exp\left[10.8Tu + 6.7(10)^{-3}Re_k\right]; \quad (dp/dx = 0) \tag{6.32}$$

In this case, an estimated turbulence level of 0.5 percent, based on Feindt's measurement of $Re_{xt,0}$, was used.

The results in Fig. 6.17 indicate that roughness has little-to-no effect on the onset of transition until the roughness Reynolds number reaches its critical value, above which the transition Reynolds number increases with increasing roughness according Eq. (6.31). The result of this correlation for the one-percent turbulence level agrees quite well with the data, as shown by the dashed line in part (a) of the figure. Since data for Re_k greater than about a hundred and thirty is not available for the higher turbulence levels, it presently remains unknown whether this or any other correlation is valid for these higher values.

Gibbings and Al-Shukri also measured the effect of roughness on the length of transition Re_{LT} at elevated turbulence levels. Again it is found that roughness has little-to-no effect on the rate until the critical Reynolds number is reached. Above $Re_{k,crit}$ however, while Re_{LT} generally decreases with increasing roughness Reynolds number, it appears to decrease more for the lower than the higher turbulence levels. A preliminary correlation of this effect, expressed in terms of the spot parameter $\hat{n}\sigma$, is given by

$$\frac{\hat{n}\sigma}{(\hat{n}\sigma)_0} = 1 + 3.5(10)^{-4}\,Tu^{-1.65}\left(Re_k/Re_{k,crit} - 1\right)^{0.55}; \quad Re_k \geq Re_{k,crit} \quad (dp/dx = 0) \tag{6.33}$$

where $(\hat{n}\sigma)_0$ is the spot parameter for a smooth surface at a given turbulence level. For $Re_k < Re_{k,crit}$, the spot parameter equals $(\hat{n}\sigma)_0$. Again, for the higher turbulence levels, this correlation is limited to roughness Reynolds numbers less than about a hundred and thirty.

Finally, it should be noted that besides offering correlations based on a regression analysis of their transition data, Gibbings and Al-Shukri also provide correlations for the standard boundary-layer parameters in a laminar boundary layer developing on a rough surface with elevated turbulence levels and a zero free-stream pressure gradient. This information is useful for calculating pre-transitional boundary layers on rough surfaces with $dp/dx = 0$.

References

Abu-Ghannam, B.J., and Shaw, R., 1980, "Natural Transition of Boundary Layers–The Effects of Turbulence, Pressure Gradient and Flow History," J. Mech. Engng. Sci., 22, pp. 213–228.

Acharya, M., 1985, "Pressure-Gradient and Free-Stream Turbulence Effects on Boundary-Layer Transition," Forschungsbericht KLR 85-127 C, Brown Boveri Forschungszentrum, Baden, Switzerland; also Feiereisen, W.J., and Acharya, M., 1986, "Modeling of Transition and Surface Roughness Effects in Boundary-Layer Flows," AIAA Journal, 24, pp. 1642–1649.

Blair, M.F., 1982, "Influence of Free-Stream Turbulence On Boundary Layer Transition in Favorable Pressure Gradients," J. Engng. for Power, 104, pp. 743–750.

Blair, M.F., 1983, "Influence of Free-Stream Turbulence on Turbulent Boundary Layer Heat Transfer and Mean Profile Development, Part I–Experimental Data," J. Heat Transfer, 105, pp. 33–40.

Blair, M.F., 1992, "Boundary Layer Transition in Accelerating Flows with Intense Free-Stream Turbulence: Part 1–Disturbances Upstream of Transition Onset, and Part 2–The Zone of Intermittent Turbulence," J. Fluids Engng., 114, pp. 313–332.

Brown, B., and Burton, R.C., 1977, "The Effects of Free-Stream Turbulence Intensity and Velocity Distribution on Heat Transfer to Curved Surfaces," ASME Paper 77-GT-48.

Burgers, J.M., 1924, "The Motion of a Fluid in the Boundary Layer along a Plane Smooth Surface," Proc. 1st Intern. Congress for Applied Mech., Delft, p. 113.

Chen, K.K., and Thyson, N.A., 1971, "Extension of Emmons' Spot Theory to Flows on Blunt Bodies," AIAA Journal, 9, pp. 821–825.

Dey, J., and Narasimha, R., 1984, "Spot Formation Rates in Incompressible Constant-Pressure Boundary Layers," Rep. 84 FM 11, Dept. Aerospace Engng., Indian Institute of Science, Bangalore, India.

Dhawan, S., and Narasimha, R., 1958, "Some Properties of Boundary Layer Flow During Transition from Laminar to Turbulent Motion," J. Fluid Mech., 3, pp. 418–436.

Dryden, H.L., 1936, "Airflow in the Boundary Layer Near a Plate," NACA Rept. 562.

Emmons, H.W., and Bryson, A.E., 1952, "The Laminar-Turbulent Transition in a Boundary Layer, Part II," Proc. 1st U.S. Natl. Cong. Theor. Appl. Mech., pp. 859–868.

Feindt, E.G., 1956, "Untersuchungen über die Abhängigkeit des Umschlages laminar-turbulent von Oberflächenrauhigkeit und der Druckverteilung," Schiffbautech. Gesellschaft, 50, pp. 180–203.

Gibbings, J.C., and Al-Shukri, S.M., 1997, "Effect of Sandpaper Roughness and Stream Turbulence on the Laminar Layer and Its Transition," Aeronautical J., 101, pp. 17–24.

Görtler, H., 1940, "Über eine dreidimensionale Instabiltät laminar Grenzschichten an konkaven Wänden," Nachr. Wiss. Ges. Göttingen, Math. Phys. Klasse, Neue Folge 2, No.1; also 1941, ZAMM 21, pp. 250–252.

Gostelow, J.P., 1989, "Adverse Pressure Gradient Effects on Boundary Layer Transition in a Turbulent Free Stream," 9th Int. Sym. on Air Breathing Engines, 2, pp. 1299–1306.

Gostelow, J.P. and Blunden, A.R., 1988, "Investigations of Boundary Layer Transition in Adverse Pressure Gradient," J. Turbomachinery, 111, pp. 366–375.

Gostelow, J.P., Blunden, A.R., and Walker, G.J., 1994, "Effects of Free-Stream Turbulence and Adverse Pressure Gradients on Boundary Layer Transition," J. Turbomachinery, 116, pp. 392–404.

Gostelow, J.P. and Ramachandran, R.M., 1983, "Some Effects of Free Stream Turbulence on Boundary Layer Transition," Proc. 8th Australasian Fluid Mech. Conference.

Gostelow, J.P., and Walker, G.J., 1990, "Similarity Behavior in Transitional Boundary Layers over a Range of Adverse Pressure Gradients and Turbulence Levels," ASME Paper 90-GT-130.

Hansen, M., 1928, "Die Geschwindigkeitsverteilung in der Grenzschicht an der längsangeströmungten ebenen Platte," ZAMM, 8, pp. 185–199; also 1930, NACA TM 585.

Hall, A.A., and Hislop, G.S., 1938, "Experiments on the Transition of the Laminar Boundary Layer on a Flat Plate," ARC RM 1843.

Hegge Zijnen, B.J. van der, 1924, "Measurements of the Velocity Distribution in the Boundary Layer along a Plane Surface," Thesis, Delft.

Hislop, G.S., 1940, "The Transition of a Laminar Boundary Layer in a Wind Tunnel," Ph.D. Thesis, Cambridge University.

Hourmouziadis, J., 1989, "Aerodynamic Design of Low Pressure Turbines," AGARD Lecture Series, 167.

Jonàs, P., Mazur, O., and Uruba, V., 2000, "On the Receptivity of the By-Pass Transition to the Length Scale of the Outer Stream Turbulence," European J. Mech. B/Fluids, 19, pp. 707–722.

Jones, W.P., and Launder, B.E., 1972, "The Prediction of Laminarization With a Two-Equation Model of Turbulence," Int. J. Heat and Mass Transfer, 15, pp. 301–314.

Klebanoff, P.S., Tidstrom, K.D., and Sargent, L.M.,1962, "The Three-Dimensional Nature of Boundary-Layer Instability," J. Fluid Mech., 12, pp. 1–34.

Kuan, C., and Wang, T., 1990, "Some Intermittent Behavior of Transitional Boundary Layers," Exp. Thermal & Fluid Sci., 3, pp. 157–170.

Liepmann, H.W., 1943, "Investigations on Laminar Boundary-Layer Stability and Transition on Curved Boundaries," NACA ACR 3H30; also NACA-WR-W-107.

Mayle, R.E., 1991, "The Role of Laminar-Turbulent Transition in Gas Turbine Engines," J. Turbomachinery, 113, pp. 509–537.

Mayle, R.E., 1999, "A Theory for Predicting the Turbulent-Spot Production Rate," J. Turbomachinery, 121, pp. 588–593.

Mick, W.J., 1987, "Transition and Heat Transfer in Highly Accelerated Rough-Wall Boundary Layers," Ph.D. Thesis, Rensselaer Polytechnic Institute.

Narasimha, R., 1957, "On the Distribution of Intermittency in the Transition Region of a Boundary Layer," J. Aero. Sci., 24, pp. 711–712.

Narasimha, R., 1984, "Subtransitions in the Transition Zone," IUTAM Symposium on Laminar-Turbulent Transition (ed., V.V. Kozlov), Springer-Verlag, Berlin, pp. 140–151.

Narasimha, R., 1985, "The Laminar-Turbulent Transition Zone in the Boundary Layer," Prog. Aerospace Sci., 22, pp. 29–80.

Narasimha, R., Devasia, K.J., Gururani, G., and Badri Narayanan, M.A., 1984, "Transitional Intermittency in Boundary Layers Subjected to Pressure Gradient," Experiments in Fluids, 2, pp. 171–176.

Ombrewski, H.G., Morkovin, M.V., and Landahl, M., 1969, "A Portfolio of Stability Characteristics of Incompressible Boundary Layers," AGARDograph No. 134.

Pinson, M.W. and Wang, T., 1997, "Effects of Leading Edge Roughness on Flow and Heat Transfer in Transitional Boundary Layers," J. Heat Mass Transfer, 40, pp. 2813–2823.

Pinson, M.W. and Wang, T., 2000, "Effect of Two-scale Roughness on Boundary Layer Transition over a Heated Flat Plate: Part 1–Surface Heat Transfer, and Part 2–Boundary Layer Structure," J. Turbomachinery, 122, pp. 301–316.

Reshotko, E., 1969, "Stability Theory as a Guide to the Evaluation of Transition Data," AIAA Journal, 7, pp. 1086–1091.

Riley, S., Johnson, M.W., and Gibbings, J.C., 1989, "Boundary Layer Transition of Strongly Concave Surfaces," ASME Paper 89-GT-321.

Roach, P.E., and Brierley, D.H., 1990, "The Influence of a Turbulent Free-Stream on Zero Pressure Gradient Transitional Boundary Layer Development Including the T3A & T3B Test Case Conditions," Proc. 1st ERCOFTAC Workshop on Numerical Simulation of Unsteady Flows, Transition to Turbulence and Combustion, Lausanne; also 1992, Numerical Simulation of Unsteady Flows and Transition to Turbulence, eds., O. Perinea et al.), Cambridge University Press, pp. 319–347, and 1993, data transmitted by J. Coupland from the Rolls-Royce Applied Science Laboratory.

Roach, P.E., and Brierley, D.H., 2000, "Bypass Transition Modeling: A New Method which Accounts for Free-Stream Turbulence and Length Scale," ASME Paper 2000-GT-278.

Roberts, S.K., and Yaras, M.I., 2003, "Measurements and Prediction of Free-Stream Turbulence Effects on Attached-Flow Boundary-Layer Transition," ASME Paper GT2003-38261.

Roberts, S.K., and Yaras, M.I., 2004, "Modeling of Boundary-Layer Transition," ASME Paper GT2004-53664.

Roberts, S.K., and Yaras, M.I., 2005, "Boundary-Layer Transition Affected by Surface Roughness and Free-Stream Turbulence," J. Fluids Eng., 127, pp. 449–457.

Roberts, S.K., and Yaras, M.I., 2006, "Effects of Surface Roughness Geometry on Separation-Bubble Transition," J. Turbomachinery, 128, pp. 349–356.

Schlichting, H., 1979, Boundary-Layer Theory, McGraw-Hill, New York.

Schraub, F.A., and Kline, S.J., 1965, Mech. Engng. Dept. Rep. MD-12, Stanford University, Palo Alto.

Schubauer, G.B., and Klebanoff, P.S., 1955, "Contribution to the Mechanism of Boundary-Layer Transition," NACA TN 3489.

Schubauer, G.B., and Skramstad, H.K., 1948, "Laminar Boundary-Layer Oscillations and Transition on a Flat Plate," NACA Rept. 909.

Sharma, O., 1987, "Momentum and Thermal Boundary Layer Development on Turbine Airfoil Suction Surfaces," AIAA Paper No. 87-1918.

Sharma, O.P., Wells, R.A., Schlinker, R.H., and Bailey, D.A., 1982, "Boundary Layer Development on Turbine Airfoil Suction Surfaces," Jour. Engng. for Power, 104, pp. 698–706.

Solomon, W.J., Walker, G.J., and Gostelow, J.P., 1996, "Transition Length Prediction for Flows with Rapidly Changing Pressure Gradients," J. Turbomachinery, 118, pp. 744–751.

Taylor, G.I., 1936, "Effect of Turbulence on Boundary Layer," Proc. Roy. Soc., London A, 156, pp. 307–317.

Thwaites, B., 1949, "Approximate Calculation of the Laminar Boundary Layer," Aero. Quart., pp. 245–280.

Walker, G.J., 1987, "Transitional Flow on Axial Turbomachine Blading," AIAA Paper 87-0010.

Walker, G.J., 1989, "Modeling of Transitional Flow in Laminar Separation Bubbles," 9th Int. Sym. Air Breathing Engines, pp. 539–548.

Walker, G.J., and Gostelow, J.P., 1990, "Effects of Adverse Pressure Gradients on the Nature and Length of Boundary Layer Transition," J. Turbomachinery, 112, pp. 196–205.

Wang, T., and Keller, F.J., 1999, "Intermittent Flow and Thermal Structures of Accelerating Transitional Boundary Layers: Part 1–Mean Quantities, and Part 2–Fluctuation Quantities," J. Turbomachinery, 121, pp. 98–112.

Wang, T., and Rice, M.C., 2003, "Effect of Elevated Free-Stream Turbulence on Transitional Heat Transfer over Dual Scaled Rough Surfaces," ASME Paper GT2003-38835.

Wang, T., and Simon, T.W., 1987, "Heat Transfer and Fluid Mechanic Measurements in Transitional Boundary Layers on Convex-Curved Surfaces," J. Turbomachinery, 109, pp. 443–452.

Wells, C.S., 1967, "Effect of Free-Stream Turbulence on Boundary-Layer Transition," AIAA Journal, 5, pp. 172–174.

Westin, K.J.A., Boiko, A.V., Klingmann, B.G.B., Kozlov, V.V., and Alfredsson, P.H., 1994, "Experiments in a Boundary Layer Subjected to Free Stream Turbulence. Part 1. Boundary Layer Structure and Receptivity," J. Fluid Mech., 281, pp.193–218.

Zhong, S., Chong, T.P., and Hodson, H.P., 2002, "On the Spreading Angle of Turbulent Spots in Non-Isothermal Boundary Layers with Favorable Pressure Gradients," ASME Paper GT-2002-

30222.

Zhou, D., and Wang, T., 1995, "Effects of Elevated Free-Stream Turbulence on Flow and Thermal Structures in Transitional Boundary Layers," J. Turbomachinery, 117, pp. 407–417.

7 Transitional Boundary Layers

Classical approach. Zonal approach. Intermittency models. Typical errors.

Here's where it all comes together – pre-transitional (laminar) flow, onset of transition, turbulent spots, intermittency, and post-transitional (turbulent) flow. The method by which models for them, or a subset of them, are assembled to predict transitional flow is simply referred to as a "transitional flow model."

The first transition model was a simple patching of the laminar and turbulent solutions at an appropriate "transition" location. Both the method of patching and the location were somewhat arbitrary, but usually transition was assumed to abruptly occur at the time-averaged position x_t. While these methods, as shown in Fig. 7.1, are satisfactory for estimating the total drag on a plate once the onset of transition is known (in this case when $Re_{xt} = 500{,}000$), they produce unrealistic jumps in the *local* wall shear stress at transition.[1] Although numerous methods were suggested to smooth out these jumps, they became outdated after Emmons [1951] discovered turbulent spots and developed his turbulent-spot theory.

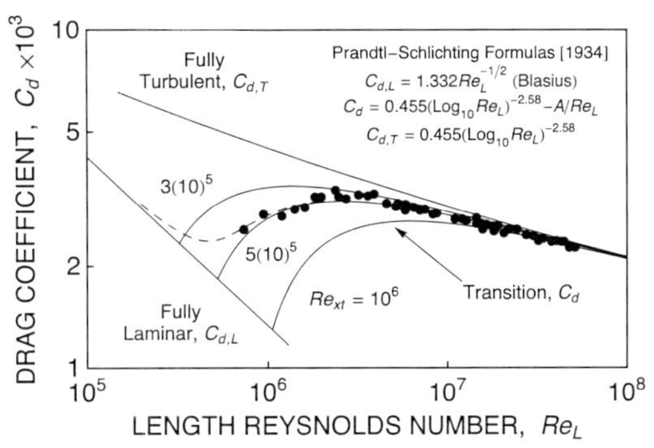

Figure 7.1.
Drag coefficient for a smooth flat plate at zero incidence to the flow. Transition curves for $Re_{xt} = 3(10)^5$, $5(10)^5$ and 10^6 obtained from Prandtl-Schlichting skin-friction formula using $A = 1050$, 1700 and 3300. Data from Gebers [1908].

Spurred by the rapid development of computational fluid dynamics in the 1970s and the realization that the production and dissipation terms in the turbulent kinetic-energy equation could be modified to *mimic* transition, a new group of transition models quickly became the mainstay of industry for calculating transitional flow. Typical schemes in this group are those proposed by Launder and Spalding [1972], McDonald and Fish [1973], Cebeci and Smith [1974], Crawford and Kays [1976], and more recently by Yang and Shih [1993], Wilcox [1994], Westin and Henkes [1997], Craft et al. [1997], Chen et al. [1998], and Hadzić and Hanjalić [2000]. A survey of the earlier schemes is given by Savill [1992, 1996], who concluded that none of them yield reliable results for various combinations of Reynolds number, free-stream turbulence level, and pressure gradient. In addition, he found that the results are sensitive to grid resolution, and like transition itself, initial and boundary conditions. It has also been the author's experience that these schemes must always be "fine tuned" for a particular application by readjusting the various model parameters to correlate measurements, and therefore should be considered more "postdictive" than predictive.[2] Consequently, the emphasis in this book will be on the other models.

---

[1] The same can be said about the calculated total and local heat transferred from the plate.

[2] My first experience with the practical methods of calculating transitional flow was using the McDonald-Fish computational program at Pratt & Whitney Aircraft in the 1970s. Frankly, without fine tuning the program for our class of turbine airfoils, the comparisons of the program with experimental measurements then looked about the same as the comparisons do today.

Broadly speaking, there are two approaches for treating intermittent transitional flow and two ways to determine the intermittency. The former we will call the "classical" and "zonal" approaches. The primary difference between them is whether they consider the interaction between the laminar and turbulent portions of the flow or not. The two ways to determine intermittency we will call "turbulent-spot" and "intermittency-transport" models. The primary difference between them is whether they use turbulent-spot theory or not. Most modern transition models are made up of various combinations of these approaches and models.

In this chapter, we will first consider the classical approach and the approximations associated with it. We will then introduce the conditional-averaging technique associated with the zonal approach and develop an appropriate set of conditionally-averaged equations of motion. Afterward, we will consider the intermittency models and finally the errors associated with various aspects of calculating transitional flow.

Some of the important results in this chapter are:

1. All transitional flow models are approximate with some more *forgiving* than others.
2. Predicting transitional flow *hasn't* improved much in the last fifty years because we still can't predict the onset and rate of transition.
3. The breakthrough in *predicting* transitional flow will arrive only when investigators begin concentrating on the basic problem – predicting the birth, birth rate, growth, and propagation of turbulent spots.

Beside the references at the end of this chapter, other relevant articles may be found in the Bibliography.

7.1 Classical Approach.

Emmons-Narasimha model. The approximations. Some calculations.

First introduced by Emmons [1951], used again by Emmons and Bryson [1952], and further developed by Narasimha and his coworkers (Narasimha [1957, 1985, 1990], Dhawan and Narasimha [1958], and Narasimha and Dey [1989]), this approach is sometimes referred to as the "linearly-combined model." Although the approach is usually associated with Emmons' turbulent-spot theory, the intermittency may also be calculated from a transport equation. Furthermore, since laminar-turbulent interactions are not accounted for in this approach, almost any reasonable analysis may be used to obtain the laminar and turbulent components of the flow.

7.1.1 Emmons-Narasimha model.

The classical approach for calculating transitional flow is to solve the laminar and turbulent flow equations independent of one another, and combine the results in proportion to the time the flow is laminar and turbulent by using the *near-wall* intermittency distribution. In his original paper, Emmons used it to calculate the drag coefficient for a smooth flat plate at zero incidence to the flow. Defined by

$$C_d = \frac{1}{L}\int_0^L \overline{c}_f(x)\,dx$$

he determined the local time-averaged skin-friction coefficient $\overline{c}_f(x)$ from its laminar and turbulent components by using

$$\overline{c}_f = \left(1-\gamma_0\right)\overline{c}_{f,L} + \gamma_0 \overline{c}_{f,T} \tag{7.1}$$

This follows directly from turbulent spot theory, the definition of the skin-friction coefficient, and the fact that the time-averaged velocity at any position in the flow is given by

$$\bar{u} = (1-\gamma)\bar{u}_L + \gamma\bar{u}_T \tag{7.2}$$

where $\gamma = \gamma_0$ near the wall.

Emmons used an expression for the near-wall intermittency that assumed a uniform distribution for the production of turbulent spots over the entire surface and a constant propagation rate.[1] In addition, he used the well known skin-friction-coefficient distributions for the laminar and turbulent boundary layer with both "starting" at the leading edge of the plate. For two-dimensional flow, this is equivalent to regarding the flow as two separate parallel streams, laminar and turbulent, of relative widths $(1-\gamma_0)$ and γ_0 respectively. While the reader may find the details of these calculations in his paper and his solution for γ_0 in §5.4.3, his result is shown by the dashed line in Fig. 7.1. The agreement with Gebers [1908] data and Prandtl-Schlichting's drag formula is quite remarkable considering that only the *product* of the production and propagation rates could be adjusted to fit the data.[2]

Dhawan and Narasimha [1958] showed that Emmons' model can be extended to calculating the displacement and momentum thickness distributions through transition by using Eq. (7.2) with the intermittency equal to the near-wall value *throughout* the boundary layer. In this case, it is relatively easy to show that the expressions for the displacement and momentum thicknesses are given by

$$\bar{\delta}_1(x) = (1-\gamma_0)\bar{\delta}_{1,L} + \gamma_0\bar{\delta}_{1,T} \tag{7.3}$$

and

$$\bar{\delta}_2(x) = (1-\gamma_0)^2\bar{\delta}_{2,L} + \gamma_0^2\bar{\delta}_{2,T} + \gamma_0(1-\gamma_0)\frac{1}{\bar{u}_\infty^2}\int_0^\infty\left[\bar{u}_L(\bar{u}_\infty - \bar{u}_T) + \bar{u}_T(\bar{u}_\infty - \bar{u}_L)\right]dy \tag{7.4}$$

Clearly, only quantities that depend linearly on the velocity can be linearly combined, otherwise the nonlinear contributions of γ_0 and $(1-\gamma_0)$ must be considered.

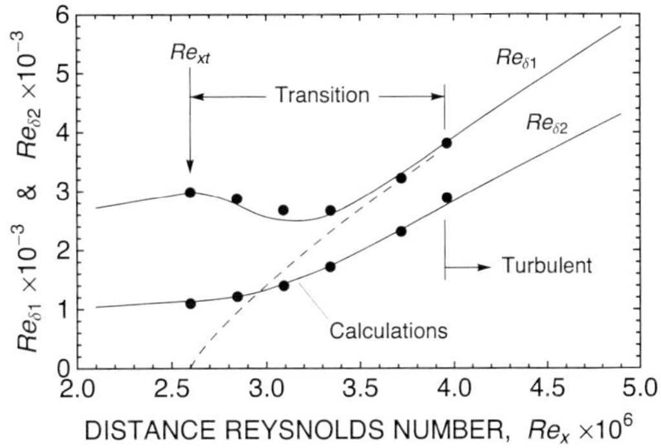

Figure 7.2.
Measured and calculated displacement and momentum thicknesses through transition on a flat plate. Data from Schubauer and Klebanoff [1955].

The results of their calculations for Schubauer and Klebanoff's [1955] data are shown in Fig. 7.2. In this case, Dhawan and Narasimha used a Blasius profile for the laminar velocity component, and the

[1] This assumption was apparently spurred by the fact that it is the simplest distribution to use and, as told me by an attendee at one of my presentations who saw Emmons' experiment, that laminar breakdowns did occur more or less uniformly over his test surface.

[2] With a uniform distribution for the production of turbulent spots beginning at the leading edge of the plate, "transition" begins at the leading edge. Consequently, Emmons defines the "transition point" as the position where the flow is turbulent half of the time. For the calculation shown in Fig. 7.1, this occurs where $Re_{xt} = 505{,}000$, which just so happens to be where the drag coefficient is a minimum. (See Emmons' paper, but note that the kinematic viscosity in the denominator of his dimensionless propagation parameter shown in Fig. 3 should be replaced by U.)

law of the wall together with Clauser's [1956] velocity defect law for the turbulent, but one could also have used a power-law profile for the turbulent component.[1] Following Emmons, they used the well known displacement and momentum thickness distributions for a laminar and turbulent boundary layer, but started the turbulent boundary layer at the onset of transition as shown by the dashed line in the figure.[2] In addition, they used Eq. (5.18) for the near-wall intermittency distribution. Its distribution is shown in Fig. 5.18 together with the corresponding values for Re_{xt} and $\hat{n}\sigma$.

The "dip" we see in the displacement thickness in Fig. 7.2 is caused by the increase in the average velocity near the wall as the flow becomes more turbulent. This behavior is seen in Fig. 7.3 where the time-averaged velocity profiles measured by Schubauer and Klebanoff and calculated by Dhawan and Narasimha for the first four locations in Fig. 7.2 are compared. The laminar and turbulent profiles used in the above calculations are respectively plotted as the dot-dashed and dashed lines in this figure. The approximate thicknesses of the laminar and turbulent portions of the flow are indicated by the long-dashed lines. Thus, as transition progresses, the time-averaged velocity profile simply morphs from a laminar to a turbulent profile.

Actually, Dhawan and Narasimha's calculations were somewhat more involved than indicated above because the velocity measurements shown in Fig. 7.3 were made using a Pitot tube. Since the Pitot tube pressure (minus the static pressure) is proportional to the square of the average velocity, they used

$$\bar{u} = \left[(1-\gamma_0)\bar{u}_L^2 + \gamma_0 \bar{u}_T^2 \right]^{1/2} \tag{7.5}$$

instead of Eq. (7.2). As we will see in §7.2.1, this expression is not exact for intermittent flows but the error turns out to be relatively small.[3]

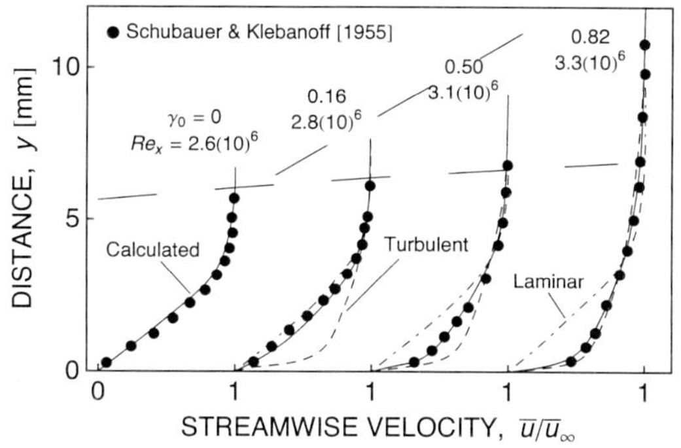

Figure 7.3.
Measured and calculated velocity profiles through transition. Data from Schubauer and Klebanoff [1955]. Calculations from Dhawan and Narasimha [1958].

The results shown in Figs. 7.2 and 7.3 are quite remarkable when we recall that *steady* boundary-layer results and the *near-wall* intermittency were used to calculate a random, unsteady, intermittent,

[1] See §3.2.2 and §3.2.3.

[2] As mentioned in §5.2.5, this result contrasts with that of Schubauer and Klebanoff who concluded that the laminar and turbulent boundary-layer thicknesses are the same at the onset of transition. In this case, the origin of the turbulent boundary layer in Fig. 7.2 would be according to Eq. (5.45) at $2.3(10)^6$.

[3] The error has to do with the time-average of the fluctuations, which in this case is given by Eq. (7.28) with $P = Q = u$. Compared to the square of the mean velocity, the time-average of the laminar and turbulent fluctuations (first and second terms on the right-hand side of Eq. (7.28)) can usually be neglected, while the component resulting from the "jumps" in velocity between the laminar and turbulent segments of the flow (third term) cannot. As it turns out, however, the error associated with this component is roughly one-eighth the magnitude of the jumps, which again compared to the square of the mean velocity can usually be neglected.

three-dimensional flow. Apparently, Emmons' turbulent spot theory adequately accounts for the flow despite its complexity because the errors associated with using approximate laminar and turbulent profiles, and the near-wall intermittency in the calculations, are small.

7.1.2 The approximations.

Laminar flow and turbulent spot approximations.

Without much thought, Emmons' approach appears quite plausible. But when we consider the physics, the leap he made is significant. In particular, he conjectured that the time-averaged laminar and turbulent skin-friction coefficients at x equal their counterparts $\bar{c}_{f,L}(x)$ and $\bar{c}_{f,T}(x)$ for *steady* laminar and turbulent boundary-layer flow over the surface, even though the flow at any one position is randomly switching between segments of laminar and turbulent flow. For the laminar portion of the flow, since turbulent spots form and grow within the developing laminar boundary layer, this might not seem such a large leap at first until we remember that the flow within the becalmed zone following a spot[1] is quite different from the surrounding laminar layer. For the turbulent portion, the leap is large; particularly when it was made before Schubauer and Klebanoff showed that a spot downstream from inception grows at roughly the same rate as a turbulent boundary layer.[2] Here the assumption concerns the turbulent flow within developing finite spots, the local time average of which is by no means obviously equal to that for a steady turbulent boundary layer.

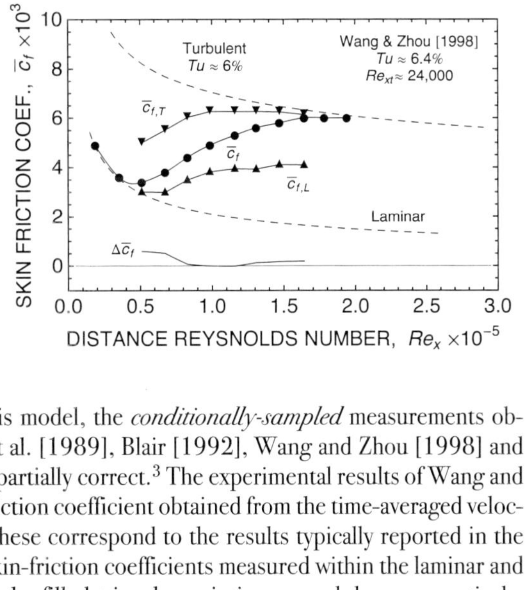

Figure 7.4.
Measured and calculated laminar and turbulent skin-friction coefficients through transition on a flat plate. Data from Wang and Zhou [1998].

About forty years after Emmons proposed his model, the *conditionally-sampled* measurements obtained by Blair and Anderson [1987], Kim et al. [1989], Blair [1992], Wang and Zhou [1998] and others, showed that his assumptions are only partially correct.[3] The experimental results of Wang and Zhou are shown in Fig. 7.4. Here, the skin-friction coefficient obtained from the time-averaged velocity profiles are plotted as the filled circles. These correspond to the results typically reported in the literature for transitional flow. The average skin-friction coefficients measured within the laminar and turbulent portions of the flow are plotted as the filled triangles pointing up and down respectively. Clearly, the measured skin-friction coefficients within the segments are not equal to those calculated for an undisturbed laminar and fully-developed turbulent boundary layer as assumed by Emmons (in-

---

[1] See §5.1.

[2] Again see §5.1. Here it is noteworthy that according to Emmons [1990], Dryden asked Schubauer and Klebanoff to conduct their wind tunnel experiments because after seeing Emmons' water channel experiment, he (Dryden) believed that the spots were not "spots of turbulence" but some surface tension effect.

[3] We will examine the measurement technique of conditional sampling and the mathematics associated with it in the following section. Conditional-sampling, however, was briefly discussed in §5.1.2 when we examined data within a single turbulent spot.

dicated by the dashed lines). With $Re_{LT} \approx 1.5(10)^5$, the spots are immature in this case, and with $Re_{\delta2,t} \approx 100$, the flow within them is more "laminar-like" than turbulent. The result is that the turbulent component of the skin-friction coefficient within the spots should be closer to the laminar than turbulent value near the beginning of transition as confirmed by the data. Since the skin-friction coefficient in the laminar becalmed zone behind a spot is significantly higher than in the surrounding laminar portion of the flow,[1] and the laminar flow near the end of transition is mostly that affected by the spots, either becalmed or something else,[2] the laminar component of the skin-friction coefficient should be higher than the extrapolated laminar flow near the end of transition, again, as confirmed by the data.[3]

The difference between using the assumed and measured laminar and turbulent components of the skin-friction coefficients in Eq. (7.1) is plotted as the $\Delta\overline{c}_f$ line in the lower portion of Fig. 7.4. Here, we see that even though the differences between the actual and assumed skin-friction coefficients are large, the error in using the linearly-combined model is surprisingly small (at most about fifteen percent). As it turns out, this is simply because the largest errors occur where the differences have the smallest effect, that is, at the beginning and end of transition.

Another source of error in the Emmons-Narasimha model is the assumption of using the near-wall intermittency rather than the actual intermittency in Eq. (7.2). As shown in Fig. 7.5, where the measured average and conditionally-sampled velocity profiles of Kim et al. are presented, this error again turns out to be surprisingly small. The intermittency distribution presented in this figure corresponds to the calculated profile shown in Fig. 5.27. The measurements were obtained using hot-wire anemometers. The time-averaged velocity measurements are plotted using filled circles. The conditionally-averaged measurements are plotted using empty triangles. Those pointing up are measurements taken only in the laminar segments of the flow averaged over a large number of laminar segments. Those pointing down are measurements taken only in the turbulent segments.

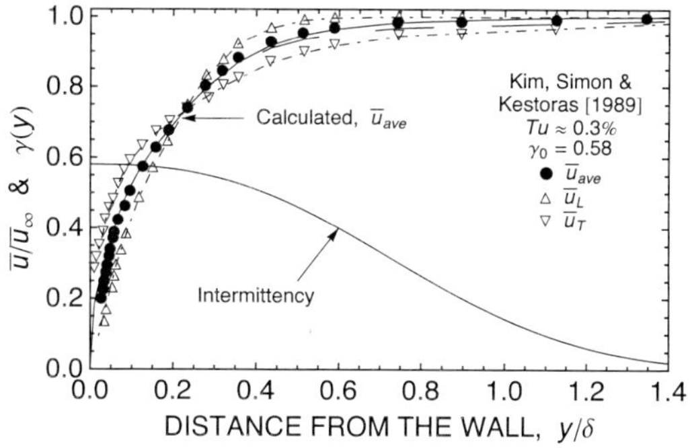

Figure 7.5.
Measured and calculated velocity profiles through transition. Data from Kim, Simon and Kestoras [1989].

As it happens, the calculated average velocity profile using Eq. (7.2) is virtually the same using either γ (solid line) or γ_0 (long-dashed line), with that using γ agreeing better with the data only in the outer portion of the boundary layer where the intermittency is obviously not equal to its near-wall value. This is easy to understand when the error between the two is considered. Using Eq. (7.2), their difference is given by

[1] See §5.1.4.

[2] See Fig. 5.50.

[3] Note here that, as far as the author knows, conditionally-averaged measurements have yet been used to develop a more comprehensive turbulent-spot model for transitional flow.

$$\Delta\bar{u} = \left(\gamma_0 - \gamma\right)\left(\bar{u}_T - \bar{u}_L\right)$$

Clearly, the largest error occurs where either γ is small, which is away from the wall, or where the laminar and turbulent velocities are most different, which is near the wall. Since this occurs at the two extremes of the region, since the laminar and turbulent velocities are equal somewhere between these extremes, and since γ is nearly equal to γ_0 over a good portion of the boundary layer, this error will always be relatively small. In fact, as seen in Fig. 7.5, the largest error occurs near the edge of the boundary layer, which in this case is only about two percent. Therefore, using $\gamma(x, y) = \gamma_0(x)$ is not only an excellent approximation for calculating transitional flow, but a convenient one as well.

Similar to the previous result, the largest differences between the approximation and reality again occur where they affect the intermittency-weighted average the least. Consequently, it appears that the Emmons-Narasimha model or any other linearly-combined model are viable transition models simply because they are "forgiving" where physically incorrect.

7.1.3 Some calculations.

The standard linearly-combined models use Eqs. (7.1)–(7.4) together with Eq. (5.18) for the near-wall intermittency distribution. Thus the major differences between them are the methods used to calculate the laminar and turbulent boundary-layer components of the flow, and/or the correlations used to obtain the onset and rate of transition. Since the choices are numerous, the interested reader is referred to Narasimha and Dey's compilation of the pre-1989 models and Solomon et al. [1996] for one of the later models. Many of them use the integral equations of motion to evaluate the laminar and turbulent components of the flow, while others use the differential equations of motion. As examples, the solutions described in §2.1 and §3.2 may be used to obtain the laminar and turbulent components for flow over a flat plate with low free-stream turbulence, while the methods described in §3.3 and §4.3.2 may be used to obtain the components for more complex flows. Many of the models using the differential approach include the effects of free-stream turbulence, pressure gradients, roughness, curvature, and compressibility.

The potential of these models is illustrated in Fig. 7.6 where some results using the simplest Emmons-Narasimha model are shown. The data are from Roach and Brierley's [1990] test cases "T3A" and "T3B," which we already examined in §4.3.3.[1] These data are for flow on a flat plate with a zero pressure gradient at the nominal incident turbulence levels of three and six percent. Clearly, if we can't calculate transition for this flow, we have little hope of calculating transition for any other.

The transition calculations (solid lines) are obtained from Eq. (7.1). Blasius' solution was used to calculate the laminar component of the skin-friction coefficient, while Eq. (3.40) was used to calculate the turbulent component.[2] These results are shown by the dot-dashed lines extending beyond the transition solutions marked "Fully Laminar" and "Fully Turbulent." The fully turbulent skin-friction coefficient was calculated to match the laminar and turbulent boundary-layer thicknesses at onset by using Eq. (5.45). As a result, the turbulent boundary layer for the six and three percent turbulence levels was calculated to begin at $Re_{x0} = 17{,}000$ and $94{,}000$ respectively. These positions are shown by the upward pointing arrows in the figure.

---

[1] The data plotted in this figure were obtained by the author from a best fit of the investigators' velocity profile data to the near-wall-laminar solution, Eq. (3.28), and Reichardt's law of the wall with $B = 5.0$, Eq. (3.32). The results generally agree with the investigators', who used $B = 5.2$, except most notably near the end of transition where some large differences were found.

[2] These components may also be obtained by using any reasonable method for calculating the pre- and post- transitional components of the model providing they correctly account for the effects of free-stream turbulence on the flow.

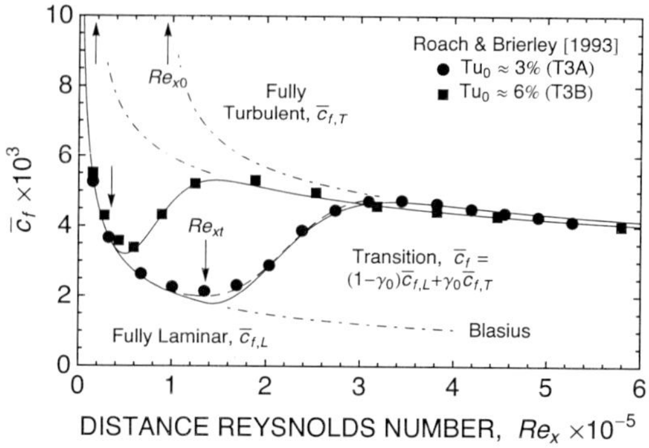

Figure 7.6.
Calculated time-averaged skin-friction coefficients using a linearly-combined transition model. Data from Roach and Brierley [1990].

Narasimha's line-source solution, given by Eq. (5.18), was used to calculate the near-wall intermittency distribution. The onset and rate of transition as determined from the best fit to the measured intermittency distribution shown in Fig. 6.2 were used to calculate the three-percent result. Since the intermittency distribution was unavailable for the six-percent test case, the values of Re_{xt} and $\hat{n}\sigma$ were adjusted to obtain the best fit of the model to data in Fig. 7.6. That is, in one case the measured intermittency distribution was used to determine the onset and rate of transition, whereas in the other the measured skin-friction coefficient distribution was used. In other words, the correlations for Re_{xt} and $\hat{n}\sigma$ given in Chapter 6 were *not* used.[1] The values used for Re_{xt} and $\hat{n}\sigma$ in these calculations are listed in the first two columns of Table 7.1. The listed transition Reynolds numbers are shown by the downward pointing arrows in the figure.

Table 7.1
Transition Reynolds numbers and spot parameters for best fit to Roach and Brierley's [1993] T3A and T3B test cases using a linearly-combined model.

Tu_0 (%)	Re_{xt}	$\hat{n}\sigma$	$(Re_{\delta 1,t})_{meas}$	$(Re_{\delta 1,t})_{Blas}$	$(Re_{\delta 2,t})_{meas}$	$(Re_{\delta 2,t})_{Blas}$	Re_{LT}	$(Re_{\delta 2,T})_{meas}$
3	136,000	$8.8\ 10^{-11}$	634	639	280	245	230,000	760
6	35,000	$2.6\ 10^{-10}$	313	324	130	124	133,000	450

Overall, the Emmons-Narasimha model provides a reasonably good method of calculating transitional flow as long as Re_{xt} and $\hat{n}\sigma$ are known a priori. In general, the calculations compare well with the data through the later stage of transition and beyond, and only reasonable through the later stage of pre-transition into early transition. The discrepancies before and near transition in these examples are partly a result of disregarding the effect of turbulence in calculating the pre-transitional boundary layer, and partly a result of using the line-source solution for the intermittency distribution instead of the distributed-source solution. If the distributed-source solution that gives the best fit to the measured intermittency distribution is used (see Fig. 6.2), the result shown by the dashed line in Fig. 7.6 is obtained which certainly improves the comparison. Any further improvement in the comparison depends on our ability to calculate pre-transitional flow.

The other columns in Table 7.1 contain the displacement- and momentum-thickness Reynolds number at transition as measured and calculated from Blasius' solution, as well as the Reynolds numbers

[1] Results of this model using these correlations will be presented and discussed in §7.4.

based on the length of transition and the measured momentum-thickness Reynolds number at the end of transition. To be specific, the length of transition was determined from Eq. (6.2) while the latter corresponds to momentum-thickness Reynolds number where $Re_x = Re_{xt} + Re_{LT}$. According to §3.2.2, the flow through and immediately after transition for these examples correspond to *very* low Reynolds-number turbulent flow. Also, according to Fig. 5.5*b*, the turbulent spots by the end of transition are still immature in contrast to those for the Schubauer and Klebanoff tests where $Re_{\delta 2,T} \approx 3000$ and $Re_{LT} \approx 1.5(10)^6$. Consequently, most of the correlations for a single fully-developed turbulent spot in §5.3.2 and §5.3.3 are beyond the range for transition with elevated turbulence levels, and perhaps also useless for many practical applications. Clearly, this can only be rectified by conducting turbulent spot experiments at elevated turbulence levels.

7.2 Zonal Approach.

Some preliminaries. Conditionally-averaged equations of motion. Conditionally-averaged Reynolds-stress equations.

Models using the zonal approach solve for the flow in each region including their interaction. Since this requires solving a set of *conditionally-averaged* differential equations of motion using the local value of the intermittency, this approach is inherently more complex than the linearly-combined models previously discussed. Despite this additional complexity, however, they do have a greater potential for predicting transition for a wider variety of flows.

Some of the earliest analyses using conditional averaging for intermittent flows are those presented by Antonia [1972], Hedley and Keffer [1974], Libby [1975] and Byggstøyl and Köllmann [1981, 1986a, and 1986b]. Originally developed for turbulent mixing layers and the wake portion of turbulent boundary layers, where large regions of turbulent flow engulf non-turbulent flow, they are also applicable to transitional boundary-layer flow. This was shown by Vancoillie [1984], and Vancoillie and Dick [1988], who used the near-wall intermittency, and later by Steelant and Dick [1996], who introduced an intermittency transport equation.

In this section, after introducing the formal concepts of intermittency and conditional or zonal averaging, we will develop the basic set of equations for the intermittent laminar and turbulent portions of the flow in a transitional boundary layer. Of particular interest to us will be the extra terms that appear in these equations as a result of the laminar-turbulent interactions.

7.2.1 Some preliminaries.

Intermittency function. Intermittency factor. Global and conditional averaging. Zonal velocities and Reynolds stresses.

We begin by considering the variation of any quantity with time at a given position within transition. As shown in Fig. 7.7, this sample typically consists of alternating segments of high frequency, large-amplitude fluctuations usually associated with turbulent flow and low frequency, small amplitude fluctuations associated with laminar flow. A sample obtained near the onset of transition will display much longer intervals of laminar than turbulent flow, whereas one obtained near the end of transition will display much longer intervals of turbulent than laminar flow. Furthermore, for a non-turbulent free stream, a sample obtained within the outer region of the boundary layer will display shorter intervals of turbulent flow than a sample obtained near the wall.

In addition, the average value of the quantity within the turbulent segments will generally be different from that in the laminar segments. For example, it follows from Fig. 7.5 that the *average velocity* of the flow within the turbulent segments of a sample near the wall will usually be greater than that in the laminar segments, whereas the reverse is found in the outer region of the boundary layer (also see Fig 5.10). From this observation we deduce that the sample of data shown in Fig. 7.7 corresponds to one obtained near the wall.

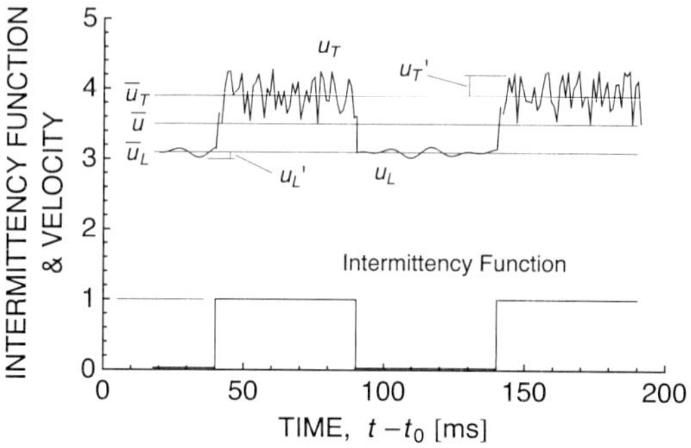

Figure 7.7.
A velocity-time signal in a transitional boundary layer and its corresponding intermittency function I.

Intermittency. To account for the intermittent behavior of the flow, we will first presume that some sort of scheme can be devised to distinguish between its turbulent and laminar parts.[1] Second we will introduce a piecewise function I, called the "intermittency function," such that its value as shown in the lower portion of Fig. 7.7 equals unity when the flow is turbulent and zero when the flow is laminar.[2] In particular, we define

$$I(x_i,t)=\begin{cases}1 \text{ ; turbulent flow}\\ 0 \text{ ; laminar flow}\end{cases} \tag{7.6}$$

Integrating the intermittency function with respect to time yields the sum of all time-intervals during which the flow is *turbulent* at the position x_i. That is

$$\int_{t_0}^{t_0+T} I(x_i,t)dt = \int_{t_1}^{t_1+(\Delta t_T)_1} dt + \int_{t_2}^{t_2+(\Delta t_T)_2} dt + \cdots = \sum_{n=1}^{N_T} (\Delta t_T)_n = T_T(x_i)$$

where t_n is the time at the beginning of the nth turbulent segment of the flow at x_i, $(\Delta t_T)_n$ is its duration, and N_T the total number of turbulent segments within the sampling time T. Within the same period, the sum of all time-intervals during which the flow is laminar is given by $T_L = T - T_T$, which is the integral of $1-I$. Otherwise, it is defined as

$$T_L(x_i) = \sum_{n=1}^{N_L} (\Delta t_L)_n$$

where the notation is obvious.

Recalling that the *intermittency* is defined as the fraction of time the flow is turbulent, we obtain

$$\gamma(x_i) = \frac{T_T(x_i)}{T} = \frac{1}{T}\int_{t_0}^{t_0+T} I(x_i,t)dt \tag{7.7}$$

---

[1] Typical discrimination schemes generally distinguish between the frequency and amplitude of velocity, vorticity, or temperature fluctuations in the laminar and turbulent parts of the flow. As might be expected, these differences are not always easy to distinguish, particularly when the free-stream turbulence is high.

[2] This assumes that the thin viscous "super layer" between the two regions, where the velocity (and other physical quantities) change rapidly, is negligibly thin. For further information about the super layer, the reader is referred to the work by Corrsin and Kistler [1955].

where it becomes clear that T must now be much larger than the characteristic time between successive turbulent segments for γ to be independent of time. Hence, the intermittency factor γ is the *time average* of I the intermittency function. As we well know, $0 \le \gamma \le 1$ where $\gamma = 0$ denotes a flow that is always laminar and $\gamma = 1$ denotes a flow that is always turbulent. In addition, since the last expression in the above string of equalities is simply the time-average of I, we have

$$\overline{I(x_i,t)} = \gamma(x_i) \tag{7.8}$$

Since I raised to any power equals I, it also follows that

$$\overline{I^n} = \gamma, \quad \overline{(1-I)^n} = 1-\gamma \quad \text{and} \quad \overline{I^n(1-I)^m} = 0 \tag{7.9}$$

where n and m are positive integers.

Noting that the limits on the integral in Eq. (7.7) are independent of x_i, we obtain

$$\frac{\partial \gamma(x_i)}{\partial x_i} = \frac{\partial \overline{I}}{dx_i} = \frac{1}{T}\int_{t_0}^{t_0+T} \frac{\partial I(x_i,t)}{\partial x_i} dt \tag{7.10}$$

where the integrand equals zero except at the leading and trailing edges of the turbulent flow segments at which points I changes from zero to one and vice versa. To evaluate these contributions, consider Fig. 7.8 where the intermittency function associated with one turbulent segment at several streamwise positions is sketched. In this figure, the temporal variation of I at the position x is shown in the upper portion, while its variation at the small distances ½δx up and downstream of x are shown in the lower portion. The lower graphs presume that the turbulent segment is moving downstream with its leading edge velocity $u_{le}(x_i)$ greater than its trailing edge velocity $u_{te}(x_i)$.[1] Consequently, $\delta t_{le} = \delta x/u_{le}$ and $\delta t_{te} = \delta x/u_{te}$, where δt_{le} and δt_{te} are the time intervals associated with the change at the leading and trailing edges.[2] Furthermore, the changes in I with respect to x are $\Delta I/\delta x = -1/\delta x = -1/u_{le}\delta t_{le}$ at the leading edge and $\Delta I/\delta x = 1/\delta x = 1/u_{te}\delta t_{te}$ at the trailing edge. These changes and subsequent variation of $\Delta I/\delta x$ are shown by the dashed line in Fig. 7.8.

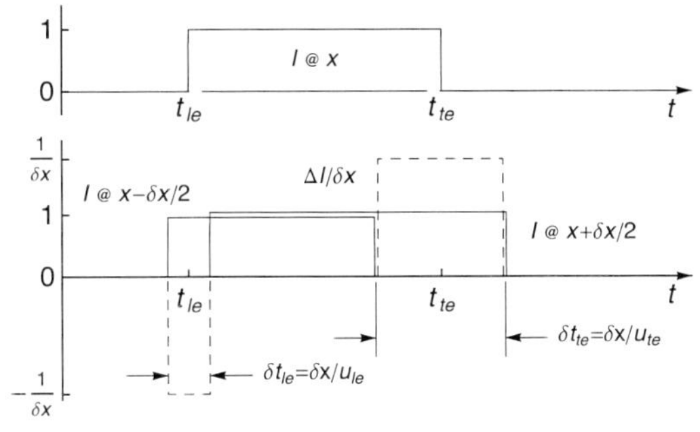

Figure 7.8.
The intermittency function and its change with respect to distance for one turbulent segment.

It then follows that the integral in Eq. (7.10) for *one* turbulent segment is

[1] See Fig. 5.30.

[2] Note that the change in the duration of the segment with respect to distance is

$$d\Delta t_T/dx = \delta(t_{te} - t_{le})/\delta x = (1/u_{te} - 1/u_{le})$$

which agrees with that we already established in Eq. (5.54) for a turbulent spot.

$$\int_{(t_{le})^-}^{(t_{le})^+} \frac{\partial I(x_i,t)}{\partial x_i} dt = \left[\left(-\frac{1}{u_{le,i}\delta t_{le}}\right)\delta t_{le} + \left(\frac{1}{u_{te,i}\delta t_{te}}\right)\delta t_{te}\right] = \left[\frac{1}{u_{te,i}(x_i)} - \frac{1}{u_{le,i}(x_i)}\right]$$

whence, by summing over all the segments in the sample, yields

$$\frac{\partial \gamma}{\partial x_i} = \frac{1}{T}\sum_1^{N_T}\left[\frac{1}{u_{te,i}(x_i)} - \frac{1}{u_{le,i}(x_i)}\right]_n \tag{7.11}$$

In similar fashion, the time average of $I\partial I/\partial x$ can be found by multiplying $\Delta I/\delta x$ in the lower graph of Fig. 7.8 by I, which is shown in the upper graph. This provides

$$\int_{(t_{le})^-}^{(t_{le})^+} I\frac{\partial I(x_i,t)}{\partial x_i} dt = \left[\left(-\frac{1}{u_{le,i}\delta t_{le}}\right)\frac{\delta t_{le}}{2} + \left(\frac{1}{u_{te,i}\delta t_{te}}\right)\frac{\delta t_{te}}{2}\right] = \frac{1}{2}\left[\frac{1}{u_{te,i}(x_i)} - \frac{1}{u_{le,i}(x_i)}\right]$$

for a single segment, and

$$\overline{I\frac{\partial I}{\partial x_i}} = \frac{1}{2}\frac{\partial \gamma}{\partial x_i} \tag{7.12}$$

when all the segments within the sample are added up.

Here we should note that the methods of dealing with the derivatives and integrals of nonlinear combinations of I generally depend on the investigator, which might be surprising, and whether I is considered a continuous function or not. Usually, experimentalists treat I as a piecewise function of x_i and t that *switches* between zero and one, whereas most analysts treat it as a *continuous function* that repeatedly varies from zero to one and back again across widths that in a limit approach zero.[1] Although both descriptions appear to be the same as long as the details at the *jumps* are not scrutinized, it turns out that the contributions from the jumps are significantly different when differentiating and integrating their nonlinear combinations. For example, choosing the experimentalists viewpoint

$$\overline{I(1-I)\partial I/\partial x_i} = 0 \tag{7.13}$$

which follows directly from Fig. 7.9, whereas using the limit of a continuous linear function for I through a jump yields $\tfrac{1}{6}(\partial\gamma/\partial x_i)$. In the following, we will take the experimentalist's point of view.

Global and conditional averaging. If we are to consider the flow in the individual portions separately, we must reconsider the averaging process introduced in §1.2. As defined there, any physical quantity Q may be represented as the sum of its time-averaged and unsteady components according to

$$Q(x_i,t) = \bar{Q}(x_i) + Q'(x_i,t) \tag{7.14}$$

where

$$\bar{Q}(x_i) = \frac{1}{T}\int_{t_0}^{t_0+T} Q(x_i,t)dt \quad \text{and} \quad \int_{t_0}^{t_0+T} Q'(x_i,t)dt = 0$$

We shall henceforth refer to this time-averaged value of Q as its *global average* since it involves the average of a quantity independent of whether the flow is laminar, turbulent, or transitional. Thus the global average of the equations of motion remain as given in §1.2.

However, if we identify the values of Q in the laminar and turbulent portions of the flow by Q_L and Q_T, we can write

[1] See for example Steelant and Dick [1996]. The interested reader may also want to study Erdélyi [1962].

$$Q(x_i,t)=\left(1-I\right)Q_L(x_i,t)+IQ_T(x_i,t) \tag{7.15}$$

That is $Q = Q_L$ when $I = 0$ and $Q = Q_T$ when $I = 1$. Expressing Q_L and Q_T as a sum of their time-averaged and fluctuating components provides

$$\begin{aligned} Q_L(x_i,t)&=\bar{Q}_L^c(x_i)+Q_L'(x_i,t)\\ Q_T(x_i,t)&=\bar{Q}_T^c(x_i)+Q_T'(x_i,t)\end{aligned} \tag{7.16}$$

where the time-averaged components with a superscript "c" are known as either the *conditional* or *zonal average*. In general, they are defined[1] by

$$\bar{Q}_L^c(x_i)=\sum_{n=1}^{N_L}\int_{t_n}^{t_n+(\Delta t_L)_n}Q(x_i,t)\,dt \Bigg/ \sum_{n=1}^{N_L}(\Delta t_L)_n=\sum_{n=1}^{N_L}\int_{t_n}^{t_n+(\Delta t_L)_n}Q_L(x_i,t)\,dt \Bigg/ \sum_{n=1}^{N_L}(\Delta t_L)_n$$

and

$$\bar{Q}_T^c(x_i)=\sum_{n=1}^{N_T}\int_{t_n}^{t_n+(\Delta t_T)_n}Q(x_i,t)\,dt \Bigg/ \sum_{n=1}^{N_T}(\Delta t_T)_n=\sum_{n=1}^{N_T}\int_{t_n}^{t_n+(\Delta t_T)_n}Q_T(x_i,t)\,dt \Bigg/ \sum_{n=1}^{N_T}(\Delta t_T)_n \tag{7.17}$$

such that

$$\sum_{n=1}^{N_L}\int_{t_n}^{t_n+(\Delta t_L)_n}Q_L'(x_i,t)\,dt=0 \quad\text{and}\quad \sum_{n=1}^{N_T}\int_{t_n}^{t_n+(\Delta t_T)_n}Q_T'(x_i,t)\,dt=0 \tag{7.18}$$

Clearly, a large number of samples in both regions must be considered in this averaging. Hence the measured values of $\bar{Q}_T^c$ near the onset of transition and $\bar{Q}_L^c$ near the end of transition, where the intermittency nearly equals zero and one, are always subject to question.[2]

The general expression for Q in terms of the conditionally-averaged and fluctuating components is then given by

$$Q(x_i,t)=\left(1-I\right)\left(\bar{Q}_L^c+Q_L'\right)+I\left(\bar{Q}_T^c+Q_T'\right) \tag{7.19}$$

The global-average of Q is

$$\bar{Q}(x_i)=\left(1-\gamma\right)\bar{Q}_L^c+\gamma\bar{Q}_T^c \tag{7.20}$$

which was taken to be obvious in Eqs. (7.1) and (7.2). Therefore the global average of any quantity for transitional flow is the time-weighted average of the conditionally-averaged values.

The conditional averages in the turbulent and laminar portions of the flow are obtained by multiplying Eq. (7.19) by I and $(1–I)$ respectively, and taking their global averages. This yields

$$\begin{aligned}\overline{IQ(x_i,t)}&=\gamma\bar{Q}_T^c(x_i)\\ \overline{(1-I)Q(x_i,t)}&=\left(1-\gamma\right)\bar{Q}_L^c(x_i)\end{aligned} \tag{7.21}$$

[1] The definitions given in these equations differ in the upper limit of the integral from those given by Libby [1975]. In particular, he integrates over the whole period T rather than summing up the integrals over the individual segments. As a result, his laminar and turbulent conditional averages are $(1-\gamma)$ and γ times ours. In this regard, the reader is invited to compare our definitions and Eqs. (7.20) and (7.21) that follow with his Eqs. (7) and (8).

[2] As discussed in §3.2.2, the same can be said about conditional averages measured near the inner and outer intermittent portion of a turbulent boundary layer where the number of samples in either the turbulent or free-stream portions of the flow are small.

which when added yields the global average of Q given by Eq. (7.20). Hence, taking the global average of a quantity after multiplying by I filters out what's happening in the laminar portion of the flow, leaving only what's happening in the turbulent, while multiplying by $(1-I)$ and averaging leaves only what's happening in the laminar portion.

From Eqs. (7.14), (7.19) and (7.20), the fluctuating component of Q is

$$Q'(x_i,t)=\left(1-I\right)Q_L'+IQ_T'+\left(I-\gamma\right)\left(\bar{Q}_T^c-\bar{Q}_L^c\right) \tag{7.22}$$

where the last term represents the fluctuation caused by the *jumps* in the conditional averages about the global mean of the fluctuations, the latter of which equals zero by definition. Clearly Q' is not equal to Q_T' when $I=1$, but equal to Q_T' superposed on $(1-\gamma)$ times the jump in the conditional averages. Also, when $I=0$, Q' equals Q_L' superposed on γ times the jump in the averages. Consequently, the conditional average of Q' is not zero, but easily shown to be

$$\overline{IQ'}=\gamma\left(1-\gamma\right)\left(\bar{Q}_T-\bar{Q}_L\right) \tag{7.23}$$

in the turbulent portion, and the negative of this in the laminar. Several of these conditionally-averaged quantities are sketched in Fig. 7.9 for a typical variation of Q'.

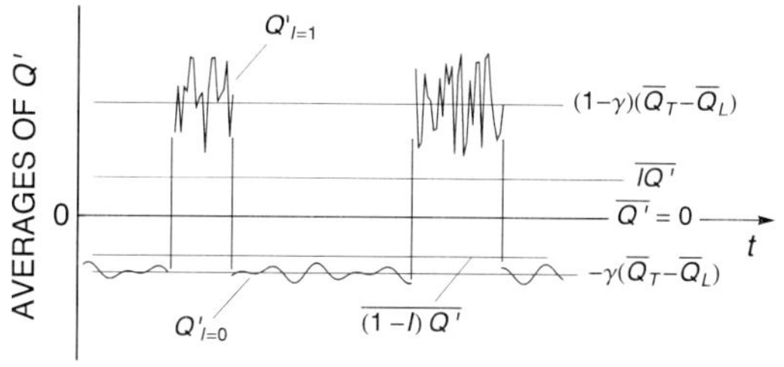

Figure 7.9.
Various averages of a fluctuating quantity in an intermittent flow.

In similar fashion, it can be shown that taking the global average of I times the spatial derivative of Q yields

$$\overline{I\frac{\partial Q(x_i,t)}{\partial x_i}}=\gamma\frac{\partial \bar{Q}_T^c}{\partial x_i}+\tfrac{1}{2}\left(\bar{Q}_T^c-\bar{Q}_L^c\right)\frac{\partial\gamma}{\partial x_i} \tag{7.24}$$

where the last term represents the contribution from all the *jumps* in Q across the laminar-turbulent interfaces.[1] Rewriting this as

$$\overline{I\frac{\partial Q(x_i,t)}{\partial x_i}}=\frac{\partial \gamma\bar{Q}_T^c}{\partial x_i}-\tfrac{1}{2}\left(\bar{Q}_L^c+\bar{Q}_T^c\right)\frac{\partial\gamma}{\partial x_i} \tag{7.25}$$

the last term now represents the amount of Q transferred across all the interfaces passing over the position x_i from the laminar to turbulent portions of the flow. In other words, it is a source of the quantity $\gamma\bar{Q}_T^c$ in the turbulent portion of the flow. The laminar counterpart to this expression is

$$\overline{(1-I)\frac{\partial Q(x_i,t)}{\partial x_i}}=\frac{\partial(1-\gamma)\bar{Q}_T^c}{\partial x_i}+\tfrac{1}{2}\left(\bar{Q}_L^c+\bar{Q}_T^c\right)\frac{\partial\gamma}{\partial x_i} \tag{7.26}$$

where the second term is the negative of that for the turbulent portion. In other words, a source from the laminar-turbulent interface in one portion of the flow becomes a sink in the other.

[1] Kovasznay et al. [1970] and Antonia [1972] referred to these contributions as "point averages."

At this point it becomes worthwhile to note that the symmetry exhibited between the last two expressions is exhibited for all conditional averages. That is, the conditional average of one can be obtained from the other by simply exchanging "T" and "L" and γ and $(1-\gamma)$. This greatly reduces our analytical efforts since we only need to obtain the conditional average for one portion of the flow (the turbulent is the easiest) and deduce that for the other.

The averages for a combination of two physical quantities, say P and Q, are obtained following the methods developed above. For example, the global average of PQ is

$$\overline{PQ} = \bar{P}\bar{Q} + \overline{P'Q'} = (1-\gamma)\left(\bar{P}_L^c \bar{Q}_L^c + \overline{P_L' Q_L'}\right) + \gamma\left(\bar{P}_T^c \bar{Q}_T^c + \overline{P_T' Q_T'}\right) \tag{7.27}$$

Unlike the global average of a fluctuation, the global average of a product of fluctuations is not zero. By using the expression given in Eq. (7.22) for a fluctuating quantity, we obtain

$$\overline{P'Q'} = (1-\gamma)\overline{P_L' Q_L'} + \gamma \overline{P_T' Q_T'} + \gamma(1-\gamma)\left(\bar{P}_T^c - \bar{P}_L^c\right)\left(\bar{Q}_T^c - \bar{Q}_L^c\right) \tag{7.28}$$

which is a sum of the contributions from the product of the fluctuations and the jumps. The coefficient of the last term in this expression is zero for either fully laminar or fully turbulent flow and a maximum of one-quarter when γ equals one-half. Subtracting the above from Eq. (7.27), yields

$$\bar{P}\bar{Q} = (1-\gamma)\bar{P}_L^c \bar{Q}_L^c + \gamma \bar{P}_T^c \bar{Q}_T^c - \gamma(1-\gamma)\left(\bar{P}_T^c - \bar{P}_L^c\right)\left(\bar{Q}_T^c - \bar{Q}_L^c\right) \tag{7.29}$$

a result that could also have been obtained from Eq. (7.20).

The conditional average of PQ and $P'Q'$ in the turbulent portion of the flow are

$$\overline{IPQ} = \gamma\left(\bar{P}_T^c \bar{Q}_T^c + \overline{P_T' Q_T'}\right)$$

and (7.30)

$$\overline{IP'Q'} = \gamma \overline{P_T' Q_T'} + \gamma(1-\gamma)^2\left(\bar{P}_T^c - \bar{P}_L^c\right)\left(\bar{Q}_T^c - \bar{Q}_L^c\right)$$

Finally, if we consider the conditional average of a quantity and the derivative of another, we obtain

$$\overline{IP\frac{\partial Q}{\partial x_i}} = \gamma\left[\bar{P}_T^c \frac{\partial \bar{Q}_T^c}{\partial x_i} + \overline{P_T' \frac{\partial Q_T'}{\partial x_i}}\right] + \frac{1}{2}\left[\bar{P}_T^c\left(\bar{Q}_T^c - \bar{Q}_L^c\right) + \overline{P_T' Q_T'} - \overline{P_T' Q_L'}\right]\frac{\partial \gamma}{\partial x_i} \tag{7.31}$$

where the correlations between the fluctuating components in the different portions of the flow are normally assumed zero.[1] Here we note as in Eq. (7.24) that the last term represents the contribution from all the jumps across the laminar-turbulent interfaces, but it is not a source term because the gradient of a quantity is not being considered. However, it is easy to show using the above result that

$$\overline{I\frac{\partial PQ}{\partial x_i}} = \frac{\partial}{\partial x_i}\left[\gamma\left(\bar{P}_T^c \bar{Q}_T^c + \overline{P_T' Q_T'}\right)\right] - \frac{1}{2}\left(\bar{P}_T^c \bar{Q}_L^c + \bar{P}_L^c \bar{Q}_T^c + \overline{P_T' Q_L'} + \overline{P_L' Q_T'}\right)\frac{\partial \gamma}{\partial x_i} \tag{7.32}$$

where now the second term does represent the transport of PQ across the interface from the laminar to turbulent portions of the flow.[2]

Expressions for the zonal velocities and Reynolds-stresses. The above relations may be applied directly to the velocity vector u_i, in which case we have

---

[1] This is physically equivalent to neglecting the fluctuating interactions taking place within the super layer between the two regions.

[2] While deriving Eq. (7.32) from Eq. (7.31) is straight forward, deriving it from first principles requires using the identities

$$\overline{I\,dI^2/dx} = d\gamma/dx, \quad \overline{I\,dI(1-I)/dx} = -\tfrac{1}{2}d\gamma/dx, \quad \text{and} \quad \overline{I\,dI(1-I)^2/dx} = 0.$$

$$\begin{aligned} u_{i,L}(x_i,t) &= \bar{u}_{i,L}(x_i) + u'_{i,L}(x_i,t) \\ u_{i,T}(x_i,t) &= \bar{u}_{i,T}(x_i) + u'_{i,T}(x_i,t) \\ u_i(x_i,t) &= (1-I)u_{i,L}(x_i,t) + Iu_{i,T}(x_i,t) \end{aligned} \tag{7.33}$$

where the notation has been simplified somewhat by dropping the superscript "c" notation. Henceforth, the barred quantities will be associated with the global averages, while the barred quantities with subscripts "L" or "T" will be associated with the conditional averages. Thus the velocity vector at any instant and position in the flow is given by

$$u_i(x_i,t) = \bar{u}_i(x_i) + u'_i(x_i,t), = (1-I)\left(\bar{u}_{i,L} + u'_{i,L}\right) + I\left(\bar{u}_{i,T} + u'_{i,T}\right) \tag{7.34}$$

while its global average and fluctuating components are

$$\bar{u}_i(x_i) = (1-\gamma)\bar{u}_{i,L} + \gamma\bar{u}_{i,T}$$

and

$$u'_i = (1-I)u'_{i,L} + Iu'_{i,T} + (\gamma - I)\left(\bar{u}_{i,L} - \bar{u}_{i,T}\right) \tag{7.35}$$

Similar expressions for the pressure and viscous stress can be written by inspection. Furthermore, it also follows from Eq. (7.28) that the Reynolds shear stress is given by

$$\overline{u'_i u'_j} = (1-\gamma)(\overline{u'_i u'_j})_L + \gamma(\overline{u'_i u'_j})_T + \gamma(1-\gamma)\left(\bar{u}_{i,L} - \bar{u}_{i,T}\right)\left(\bar{u}_{j,L} - \bar{u}_{j,T}\right) \tag{7.36}$$

For two-dimensional flow the above equations yield

$$\begin{aligned} \bar{u} &= (1-\gamma)\bar{u}_L + \gamma\bar{u}_T \\ \bar{v} &= (1-\gamma)\bar{v}_L + \gamma\bar{v}_T \end{aligned} \tag{7.37}$$

and

$$\overline{u'v'} = (1-\gamma)(\overline{u'v'})_L + \gamma(\overline{u'v'})_T + \gamma(1-\gamma)\left(\bar{u}_L - \bar{u}_T\right)\left(\bar{v}_L - \bar{v}_T\right) \tag{7.38}$$

7.2.2 Conditionally-averaged equations of motion.

Zonal continuity and momentum equations. Exchanged mass flux, momentum flux and net interfacial forces.

Given by Eqs. (1.3), (1.4) and (1.5), and repeated here for convenience, the equations of motion are

$$\frac{\partial u_i}{\partial x_i} = 0 \tag{7.39}$$

and

$$\rho\frac{\partial u_i}{\partial t} + \rho u_j\frac{\partial u_i}{\partial x_j} = -\frac{\partial p}{\partial x_i} + \frac{\partial \tau_{ij}}{\partial x_j} \tag{7.40}$$

where

$$\tau_{ij} = \mu\left(\frac{\partial u_i}{\partial x_j} + \frac{\partial u_j}{\partial x_i}\right)$$

and the body force f_i has been set equal to zero.

Given by Eqs. (1.20) and (1.22), their global averages are

$$\frac{\partial \bar{u}_i}{\partial x_i}=0 \tag{7.41}$$

and

$$\rho\frac{\partial \bar{u}_i\bar{u}_j}{\partial x_j}=-\frac{\partial \bar{p}}{\partial x_i}+\frac{\partial}{\partial x_j}\left(\bar{\tau}_{ij}-\rho\overline{u_i'u_j'}\right) \tag{7.42}$$

Zonal continuity equations. To obtain the average continuity equation for the turbulent and laminar portions of the flow, we introduce the definition of the velocity given in Eq. (7.34) into Eq. (7.39), multiply the result by I and take its global average, and then multiply by $(1-I)$ and take its average, or simply use Eqs. (7.25) and (7.26). Either approach yields

$$\frac{\partial \gamma\bar{u}_{i,T}}{\partial x_i}=\dot{m} \quad \text{and} \quad \frac{\partial(1-\gamma)\bar{u}_{i,L}}{\partial x_i}=-\dot{m} \tag{7.43}$$

where

$$\dot{m}=\tfrac{1}{2}\left(\bar{u}_{i,T}+\bar{u}_{i,L}\right)\frac{\partial \gamma}{\partial x_i} \tag{7.44}$$

is the mass flux per unit mass transferred from the laminar to turbulent portion of the flow.[1]

The global continuity equation is recovered by taking the sum of these and using Eq. (7.35). It should also be noted that when $\gamma = 1$, $\dot{m}$ equals zero, and the first relation in Eq. (7.43) becomes the global continuity equation for turbulent flow. In this case, the second reduces identically to zero. Similarly, when $\gamma = 0$, $\dot{m}$ also equals zero, and the roles of the first and second relations reverse. This will be true for all of our conditionally averaged conservation equations as long as $\dot{m}$ is replaced by the exchanged flux of the quantity being conserved.

For two-dimensional flow the above equations become

$$\frac{\partial \gamma\bar{u}_T}{\partial x}+\frac{\partial \gamma\bar{v}_T}{\partial y}=\dot{m} \quad \text{and} \quad \frac{\partial(1-\gamma)\bar{u}_L}{\partial x}+\frac{\partial(1-\gamma)\bar{v}_L}{\partial y}=-\dot{m} \tag{7.45}$$

with

$$\dot{m}=\tfrac{1}{2}\left(\bar{u}_L+\bar{u}_T\right)\frac{\partial \gamma}{\partial x}+\tfrac{1}{2}\left(\bar{v}_L+\bar{v}_T\right)\frac{\partial \gamma}{\partial y} \tag{7.46}$$

By expanding the left-hand side of the turbulent component in Eq. (7.43), we obtain

$$\frac{\partial \bar{u}_{i,T}}{\partial x_i}=-\tfrac{1}{2}\left(\bar{u}_{i,T}-\bar{u}_{i,L}\right)\frac{1}{\gamma}\frac{\partial \gamma}{\partial x_i} \tag{7.47}$$

Unlike the global average velocity, the divergence of the conditionally-averaged turbulent velocity is no longer zero, unless the flow is completely turbulent. Furthermore, its divergence increases without bound as γ approaches zero from within the transition region, unless the difference between the laminar and turbulent velocity components also approaches zero. This suggests that the turbulent velocity profile must approach the laminar profile at onset at least as fast as $(d\ \ln\gamma/dx)^{-1}$ for a divergence-free flow at the beginning of transition. Examining the result obtained for the laminar

[1] The similarity between these equations and the conservation of mass equation for one-dimensional flow with a variable area and mass either injected or ejected through its stream tube should not be overlooked. In this case the quantities γ and $(1-\gamma)$ represent the areas of the stream tubes of the turbulent and laminar portions of the flow respectively. As we shall see, the similarity extends to the conditionally averaged momentum equation as well.

component of velocity suggests that the laminar velocity profile must approach the turbulent profile near the end of transition. Neither of these results seem implausible, especially when the trends in Fig. 7.4 are considered.

By definition, all the velocity *fluctuations* are divergence free, and their global and conditional averages are zero. The conditional average of a velocity fluctuation and its divergence, however, is not. For example, using Eq. (7.31) with $P = u_j'$ and $Q = u_i'$, we obtain

$$\overline{u'_{j,T}\frac{\partial u'_{i,T}}{\partial x_i}} = -\tfrac{1}{2}\overline{u'_{j,T}\left(u'_{i,T}-u'_{i,L}\right)}\frac{\partial \gamma}{\partial x_i} \tag{7.48}$$

As we will see later, this relation and its laminar counterpart are needed when rearranging derivatives of the conditionally-averaged Reynolds stresses.

Zonal momentum equations. Following the same procedure used to determine the zonal continuity equations, the momentum equations for the turbulent and laminar portions of the flow are

$$\rho\frac{\partial(\gamma \bar{u}_{i,T}\bar{u}_{j,T})}{\partial x_j} = -\frac{\partial \gamma \bar{p}_T}{\partial x_i} + \frac{\partial}{\partial x_j}\left[\gamma\left(\bar{\tau}_{ij,T} - \rho(\overline{u_i'u_j'})_T\right)\right] + \dot{S}_i \tag{7.49}$$

and

$$\rho\frac{\partial[(1-\gamma)\bar{u}_{i,L}\bar{u}_{j,L})]}{\partial x_j} = -\frac{\partial(1-\gamma)\bar{p}_L}{\partial x_i} + \frac{\partial}{\partial x_j}\left[(1-\gamma)\left(\bar{\tau}_{ij,L} - \rho(\overline{u_i'u_j'})_L\right)\right] - \dot{S}_i \tag{7.50}$$

where

$$\dot{S}_i = \tfrac{1}{2}\rho\left(\bar{u}_{i,T}\bar{u}_{j,L} + \bar{u}_{i,L}\bar{u}_{j,T}\right)\frac{\partial \gamma}{\partial x_j} + \tfrac{1}{2}\left[\left(\bar{p}_T + \bar{p}_L\right)\delta_{ij} - \left(\bar{\tau}_{ij,T} + \bar{\tau}_{ij,L}\right)\right]\frac{\partial \gamma}{\partial x_j} \tag{7.51}$$

is the momentum flux transferred from the laminar to turbulent portion of the flow. The first term in the above expression is the mean momentum flux exchanged between the two flows when the correlations between the laminar and turbulent velocity fluctuations are zero,[1] while the second is the net force resulting from the pressure and viscous stresses acting on the interface between the two flows. Again, the global equation is recovered by adding the laminar and turbulent components of the equation.

For a two-dimensional boundary layer, these equations become

$$\frac{\partial \gamma \bar{u}_T \bar{u}_T}{\partial x} + \frac{\partial \gamma \bar{v}_T \bar{u}_T}{\partial y} = -\gamma\frac{1}{\rho}\frac{d\bar{p}_\infty}{dx} + \frac{\partial}{\partial y}\left[\gamma\left(\nu\frac{\partial \bar{u}_T}{\partial y} - (\overline{u'v'})_T\right)\right] + \frac{1}{\rho}\dot{S}_x \tag{7.52}$$

and

$$\frac{\partial(1-\gamma)\bar{u}_L\bar{u}_L}{\partial x} + \frac{\partial(1-\gamma)\bar{v}_L\bar{u}_L}{\partial y} = -\left(1-\gamma\right)\frac{1}{\rho}\frac{d\bar{p}_\infty}{dx} + \frac{\partial}{\partial y}\left[\left(1-\gamma\right)\left(\nu\frac{\partial \bar{u}_L}{\partial y} - (\overline{u'v'})_L\right)\right] - \frac{1}{\rho}\dot{S}_x \tag{7.53}$$

where

$$\dot{S}_x = \rho\bar{u}_L\bar{u}_T\frac{\partial \gamma}{\partial x} + \left\langle \tfrac{1}{2}\rho\left(\bar{u}_T\bar{v}_L + \bar{u}_L\bar{v}_T\right)\frac{\partial \gamma}{\partial y} - \mu\frac{\partial \bar{u}_T}{\partial y}\frac{\partial \gamma}{\partial y}\right\rangle$$

In this case, we have presumed that $\bar{p}_T = \bar{p}_L = \bar{p}_\infty$ and the correlations between the laminar and tur-

---

[1] Otherwise, the parenthesis in the first term of the equation would also include $\overline{u'_{i,L}u'_{j,T}} + \overline{u'_{i,T}u'_{j,L}}$.

bulent fluctuating components equal zero. The terms within the angled brackets are about an order of magnitude smaller than the first and can usually be neglected by comparison. It also turns out that, since each of the terms within the brackets are about the same size, they more or less cancel each other out.

Finally, if Eq. (7.49) is rearranged into the form we usually associate with the momentum equation, we obtain

$$\rho\frac{\partial \bar{u}_{i,T}\bar{u}_{j,T}}{\partial x_j}=-\frac{\partial \bar{p}_T}{\partial x_i}+\frac{\partial}{\partial x_j}\Big[\bar{\tau}_{ij,T}-\rho(\overline{u_i'u_j'})_T\Big]-\tfrac{1}{2}\rho\Big[\bar{u}_{i,T}\big(\bar{u}_{j,T}-\bar{u}_{j,L}\big)$$
$$+\bar{u}_{j,T}\big(\bar{u}_{i,T}-\bar{u}_{i,L}\big)+2\overline{u_{i,T}'u_{j,T}'}\Big]\frac{1}{\gamma}\frac{\partial\gamma}{\partial x_j}-\tfrac{1}{2}\Big[\big(\bar{p}_T-\bar{p}_L\big)\delta_{ij}-\big(\bar{\tau}_{ij,T}-\bar{\tau}_{ij,L}\big)\Big]\frac{1}{\gamma}\frac{\partial\gamma}{\partial x_j} \qquad (7.54)$$

which simplifies somewhat for boundary-layer flows when the pressures in the individual portions of the flow are set equal to the free-stream value. As noted before, the conditionally-averaged momentum equation in the laminar portion of the flow can be obtained by exchanging "T" and "L" and γ and $(1-\gamma)$.

The situation here is the same found for the conditioned continuity equation when the intermittency approaches the values of zero and one. In this case, however, not only must the laminar and turbulent velocity profiles be equal at the onset and end of transition for a continuous solution, but the laminar and turbulent mean pressure and viscous-stress profiles must also.

7.2.3 Conditionally-averaged Reynolds-stress and turbulent-kinetic-energy equations.

Reynolds-stress equations. Turbulent-kinetic-energy equation.

The globally-averaged turbulent-kinetic-energy and Reynolds-stress equations were developed in §1.2. Given by Eq. (1.28) and repeated here for convenience, the global turbulent-kinetic-energy equation is

$$\rho\bar{u}_j\frac{\partial k}{\partial x_j}=-\rho\overline{u_i'u_j'}\frac{\partial \bar{u}_i}{\partial x_j}+\frac{\partial}{\partial x_j}\left[\mu\frac{\partial k}{\partial x_j}-\rho\overline{u_j'\left(\tfrac{1}{2}u_k'u_k'\right)}-\overline{u_i'p'}\delta_{ij}\right]-\rho\hat{\varepsilon} \qquad (7.55)$$

where

$$\hat{\varepsilon}=\nu\overline{\frac{\partial u_i'}{\partial x_j}\frac{\partial u_i'}{\partial x_j}} \qquad (7.56)$$

and the body forces have been set equal to zero. Likewise, given by Eq.(1.32), the global Reynolds-stress equation is

$$\rho\bar{u}_k\frac{\partial \overline{u_i'u_j'}}{\partial x_k}=-\rho\overline{u_i'u_k'}\frac{\partial \bar{u}_j}{\partial x_k}-\rho\overline{u_j'u_k'}\frac{\partial \bar{u}_i}{\partial x_k}+\frac{\partial}{\partial x_k}\left(\mu\frac{\partial \overline{u_i'u_j'}}{\partial x_k}-\rho\overline{u_k'u_i'u_j'}\right)-\overline{u_i'\frac{\partial p'}{\partial x_j}}-\overline{u_j'\frac{\partial p'}{\partial x_i}}-\rho\hat{\varepsilon}_{ij} \qquad (7.57)$$

where

$$\hat{\varepsilon}_{ij}=2\nu\overline{\frac{\partial u_i'}{\partial x_k}\frac{\partial u_j'}{\partial x_k}} \qquad (7.58)$$

The Reynolds-stress equations. While the conditionally-averaged Reynolds-stress equations are not particularly easy to obtain, the initial step in their derivation begins with the instantaneous momentum equation for u_i'. Given by Eq. (1.21), it is rewritten here as

$$\rho\frac{\partial u_i'}{\partial t}+\rho\bar{u}_k\frac{\partial u_i'}{\partial x_k}=-\rho u_k'\frac{\partial \bar{u}_i}{\partial x_k}-\rho u_k'\frac{\partial u_i'}{\partial x_k}+\rho\overline{u_k'\frac{\partial u_i'}{\partial x_k}}-\frac{\partial p'\delta_{ik}}{\partial x_k}+\frac{\partial \tau_{ik}'}{\partial x_k}$$

where

$$\tau_{ij}'=\mu\left(\frac{\partial u_i'}{\partial x_j}+\frac{\partial u_j'}{\partial x_i}\right)$$

Similar to the method we used in §1.2 to derive the global Reynolds-stress equation, the above is multiplied by the quantity Iu_j' (for the turbulent component of the equation), and added to the instantaneous momentum equation for u_j' multiplied by Iu_i' before taking its average. Furthermore, before averaging, all the fluctuating quantities must be expanded in a fashion similar to that given for the fluctuating velocity in Eq. (7.35). If this order of substitution and integration is not followed, some terms are lost in the process and the sum of the conditioned laminar and turbulent equations will not equal the globally-averaged equation.

The result is voluminous and may be cast into many forms. Its rawest form, which appears useless, is presented in Appendix L. If we consider that the *mean pressure* in each portion of the flow is equal and constant, and that the correlations between the fluctuating quantities in the *different portions* of the flow are zero, then a somewhat reduced form of the equation is given by

$$\rho\bar{u}_k\frac{\partial\overline{(u_i'u_j')_T}}{\partial x_k}=-\rho\overline{(u_i'u_k')_T}\frac{\partial\bar{u}_j}{\partial x_k}-\rho\overline{(u_j'u_k')_T}\frac{\partial\bar{u}_i}{\partial x_k}+\frac{\partial}{\partial x_k}\left[\mu\frac{\partial\overline{(u_i'u_j')_T}}{\partial x_k}-\rho\overline{(u_k'u_i'u_j')_T}\right]$$

$$-\overline{u_{i,T}'\frac{\partial}{\partial x_j}\left(p'-\tfrac{1}{2}\tau_{kk}'\right)_T}-\overline{u_{j,T}'\frac{\partial}{\partial x_i}\left(p'-\tfrac{1}{2}\tau_{kk}'\right)_T}-\rho\hat{\varepsilon}_{ij,T}\qquad\text{(7.59, Group 1)}$$

$$-(1-\gamma)^2\rho\Delta\bar{u}_k\left(\Delta\bar{u}_i\frac{\partial\bar{u}_j}{\partial x_k}+\Delta\bar{u}_j\frac{\partial\bar{u}_i}{\partial x_k}\right)-(1-\gamma)^2\rho\bar{u}_k\frac{\partial\Delta\bar{u}_i\Delta\bar{u}_j}{\partial x_k}$$

$$-(1-\gamma)\rho\Delta\bar{u}_k\left[\frac{\partial\Delta\bar{u}_i\Delta\bar{u}_j}{\partial x_k}+\frac{\partial\overline{(u_i'u_j')_T}}{\partial x_k}\right]+\gamma(1-\gamma)^2\left[\rho\Delta\bar{u}_i\Delta\bar{u}_k\frac{\partial\Delta\bar{u}_j}{\partial x_k}+\rho\Delta\bar{u}_j\Delta\bar{u}_k\frac{\partial\Delta\bar{u}_i}{\partial x_k}\right]$$

$$-(1-\gamma)\left[\rho\overline{(u_j'u_k')_T}\frac{\partial\Delta\bar{u}_i}{\partial x_k}+\rho\overline{(u_k'u_i')_T}\frac{\partial\Delta\bar{u}_j}{\partial x_k}\right]$$

$$-(1-\gamma)^2\left\{\rho\Delta\bar{u}_i\frac{\partial}{\partial x_k}\left[\overline{(u_j'u_k')_T}-\overline{(u_j'u_k')_L}\right]+\rho\Delta\bar{u}_j\frac{\partial}{\partial x_k}\left[\overline{(u_i'u_k')_T}-\overline{(u_i'u_k')_L}\right]\right\}$$

$$+(1-\gamma)^2\frac{\partial}{\partial x_k}\left(\Delta\bar{u}_i\Delta\bar{\tau}_{jk}+\Delta\bar{u}_j\Delta\bar{\tau}_{ik}\right)-(1-\gamma)^2\left[\Delta\bar{\tau}_{ik}\frac{\partial\Delta\bar{u}_j}{\partial x_k}+\Delta\bar{\tau}_{jk}\frac{\partial\Delta\bar{u}_i}{\partial x_k}\right]$$

$$-(1-\gamma)^2\left[\Delta\bar{u}_i\frac{\partial\Delta\bar{p}\delta_{jk}}{\partial x_k}+\Delta\bar{u}_j\frac{\partial\Delta\bar{p}\delta_{ik}}{\partial x_k}\right]\qquad\text{(7.59, Group 2)}$$

$$-2(1-\gamma)(\tfrac{1}{2}-\gamma)\rho\bar{u}_k\Delta\bar{u}_i\Delta\bar{u}_j\frac{1}{\gamma}\frac{\partial\gamma}{\partial x_k}-\tfrac{1}{2}\rho\bar{u}_k\left(\overline{u'_{i,T}\Delta u'_j}+\overline{u'_{j,T}\Delta u'_i}\right)\frac{1}{\gamma}\frac{\partial\gamma}{\partial x_k}$$

$$-(\gamma-\tfrac{1}{2})\rho(\overline{u'_iu'_j})_T\Delta\bar{u}_k\frac{1}{\gamma}\frac{\partial\gamma}{\partial x_k}-(1-\gamma)(1-2\gamma)^2\rho\Delta\bar{u}_i\Delta\bar{u}_j\Delta\bar{u}_k\frac{1}{\gamma}\frac{\partial\gamma}{\partial x_k}$$

$$+\gamma(1-\gamma)\left\{\rho\Delta\bar{u}_i\left[(\overline{u'_ju'_k})_T-(\overline{u'_ju'_k})_L\right]+\rho\Delta\bar{u}_j\left[(\overline{u'_ku'_i})_T-(\overline{u'_ku'_i})_L\right]\right\}\frac{1}{\gamma}\frac{\partial\gamma}{\partial x_k}$$

$$-(1-\tfrac{3}{2}\gamma)\left[\rho(\overline{u'_iu'_j})_T\Delta\bar{u}_k+\rho(\overline{u'_ju'_k})_T\Delta\bar{u}_i+\rho(\overline{u'_ku'_i})_T\Delta\bar{u}_j\right]\frac{1}{\gamma}\frac{\partial\gamma}{\partial x_k}$$

$$-\tfrac{1}{2}\rho(\overline{u'_iu'_ju'_k})_T\frac{1}{\gamma}\frac{\partial\gamma}{\partial x_k}+(1-\gamma)(\tfrac{1}{2}-\gamma)\left(\Delta\bar{u}_i\Delta\bar{\tau}_{jk}+\Delta\bar{u}_j\Delta\bar{\tau}_{ik}\right)\frac{1}{\gamma}\frac{\partial\gamma}{\partial x_k}$$

$$+(1-\gamma)(\tfrac{1}{2}-\gamma)\left(\Delta\bar{u}_i\Delta\bar{\tau}_{jk}+\Delta\bar{u}_j\Delta\bar{\tau}_{ik}\right)\frac{1}{\gamma}\frac{\partial\gamma}{\partial x_k}+\tfrac{1}{2}\left[\overline{u'_i\Delta\tau'_{jk}}+\overline{u'_j\Delta\tau_{ik}}\right]\frac{1}{\gamma}\frac{\partial\gamma}{\partial x_k}$$

$$-(1-\gamma)(\tfrac{1}{2}-\gamma)\Delta\bar{p}\left(\Delta\bar{u}_i\delta_{jk}+\Delta\bar{u}_j\delta_{ik}\right)\frac{1}{\gamma}\frac{\partial\gamma}{\partial x_k}$$

$$-\tfrac{1}{2}\left[\overline{u'_i\Delta p'}\delta_{jk}+\overline{u'_j\Delta p'}\delta_{ik}\right]\frac{1}{\gamma}\frac{\partial\gamma}{\partial x_k} \qquad (7.59,\text{Group 3})$$

where

$$\hat{\varepsilon}_{ij,T}=2\nu\overline{\frac{\partial u'_{i,T}}{\partial x_k}\frac{\partial u'_{j,T}}{\partial x_k}} \qquad (7.60)$$

and $\Delta\bar{u}_i=\bar{u}_{i,T}-\bar{u}_{i,L},\quad \Delta u'_i=u'_{i,T}-u'_{i,L},\quad$ etc.

The terms of this equation have been purposely grouped according to averages of velocity fluctuations, velocity differences, and gradients of the intermittency. The first group contains the terms related to the production, diffusion, and dissipation of Reynolds-stresses in the turbulent portion of the flow caused by the fluctuations, while the second and third groups contain those related to the interfacial jumps. Since some of the terms in the last two groups may be rewritten and interchanged using both Eq. (7.48) and

$$\gamma(1-\gamma)\frac{\partial\Delta\bar{u}_i}{\partial x_i}=-(\tfrac{1}{2}-\gamma)\Delta\bar{u}_i\frac{\partial\gamma}{\partial x_i} \qquad (7.61)$$

which is a consequence of the conditioned continuity equations given in Eq. (7.47), the exact form of these groups are subject to one's outlook.[1]

As it happens, most investigators neglect all the terms in these last two groups, at least when considering the kinetic energy equation. Because spot growth depends directly on the production of turbulent shear stress at the laminar-turbulent interface (see §5.1), however, this is somewhat surprising. By keeping only the production terms associated with the interfacial jumps and neglecting all the others, it appears that a reasonable approximation to the Reynolds-stress equation might be

[1] Frankly, after ensuring that the production terms in these groups were correct, the author became tired of searching for a meaningful display of the other terms.

$$\rho\bar{u}_k\frac{\partial\overline{(u_i'u_j')}_T}{\partial x_k}=-\rho\overline{(u_i'u_k')}_T\frac{\partial\bar{u}_j}{\partial x_k}-\rho\overline{(u_j'u_k')}_T\frac{\partial\bar{u}_i}{\partial x_k}+\frac{\partial}{\partial x_k}\left[\mu\frac{\partial\overline{(u_i'u_j')}_T}{\partial x_k}-\rho\overline{(u_k'u_i'u_j')}_T\right]$$
$$+\frac{\partial}{\partial x_k}\left[\mu\frac{\partial\overline{(u_i'u_j')}_T}{\partial x_k}-\rho\overline{(u_k'u_i'u_j')}_T\right]-\overline{u_{i,T}'\frac{\partial p_T'}{\partial x_j}}-\overline{u_{j,T}'\frac{\partial p_T'}{\partial x_i}}-\rho\hat{\varepsilon}_{ij,T}$$
$$-\left(1-\gamma\right)^2\rho\Delta\bar{u}_k\left[\Delta\bar{u}_i\frac{\partial\bar{u}_j}{\partial x_k}+\Delta\bar{u}_j\frac{\partial\bar{u}_i}{\partial x_k}\right]-\left(1-\gamma\right)^2\rho\bar{u}_k\frac{\partial\Delta\bar{u}_i\Delta\bar{u}_j}{\partial x_k}\tag{7.62}$$

where the normal viscous-stress terms in the second line of Eq. (7.59, Group 1) are easily shown to be negligible when compared to the other terms in the equation.[1] In this case, writing the laminar counterpart of the above equation and adding recovers the global Reynolds-stress equation to within the error of the neglected terms.

Since all but the terms associated with the interfacial jumps are similar to those in the global Reynolds-stress equation, they may be modeled as discussed in §3.3 or by using any other suitable method. Furthermore, since the terms associated with the interfacial jumps contain only the average velocities and intermittency, no modeling is required. Consequently, these equations together with the continuity, momentum, and dissipation equations form a closed set which can presumably be solved. The same can be said about the laminar counterpart of the equation, which is obtained from Eq. (7.62) by replacing the subscripts "T" with "L" and the coefficients $(1-\gamma)^2$ with γ^2. In this case, the terms similar to those in the global Reynolds-stress equation may be modeled as discussed in §4.3 or by using any other suitable method.

For a two-dimensional boundary layer, the equations for the streamwise, wall-normal, and turbulent shear stresses reduce to

$$\rho\bar{u}\frac{\partial\overline{(u'u')}_T}{\partial x}+\rho\bar{v}\frac{\partial\overline{(u'u')}_T}{\partial y}=-2\rho\overline{(u'v')}_T\frac{\partial\bar{u}}{\partial y}-2\overline{u_T'\frac{\partial p_T'}{\partial x}}$$
$$+\frac{\partial}{\partial y}\left(\mu\frac{\partial\overline{(u'u')}_T}{\partial y}-\rho\overline{(v'u'u')}_T\right)-\rho\hat{\varepsilon}_{11,T}-2\left(1-\gamma\right)^2\rho\Delta\bar{u}\Delta\bar{v}\frac{\partial\bar{u}}{\partial y}-\left(1-\gamma\right)^2\rho\bar{v}\frac{\partial\Delta\bar{u}^2}{\partial y}\tag{7.63}$$

$$\rho\bar{u}\frac{\partial\overline{(v'v')}_T}{\partial x}+\rho\bar{v}\frac{\partial\overline{(v'v')}_T}{\partial y}=-2\overline{v_T'\frac{\partial p_T'}{\partial y}}+\frac{\partial}{\partial y}\left(\mu\frac{\partial\overline{(v'v')}_T}{\partial y}-\rho\overline{(v'v'v')}_T\right)-\rho\hat{\varepsilon}_{22,T}$$
$$-2\left(1-\gamma\right)^2\rho\Delta\bar{v}\Delta\bar{v}\frac{\partial\bar{v}}{\partial y}-\left(1-\gamma\right)^2\rho\bar{v}\frac{\partial\Delta\bar{v}^2}{\partial y}\tag{7.64}$$

$$\rho\bar{u}\frac{\partial\overline{(u'v')}_T}{\partial x}+\rho\bar{v}\frac{\partial\overline{(u'v')}_T}{\partial y}=-\rho\overline{(v'v')}_T\frac{\partial\bar{u}}{\partial y}-\overline{v_T'\frac{\partial p_T'}{\partial x}}-\overline{u_T'\frac{\partial p_T'}{\partial y}}$$
$$+\frac{\partial}{\partial y}\left(\mu\frac{\partial\overline{(u'v')}_T}{\partial y}-\rho\overline{(v'u'v')}_T\right)-\rho\hat{\varepsilon}_{12,T}-\left(1-\gamma\right)^2\rho\Delta\bar{v}\Delta\bar{v}\frac{\partial\bar{u}}{\partial y}-\left(1-\gamma\right)^2\rho\bar{v}\frac{\partial\Delta\bar{u}\Delta\bar{v}}{\partial y}\tag{7.65}$$

[1] Note here that the normal viscous-stress terms in the global Reynolds-stress equation are exactly equal to zero because the divergence of the fluctuating velocity u_k' is zero.

where

$$\hat{\varepsilon}_{11,T} = 2\nu\overline{\left(\frac{\partial u'}{\partial y}\right)^2_T}, \quad \hat{\varepsilon}_{22,T} = 2\nu\overline{\left(\frac{\partial v'}{\partial y}\right)^2_T}, \quad \hat{\varepsilon}_{12,T} = 2\nu\overline{\left(\frac{\partial u'}{\partial y}\frac{\partial v'}{\partial y}\right)_T}$$

$$\Delta\bar{u} = \bar{u}_T - \bar{u}_L, \quad \Delta\bar{v} = \bar{v}_T - \bar{v}_L,$$

and the usual smaller order angled-bracketed terms have been neglected.

The turbulent-kinetic-energy and dissipation equations. At this point, the turbulent-kinetic-energy equation is most quickly obtained by contracting Eq. (7.59). The approximate version corresponding to Eq. (7.62) for the Reynolds-stress equation is

$$\rho\bar{u}_k\frac{\partial k_T}{\partial x_k} = -\rho\overline{(u_i'u_k')}_T\frac{\partial \bar{u}_i}{\partial x_k} + \frac{\partial}{\partial x_k}\left[\mu\frac{\partial k_T}{\partial x_k} - {}^1\!/\!_2\,\rho\overline{(u_k'u_i'u_i')}_T\right]$$
$$-\overline{u_{i,T}'\frac{\partial p_T'}{\partial x_i}} - \rho\hat{\varepsilon}_T - (1-\gamma)^2\rho\Delta\bar{u}_k\Delta\bar{u}_i\frac{\partial \bar{u}_i}{\partial x_k} - {}^1\!/\!_2(1-\gamma)^2\rho\bar{u}_k\frac{\partial \Delta\bar{u}_i\Delta\bar{u}_i}{\partial x_k} \tag{7.66}$$

where

$$k_T = {}^1\!/\!_2\overline{(u_i'u_i')}_T \quad \text{and} \quad \hat{\varepsilon}_T = 2\nu\overline{\left(\frac{\partial u_i'}{\partial x_k}\frac{\partial u_i'}{\partial x_k}\right)_T}$$

For a two-dimensional boundary layer, this equation reduces to

$$\rho\bar{u}\frac{\partial k_T}{\partial x} + \rho\bar{v}\frac{\partial k_T}{\partial y} = -\rho\overline{(u'v')}_T\frac{\partial \bar{u}}{\partial y} - \left[\overline{u_T'\frac{\partial p_T'}{\partial x}} + \overline{v_T'\frac{\partial p_T'}{\partial y}}\right] + \frac{\partial}{\partial y}\left[\mu\frac{\partial k_T}{\partial y} - \rho\overline{(v'k)}_T\right] - \rho\hat{\varepsilon}_T$$
$$-(1-\gamma)^2\rho\Delta\bar{u}\Delta\bar{v}\frac{\partial \bar{u}_T}{\partial y} - {}^1\!/\!_2(1-\gamma)^2\rho\bar{v}_T\frac{\partial \Delta\bar{u}^2}{\partial y} \tag{7.67}$$

where again the usual smaller order terms have been neglected. As mentioned above, this equation needs no further modeling than that used for standard turbulent boundary layers. As a result, the dissipation transport equation developed in §3.3.2 should be sufficient to close the set. For a two-dimensional boundary layer, the conditioned equation corresponding to the globally-averaged version given by Eq. (3.96) is

$$\bar{u}\frac{\partial \tilde{\varepsilon}_T}{\partial x} + \bar{v}\frac{\partial \tilde{\varepsilon}_T}{\partial y} = -c_{\varepsilon 1}\frac{\tilde{\varepsilon}_T}{k_T}\overline{(u'v')}_T\frac{\partial \bar{u}}{\partial y}$$
$$+\frac{\partial}{\partial y}\left[\left(\nu + c_\varepsilon\overline{(v'v')}_T\frac{k_T}{\hat{\varepsilon}_T}\right)\frac{\partial \tilde{\varepsilon}_T}{\partial y}\right] - c_{\varepsilon 2}f_{\varepsilon 2}\frac{\tilde{\varepsilon}_T^2}{k} + c_{\varepsilon 3}\nu\overline{(v'v')}_T\frac{k_T}{\tilde{\varepsilon}_T}\left(\frac{\partial^2\bar{u}}{\partial y^2}\right)^2 \tag{7.68}$$

where

$$\hat{\varepsilon}_T = \tilde{\varepsilon}_T + \nu f_{\varepsilon,w}\left|\frac{1}{y}\frac{\partial k_T}{\partial y}\right|$$

The constants c's and functions f's for this model are given in Table 3.3 and Eq. (3.72). Of course, any other turbulence modeling may be used.

7.3 Intermittency Models.

Turbulent-spot models. Transport models.

As mentioned before, there are two basic types of intermittency models – the *turbulent-spot* models and *intermittency-transport* models. While the former represents the classical approach, the latter represents the more modern, albeit not necessarily better approach. In this section, we will first examine the classical turbulent-spot models and then the intermittency-transport models. As also mentioned before, either of these may be coupled to either of the approaches described in the previous section for calculating transitional flow.

7.3.1 Turbulent-spot models.

Early models. A different tack. Some calculations.

These models date back to Emmons' 1951 paper on turbulent spots. In their simplest form, they provide an algebraic expression for the near-wall intermittency distribution which can easily be incorporated into any boundary-layer code for calculating transitional flow. In their most general form, they yield an integral expression which is still rather easy to incorporate into these codes. A criticism generally lodged against them, however, is that they are difficult to incorporate into modern three-dimensional Navier-Stokes codes using multiple parallel processors and unstructured grids, particularly because it becomes difficult to separate the processing (slowing computation time) and determine the required boundary-layer parameters because the wall-normal isn't necessarily part of the grid. But, as we will see in §7.4, this is the least of our problems.

In the following, we will first summarize the early models, consider a new approach to obtaining the near-wall intermittency for flows with an arbitrarily varying free-stream velocity, and finally provide some calculations of the intermittency distribution for flows with different piecewise-varying streamwise pressure gradients.

Early models. As explained in §6.1.2, the central portion of almost any measured near-wall intermittency distribution can be fitted by Eq. (5.18), namely,

$$\gamma_0(x)=1-\exp\left[-\hat{n}\sigma\left(Re_x-Re_{xt}\right)^2\right]; \quad (Re_x \geq Re_{xt}) \tag{7.69}$$

even though it strictly applies only when the free-stream velocity is constant, turbulent spots expand linearly as they propagate downstream, and laminar breakdowns occur over a relatively narrow range of distances compared to the length of transition. As a result, Eq. (7.69), which was used to calculate the results in Fig. 7.6, became the earliest intermittency model for transitional flow.

The first to extend Emmons turbulent spot theory to transition with a variable free-stream velocity were Chen and Thyson [1971]. Discussed in §5.5.1, they considered spots with triangular shaped footprints and assumed that all spots were produced at one streamwise location. Their general result, given previously by Eq. (5.111), is

$$\gamma_0(x)=1-\exp\left\{-n'f_s\int_{x_t}^{x}\tan\alpha\, dx'\int_{x_t}^{x}\frac{u_{le}-u_{te}}{u_{le}u_{te}}dx'\right\} \tag{7.70}$$

Using Schubauer and Klebanoff's [1955] information, Chen and Thyson assumed that the ratios of the leading- and trailing-edge velocities to the local free-stream velocity, and the lateral half-angle α remain constant as the spot propagates downstream. We now know that none of these assumptions are strictly true.[1] Nevertheless, using their assumptions, the above expression reduces to

---

[1] See the velocity ratios for flows with favorable and adverse pressure gradients plotted against distance in Figs. 5.40 and 5.44, and Eq. (5.82) for the lateral half-angle in a flow with an adverse pressure gradient,

$$\gamma_0(x)=1-\exp\left[-\dot{n}'f_s\frac{\bar{u}_\infty}{u_{le}}\frac{u_{le}-u_{te}}{u_{te}}\left(x-x_t\right)\tan\alpha\int_{x_t}^{x}\frac{1}{\bar{u}_\infty(x')}dx'\right];\quad (u_{le}/\bar{u}_\infty, u_{te}/\bar{u}_\infty \text{ and } \alpha \text{ constant}) \tag{7.71}$$

The integral in this expression may be recognized as the "time of flight" of a particle moving at the free-stream velocity, which is often used to transpose a solution for a constant free-stream velocity into one for a variable free-stream velocity. Along with this equation, they also proposed the model described in §6.2.3 for the production rate $\dot{n}'$, which has the added advantage of canceling the $\tan\alpha$ factor in the above expression.[1] However, as shown by Walker et al. [1988], Eq. (7.71) grossly over estimates the length of transition in adverse pressure gradients. Nevertheless, Chen and Thyson's equation is still widely used as the intermittency model when the free-stream velocity varies.

In a pair of papers, Gostelow et al. [1996] and Solomon et al. [1996] developed a somewhat convoluted method to evaluate Eq. (7.70) for flows with a variable free-stream velocity by using the *local* value of the pressure-gradient parameter $\lambda_{\delta 2}$. It basically consists of using Eq. (5.106), which by the way assumes that the spot expands linearly across the surface, to eliminate the velocity ratios in terms of α and σ, and then using the correlations of Gostelow's et al. [1994] data among others to obtain the local functions $\alpha(\lambda_{\delta 2})$ and $\sigma(\lambda_{\delta 2})$. Although their model is a step in the right direction, we now know that neither α nor σ are constant for flows with a constant pressure-gradient parameter, and especially not when the pressure gradient is adverse.

A different tack. Since α is the angle tangent to the path line of the spot's tip at any position x, Eq. (7.70) may be expressed completely in terms of velocities associated with the spot. Introducing the spanwise and streamwise velocity components of the spot's outboard tip w_{tp} and u_{tp}, we obtain

$$\gamma_0(x)=1-\exp\left[-\dot{n}'f_s\int_{x_t}^{x}\frac{w_{tp}}{u_{tp}}dx'\int_{x_t}^{x}\frac{u_{le}-u_{te}}{u_{le}u_{te}}dx'\right];\quad (x\geq x_t) \tag{7.72}$$

where all the velocities are to be evaluated at the position x', and where, for simplicity, it has been assumed that the size of a nascent spot can be neglected compared to its size farther downstream. Here, a nascent spot of finite size can easily be taken into account by introducing an initial spot size in the lower limits of the integrals. A good guess for the initial size would be the thickness of the laminar boundary layer at the onset of transition, or some factor times it. The idea behind using the above expression is that it *may* be easier to relate the velocity associated with the spot's tip to the parameters describing the physics of the surrounding flow than using either the breadth b or angle α, simply because it together with the leading- and trailing-edge velocities are the fundamental kinematic quantities describing the motion of the spot. Thus, the above equation could be called a "kinematic model" for the propagation of a turbulent spot.

Since u_{tp} nearly equals the spot's trailing edge velocity, the near-wall intermittency can be determined once the leading-edge, trailing-edge, and transverse trailing-edge-tip velocities are known. As shown in Figs. 5.31, 5.40 and 5.43, these depend at the very least on the transition Reynolds number, the pressure-gradient parameter, and the distance downstream from the onset of transition. In addition, they probably depend on turbulence level and everything else that affects the stability of the surrounding laminar boundary layer. Nevertheless, bearing this in mind, a plot of the leading- and trailing-edge velocity data in Figs. 5.40 and 5.43 as a function of the distance Reynolds number downstream from onset is shown in Fig. 7.10a. Here, as before, the ratios are taken using the *local* free-stream velocity (which varies with distance). The distance Reynolds number, however, is formed using the free-

[1] See Eq. (6.24).

stream velocity at the spot source. In this format, the variation of the leading-edge velocity ratios for both the favorable and adverse pressure gradients are about the same. Although not reported by Wygnanski et al. [1982], the variation for zero pressure gradient presumably lies between them. The equation for the line drawn through the data is

$$\frac{u_{le}}{\overline{u}_\infty}=0.68+0.30\left\{1-\exp\left[-3.5(10)^{-6}\,\overline{u}_{\infty,t}\left(x-x_t\right)/\nu\right]\right\} \tag{7.73}$$

The asymptotic value of this ratio for large distances is 0.98, which according to Schubauer and Klebanoff [1955], is closer to the velocity of the leading edge nose of the spot than of its footprint. Nevertheless, for distance Reynolds numbers greater than $5(10)^5$, which is, according to Fig. 5.5b, when the spot is nearly fully-developed, the above expression yields a leading-edge velocity that is greater than about ninety percent of the free-stream velocity.

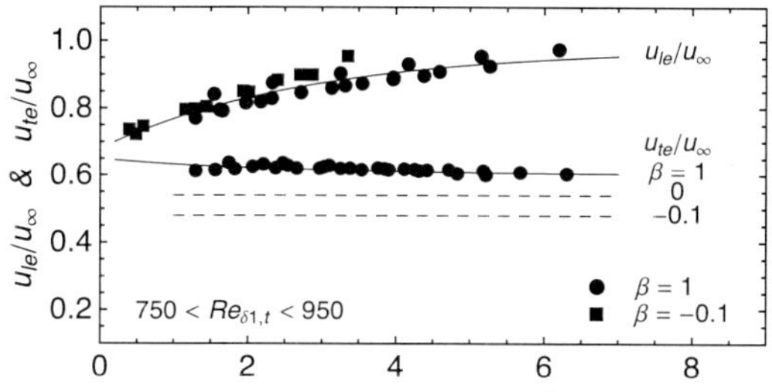

(a)

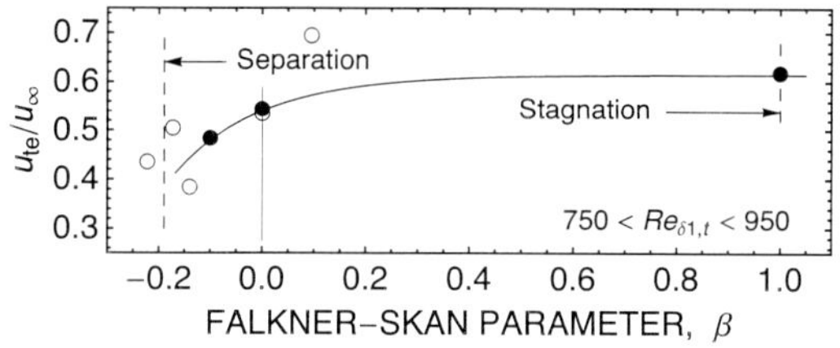

(b)

Figure 7.10a,b.
Variation of (*a*) the leading- and trailing-edge velocity ratios with distance Reynolds number, and (*b*) the average trailing-edge velocity ratio with the pressure-gradient parameter. Data from Katz et al. [1990] and Seifert and Wygnanski [1995].

The average trailing-edge velocities plotted in the figure are 0.48, 0.54 and 0.61, which roughly correspond to the values measured for the adverse, zero and favorable pressure gradients with $\beta = -0.1$, 0 and 1 ($m = -0.048$, 0 and 1; $\lambda_{\delta 2} = -0.027$, 0 and 0.085).[1] Although the trailing-edge velocity decreases slightly with increasing distance, its variation within the range shown in Fig. 7.10a (which is admittedly small) is less than ten percent of the average. The variation of the average trailing-edge velocity with the Falkner-Skan parameter is plotted in Fig. 7.10b, where the filled circles are the results of Wygnanski and his coworkers and the empty circles are data from Table 2 in Gostelow et al. [1996]. A curve fit to this data is given by

$$\frac{u_{te}}{\overline{u}_\infty}=0.54+0.075\left(1-e^{-5.9\beta}\right) \tag{7.74}$$

Clearly, an adverse pressure gradient affects the spot's trailing-edge velocity more than a favorable

[1] The value for the adverse pressure gradient is an average of a few largely scattered data within a limited range of Reynolds numbers. For further information, the reader is referred to Fig. 7 in Seifert and Wygnanski [1995].

pressure gradient. As a consequence, since the leading-edge velocity ratio is about the same for both, the duration of a spot over any given position is greater for the adverse than favorable pressure gradient. The effect of turbulence level on either the spot's leading or trailing edge velocity is unknown.

Correlations for the asymptotic variation of the spot's tip with distance for $\beta = -0.1$, 0, and 1 are provided respectively in Eqs. (5.81), (5.68) and (5.78) for very low free-stream turbulence levels. A general expression for all of these that yields a linear asymptotic state of growth far downstream and satisfies $z_{tp} = 0$ at the source of the spot is

$$\frac{z_{tp}}{\delta_{1,t}} = a\left\{\left(Re_x - Re_{xt}\right) - \frac{1}{b}\left\{1 - \exp\left[-b\left(Re_x - Re_{xt}\right)\right]\right\}\right\} \tag{7.75}$$

where a and b are functions of $Re_{\delta 1,t}$, β, and other flow parameters. For large distances from the source, the half-width of the spot varies linearly with distance which matches the asymptotic correlations for the zero and favorable pressure gradient cases. For small distances, it varies according to

$$\frac{z_{tp}}{\delta_{1,t}} = \frac{1}{2}ab\left(Re_x - Re_{xt}\right)^2 + \cdots$$

which matches the correlation for the case with an adverse pressure gradient. The actual values of a and b that provide these matches are given in Table 7.2.[1]

Table 7.2
Coefficients a and b for use in Eq. (7.75) when $Tu < 0.005$.

β	a	b
−0.1	$56.3/Re_{\delta 1,t}$	$1.96\,(10)^{-5}/Re_{\delta 1,t}$
0	$0.0117/Re_{\delta 1,t}^{\ 0.6}$	$4.57\,(10)^{-4}/Re_{\delta 1,t}^{\ 0.6}$
1	$1.49\,(10)^{-4}$	$1.07\,(10)^{-5}$

The derivative of z_{tp} with respect to x yields the expression for the ratio of the spot's outboard-tip velocity components, namely,

$$\frac{w_{tp}}{u_{tp}} = aRe_{\delta 1,t}\left\{1 - \exp\left[-b\left(Re_x - Re_{xt}\right)\right]\right\} \tag{7.76}$$

This relation is plotted as a function of streamwise distance in Fig. 7.11a for a spot-source Reynolds number of 850, which is about halfway within the range of the correlated data. The trend of the velocity ratio with distance clearly depends on whether β, and therefore the pressure-gradient parameter $\lambda_{\delta 2}$, is less than or greater than zero. For $\beta \geq 0$, the ratio becomes nearly constant (corresponding to an asymptotic state of linear spot expansion) when the distance Reynolds number is roughly $5(10)^5$.[2] As noted before, this also corresponds to where a spot developing in a flow with $\beta = 0$ becomes fully developed. For $\beta = -0.1$, however, the ratio increases linearly with distance indicating that an asymptotic state of linear spot growth is not attained within the distance plotted in the figure. Presumably, the curves for intermediate values of β in this figure continuously morph between those shown. While

[1] This equation with the values for a and b given in Table 7.2 was used to plot the dashed curves in Figs. 5.34 and 5.41 for the flows with $\beta = 0$ and 1.

[2] This distance depends on the parameter b in Eq. (7.75), which in turn depends on the source Reynolds number.

these would not be too difficult to estimate for favorable pressure gradients, those for adverse gradients would be a challenge. Clearly, more data would be helpful in this effort, including data at elevated free-stream turbulence levels.

The ratio of the tip velocity components varies with Reynolds number as shown in Fig. 7.11b. The solid lines represent the variation within the data range, while the dashed lines provide their extrapolation into a lower range. Except for the case with the adverse gradient, the variation is slight, particularly within the range of data. The extrapolation for $\beta = -0.1$, which is tentative, indicates that a value greater than one might be possible; that is, the tip might move faster in the spanwise direction for lower spot Reynolds numbers than it moves streamwise.

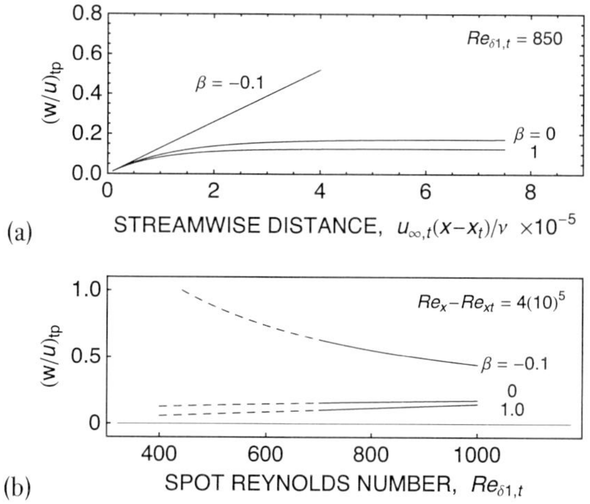

Figure 7.11a,b.
Calculated variation of the spot's outboard-tip velocity ratio with (a) distance Reynolds number, and (b) spot-source Reynolds number for three values of the pressure-gradient parameter β.

Returning to the problem at hand, Eq. (7.72) for the near-wall intermittency requires either correlating or predicting the behavior shown in Figs. 7.10 and 7.11. While correlating the results in Figs. 7.10a,b might be relatively easy, correlating those in Figs. 7.11a,b appear daunting. Nevertheless, a promising approach in this effort might be to use Eq. (7.76) and consider only b to be a function of the pressure gradient parameter. In this case, we would expect b to increase with a decreasing β. Before this can be attempted, however, more data in the region $\beta < 0$ is required.

Nevertheless, if we use the correlations presented in Figs. 7.10 and 7.11 to calculate the near-wall intermittency distributions for the three pressure gradient parameters $\beta = -0.1$, 0 and 1, we obtain the results shown in Fig. 7.12a. In this case, Eq. (7.72) was used together with the correlations given by Eqs. (7.73), (7.74), (7.76) and Table 7.2. The free-stream velocity distribution was obtained from

$$\frac{\bar{u}_\infty}{\bar{u}_{\infty,t}} = \left[1 + \frac{\bar{u}_{\infty,t}(x - x_t)/\nu}{Re_{xt}}\right]^{\frac{\beta}{2-\beta}}$$

which is equivalent to the Falkner-Skan distribution given by Eq. (4.19). The distance Reynolds number Re_{xt}, which is measured from the virtual origin of the flow, depends on $Re_{\delta 1,t}$ as well as β. This relation can be derived from Eq. (4.19) and evaluated using Table D.3 in Appendix D. The results for $\beta = -0.1$, 0 and 1 are respectively given by

$$Re_{xt} = 0.229\,Re_{\delta 1,t}^2,\quad 0.337\,Re_{\delta 1,t}^2,\quad \text{and}\quad 2.4\,Re_{\delta 1,t}^2$$

The calculated results in Fig. 7.12 are shown for $Re_{\delta 1,t} = 850$, which yields the distance Reynolds numbers based on the source's location of 165,000, 243,000 and 1,730,000 respectively.

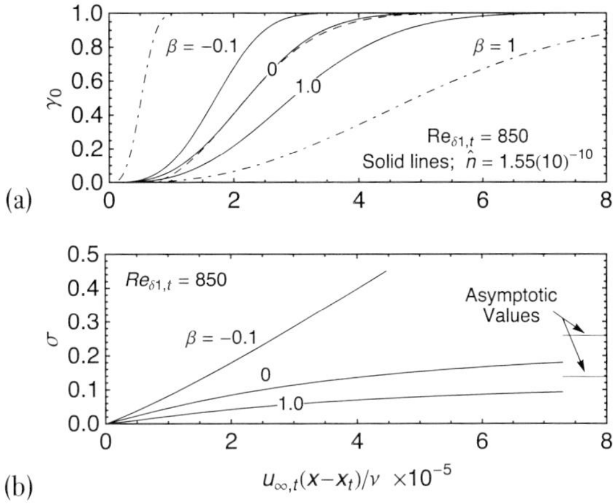

Figure 7.12a,b.
Calculated distributions of (*a*) the near-wall intermittency, and (*b*) the propagation parameter for an adverse, zero, and favorable pressure gradient with a very low free-stream-turbulence level and $Re_{\delta 1,t} = 850$.

The value of $\hat{n}'$ in Eq. (7.72) for the $\beta = 0$ case was obtained from the correlation for $\hat{n}\bar{\sigma}$ given by Eq. (6.18). For $Re_{\delta 2,t} = 850/2.6 = 328$, one obtains $\hat{n}\bar{\sigma} = 2.62(10)^{-11}$. Since σ varies with distance and depends upon the pressure-gradient parameter, $\bar{\sigma}$ was determined from $\bar{\sigma} = f_\sigma \sigma_{1/2}$ where $\sigma_{1/2}$ is the propagation parameter when the intermittency equals one-half and f_σ is the factor that ensures the intermittency distribution obtained from Eq. (7.72) matches Narasimha's solution in the central portion of the distribution. As a consequence, the production parameter was obtained from

$$\hat{n} = 2.62(10)^{-11} \Big/ f_\sigma \sigma_{1/2}$$

The spot propagation parameter for these calculations was determined by using

$$\sigma(x) = \frac{\bar{u}_\infty f_s}{(x - x_s)^2} \int_{x_t}^{x} \frac{w_{tp}}{u_{tp}} dx' \int_{x_t}^{x} \frac{u_{le} - u_{te}}{u_{le} u_{te}} dx' \tag{7.77}$$

which follows from Eq. (5.110) after introducing the tip-velocity ratio.

In this case, the propagation parameter $f_\sigma \sigma_{1/2}$ was found equal to 0.169, which in turn provided $\hat{n}$ = $1.55(10)^{-10}$. The fit to Narasimha's solution is drawn for comparison as the dashed line in the figure after shifting its transition Reynolds number downstream by about 58,000.

The results for the two other pressure-gradient parameters were calculated using the same production rate. These are shown by the solid lines in Fig. 7.12a. Thus, only the effect of pressure gradient on *spot propagation rate* is displayed by the solid lines in this figure. To determine the entire effect of pressure gradient on the intermittency, $\hat{n}\bar{\sigma}$ was obtained from Eq. (6.31) for the adverse gradient and Fig. 6.15 for the favorable. This provided $\hat{n}\bar{\sigma} = 5.7(10)^{-10}$ for $\beta = -0.1$ ($\lambda_{\delta 2} = -0.027$), which is about twenty times larger than that for the zero pressure-gradient case, and $\hat{n}\bar{\sigma} \approx 4(10)^{-12}$ for $\beta = 1.0$ ($\lambda_{\delta 2} = 0.085$) which is about seven times smaller. The corresponding values of $\hat{n}$ are $6.5(10)^{-9}$ and $4.2(10)^{-11}$ respectively. That is, the production of turbulent spots decrease as the pressure-gradient parameter increases. The calculated intermittency distributions for these conditions are shown by the dot-dashed lines to the far left and right of the figure. When Narasimha's solution is used to match these results, the transition Reynolds numbers for the adverse and favorable pressure-gradient solutions must be shifted downstream by about 16,500 and 70,000 respectively. Nevertheless, we see that the effect of the pressure gradient on the intermittency distribution is significant, with its effect on both the spot production and propagation rates being important.[1]

The corresponding variations of the spot propagation parameter are shown in part (*b*) of Fig. 7.12. A point to be noted here is that the flow becomes completely turbulent before the propagation parameters attain their asymptotic values. For the zero and favorable pressure-gradient cases these values are 0.27 and 0.14 respectively. Using Eq. (6.7), transition occurs at $Re_{\delta 2,t} = 328$ for flow without a pressure gradient when the turbulence level is about 1.4 percent. Consequently, in agreement with the observation made in §5.1.1, transition with moderate-to-high free-stream turbulence levels occur before the spot is fully developed, at least as defined by the low free-stream turbulence "single-spot" investigations discussed there. This again raises questions regarding spot growth in a turbulent free stream. Since turbulence rather than the instability of the flow probably controls spot growth, the result could be significantly different and maybe even simpler. Clearly, single-spot investigations with higher free-stream turbulence levels must be conducted before these questions can be answered. When they *are* conducted, data must be obtained both closer to the source and for lower source Reynolds numbers.

Finally, it must be emphasized in Eq. (7.72) that $\dot{n}'$ is the total number of spots produced per unit time per unit span by a line source at $x = x_t$ and is, therefore, a constant. If a distributed breakdown is considered, we must then return to the basic analysis carried out in §5.2.3 for distributed sources and apply it to any one of Eqs. (7.70)–(7.72). If we use Eq. (7.72), the result is

$$\gamma_0(x;x_l) = 1 - \exp\left\{ -\int_0^{x_l} \dot{n}''(\xi)\mathrm{U}(x-\xi)\,d\xi \int_\xi^x \frac{w_{tp}}{u_{tp}}\,dx' \int_\xi^x \frac{u_{le}-u_{te}}{u_{le}u_{te}}\,dx' \right\} \tag{7.78}$$

where $\dot{n}''(\xi)$ is the source strength per unit time per unit surface area and U(z) is the unit step function of z. When the breakdown is Gaussian-like, the source distribution is given by Eq. (5.28) and the limits on ξ in the integral above may be extended to ±infinity, with but few exceptions. While the above expression for γ_0 represents the most general of the turbulent-spot models, it clearly requires more information about the breakdown of laminar flow and the becalmed zones than is presently available, namely, the appropriate transition Reynolds number Re_{xt} and spot-production distribution $\dot{n}''(x)$.

Some calculations. Once the kinematic conditions defining the propagation of a turbulent spot are known, we can use Eq. (7.72) to calculate the intermittency distribution through transition for any free-stream velocity variation providing we assume that $\dot{n}'$ depends only on the conditions at the onset of transition. We will now consider calculating these intermittency distributions for two typical velocity variations. In particular, using the information we know about spot kinematics for flows with $\beta = -0.1$, 0, and 1, we will consider transition when this parameter varies from one of these values to another using the same conditions we used to calculate the results in Fig. 7.12. In this case, the calculated distribution before the change in β will be identical to that shown in Fig. 7.12, while after the change, since the spots expand differently across the surface than before, the intermittency distribution will be different. To calculate the new expansion rates we again use Eqs. (7.73), (7.74) and (7.76) with the new value of β. Since the size of the spot must be the same immediately before and after the change, the breadth and duration (not length) of the spot must match where the change occurs. This is done by adjusting the virtual origins for both the width and duration of the spot developing in the downstream region.

---

[1] Here it is worthwhile to repeat that Chen and Thyson's model using Eq. (7.71) has nowhere near the same effect on the intermittency distribution.

For the first example consider transition occurring with an abrupt change in β from 0 to –0.1, that is, from a zero to an adverse pressure gradient.[1] If we arbitrarily consider that the change occurs at the position corresponding to a Reynolds number of 200,000 downstream from the onset of transition (see Fig. 7.12, β = 0), then the resulting free-stream velocity distribution is shown in Fig. 7.13a. (Note: We could have used any free-stream velocity distribution if we had a general correlation for the tip-velocity ratio.) Assuming the same conditions for the onset of transition as in Fig. 7.12 and matching the size of the spot at the change in β yields the intermittency distribution given by the solid line in part (b) of the figure. The dashed line marked β = 0 is that shown in Fig. 7.12a for a zero pressure gradient throughout transition. The difference between these results is caused solely by the spots expanding at different rates in both directions as they encounter the adverse pressure gradient. This effect is shown in parts (*c*) and (*d*) of Fig. 7.13 where the calculated variations of the spot's half-width and duration with distance are shown. The dramatic increase of the spot's width in the adverse pressure gradient, which is almost twice its increase in length, is obvious.

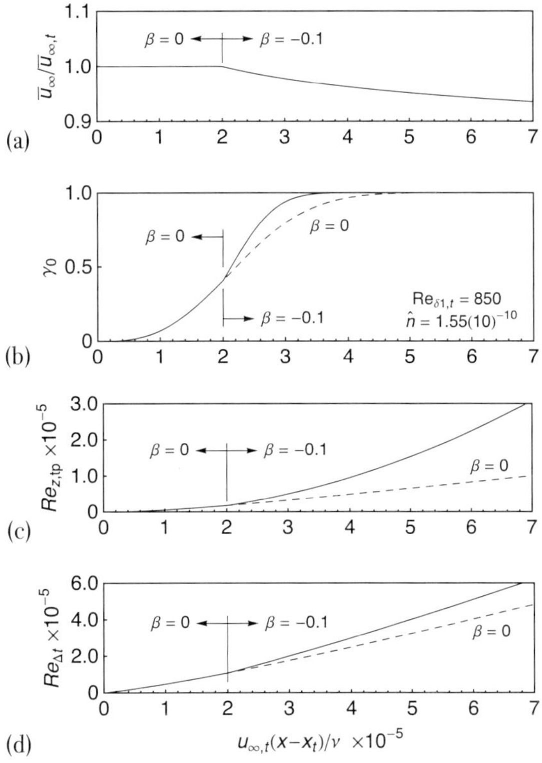

Figure 7.13a,b,c,d.
Calculated distributions of the (*a*) free-stream velocity, (*b*) near-wall intermittency, (*c*) width, and (*d*) duration, for a change from a zero to adverse pressure gradient through transition with $Re_{\delta 1,t}$ = 850 and $\hat{n} = 1.55(10)^{-10}$.

For the second example consider transition occurring with an abrupt change in β from 1 to 0, that is, from a favorable to zero pressure gradient. Since the pressure gradient decreases as in the previous example, the rate of expansion and intermittency should increase with this change also. Again considering that the change occurs at the position corresponding to a Reynolds number of 200,000 downstream of the onset of transition (see Fig. 7.12, β = 1), the resulting free-stream velocity distribution

[1] This roughly corresponds to the state of affairs on the suction side of an airfoil aft the position of minimum pressure.

is shown in Fig. 7.14a. The intermittency distribution is given by the solid line in part (b) of the figure for the same conditions at onset as in the previous figures. In this figure, however, the dashed line corresponds to that shown in Fig. 7.12a for $\beta = 1$. Although the spot size is not plotted for this example, the width and duration at the far right of the figure are about thirty and fifty percent above those calculated for the $\beta = 1$ case. Consequently, the duration of the spots in this example increases more than their width which is in contrast with the results shown in parts (c) and (d) of Fig. 7.13.

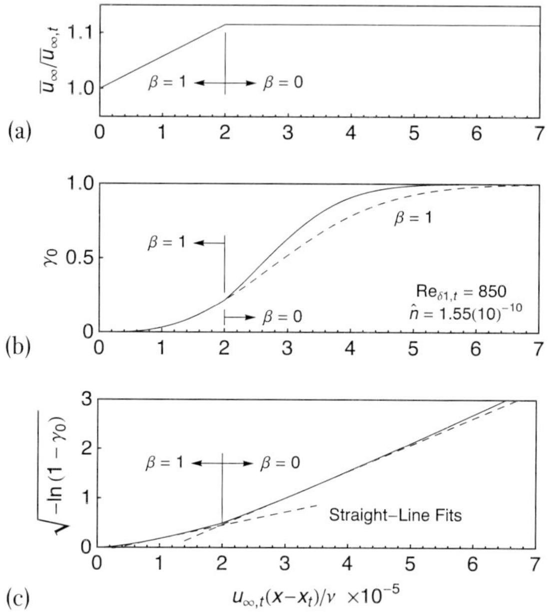

Figure 7.14a,b,c.
Calculated distributions of the (*a*) free-stream velocity, (*b*) near-wall intermittency, and (*c*) Narasimha's function, for a change from a favorable to zero pressure gradient through transition with $Re_{\delta 1,t}$ = 850 and $\hat{n}$ = 1.55(10)$^{-10}$.

The subsequent increase in the intermittency with the decrease in the pressure gradient as shown here in Figs. 7.13 and 7.14 qualitatively agrees with the measurements of Narasimha [1984, 1985] and Narasimha et al. [1984]. In their experiments, they found that the data fell on two, rather than one straight line when plotted as the square root of $-\ln(1-\gamma_0)$ against distance (see §6.1.2, Fig. 6.1) with the slope of the line nearest the onset of transition smaller than that farther downstream. A plot of this function for our second example is provided in part (c) of Fig. 7.14 and, as found by Narasimha and his coworkers, it can also be approximated by two straight lines. The intersection of these lines occurs about where the pressure gradient and spot-expansion rates change. Narasimha views this change in behavior as a "subtransition" and ascribes it to a change in the stability of the surrounding laminar boundary layer. While definitely not associated with another sort of transition, it is definitely caused by a change in the spot-expansion rates which in turn depends on the stability of the surrounding boundary layer. In fact, as discussed in §5.1.1 and shown in Fig. 7.12b, the rate of growth dramatically increases as the laminar boundary layer becomes less stable. So far, however, a definite connection has yet to be made.

7.3.2 Intermittency transport models.

Prescribed production models. Turbulence transport models.

As implied by their name, intermittency-transport models use a transport equation for the intermittency. In general, these models can be divided into two groups – those that use a prescribed near-wall intermittency distribution, generally based on Emmons' turbulent-spot theory, which we will call the "prescribed production models," and those that use the standard transport equation with the intermittency production and dissipation terms developed specifically for the turbulence caused by the spots, which we will call the "turbulent production models." While the former group of models depend on correlations for the onset of transition and production of turbulent spots (usually similar to those presented in §6.2 and §6.3), the latter models generally use a correlation for the onset of transition and a variety of model constants to correlate intermittency data. However, since both are relatively easy to incorporate into most numerical computational schemes, either steady or unsteady, and allow modeling of unbounded intermittent flows such as those encountered in separation bubbles and confluent wakes, they have recently become the vogue in transitional flow modeling.

With the oldest dating from the 1980s, intermittency transport models are relatively new. Generally based on the standard transport equation, they take the form

$$\rho\bar{u}_i\frac{\partial\gamma}{\partial x_i}=\rho P_\gamma+\frac{\partial}{\partial x_i}\left[\left(\mu+\mu_{t,\gamma}\right)\frac{\partial\gamma}{\partial x_i}\right]+\rho D_\gamma \tag{7.79}$$

where P_γ and D_γ are the production and dissipation of intermittency, and $\mu_{t,\gamma}$ is the eddy viscosity for the turbulent diffusion of the intermittency. Expressions for P_γ and D_γ vary according to the investigator with some more complicated than others, and some more suitable for one computational scheme than another. Since most of the models are "tuned" to the equations being solved (global or conditional, steady or unsteady) and the turbulence model being used, the reader is strongly advised to examine the papers referenced in the following discussion for details. Nevertheless, it should be kept in mind that these models must account for the effects of the major independent variables of the flow, such as turbulence level and pressure gradients, on the production of turbulent spots, the becalmed zones, and the growth of the turbulent spots as they propagate downstream. This appears to be even more daunting than correlating these effects for a single turbulent spot and its becalmed zone as proposed in the previous section.[1]

Prescribed production models. Transport equations using a production term that exploits the standard transition correlations are those developed by Steelant and Dick [1996, 2001] where the conditioned equations of motion are solved, and Suzen and Huang [2000], Suzen et al. [2002, 2003], Pecnik et al. [2003], Lodefier et al. [2006], and Piotrowski et al. [2008] where various Reynolds-averaged equations of motion are solved. In these models, the dissipation term in Eq. (7.79) is usually neglected such that the intermittency transport equation for boundary-layer flow becomes

$$\rho\bar{u}\frac{\partial\gamma}{\partial x}+\rho\bar{v}\frac{\partial\gamma}{\partial y}=\rho P_\gamma+\frac{\partial}{\partial y}\left[\left(\mu+\mu_t\right)\frac{\partial\gamma}{\partial y}\right] \tag{7.80}$$

where μ_t is the eddy viscosity.

Since $\partial\gamma/\partial y$ and the normal velocity and its derivative with respect to y are negligibly small near the wall, the above equation near the wall reduces to

$$P_{\gamma,0}=\bar{u}\frac{d\gamma_0}{dx}$$

[1] Frankly, the author is not enamored by any attempt to model transitional flow without considering turbulent spots and the physics associated with them and their motion. Since turbulent spots with their attending becalmed zones are the basic ingredients of transitional flow, any method predicting the flow without considering them is bound to be erroneous, and sometimes significantly so. To be fair, however, of the two following models, the prescribed production model has a greater potential for predicting transitional flow because it at least recognizes that turbulent spots exist.

where $P_{\gamma,0}$ is the *near-wall* production rate.

If we consider transition on a flat plate with a constant free-stream velocity, assume that laminar breakdowns occur within a relatively narrow band, and further assume that the product $\hat{n}\sigma$ is a constant, then the derivative of γ_0 can be obtained from Eq. (7.69). In this case, the near-wall production of intermittency is given by

$$P_{\gamma,0} = 2\bar{u}\sqrt{\hat{n}\sigma}\,(\bar{u}_\infty/\nu)\left(1-\gamma_0\right)\sqrt{-\ln(1-\gamma_0)} \tag{7.81}$$

Although the production term has taken a variety of forms through its development, and is still evolving, P_γ is generally assumed proportional to the right-hand side of the above expression after setting $\gamma_0 = \gamma$. As a result, the production term in Eq. (7.80) becomes

$$P_\gamma = 2F_s\bar{u}\sqrt{\hat{n}\sigma}\,(\bar{u}_\infty/\nu)\left(1-\gamma\right)\sqrt{-\ln(1-\gamma)} \tag{7.82}$$

where the function F_s is used to correct for distributed breakdowns, turning off production in the free stream, and fixing numerical starting problems at the beginning of transition. Furthermore, to account for a variable free-stream velocity, the *local* values of the free-stream velocity and product of the production and propagation parameters are used. That is, they use the correlations provided in §6.2.2 and §6.3.2 (or their equivalent) for the latter, but apply them locally. In addition, some correlations are corrected to include the effect of Mach number. [1]

Steelant and Dick [2001] use the conditioned equations of motion described in §7.2 with Yang and Shih's [1993] low-Reynolds number k-ε turbulence model and two intermittency transport equations – one for the intermittency γ and the other for the free-stream turbulence. On the other hand, Lodefier et al. use the global equations of motion with Menter's [1994] high Reynolds number k-ω turbulence model, where ω is the vorticity or inverse-time-scale of turbulence, and one transport equation for γ. The production term for both is provided by Eq. (7.82) with somewhat different expressions for the function F_s.

For low Mach numbers, Steelant and Dick suggest

$$F_s = 1-\exp\left[-1.74\tan\left(5.45\gamma-0.954\right)-2.2\right]$$

which corrects Eq. (7.82) for a prescribed type of distributed breakdown (see Steelant and Dick [1996]).

To account for a variable pressure gradient, they fit Mayle's [1991] plot of the spot parameter $\hat{n}\sigma$ against the acceleration parameter K. This fit is given by

$$\hat{n}\sigma = \begin{cases} 1.25(10)^{-11}\,Tu^{7/4}(474Tu^{-2.9})^{1-\exp[2(10)^6 K]} & ;K<0 \\ 1.25(10)^{-11(1+293K^{0.598})}\,Tu^{7/4} & ;K\geq 0 \end{cases}$$

where Tu is the *local* free-stream-turbulence level in percent. For zero pressure gradient, $K = 0$ and the above correlation reduces to the value given by Eq. (6.17) using a coefficient that fits the intermittency data best.

To determine the onset of transition, Steelant and Dick use Mayle's original version of Eq. (6.7) that fits the intermittency data best, modified to account for a distributed breakdown. This shifts their onset upstream from that calculated using Blasius' relation by an amount equal to $Re_{\Delta x}/2$. For their F_s function, they determined that $Re_{\Delta x}$ is given by

[1] The effect of Mach number on the onset of transition and spot growth are also examined in the papers by Chen and Thyson [1971], Narasimha [1985], Mayle [1991], and Walters and Leylek [2004].

$$Re_{\Delta x} = 0.744 \big/ \sqrt{(\hat{n}\sigma)_{xt}}$$

where $(\hat{n}\sigma)_{xt}$ is evaluated at the onset of transition.

For low Mach numbers, Lodefier et al. multiplies Steelant and Dick's expression for F_s by the factor

$$\min\left[20\max\left(1.12\frac{S}{2.2Re_{\delta 2,t}}+0.1(1-a)\gamma\frac{\mu_t}{\mu}-1;0\right);1\right]; \quad (0 < a \leq 1)$$

where $Re_{\delta 2,t}$ is determined from Eq. (6.7) using Steelant and Dick's version, and S is the absolute value of the strain rate. For a boundary layer, the latter is given by

$$S = \sqrt{2}\frac{y^2}{\nu}\frac{\partial \bar{u}}{\partial y}$$

which is the square root of two times Driest and Bummer's [1963] vorticity parameter.

The main purpose of their multiplicative factor is to "turn on" transition when the quantity $S/2.2Re_{\delta 2,t}$ reaches unity within the boundary layer and keep it on through transition. In addition, it turns the production term off in the free stream and corrects for the upstream shift in the onset of transition because of a distributed breakdown. Much of this is handled by the min-max makeup of the factor, and the factor a which detects the edge of the boundary layer by switching rapidly from a value near zero within it to a value of unity outside. For an explicit explanation of all these factors and the method of correcting for elevated Mach numbers, the reader is referred to their original paper.

Unlike Steelant and Dick, however, who didn't alter the turbulence model, Lodefier et al. multiplies the eddy viscosity in the equations of motion by γ, and the production terms of the k and ω turbulence equations by

$$\left\{a+(1-a)\left[\gamma+(1-\gamma)\max\left(\mu/\mu_t;0.1\right)\right]\right\}$$

This produces small but nonzero values of turbulence within the boundary layer at the start of transition to let the turbulence grow sufficiently fast in the transition zone without affecting production in the free stream. The dissipation terms in these equations they leave untouched.

Their calculated results are compared to Roach and Brierley's data in Fig. 7.15. The dashed lines in the figures are the results shown in Fig. 7.6 calculated by using the classical linearly-combined model. Clearly, Steelant and Dick's result, which solves two intermittency transport equations, overestimates the effect of free-stream turbulence on the laminar boundary layer and transitions too early, which also explains their higher than measured skin-friction coefficients throughout transition and beyond. On the other hand, Lodefier's et al. pre-transitional results, which are virtually the same as Blasius', underestimates the effect of free-stream turbulence yet transitions about as measured. While this probably explains their lower than measured results through the beginning of transition for the case with the three percent turbulence level, their lower post-transitional result for the highest turbulence level, shown in part (b) of the figure, is probably caused by an error in their production and/or propagation rate model.

In conclusion, if we use the kinematic spot-propagation model presented in the previous section, specifically Eq. (7.72), a more general expression for the near-wall production term than that given by Eq. (7.81) can be determined. Specifically, if we assume that the onset of transition occurs at $x = x_t$, we obtain

$$P_{\gamma,0} = \bar{u}_\infty \hat{n}' f_s (1-\gamma_0)\left\{\frac{w_{tp}}{u_{tp}}\int_{x_t}^{x}\frac{u_{le}-u_{te}}{u_{le}u_{te}}dx' + \frac{u_{le}-u_{te}}{u_{le}u_{te}}\int_{x_t}^{x}\frac{w_{tp}}{u_{tp}}dx'\right\} \tag{7.83}$$

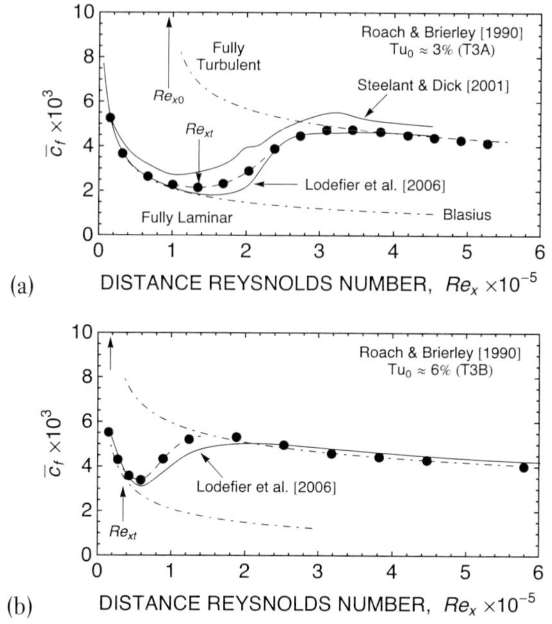

Figure 7.15a,b.
Calculated and measured distributions of the skin-friction coefficient. Data from Roach and Brierley [1990]. Calculations from Steelant and Dick [2001] and Lodefier et al. [2006].

Furthermore, if we consider a distributed breakdown, then the correct expression for the near-wall production term, obtained from Eq. (7.78), is

$$P_{\gamma,0}(x;x_l)=\bar{u}_\infty f_s\left(1-\gamma_0\right)\left\{\frac{w_{tp}}{u_{tp}}\int_0^{x_l}\dot{n}''(\xi)\mathrm{U}(x-\xi)\,d\xi\int_\xi^x\frac{u_{le}-u_{te}}{u_{le}u_{te}}dx'\right.$$
$$\left.+\frac{u_{le}-u_{te}}{u_{le}u_{te}}\int_0^{x_l}\dot{n}''(\xi)\mathrm{U}(x-\xi)\,d\xi\int_\xi^x\frac{w_{tp}}{u_{tp}}dx'\right\} \tag{7.84}$$

where as noted before $\dot{n}''(\xi)$ is the source strength per unit time per unit surface area.

Turbulence production models. Developed initially for free-shear layers, turbulent boundary layers, and then transitional flow, the evolution of these models are described in the papers by Byggstøyl and Kollmann [1981, 1986a, 1986b], Cho and Chung [1990, 1992], Savill [1995, 1996], and Vicedo et al. [2003]. Specifically, Byggstøyl and Kollmann derived an intermittency transport equation and coupled it with the conditionally-averaged Navier-Stokes equations to compute free-shear wakes. Considering this work, Cho and Chung developed a global set of high-Reynolds k-ε-γ equations to predict turbulence in the far wake and various jet flows. Savill extended the approach to calculate boundary-layer transition using a low-Reynolds number turbulence model, while Vicedo et al. examined transition in separated flows by using the globally-averaged Navier-Stokes equations. One advantage of these models is their dependence on locally-determined quantities throughout the flow, which are well suited for modern Navier-Stokes codes. The disadvantages are that they disregard any fundamentals about the production and propagation of turbulent spots and their becalmed zones, and as all other models require correlations for the onset of transition.

Because it is one of the latest, useful for both attached and separated transitional flows, typical, and rather straightforward to describe, we will concentrate on the model proposed by Vicedo's et al. Starting with the diffusion term in Eq. (7.79), they use the version proposed by Byggstøyl and Kollmann [1981], which in turn is based on the velocity-jump model of Lumley [1980]. This yields an intermittency-transport equation of the form

$$\rho\bar{u}\frac{\partial\gamma}{\partial x}+\rho\bar{v}\frac{\partial\gamma}{\partial y}=\rho P_\gamma+\frac{\partial}{\partial y}\left\{\left[\mu+\gamma(1-\gamma)\mu_t\right]\frac{\partial\gamma}{\partial y}\right\}+\rho D_\gamma \tag{7.85}$$

where the expression for μ_t depends on the turbulence model.

For boundary layers, the production term is expressed as the sum

$$P_\gamma=C_{\gamma1}\gamma(1-\gamma)\frac{P_k}{k}+C_{\gamma2}\gamma\nu_t\left(\frac{\partial\gamma}{\partial y}\right)^2-C_{\gamma3}\gamma(1-\gamma)\nu_t\frac{1}{\bar{u}}\frac{\partial\bar{u}}{\partial y}\frac{\partial\gamma}{\partial y}$$

where P_k is the trace of the turbulent-kinetic-energy production tensor P_{ij} given in Eq. (3.45). The terms on the right-hand side of the above equation respectively represent the production of intermittency by the production of turbulent kinetic energy, the transport of mass and momentum caused by spatial inhomogeneity, and the entrainment of intermittency caused by large-scale turbulent motion. The first term is the rate of turbulent energy produced per unit energy, modified to turn production off when the intermittency equals either zero or one. The second term is a modified version of the term proposed by Byggstøyl and Kollmann to account for the increased inhomogeneity of the flow as it becomes turbulent. The third is based on the model of Cho and Chung [1990] to account for the entrainment, which is independent of the fluid viscosity because it depends only on the large-scale turbulent motions.

The dissipation term, again proposed by Byggstøyl and Kollmann, is given by

$$D_\gamma=-C_{\gamma4}\gamma(1-\gamma)\frac{\varepsilon}{k}$$

which, similar to the production term, is the rate of turbulent energy dissipated per unit energy, modified to turn the dissipation off when the intermittency equals either zero or one.

Substituting the above expressions into Eq. (7.85), we obtain

$$\begin{aligned}\rho\bar{u}\frac{\partial\gamma}{\partial x}+\rho\bar{v}\frac{\partial\gamma}{\partial y}&=C_{\gamma1}\rho\gamma(1-\gamma)\frac{P_k}{k}+C_{\gamma2}\gamma\mu_t\left(\frac{\partial\gamma}{\partial y}\right)^2-C_{\gamma3}\gamma(1-\gamma)\mu_t\frac{1}{\bar{u}}\frac{\partial\bar{u}}{\partial y}\frac{\partial\gamma}{\partial y}\\&\quad+\frac{\partial}{\partial y}\left\{\left[\mu+\gamma(1-\gamma)\mu_t\right]\frac{\partial\gamma}{\partial y}\right\}-C_{\gamma4}\rho\gamma(1-\gamma)\frac{\varepsilon}{k}\end{aligned} \tag{7.86}$$

which depends only on local values of the quantities involved. Based on the results of Savill [1995], the constants C_γ in this equation are given by

$$C_{\gamma1}=1.60,\quad C_{\gamma2}=0.15,\quad C_{\gamma3}=0.16,\quad\text{and}\quad C_{\gamma4}=0.10$$

Vicedo et al. used this equation together with a modified version of Yang and Shih's k-ε model to account for the intermittent behavior of the flow through transition.[1] Recalling that they were interested in calculating *separated-flow* transition, they used the author's [1991] correlation for the onset of transition in separated flows. In general, when compared with the experiments performed by Rolls Royce (see Coupland [1995]) for separated-flow transition, their calculations using Mayle's [1991]

[1] Briefly, the major changes to Yang and Shih's model involves forcing the eddy viscosity to be zero in regions inside the boundary layer where the flow is laminar. This is accomplished by multiplying all the source terms in both the turbulence model and intermittency transport equation by the intermittency function, $I(x,y)$.

correlations for separated-flow transition (see §8.3.3) fair better than those that don't use an intermittency-transport equation. For the details concerning these comparisons, the intermittency equation, and the modifications made to the k-ε equations, the reader is referred to their paper.

In any case, the most recent models of this genre are those that also solve a "mock" transport equation for the transition Reynolds number $Re_{\delta 2,t}$. The idea behind this approach is to determine the onset of transition using a *local* transition criterion which is preferred simply as a matter of convenience by some who use the modern Navier-Stokes codes. Despite the fact that there isn't any physical basis for a transition Reynolds number transport equation, a disadvantage of this approach is that the two transport equations must be well tuned.

As an example of their ability to calculate transitional flow, the results from two of these models are compared to Roach and Brierley's data in Fig. 7.16. The calculations by Piotrowski et al. [2008] use an algorithm similar to Menter et al. [2006], but tuned to a more extensive set of correlations. Neither claim to model the physics of the transition process (unlike turbulence models), but attempt to introduce "correlation-based models into general purpose computational fluid dynamic methods." Presumably for propriety reasons, these correlations are not available in their paper; a practice that should *not be allowed* in professional publications. In any case, it doesn't matter. While the calculated results for the three percent turbulence level shown in part (a) of the figure are reasonably good through transition, the results for the six percent data shown in part (b) are exceptionally poor. Apparently, both models were "tuned" to the three percent transition data. Note that the dashed line in part (b) of this figure correspond to the result shown in Fig. 7.6 using the linearly-combined model.

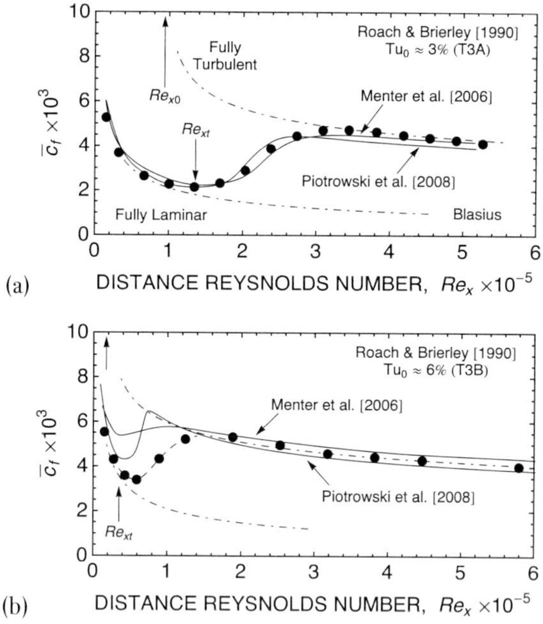

Figure 7.16a,b.

Calculated and measured distributions of the skin-friction coefficient for Roach and Brierley's [1990] (a) T3A and (b) T3B test cases. Calculations from Menter et al. [2006] and Piotrowski et al. [2008].

Unfortunately, the results shown in Figs. 7.15 and 7.16 represent the status quo of calculating transitional flow using an intermittency-transport equation. Although some investigators might disagree, this author believes that without considering turbulent spots this method will never be able to *predict* transitional flow, and is therefore relegated to *calculating* transition only within a very select subset of geometries and flows. Consequently, a turbulent-spot intermittency model, whether cobbled up from single-spot data as described in the previous subsection or obtained from a yet undeveloped physical spot theory, appears to be the only path to a truly predictive transition model.[1]

7.4 Some Typical Errors.

Pre-transitional flow errors. Correlation errors. Turbulent flow errors.

When a turbulent-spot intermittency model together with either the classical or zonal approach is used to calculate transitional flow, errors in calculating any one of the three major components of the model will affect the result. In this section, we will use the Emmons-Narasimha model to assess some of the errors associated with incorrectly calculating the flow before transition, the intermittency distribution, and the flow after transition.

Pre-transitional flow errors. As noted previously, Blasius solution, or its laminar equivalent for a variable free-stream velocity, cannot always be used to calculate pre-transitional laminar flow. This is especially true for elevated turbulence levels, highly curved surfaces, and large surface roughness. Besides yielding incorrect values for the boundary-layer thickness Reynolds numbers and skin-friction coefficient, calculating pre-transitional flow incorrectly also affects the transitional-flow calculation because 1) the laminar component in the transitional region is an extension of the pre-transitional solution, and 2) the correlations or methods used to determine the onset and rate of transition depend on a transition Reynolds number.

To assess the errors associated with incorrectly calculating pre-transitional flow, we will consider Roach and Brierley's [1990] test for $dp/dx = 0$ and a nominal turbulence level of three percent,[2] and evaluate the laminar or pre-transitional component of the flow using both Blasius' solution and the numerical results obtained from the pre-transition calculations described in §4.3.3. We will evaluate the turbulent or post-transitional component using Eqs. (3.40) and (5.45) with $r = 1$ to match the laminar and turbulent boundary-layer thicknesses at onset. Furthermore, we will use the best fit to their intermittency data (Fig. 6.2) using the line-source solution given by Eq. (7.69).

The calculated skin-friction results are shown in Fig. 7.17. The transitional-flow results, which were obtained from Eq. (7.1), are shown by the solid lines bordering the grey error bands. The extensions of the fully-laminar and turbulent results are drawn using dot-dashed lines, while that for the pre-transition result is drawn using a dashed line. The virtual origins of the turbulent boundary layer are shown by the upward pointing arrows, while the transition Reynolds numbers are indicated by the downward pointing arrows.

The calculations were obtained for three different matching conditions. In part (a) of the figure, onset for both pre-transition models was assumed to occur where the distance Reynolds number equaled that obtained by fitting the intermittency measurements, that is, $Re_{xt} = 136{,}000$ according to the results shown in Fig. 6.2. The virtual origin of the turbulent boundary layer in this case is $Re_{x0} = 94{,}000$. In part (b), onset was assumed to occur at the position where the momentum-thickness Reynolds number for both models equals its measured value at $Re_{xt} = 136{,}000$. According to Table 7.1, this occurs where $Re_{\delta 2,t} = 280$. In this case, the transition Reynolds numbers based on distance are then $Re_{xt} = 148{,}000$ and $178{,}000$, while $Re_{x0} = 104{,}000$ and $128{,}000$. Finally, in part (c) of the figure, onset was assumed to occur at the distance where the displacement-thickness Reynolds number for both pre-transition models equals its measured value at $Re_{xt} = 136{,}000$. Again from Table 7.1, this occurs where $Re_{\delta 1,t} = 634$. The corresponding distance Reynolds numbers in this case are $Re_{xt} = 149{,}000$ and $134{,}500$, and $Re_{x0} = 105{,}000$ and $93{,}000$. In all cases, the rate parameter was

[1] My guess here is that all those investigating transitional flow using DNS methods will disagree.

[2] Similar conclusions are also attained by considering their one and six percent turbulence level results.

assumed to be that obtained by fitting the intermittency measurements, that is, $\hat{n}\sigma = 8.8(10)^{-11}$. The effect of an elevated turbulence level on the pre-transitional flow is obvious. Increasing the turbulence level to three percent, increases the skin-friction coefficient. In this case, the calculated increase is slightly greater than that measured, which implies that the near-wall modeling described in §4.3.2 is not quite right. At the onset of transition, the error in the calculated skin-friction coefficient in part (a) of the figure is about seven percent, while the error associated with using Blasius' solution is about minus twelve percent. In other words, the maximum width of the error band is about twenty percent when onset is based on the distance to transition of Re_{xt} = 136,000. When onset is based on the measured momentum thickness at Re_{xt} = 136,000, which occurs at different distances because the momentum thickness for the undisturbed laminar boundary layer grows at a lower rate than the disturbed, the error as shown in part (b) of the figure is significantly larger than that shown in part (a). On the other hand, when onset is based on the measured displacement thickness at Re_{xt} = 136,000, the error as shown in part (c) is again about twenty percent at onset, but now limited to the late pre-transition and early transition regions of the flow.

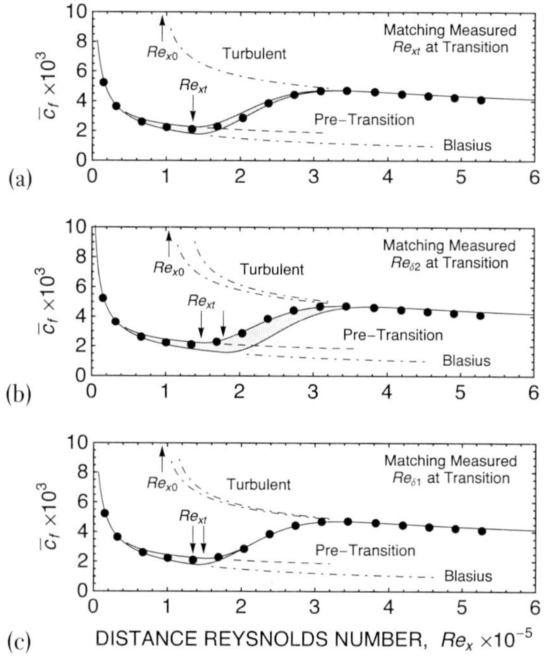

Figure 7.17a,b,c.
Errors associated with calculating pre-transitional flow. Data from Roach and Brierley's [1990] T3A test case with an incident turbulence level of 3 percent.

Two conclusions based on this example are obvious. First, pre-transitional flow must be calculated correctly to accurately predict the onset of transition. Second, if pre-transitional flow is not correctly calculated, a larger error is incurred when onset is assumed to occur where the momentum-thickness Reynolds number equals its measured value at transition than where the displacement-thickness Reynolds number equals its measured value. The latter conclusion follows from the fact, as mentioned in §6.2.2, that the growth in displacement thickness is relatively unaffected by free-stream turbulence compared to the growth in momentum thickness. This result also agrees with the conclusion reached in the same section suggesting that smaller errors are obtained in the correlations when based on the transition displacement-thickness Reynolds number instead of the momentum-thickness Reynolds number.

One conclusion that's not so obvious concerns the correlations. Since many investigators have presumed that pre-transitional flow is synonymous with undisturbed laminar flow and correlated their data accordingly, and since many investigators (including the author) have used laminar flow solutions to convert correlations based on distance, momentum-thickness, and displacement-thickness Reynolds numbers to one based on another, all of the correlations in Chap. 6 are prone to large errors; at least as large as those shown in Fig. 7.17. Unfortunately, most of these errors are irreversible.

Correlation errors. An example of errors associated with the correlations is shown in Fig. 7.18. Again the results of Roach and Brierley are used. Here a comparison between the measured and calculated skin-friction coefficients using two different correlations are shown in part (a) of the figure, while the corresponding intermittency distributions are presented in part (b). The best fit to the measured intermittency distribution, obtained in §6.1.2, is also plotted in part (b).

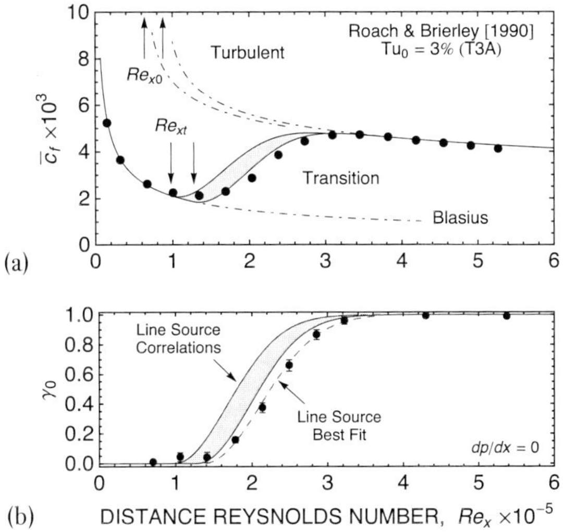

Figure 7.18a,b.
Errors associated with using present-day transition correlations for elevated free-stream turbulence levels. Data from Roach and Brierley [1990].

The calculated results, plotted as the solid lines, were obtained using the Emmons-Narasimha model with the pre- and post-transition components determined as described in the previous section using Blasius' solution for the former. In this case, however, the intermittency was determined from Eq. (7.69) using the correlations given by Eqs. (6.7) and (6.8) to obtain Re_{xt}, and Eq. (6.17) for $\hat{n}\sigma$. Here we should recall that these correlations represent the best fit to the intermittency data alone, and the best fit to both the intermittency and minimum wall-shear-stress data. Using the nominal free-stream turbulence level, these correlations yield Re_{xt} = 98,000 and 128,000 respectively, and $\hat{n}\sigma$ = $1.03(10)^{-10}$. The corresponding virtual origins of the turbulent boundary layer are Re_{x0} = 64,000 and 88,000. Using the free-stream turbulence level at transition of 1.7 percent, shifts the predicted transition Reynolds numbers to Re_{xt} = 200,000 and 260,000, or roughly twice the measured result, which are obviously incorrect.

Unfortunately, the results shown in Fig. 7.18, being about the same as those that would be obtained by using any of the other correlations mentioned in §6.2 and §6.3, are typical of our present inability even after a half century of research to correlate, calculate, or predict the onset of transition, and consequently transitional flow. In most cases, errors between the calculated and measured distributions of this magnitude are completely unacceptable, particularly for flows with a variable free-stream velocity where the flow might transition before it separates. Nonetheless, it is noteworthy that the results shown in this figure also explain why many investigators compare their transition calculations to data only *after* matching their calculated transition Reynolds number to that measured. In the present set of comparisons, this technique would provide the comparison shown in Fig. 7.17a, which is obviously better than that shown in Fig. 7.18a.

Turbulent-flow errors. Although calculating the turbulent component of the model appears straightforward, it isn't. This is particularly true for moderate-to-high turbulence levels where transition lengths are short, the spots are immature, and the momentum-thickness Reynolds numbers are low. Nevertheless, an example of the errors associated with incorrectly calculating the turbulent component of the flow are presented in Fig. 7.19. The data in this example were obtained by Wang and Zhou [1998] for flow over a flat plate with a zero pressure gradient and a turbulence level of about 6.4 percent.

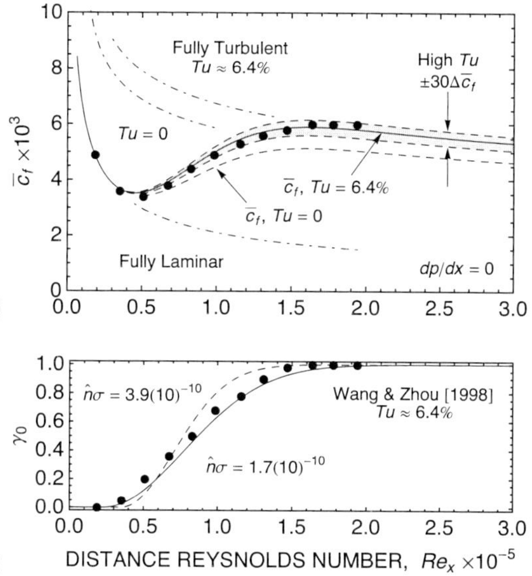

Figure 7.19a,b.
The effect of incorrectly calculating the turbulent component on the skin-friction coefficient for flow on a flat plate with an elevated free-stream turbulence level. Data from Wang and Zhou [1998].

The measured and calculated time-averaged skin-friction coefficients are plotted in part (a) of the figure. The corresponding intermittency distributions are shown in part (b). The transition calculation (solid line) was again obtained using Eqs. (7.1) and (7.69). Since the transition Reynolds number is so low and the turbulent length scale is small, the laminar component of the skin-friction coefficients using either Blasius' solution or the pre-transitional analysis for an elevated turbulence level are nearly the same. For the turbulent component, Eqs. (3.40) and (5.45) were again used but corrected for the effect of free-stream turbulent intensity and length scale using Blair's [1983] correlation.[1] Because the turbulent length scale is relatively small in this test, the effect of free-stream turbulence on the developing turbulent boundary layer increases the turbulent skin-friction coefficient by about fifteen percent. This contrasts with the increase of less than one percent in Roach and Brierley's T3B test (see Fig. 7.6) where the length scale was relatively large for roughly the same turbulence level; a

[1] A k–ε or Reynolds-stress analysis could also have been used to calculate this component.

result of placing the turbulence grid in the test section downstream of the contraction nozzle and relatively far upstream of the test plate, rather than upstream of the nozzle in front of the test plate. Nevertheless, the estimated error associated with the turbulent flow calculation is about ±30 times the increase in skin-friction coefficient *above* the value for $Tu = 0$, which was obtained from Eq. (3.40). This error is shown by the gray area in Fig. 7.19a. Clearly, the error would be much greater if the effect of free-stream turbulence on the turbulent boundary layer was not taken into account. In either case, the effect of the post-transitional calculations is limited to the flow downstream of the onset of transition as should be expected.

Again, the Emmons-Narasimha or linearly-combined model is sufficiently accurate to calculate transitional flow as long as the onset and rate of transition are known. In this case, $Re_{xt} = 23{,}000$ and $\hat{n}\sigma = 1.7(10)^{-10}$ provide the best fit to the data through transition. The corresponding value for Re_{x0} is 9,200. Since Blasius' solution just happens to be a good approximation for the laminar flow in this case, we obtain $Re_{\delta 2,t} \approx 100$. According to the correlations given in Eqs. (6.7) and (6.17), however, $Re_{\delta 2,t} = 125$ and $\hat{n}\sigma = 3.9(10)^{-10}$. That is, the correlations provide $Re_{xt} = 36{,}000$, which is one-and-a-half times greater than that for the best fit, and a value for $\hat{n}\sigma$ that is about twice as large. The intermittency distribution using these values is shown by the dashed line in part (b) of Fig. 7.19. The corresponding skin-friction-coefficient distribution just happens to lie close to the upper bound of the gray area in part (a) of the figure and therefore is not shown.

References

Antonia, R.A., 1972, "Conditionally Sampled Measurements Near the Outer Edge of a Turbulent Boundary Layer," J. Fluid Mech., 56, pp. 1–18.

Blair, M.F., 1983, "Influence of Free-Stream Turbulence On Turbulent Boundary Layer Heat Transfer and Mean Profile Development: Part 1–Experimental Data, and Part 2–Analysis of Results," J. Heat Transfer, 105, pp. 33–47.

Blair, M.F., 1992, "Boundary Layer Transition in Accelerating Flows with Intense Free-Stream Turbulence: Part 1–Disturbances Upstream of Transition Onset, and Part 2–The Zone of Intermittent Turbulence," J. Fluids Engng., 114, pp. 313–332.

Blair, M.F., and Anderson, O.L., 1987, "Study of the Structure of Turbulence in Accelerating Transitional Boundary Layers," United Technologies Research Center Report R87-956900-1.

Byggstøyl, S., and Kollmann W., 1981, "Closure Model for Intermittent Turbulent Flows," Int. J. Heat Mass Transfer, 24, pp. 1811–1822.

Byggstøyl, S., and Kollmann W., 1986a, "Stress Transport in the Rotational and Irrotational Zones of Turbulent Shear Flows," Phys. Fluids, 29, pp. 1423–1429.

Byggstøyl, S., and Kollmann W., 1986b, "A Closure Model for Conditioned Stress Equations and its Application for Turbulent Shear Flows," Phys. Fluids, 29, pp. 1430–1440.

Cebeci, T., and Smith, A.M.O., 1974, Analysis of Turbulent Boundary-Layers, Academic Press, New York.

Chen, K.K., and Thyson, N.A., 1971, "Extension of Emmons' Spot Theory to Flows on Blunt Bodies," AIAA Journal, 9, pp. 821–825.

Chen, W.A., Lien, F.S., and Leschziner, M.A., 1998, "Non-Linear Eddy Viscosity Modelling of Transitional Boundary Layer Pertinent to Turbomachinery Aerodynamics," Int. J. Heat and Fluid Flow," 19, pp. 297–306.

Cho, J.T., and Chung, M.K., 1990, "Intermittency Transport Modeling Based on Interactions Between Intermittency and Mean Velocity Gradients," Engineering Turbulence Modelling and Experiments (eds., W. Rodi and E. Ganic), Elsevier.

Cho, J.T., and Chung, M.K., 1992, "A k–ε–γ Equation Turbulence Model," J. Fluid Mech., 237, pp. 301–322.

Clauser, F.H., 1956, "The Turbulent Boundary Layer," Adv. Appl. Mech., 4, pp. 1–51.

Corrsin, S., and Kistler, A.L., 1955, "Free-Stream Boundaries of Turbulent Flows," NACA TR 1244; also 1954, NACA TN 3133.

Coupland, J., 1995, "Transition Modelling for Turbomachinery Flows," ERCOFTAC Bulletin, 24, pp. 5–8.

Craft, T., Launder, B.E., and Suga, K., 1997, "Prediction of Turbulent Transition Phenomena with a Non-Linear Eddy-Viscosity Model," Int. J. Heat and Fluid Flow," 18, pp. 15–28.

Crawford, M.E., and Kays, W.M., 1976, "STAN-5 – A Program for Numerical Computation of Two-Dimensional Internal and External Boundary-Layer Flows," NASA CR-2742.

Dhawan, S., and Narasimha, R., 1958, "Some Properties of Boundary Layer Flow During Transition from Laminar to Turbulent Motion," J. Fluid Mech., 3, pp. 418–436.

Driest, E.R. van, and Bummer, C.B., 1963, "Boundary Layer Transition: Freestream Turbulence and Pressure Gradient Effects," AIAA Journal, 1, pp. 1303–1306.

Emmons, H.W., 1951, "The Laminar-Turbulent Transition in a Boundary Layer, Part I," J. Aero. Sci., 18, pp. 490–498.

Emmons, H.W., and Bryson, A.E., 1952, "The Laminar-Turbulent Transition in a Boundary Layer, Part II," Proc. 1st U.S. Natl. Cong. Theor. Appl. Mech., pp. 859–868.

Emmons, H.W., 1990, personal communication.

Erdélyi, A., 1962, Operational Calculus and Generalized Functions, Holt, Rinehart, and Winston, New York; also 2013, Dover.

Gebers, F., 1908, "Ein Beitrag zur experimentellen Ermittlung des Wasserwiderstandes gegen bewegte Körper," Schiffbau, 9, pp. 435–452 and pp. 475–485; also 1920/21, "Das Ähnlichkeitsgesetz für den Flächenwiderstand in Wasser geradlinig fortbewegter polierter Platten," Schiffbau, 22, pp. 689–930.

Gostelow, J.P., Blunden, A.R., and Walker, G.J., 1994, "Effects of Free-Stream Turbulence and Adverse Pressure Gradients on Boundary Layer Transition," J. Turbomachinery, 116, pp. 392–404.

Gostelow, J.P., Melwani, N., and Walker, G.J., 1996, "Effects of Streamwise Pressure Gradient on Turbulent Spot Development," J. Turbomachinery, 118, pp. 737–743.

Hadzić, I., and Hanjalić, K., 2000, "Separation-Induced Transition to Turbulence: Second-Moment Closure Modeling," Flow, Turbulence and Combustion, 63, pp. 153–173.

Hedley, T.B., and Keffer, J.F., 1974, "Turbulent/Non-Turbulent Decisions in an Intermittent Flow," J. Fluid Mech., 64, pp. 625–644.

Katz, Y., Seifert, A., and Wygnanski, I., 1990, "On the Evolution of a Turbulent Spot in a Laminar Boundary Layer in a Favorable Pressure Gradient," J. Fluid Mech., 221, pp. 1–22.

Kim, J., Simon, T.W., and Kestoras, M., 1989, "Fluid Mechanics and Heat Transfer Measurements in Transitional Boundary Layers Conditionally Sampled on Intermittency," Heat Transfer in Convective Flows, ASME Publication HTD-107, pp. 69–81.

Kovasznay, L.S.G., Kibens, C., and Blackwelder, R.F., 1970, "Large-Scale Motion in the Intermittent Region of a Turbulent Boundary Layer," J. Fluid Mech., 41, pp. 283–325.

Launder, B.E., and Spalding, B.D., 1972, Mathematical Models of Turbulence, Academic Press, New York.

Libby, P.A., 1975, "On the Prediction of Intermittent Turbulent Flows," J. Fluid Mech., 68, pp.

273–295.

Lodefier, K., Merci, B., De Langhe, C., and Dick, E., 2006, "Intermittency Based RANS Bypass Transition Modelling," Progress in Computational Fluid Dynamics, 6, pp. 68–78.

Lumley, J.L., 1980, "Second Order Modelling of Turbulent Flows," Prediction Methods for Turbulent Flows (ed., W. Kollmann), 1, Hemisphere.

Mayle, R.E., 1991, "The Role of Laminar-Turbulent Transition in Gas Turbine Engines," J. Turbomachinery, 113, pp. 509–537.

McDonald, H., and Fish, R.W., 1973, "Practical Calculations of Transitional Boundary Layers," Int. J. Heat and Mass Transfer, 16, pp. 1729–1744.

Menter, F.R., 1994, "Two-Equations Eddy-Viscosity Turbulence Models for Engineering Applications," AIAA Journal, 32, pp. 1598–1605.

Menter F.R., Langtry R.B., Likki S.R., Suzen Y.B., Huang P.G., Völker S., 2006, "A Correlation-Based Transition Model Using Local Variables: Part I–Model Formation," J. Turbomachinery, 128, pp. 413–422.

Narasimha, R., 1957, "On the Distribution of Intermittency in the Transition Region of a Boundary Layer," J. Aero. Sci., 24, pp. 711–712.

Narasimha, R., 1984, "Subtransitions in the Transition Zone," IUTAM Symposium on Laminar-Turbulent Transition (ed., V.V. Kozlov), Springer-Verlag, Berlin, pp. 140–151.

Narasimha, R., 1985, "The Laminar-Turbulent Transition Zone in the Boundary Layer," Progress Aerospace Sci., 22, pp. 29–80.

Narasimha, R., 1990, "Modeling the Transitional Boundary Layer," NASA CR 187487.

Narasimha, R., Devasia, K.J., Gururani, G., and Badri Narayanan, M.A., 1984, "Transitional Intermittency in Boundary Layers Subjected to Pressure Gradient," Experiments in Fluids, 2, pp. 171–176.

Narasimha, R. and Dey, J., 1989, "Transition-Zone Models for Two-Dimensional Boundary Layers: A Review," Sadhana, 14, pp. 93–120; also 1990, Narasimha, R., NASA CR 187487.

Pecnik, R., Sanz, W., Geher, A., and Woisetschläger, J., 2003, "Transition Modeling Using Two Different Intermittency Transport Equations," Flow, Turbulence and Combustion, 70, pp. 299–323.

Piotrowski, W., Elsner, W., and Drobniak, S., 2008, "Transition Prediction on Turbine Blade Profile with Intermittency Transport Equation," ASME Paper GT–2008–50796.

Prandtl, L., and Schlichting, H., 1934, "Das Widerstandsgesetz rauher Platten," Werft, Reederei, Hafen, 15, pp. 1–4.

Roach, P.E., and Brierley, D.H., 1990, "The Influence of a Turbulent Free-Stream on Zero Pressure Gradient Transitional Boundary Layer Development Including the T3A & T3B Test Case Conditions," Proc. 1st ERCOFTAC Workshop on Numerical Simulation of Unsteady Flows, Transition to Turbulence and Combustion, Lausanne; also 1992, Numerical Simulation of Unsteady Flows and Transition to Turbulence, eds., O. Perinea et al.), Cambridge University Press, pp. 319–347, and 1993, data transmitted by J. Coupland from the Rolls-Royce Applied Science Laboratory.

Savill, A.M., 1992, "A Synthesis of T3 Test Cases Predictions," Numerical Simulation of Unsteady Flows and Transition to Turbulence (eds., O. Perinea et al.), Cambridge University Press, pp. 404–442.

Savill, A.M., 1995, "The Savill-Launder-Younis (SLY) RST Intermittency Model for Predicting Transition," ERCOFTAC Bulletin No. 24, pp. 37–41.

Savill, A.M., 1996, "One-Point Closures Applied to Transition," Turbulence and Transition Modeling (eds., M. Hallback et al.), Kluwer, London, pp. 233–268.

Schubauer, G.B., and Klebanoff, P.S., 1955, "Contribution to the Mechanism of Boundary-Layer Transition," NACA TN 3489.

Seifert, A., and Wygnanski, I.J., 1995, "On Turbulent Spots in a Laminar Boundary Layer Subjected to a Self-Similar Adverse Pressure Gradient," J. Fluid Mech., 296, pp. 185–209.

Solomon, W.J., Walker, G.J., and Gostelow, J.P., 1996, "Transition Length Prediction for Flows with Rapidly Changing Pressure Gradients," J. Turbomachinery, 118, pp. 744–751.

Steelant, J., and Dick, E., 1996, "Modeling of Bypass Transition with Conditioned Navier-Stokes Equations Coupled to an Intermittency Transport Equation," Int'l Jour. Numerical Methods in Fluids, 23, pp. 193–220.

Steelant, J., and Dick, E., 2001, "Modeling of Laminar-Turbulent Transition for High Freestream Turbulence," J. Fluids Engng., 123, pp. 22–30.

Suzen, Y.B., and Huang, P.G., 2000, "Modelling of Flow Transition Using an Intermittency Transport Equation," J. Fluids Engng., 122, pp. 273–284.

Suzen, Y.B., Xiong, G., and Huang, P.G., 2002, "Predictions of Transition Flows in Low-Pressure Turbines Using Intermittency Transport Equation," AIAA Journal, 40, pp. 254–266.

Suzen, Y.B., Huang, P.G., Hultgren, L.S., Ashpis, D.E., 2003, "Predictions of Separated and Transitional Boundary Layers Under Low-Pressure Turbine Airfoil Conditions Using an Intermittency Transport Equation," J. Turbomachinery, 125, pp. 455–464.

Vancoillie, G., 1984, "A Turbulence Model for the Numerical Simulation of Transitional Boundary Layers," Proc. 2nd IUTAM Symp. Laminar-Turbulent Transition (ed., V.V. Kozlov), Springer-Verlag, Berlin, pp. 87–92.

Vancoillie, G., and Dick, E., 1988, "A Turbulence Model for the Numerical Simulation of the Transition Zone in a Boundary Layer," J. Engng. Fluid Mech., 1, pp. 28–49.

Vicedo, J., Vilman, S., Dawes, W.N., and Savill, A.M., 2003, "Intermittency Transport Modeling of Separated Flow Transition," ASME Paper GT2003-38719.

Walker, G.J., Subroto, P.H, and Platzer, M.F., 1988, "Transition Modelling Effects on Viscous/Inviscid Interaction Analysis of Low Reynolds Number Airfoil Flows Involving Laminar Separation Bubbles," ASME Paper 88-GT-32.

Walters, D.K., and Leylek, J.H., 2004, "A New Model for Boundary Layer Transition Using a Single-Point RANS Approach," J. Turbomachinery, 126, pp. 193–202.

Wang, T., and Zhou, D., 1998, "Conditionally Sample Flow and Thermal Behavior of a Transitional at Elevated Free-Stream Turbulence," Int. J. Heat and Fluid Flow, 19, pp. 348–357.

Westin, K.J.A., and Henkes, R.A.W.M., 1997, "Application of Turbulence Models to Bypass Transition," J. Fluids Engng., 119, pp. 859–866.

Wilcox, D.C., 1994, "Simulation of Transition with a Two Equation Turbulence Model," AIAA Journal, 32, pp. 247–255.

Wygnanski, I., Zilberman, M., and Haritonidis, J.H., 1982, "On the Spreading of a Turbulent Spot in the Absence of a Pressure Gradient," J. Fluid Mech., 123, pp. 69–90.

Yang, Z., and Shih, T.H., 1993, "New Time-Scale Based k-ε model for Near-Wall Turbulence," AIAA Journal, 31, pp. 1191–1198; also 1992, "A k, ε Calculation of Transitional Boundary Layers," NASA TM 105604.

Part III

And Beyond

8 The Other Modes of Transition

Wake-induced transition. Multimode transition. Separated-flow transition. Post- and reverse transition.

In this chapter we will consider some special features of laminar-turbulent transition. Considered by most as the cutting edge of transition research for the past decade or two, we will examine the basic concepts associated with *wake-induced*, *multimode*, and *separated-flow* transition in the first three sections. In the fourth, we will simply note the role of turbulent spots in early post-transitional flow, and provide the general rule of thumb for the "onset" of reverse transition in turbulent flow. More introductory than inclusive, the latest developments in these topics will be left to the reader to research. While other issues remain, such as the effects of compressibility, sound, noise, and surface vibrations on pre-transitional and transitional flow, they will not be discussed mainly because of insufficient data.[1]

Some of the important results in this chapter are:

1. Periodic turbulent wakes or disturbances in the free stream can cause an earlier than usual periodic transition to turbulent flow.
2. Transition with periodic turbulent wakes can occur through multiple modes of transition at different locations on the same surface at the same time.
3. Periodic turbulent wakes or disturbances can cause an unsteady formation and collapse of separation bubbles.
4. The role of interacting turbulent spots in the developing turbulent boundary layer of post-transitional flow is virtually unknown.

Beside the references at the end of this chapter, other relevant articles may be found in the Bibliography.

8.1 Wake-Induced Transition.

Mayle-Dullenkopf theory. Comparisons with data. Distributed wake-induced sources. Onset and production rate.

To those working in the field of turbomachinery, the effect of periodic wakes on airfoil losses, heat transfer, and transition is of particular interest. Extensively examined in the last twenty to thirty years, the literature concerning this topic is enormous and for the most part will not be covered. Instead, we will set the groundwork for those unfamiliar with *periodic, wake-induced transition* by considering one of the basic analyses of the phenomenon.

The seminal work on this subject was done by Pfeil and Herbst [1979] who after conducting an innovative set of experiments, qualitatively described the physics generally accepted today as correct for what is now known as *wake-induced transition*. In particular, Pfeil and Herbst and later Pfeil et al. [1983] examined the laminar-turbulent transition on a flat-plate boundary layer periodically disturbed by passing turbulent wakes. After measuring the time-averaged growth of the boundary layer along the plate, they concluded that: 1) passing wakes cause an unsteady boundary-layer transition, 2) the length of transition decreases as the wake frequency increases, and 3) wake-induced transition is essentially a quasi-steady phenomenon. In addition, they determined that when the wakes "impinge" on the plate, they produce a series of turbulent "strips" that span and propagate along the surface *independent* of the wake passing velocity. Furthermore, they pointed out that these strips grow as they move downstream eventually coalescing and making the boundary layer completely turbulent. Although not formally stated, they also implied that any time-averaged quantity within the boundary

[1] The exception here, which we briefly mentioned in previous chapters, is the effect of compressibility on transition.

layer may be obtained by using

$$\tilde{Q}=\left(1-\tilde{\gamma}\right)Q_L+\tilde{\gamma}Q_T \tag{8.1}$$

where Q_L is its laminar value, Q_T is its fully-turbulent value, and $\tilde{\gamma}$ is the time-averaged intermittency. Other experiments examining wake-induced transition without a pressure gradient have been conducted by Liu and Rodi [1991], Funazaki [1996], and Funazaki and Aoyama [2000]. Experiments on the effects of moving wakes on gas-turbine airfoils are numerous. Some of these are referred to herein, while others are listed in the Bibliography.

Theories for unsteady wake-induced transition have been proposed by several investigators; namely, Walker [1974], Doorly [1988], Mayle and Dullenkopf [1989, 1990], Hodson [1990], and Hodson et al. [1992]. Walker described a theory for unsteady transition on compressor blades based on the unsteady formation and collapse of separation bubbles. Doorly proposed an intermittency-based theory for the effect of a passing wake on heat transfer to the suction side of a gas turbine airfoil. In particular, he assumed that the high turbulence content of the wake produces a turbulent strip in an otherwise laminar boundary layer during each passing. To obtain the time-averaged heat flux, he used Eq. (8.1) and an intermittency determined by computing the temporal position of the wake next to the surface. The intermittency, therefore, depended on the *propagation and growth of the wake* around the airfoil. A comparison to measurements showed only a qualitative agreement. The theory proposed by Mayle and Dullenkopf was more inline with the conclusions reached by Pfeil and his co-workers. Considering Emmons' [1951] idea for transition, they assumed that each wake produces turbulent strips that subsequently propagate and grow downstream independently of the wake itself. As a result, their analysis agreed well with Pfeil and Herbst's flat plate data and the data from both large-scale and full-scale turbine airfoil measurements. The early theory proposed by Hodson is similar to Doorly's, while his later is similar to Mayle and Dullenkopf's except that it ignores the mutual exclusivity of turbulent regions. Other wake-induced transition models have been proposed by Funazaki [1996], Funazaki et al. [1997], Halstead et al. [1997], and Schulte and Hodson [1998], the latter of which is similar to that developed by Mayle and Dullenkopf. Numerical models have been proposed by Cho et al. [1993], Chakka and Schobeiri [1999], Michelassi et al. [1999], Schobeiri [2005], and others.[1]

In this section we will examine Mayle and Dullenkopf's theory of wake-induced transition. We will then consider the effect of a wake producing strips distributed over a portion of the surface, and finally discuss the effect of wakes on the onset of transition, and the turbulent-strip production and propagation rates.

8.1.1 Mayle-Dullenkopf Theory.

Time-averaged intermittency. Turbulent strips. The production switch function. Some calculations.

In this and the next few sections, we shall consider transition of a two-dimensional boundary layer on a flat plate with a free-stream flow in the x direction. Furthermore, as sketched in Fig. 8.1, we shall consider that transition is caused by a periodic series of two-dimensional turbulent wakes or disturbances[2] embedded in the free stream such that their "footprints" on the surface resemble a traveling series of parallel spanwise bands. Presuming that these wakes produce periodic regions of turbulence within the boundary layer that propagate and expand in the streamwise direction, eventually making the boundary layer completely turbulent, our task is to determine the time-averaged intermittency as a function of the distance along the surface.

---

[1] Some of these "others" have already been referenced in §7.3.2.

[2] As will become evident, any periodically propagating free-stream disturbance that causes transition, such as a pressure pulse or shock wave, may be treated similarly. See the results of Doorly and Oldfield [1985] for example.

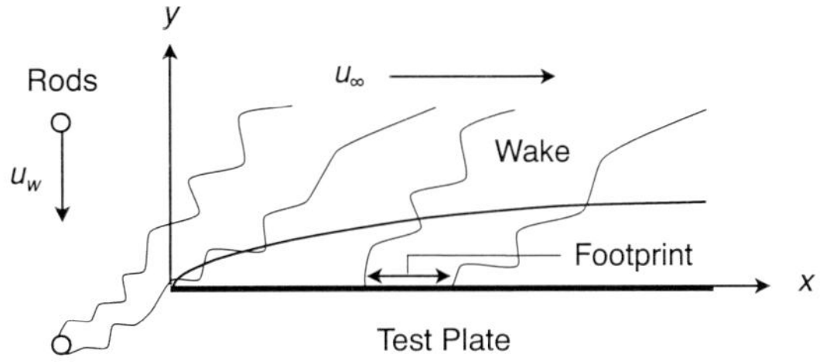

Figure 8.1.
Wakes from moving circular rods passing over a flat test plate.

Following Emmons, first suppose that the disturbances within the wakes cause a random formation of turbulent spots on the surface. Then, according to Eq. (5.12), the intermittency at any point on the surface is given by

$$\gamma(P)=1-\exp\left\{-\int_R g(P')dx'\,dz'\,dt'\right\}$$

where $g(P')$ is the rate of turbulent spots produced per unit surface area at a point P' within the influence volume R. The influence volume, as shown in Fig. 5.17, is defined as the volume of the dependence cone in an xzt-space containing all the upstream sources of turbulent spots that could cause the flow to be turbulent at the point P.

The unsteady effects of the wakes can be taken into account by considering that the production function depends on both *position* and *time*. This is different from the analysis in §5.2.2 where we considered $g(P')$ to be only a function of position. Consequently, the time-averaged intermittency at any point on the surface is obtained by integrating the above expression over one period. Denoting the period of the wakes by τ, this provides

$$\tilde{\gamma}_w(x,z)=\frac{1}{\tau}\int_{t_0}^{t_0+\tau}\gamma(x,z,t)\,dt=1-\frac{1}{\tau}\int_0^{\tau}\exp\left[-\int_R g(x',z',t')dx'\,dz'\,dt'\right]dt \tag{8.2}$$

where the subscript "w" indicates the intermittency caused by wake-induced transition, and t_0 has been arbitrarily set to zero because of the periodicity. Since nothing has yet been said about the source distribution of spots, the time-averaged intermittency may still be a function of z even though both the boundary layer and wakes have been assumed on average to be two dimensional.

The problem is greatly simplified when we suppose the wakes are so intense that all the spots formed by any spanwise portion of a wake coalesce immediately into a spanwise oriented "turbulent strip."[1] The word "immediately" is important here because it renders $g(P)$ and, consequently, the intermittency independent of z. Otherwise, we must solve Eq. (8.2) for a line source of turbulent spots as considered in §5.2.2, or a line of point sources as considered in Appendix G. Physically, "immediately" requires the spots to be produced both quickly and close together. For example, if the wakes move at the free-stream velocity, τ_w is the transit time for the wake over a point on the surface, τ_s is the time it takes a spot to be produced, and d_s is average distance between spots, then turbulent strips will rapidly form only when $\tau_s << \tau_w$ and $d_s << u_\infty\tau_w$. Assuming this is the case, Eq. (8.2) reduces to

$$\tilde{\gamma}_w(x)=1-\frac{1}{\tau}\int_0^{\tau}\exp\left[-\int_A g(x',t')dx'\,dt'\right]dt \tag{8.3}$$

[1] Mathematically, these become lines of turbulence, i.e., strips of zero thickness.

where $g(x', t')$ is now interpreted as the production function for turbulent strips, and A is the influence area in the xt-plane for the point $P'(x', t')$.[1] That is, A is the area containing all the upstream sources of turbulent strips that could cause the flow at the point P to be turbulent. To evaluate Eq. (8.3), we need to determine both A and $g(x', t')$.

First consider the influence area A. If we consider for simplicity that the free stream and turbulent strip-propagation velocities are constant, then the influence area for point P in Fig. 8.2 will appear as the light-gray wedge marked A, while the traces of the *wake* footprints will appear as the medium-gray bands marked B. For wakes traveling at the free-stream velocity, the bands in this distance-time diagram have a slope equal to $1/u_\infty$. The upper and lower slopes of the *influence wedge* are $1/u_{le}$ and $1/u_{te}$, where u_{le} and u_{te} are the leading- and trailing-edge strip-propagation velocities. Measurements[2] show that the leading- and trailing-edge velocities of these strips are about $0.88u_\infty$ and $0.5u_\infty$, which are about the same as those for turbulent spots[3] without a pressure gradient. When their leading- and trailing-edge velocities are a constant fraction of the free-stream velocity, it is always possible to represent their boundaries by straight lines in the xt-plane even when the free-stream velocity varies. This may be done by using the transformed streamwise coordinate $x = u_\infty(x)\int dx'/u(x')$. Figure 8.2 is drawn with this in mind. Nevertheless, the slopes of the wedge boundaries are greater than that of the bands with the consequence that the bands will always cut across the influence area.

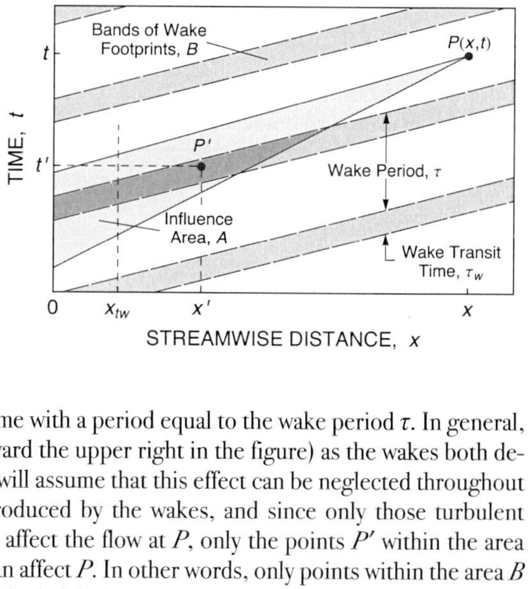

Figure 8.2.
Influence area and bands of wake footprints in the x, t solution plane.

Of course, the bands of footprints repeat in time with a period equal to the wake period τ. In general, these bands will also expand downstream (toward the upper right in the figure) as the wakes both decay and broaden. However, for simplicity, we will assume that this effect can be neglected throughout transition. Since turbulent strips are only produced by the wakes, and since only those turbulent strips produced within the influence area can affect the flow at P, only the points P' within the area common to both the bands B and the area A can affect P. In other words, only points within the area $B \cap A$ can affect P. One of these areas has been shaded dark gray in Fig. 8.2. Since the integration over A in Eq. (8.3) pertains to any x with $0 \le t \le \tau$, the shape and position of these areas change with x and t. This is easily seen by drawing the influence area for another point P in the figure.

Turning our attention to the production function $g(x', t')$, it is far easier and more instructive for us to consider, at least to begin with, that all turbulent strips are produced at the streamwise position x_{tw} as

---

[1] An interesting point to make here is that this analysis is also correct in the time-averaged sense when disturbances cause broken sections of strips or sinuous strips across the span, just as long as they are random.

[2] In particular, Pfeil et al. [1983], Orth [1991, 1992], Schobeiri and Radke[1994], and Halstead et al. [1997].

[3] See §5.1.1 and §5.3.1.

shown in the figure.[1] Then x_{tw} becomes the onset of wake-induced transition. Let $\dot{n}_w$ be the number of strips that can be produced per unit time anywhere on the line x_{tw} and $g^*(t')$ be the fraction of this number actually produced. Then the production function may be expressed as

$$g(x',t')=\dot{n}_w\delta(x'-x_{tw})g^*(t') \tag{8.4}$$

where $\delta(z)$ is the Dirac delta function of z, and $g^*(t')=0$ for $t'\notin B$. Clearly, when P' doesn't lie within a band of footprints on the line $x=x_{tw}$, $g(P')=0$ because either $\delta(x'-x_{tw})$ or $g^*(t')$ equals zero. Since a point downstream of x_{tw} can only become turbulent when the intersection of the bands B and the influence area A include the line $x=x_{tw}$, that is, where $B\cap A\cap x_{tw}$, the distance x' in Eq. (8.3) must equal x_{tw}, and the limits for A in the integral (based on the geometry shown in Fig. 8.2) become

$$t'\geq t-\left(x-x_{tw}\right)/u_{te}$$
$$t'\leq t-\left(x-x_{tw}\right)/u_{le}$$

Hence, upon substituting Eq. (8.4) into Eq. (8.3), we obtain

$$\tilde{\gamma}_w(x;x_{tw})=1-\frac{1}{\tau}\int_0^{\tau}\exp\left[-\dot{n}_w\int_{t-\left(x-x_{tw}\right)/u_{te}}^{t-\left(x-x_{tw}\right)/u_{le}} g^*(t')dt'\right]dt;\quad g^*(t'\notin B)=0 \tag{8.5}$$

where $\tilde{\gamma}_w(x;x_{tw})$ is to be interpreted as the time-averaged, near-wall intermittency at x resulting from periodically-induced transition beginning at x_{tw}.

Without any experimental information about the wake-induced production function $g^*(t')$, we will simply assume that the wakes turn the production of turbulent strips "on-and-off" like a switch as they pass over x_{tw}. This is also in line with the assumption we previously made about the wakes being "intense." For our periodic wakes then, where τ is their period and τ_w is their transit time, $g^*(t')$ is given by

$$g^*\left(t'\right)=\begin{cases}0; & t'\notin B\rightarrow\left(n\tau+\tau_w\leq t'\leq(n+1)\tau\right)\\ 1; & t'\in B\rightarrow\left(n\tau\leq t'\leq n\tau+\tau_w\right)\end{cases};\quad (n=\ldots,-2,-1,0,1,\ldots)$$

which is expressed more compactly as

$$g^*(t')=\sum_{n=-\infty}^{\infty}\left[\mathrm{U}(t'-n\tau)-\mathrm{U}(t'-\tau_w-n\tau)\right] \tag{8.6}$$

where $\mathrm{U}(z)$ is the unit step function of z. Substituting Eq. (8.6) into Eq. (8.5) and realizing that the order of integration and summation may be interchanged, yields

$$\tilde{\gamma}_w(x;x_{tw})=1-\frac{1}{\tau}\int_0^{\tau}\exp\left[-\dot{n}_w\sum_{n=-\infty}^{\infty}\left\{\int_{t-(x-x_{tw})/u_{te}}^{t-(x-x_{tw})/u_{le}}\left[\mathrm{U}(t'-n\tau)-\mathrm{U}(t'-\tau_w-n\tau)\right]dt'\right\}\right]dt \tag{8.7}$$

Defining the integral within the sum as $G_n(x,t;x_{tw})$, the above may be written as

$$\tilde{\gamma}_w(x;x_{tw})=1-\frac{1}{\tau}\int_0^{\tau}\exp\left[-\dot{n}_w\sum_{n=-\infty}^{\infty}G_n(x,t;x_{tw})\right]dt \tag{8.8}$$

[1] As we found in §5.2.1, we may then use the intermittency solution for this simple problem to construct solutions for more the difficult wake-induced transition problems that include transition with distributed sources of turbulent strips and becalmed zones. This feature will be examined further in §8.1.3, §8.2.2 and §8.2.3.

where, by noting that the integral of U(z) with respect to z equals zU(z), we obtain

$$\begin{aligned}G_n(x,t;x_{tw}) = \Big\{ & \big[t-(x-x_{tw})/u_{le}-n\tau\big]\mathrm{U}\big(t-(x-x_{tw})/u_{le}-n\tau\big) \\ & -\big[t-(x-x_{tw})/u_{le}-\tau_w-n\tau\big]\mathrm{U}\big(t-(x-x_{tw})/u_{le}-\tau_w-n\tau\big) \\ & -\big[t-(x-x_{tw})/u_{te}-n\tau\big]\mathrm{U}\big(t-(x-x_{tw})/u_{te}-n\tau\big) \\ & +\big[t-(x-x_{tw})/u_{te}-\tau_w-n\tau\big]\mathrm{U}\big(t-(x-x_{tw})/u_{te}-\tau_w-n\tau\big)\Big\}\end{aligned} \tag{8.9}$$

This equation, which was first obtained by Mayle and Dullenkopf [1989], provides the time-averaged intermittency for a periodically-induced, two-dimensional transition.

Evaluating Eq. (8.8) is numerically straightforward as long as one considers enough terms in the summation to ensure that the integrand is periodic over the period being considered.[1] If we choose the period τ as the appropriate time scale for the problem, then the obvious dimensionless distance is $(x-x_{tw})/u_\infty\tau$. In addition, upon examining Eqs. (8.8) and (8.9), the dimensionless parameters of the problem become either τ_w/τ, $\dot{n}_w\tau$, u_∞/u_{le}, and u_∞/u_{te}, the most obvious set, or some combination thereof.

Two asymptotic solutions for Eq. (8.8) are known. When $\tau_w = \tau$, the equation for $\tilde{\gamma}_w(x;x_{tw})$ is easily evaluated. In this case, $g^*(t') = 1$ and Eq. (8.5) immediately yields

$$\tilde{\gamma}_w(x;x_{tw}) = 1-\exp\left[-\dot{n}_w\sigma_w\frac{(x-x_{tw})}{u_\infty}\mathrm{U}(x-x_{tw})\right]; \quad (\tau_w/\tau\approx 1) \tag{8.10}$$

where

$$\sigma_w = u_\infty\left(u_{le}-u_{te}\right)/u_{le}u_{te}$$

is the *strip-propagation* parameter. For the measured values $u_{te}/u_\infty = 0.5$, and $u_{le}/u_\infty = 0.88$, we obtain $\sigma_w = 0.86$. This solution was obtained by Narasimha [1985] for a *continuous* disturbance that randomly produces turbulent strips (not spots) at $x = x_{tw}$. In this case, the characteristic streamwise length is $u_\infty/\dot{n}_w$.

The other asymptotic solution was obtained by Mayle and Dullenkopf in their original paper. Specifically, they found a solution for $\tilde{\gamma}_w(x;x_{tw})$ when $\dot{n}_w\tau_w << 1$, that is, when a small number of turbulent strips are produced during any one period. Without going into any details of their analysis, they obtained

$$\tilde{\gamma}_w(x;x_{tw}) \approx 1-\exp\left[-\hat{n}_w\sigma_w\frac{(x-x_{tw})}{u_\infty\tau}\mathrm{U}(x-x_{tw})\right]; \quad (\hat{n}_w << 1) \tag{8.11}$$

where

$$\hat{n}_w = \dot{n}_w\tau_w \tag{8.12}$$

is the *strip-production* parameter, or the total number of strips produced on average with each passing wake. As might be expected, the intermittency increases with the distance from onset, strip-propagation parameter σ_w, strip-production parameter $\hat{n}_w$, and decreases as distance between the wakes $u_\infty\tau$ increases; the latter of which is the characteristic length of the problem. Furthermore, in this solution, $(\hat{n}\sigma)_w$ is the wake-induced turbulent "strip parameter" equivalent of the parameter $\hat{n}\sigma$

[1] For the examples to follow, the integrand becomes periodic over the interval $0 \le t \le 1$ when $-11 \le n \le 1$. For solutions farther downstream, the range of n must be extended at its lower limit to include even "earlier" wakes.

for turbulent spots. Finally, we note that Eqs. (8.10) and (8.11) become identical when $\tau_w = \tau$.

Some calculations. Some calculated wake-induced transition results are presented in the next two figures. In Fig. 8.3, the results from Eq. (8.8) are plotted against the dimensionless streamwise distance $(x-x_{tw})/u_\infty\tau$ for different values of the strip-production parameter $\hat{n}_w$ with $\tau_w/\tau = 0.2$, $u_{te}/u_\infty = 0.5$, and $u_{le}/u_\infty = 0.9$. In Fig. 8.4, the results are plotted for several values of τ_w/τ, that is, several wake transit ratios, with $\hat{n}_w = 2$, $u_{te}/u_\infty = 0.5$, and $u_{le}/u_\infty = 0.9$. As a result, the curve for $\hat{n}_w = 2$ in the first figure is the same as the $\tau_w/\tau = 0.2$ curve (that farthest to the right) in the second. In contrast to the results shown in Fig. 5.18 for transition caused by line source of turbulent spots, the intermittency for transition from a line source of wake-induced turbulent strips increases immediately from onset. This is because the wake-induced turbulent *spots* were considered to coalesce immediately into strips, whereas spots in reality must first grow before they coalesce.

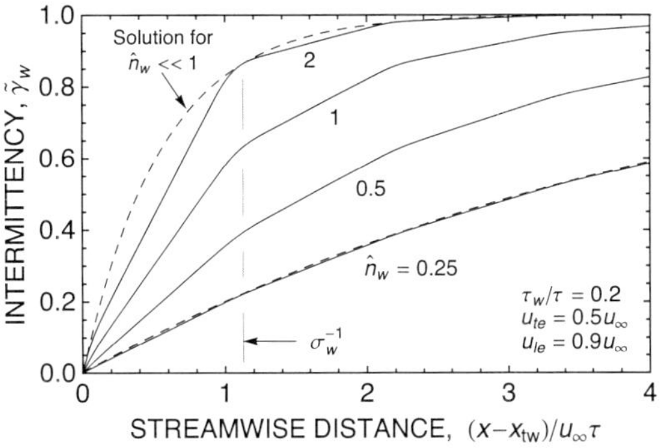

Figure 8.3.
Time-averaged intermittency for several turbulent strip production rates with $\tau_w/\tau = 1/5$, $u_{te}/u_\infty = 0.5$, and $u_{le}/u_\infty = 0.9$.

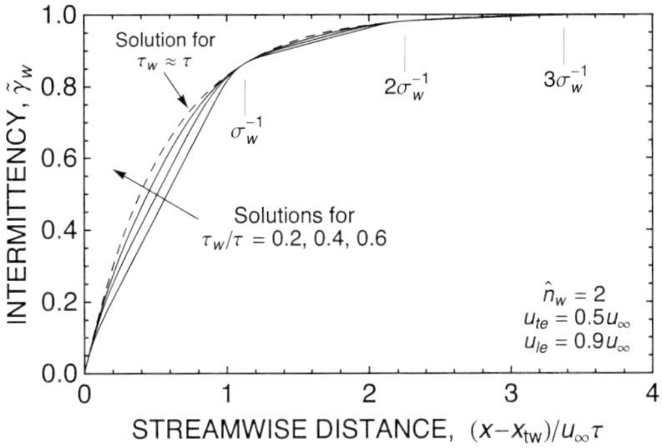

Figure 8.4.
Time-averaged intermittency for several wake transit ratios with $\hat{n}_w = 2$, $u_{te}/u_\infty = 0.5$, and $u_{le}/u_\infty = 0.9$.

Nevertheless, as expected, increasing the strip production rate decreases the transition length. Perhaps a result not expected is that the solution consists of a series of "curved kinks" and "straight-line" segments. The kinks occur where "patches" of turbulent strips within adjacent wake footprints overlap. From geometry it is easy to show that these occur at streamwise intervals of $1/\sigma_w$, which for these figures equals 9/8. For large production rates and relatively small wake transit times τ_w, the distribution simply becomes a series of connected straight lines. The initial curvature of the solution at $x = x_{tw}$, which is more clearly seen in Fig. 8.4, and the "roundness" at the kinks are the result of the individual turbulent strips coalescing within each patch of strips. This also accounts for the curves becoming smoother in Fig. 8.4 as τ_w increases, that is, as the time available for the production of strips within each wake passage increases.

The two dashed curves in Fig. 8.3 are the results obtained from Eq. (8.11) for the two extreme values of $\hat{n}_w$. For $\hat{n}_w = 0.25$, the asymptotic solution is virtually the same as the exact. Thus, the value $\hat{n}_w = 0.25$ is apparently small enough for Eq. (8.11) to be useful. This observation is independent of any other parameter as long as the quantity $\hat{n}_w$ is small. For the other extreme, that is, when $\hat{n}_w = 2$ (which is definitely not small), it just so happens that Mayle and Dullenkopf's solution provides the locus of all the "kinks" in the exact solution and hence provides an upper bound of all solutions to Eq. (8.8). Thus, Eq. (8.11) is not only the asymptotic solution to Eq. (8.8) when $\hat{n}_w << 1$, but also the upper bound to *all* solutions for any value of τ_w/τ, u_∞/u_{le}, and u_∞/u_{te}. This feature allowed Mayle and Dullenkopf to correlate all the data with $\hat{n}_w$ greater than two in their first paper. Finally, the dashed line in Fig. 8.4 is the asymptotic solution to Eq. (8.8) for $\tau_w/\tau \approx 1$ given by Eq. (8.10). As pointed out earlier, it also corresponds to the solution given by Mayle and Dullenkopf for $\tau_w/\tau \approx 1$. Nevertheless, its asymptotic nature for increasing wake transit ratios is quite apparent.

8.1.2 Comparison with Data.

In their experiment, Pfeil and Herbst [1979] measured the unsteady and time-averaged velocity components in the flow along a plate positioned downstream of a large rotating cylindrical-cage of circular wake-generating rods aligned parallel to the leading edge of the plate. The time-averaged shape factor distributions obtained by them are plotted in Fig. 8.5 for a cage with different number of rods. The data were obtained for a value of u_∞/ω equal to 0.40m, where ω is the angular frequency of the cage. Thus the streamwise distance between wakes is $u_\infty\tau = 2.52/n_r$ meters, where n_r is the number of rods in the generator. Without the generator, the flow over the plate is laminar. As the number of rods (wakes) increase, the flow transitions to turbulent as more of the plate becomes covered by propagating, expanding turbulent strips.

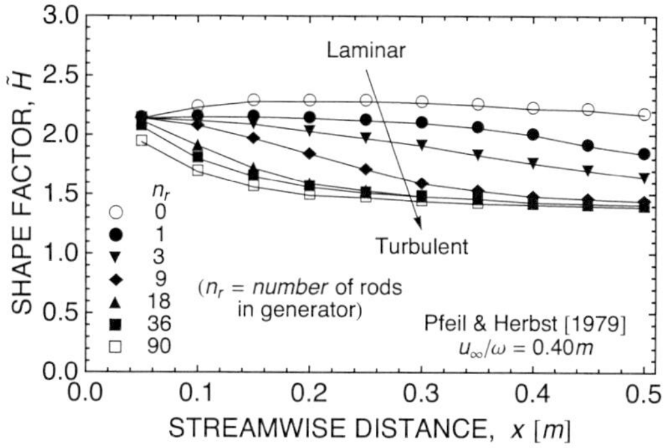

Figure 8.5.
Time-averaged shape factor distributions on a flat plate downstream from a wake-generator with different number of rods. Data from Pfeil and Herbst [1979]

To compare their theory with experiments, Mayle and Dullenkopf [1989] determined the intermittency for this data using the relation

$$\tilde{\gamma}_w = \left(\tilde{H} - \tilde{H}_L\right) / \left(\tilde{H}_T - \tilde{H}_L\right)$$

where $\tilde{H}_L$ and $\tilde{H}_T$ are the time-averaged shape factors for fully-laminar and fully-turbulent flows. While not the best quantity to use, as explained in §7.1.1, the shape factor was the only quantity that could be used to obtain an intermittency distribution for this experiment. Nevertheless, consistent with Pfeil and Herbst's conclusions, the data without the wake generator was used for the fully-laminar shape factors, while that for 90 rods was used for the fully-turbulent values.

The resulting intermittency distributions compared with theory are plotted in Fig. 8.6. In this format, the data had to be shifted streamwise by an amount equal the onset of wake-induced transition x_{tw}. These shifts amounted to the distances x_{tw} = –0.25m and –0.13m for the 1 and 3 rod data, and 0.03m for the others. In their original paper, Mayle and Dullenkopf used x_{tw} = 0.04m for all. The calculated results were obtained using the measured values of $u_{te}/u_\infty = 0.5$, and $u_{le}/u_\infty = 0.88$, which yields $\sigma_w = 0.86$ and $1/\sigma_w = 1.16$. In addition, the data had to be matched by adjusting the strip-production parameter $\hat{n}_w$. The calculated curves in the figures were obtained by using $\hat{n}_w = 2.2$. Except for a few points, notably at the beginning and end of transition, the agreement is very good.

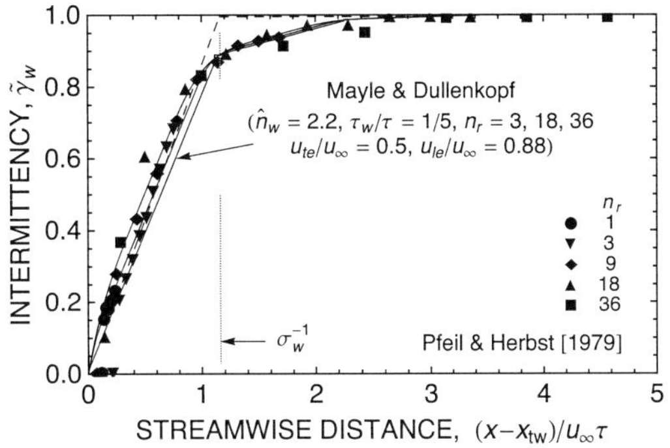

Figure 8.6.
Comparison of the Mayle-Dullenkopf wake-induced transition theory with the experiments of Pfeil and Herbst [1979].

At this point we could say that it is quite remarkable for a theory to explain so many aspects of such a complicated process. However, this is simply a direct outcome of Emmons' theory for transitional flow which reduces the problem in this case to prescribing a production rate for the turbulent strips, a strip propagation parameter, and a fraction of time that the flow is periodically disturbed. For wake-induced transition, or any periodically-forced transition, the physical picture should now be obvious. Each wake initiates a patch of random turbulent strips within its footprint that subsequently propagates downstream and grows independently of the wake itself. In most cases, the strips within each patch will coalesce a short distance downstream from where they form, long before the patches from individual wakes coalesce (as in Fig. 8.4 for $\tau_w/\tau = 0.2$). However, if the transit time for a wake is a significant fraction of their period, this will not necessarily be true (as in Fig. 8.4 for $\tau_w/\tau = 1$). In either case, however, Emmons' theory ensures that overlapping turbulent strips at any position on the surface, whether occurring within or between the patches, are not counted twice. Far downstream, of course, all turbulent strips coalesce making the flow completely turbulent.

As it turns out, the initial linear increase in intermittency displayed by the data in Fig. 8.6 and other experiments have led most investigators[1] to develop linear wake-induced transition models. If turbulent strips expand linearly as they propagate downstream, the result of these models for a constant free-stream velocity is simply

$$\tilde{\gamma}_w(x;x_{tw}) = \min\left[\sigma_w \frac{(x-x_{tw})}{u_\infty \tau}, 1\right]$$

or a slightly different version thereof. This expression is plotted as the dashed line in the figure using $\sigma_w = 0.86$. We now see that these models are limited to a distance from onset less than that to the first kink, that is, to where the previously generated turbulent strips begin to interfere with those just gen-

[1] See for example Hodson et al. [1992], Funazaki [1996], and Funazaki et al. [1997].

erated. Since the first kink occurs where $x = x_{tw} + u_\infty \tau/\sigma_w$, and since Eq. (8.11) defines the locus of all the kinks, these models are good as long as the intermittency is less than about $1-\exp(-\hat{n}_w)$, or 0.89 in this case. Beyond this kink, either the model must be modified to account for the mutual-exclusiveness of the individual turbulent strips or Mayle-Dullenkopf's theory must be used, which automatically accounts for this. One way to extend the linear models is to use a piecewise-linear model. An example of one that approximates the first two segments of the distribution is

$$\tilde{\gamma}_w(x;x_{tw}) = \begin{cases} \sigma_w x^* & ;\ 0 \le x^* \sigma_w \le 1 \\ \sigma_w\left[(1-e^{-\hat{n}_w}) + (e^{-\hat{n}_w} - e^{-2\hat{n}_w})x^*\right] & ;\ 1 \le x^* \sigma_w \le 2 \\ 1 & ;\ x^* \sigma_w > 2 \end{cases}$$

where $x^* = (x - x_{tw})/u_\infty \tau$. Its extension to more segments is straightforward.

Unlike the Mayle-Dullenkopf theory, however, these models are usually unable to describe other interactive effects associated with wake-induced transition without major alterations.[1] We will investigate one of the more interesting of these effects now.

8.1.3 Effect of Distributed Wake-Induced Sources.

In the previous sections, we assumed that all turbulent strips were produced by a single line source located at x_{tw}, as long as the wake (disturbance) was passing over it. Since it seems highly unlikely that periodic wakes would force transition to begin at only one position on the surface, we shall now consider that transition occurs *most likely* when the wakes pass over this position but also before and after it.

A solution to this problem may be obtained by using the method of superposition described in §5.2.3. In this case, however, we begin with Eq. (8.8) written for $\tilde{\gamma}_w(x;x_{tw,i})$ and use Eq. (5.7) to determine the probability the flow is laminar for any number of sources. The analysis is somewhat more difficult for a distribution of wake-induced sources than for turbulent spots since Eq. (8.9) must be written for $G_n(x, t; x_{tw,i})$ and appropriately summed. For a continuous distribution, the sum becomes an integral as we found in Eq. (5.27) for turbulent spots. Nevertheless, following the same methods, we can calculate the time-averaged intermittency for any probability distribution of sources. For an approximate solution, the analytical result for $\tilde{\gamma}_w(x;x_{tw,i})$ given by Eq. (8.11) may be used.

If we suppose that turbulent strips are produced according to a normalized Gaussian distribution centered at x_{tw} when the wake passes over, then the production rate is given by

$$\dot{n}'_w(\xi) = \frac{\dot{n}_w}{\sqrt{2\pi}\chi} \exp\left[-\frac{1}{2}\left(\frac{\xi - x_t}{\chi}\right)^2\right] \qquad (8.13)$$

where $\dot{n}'_w$ is the total number of strips produced over the surface per unit time per unit distance, and χ is the distribution's standard deviation.

The exact solution for the time-averaged intermittency using Eq. (8.13) with $\chi = u_s\tau/2$ is shown by the solid line in Fig. 8.7. This curve, as the others in the figure, was evaluated for $\hat{n}_w = 2$, $\tau_w/\tau = 0.2$, $u_{te}/u_\infty = 0.5$, and $u_{le}/u_\infty = 0.9$. The solution given by Eq. (8.8) for the line source is shown by one of the dot-dashed lines. The other is the approximate result for the Gaussian production distribution using Eq. (8.11). This is given by the closed form solution

[1] See Schulte and Hodson [1998] for example.

$$\tilde{\gamma}_w = 1 - \exp\left\{ -\frac{1}{2}\hat{n}_w \sigma_w \frac{1}{u_\infty \tau}\left[(x - x_{tw}) + \sqrt{\frac{2}{\pi}}\chi \exp\left[-\frac{(x-x_{tw})^2}{2\chi^2} \right] + (x - x_{tw})\,\mathrm{erf}\left[\frac{x - x_{tw}}{\sqrt{2}\chi} \right] \right] \right\} \quad (8.14)$$

Clearly, the effect of a distributed source of wake-induced turbulent strips can be substantial, particularly before peak production where distributing the sources over some distance completely changes the initial trend of the intermittency from one with a sudden increase in the intermittency to one with a gradual increase. In any case, we find that the exact solution lies between the line-source and approximate solutions, the latter of which again provides the upper bound to the exact solution.

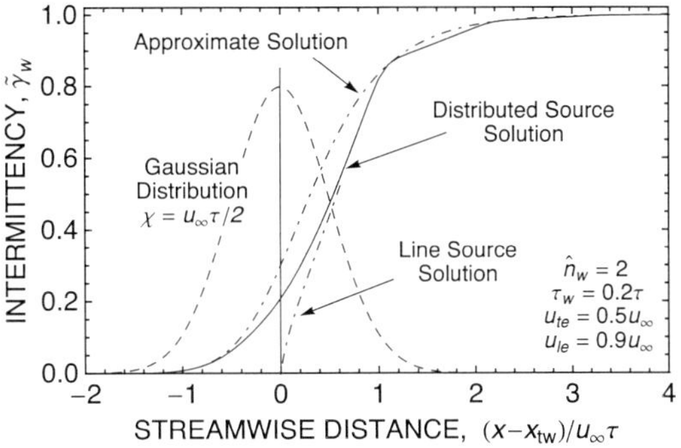

Figure 8.7.
Exact and approximate solutions for the intermittency with a Gaussian source of turbulent strips compared with that for a simple line source of strips.

8.1.4 Wake-Induced Onset and Production Parameter.

There is very little information about the onset of wake-induced transition x_{tw}. Presently, most calculations use the correlations for bypass transition like those given by Eq. (6.7), or from curve fits like those shown in Fig. 6.14, or given by Abu-Ghannam and Shaw [1980], Gostelow et al. [1994], Solomon et al. [1996], Steelant and Dick [2001], or modified versions thereof. Furthermore, most substitute the maximum turbulence level in the wake for the turbulence level in the correlation. This approach was recently confirmed by Funazaki and Koyabu [2006], who conducted experiments on a flat plate for several pressure gradient distributions. In their case, the maximum wake turbulence level was about eight percent.

However, some earlier experiments indicate otherwise. The correlation shown in Fig. 8.6 for Pfeil and Herbst's [1979] three highest frequency data (with x_{tw} = 0.03m) suggests that the transition Reynolds number based on the momentum thickness is about $Re_{\delta 2,tw}$ = 125. This is greater than the value predicted by Eq. (6.7) using the maximum intensity of turbulence in their wake. In fact, the value 125 corresponds to $Tu \approx 7$ percent which is about one-half the maximum turbulence level in their wakes. Hodson et al. [1992], based on the work by Addison and Hodson [1990], presents a theory assuming that x_{tw} varies with time according to the convected turbulence level in the wake. While this is undoubtedly true for wakes with very low turbulence levels, data for wakes with high turbulence levels seem to contradict this viewpoint. In the end, it might be that Hodson's et al. theory applies to wakes with low turbulence levels while some version of the bypass transition correlation applies to wakes with high. Obviously, more information is required here.

When a wake passes over a stationary object in the flow, the deficit of the mean flow within it causes a relative motion either toward or away from the surface of the object depending on the direction the wake generator moves relative to the incident free stream (see Kerrebrock and Mikolajczak [1970]).

The motion toward or away, which is relatively easy to determine by considering the velocity triangles associated with the motion, causes an unsteady pulse of either "converging" or "diverging" mean flow near the surface as the wake passes over. Dong and Cumpsty [1990] found that the direction of the wake source had little effect in their experiments, while Funazaki and Aoyama [2000], and Funazaki and Koyabu [2006] found that it affects the structure of the turbulent strip produced by the wake and, one might therefore presume, the onset and length of wake-induced transition. Again, more information is required regarding this effect.

The strip-production parameter $\hat{n}_w$ is somewhat easier to evaluate. By fitting Eq. (8.11) to the wake-induced transition data of Pfeil and Herbst [1979], Wittig et al. [1988], and Dring et al. [1986], Mayle and Dullenkopf [1990] determined that $(\hat{n}\sigma)_w \approx 1.9 \pm 0.05$. Using the previously determined value for the propagation parameter $\sigma_w = 0.86$, this provides $\hat{n}_w \approx 2.2$, which is the same value that correlates the curves in Fig. 8.6. Since the first segments of these curves agree with the linear models used to correlate all the other wake-induced transition data, we may conclude that $\hat{n}_w \approx 2.2$ is probably a good "all around" value for the strip-production parameter. Furthermore, since $\hat{n}_w$ is also the time-averaged number of strips produced with each passing wake, it now appears that only about two turbulent strips are produced per wake, which in turn suggests that a strip is produced as both the downstream and upstream edge of the wake passes over the induced-transition location x_{tw}. The time-averaged intermittency distribution in this case is easily obtained from Eq. (8.5) by using $g^*(t') = \sum_n \left[\delta(t' - n\tau) - \delta(t' - \tau_w - n\tau)\right]$, where τ_w is an appropriate wake "width."

8.2 Multimode Transition.

Preliminaries. Becalmed-controlled transition. Multimode transition theory. Some examples.

While many experiments clearly display the effects of "wake-induced" transition, some of them also reveal that transition by any other mode can occur between the passing turbulent strips. This multiple transitional behavior is know as "multimode transition."

8.2.1 Preliminaries.

Bypass and wake-induced transition. The becalming effect.

To the author's knowledge the first evidence of multimode transition was presented by Pfeil and Herbst [1979] and later described in a series of articles by Herbst [1980], Pfeil et al. [1983], and Schröder [1985]. Besides discovering that wakes produce turbulent strips, they discovered that a periodically unsteady transition could occur between the strips. In this regard, they were the first to establish the basic features of the multimode transition process as we know them today. Later, Wittig et al. [1988], Dullenkopf et al. [1991] and Dullenkopf [1992] measured the time-averaged heat transfer on a gas turbine airfoil in an unsteady wake-disturbed incident flow which clearly showed a multimode transitional behavior. The time-averaged intermittency distributions derived from their data are shown in Fig. 8.8, where

$$S = f_w c / u_\infty \tag{8.15}$$

is the Strouhal number based on the wake-passing frequency f_w, and c is the chord of the airfoil. In this figure, wake-induced transition occurs over the forward portion of the airfoil's surface while a combination of wake-induced and bypass transition occurs over the aft portion of the surface.

About the same time, Dong and Cumpsty [1990] presented their measurements on the unsteady interaction between wake-induced turbulent strips and a separated-flow transition bubble. In this case, laminar separation on a compressor blade was suppressed every time a turbulent strip passed over the separation location. In fact, for certain conditions, it was found that the wakes prevented both separation and transition from occurring such that the flow between the turbulent strips remained laminar all the way to the airfoil's trailing edge.[1] In any case, all of these experiments indicate that the transitional flow taking place between the strips is a periodically disturbed version of that which would have occurred there if the mainstream flow had been steady. Thus, any transition between passing turbulent strips may be called either "wake-disturbed normal" or "wake-affected normal" transition.

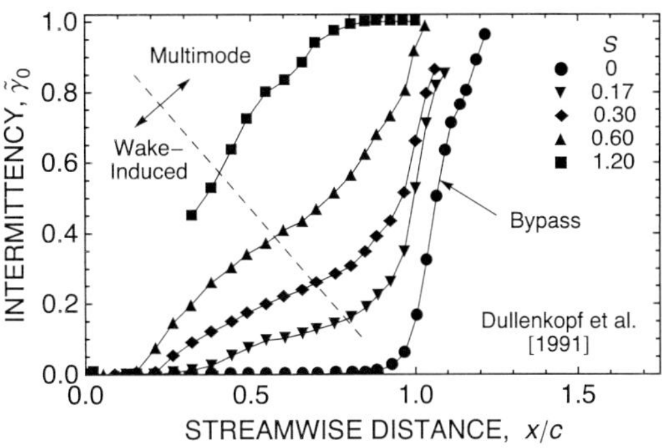

Figure 8.8.
An example of multimode transition on the suction side of a gas turbine airfoil. Data from Dullenkopf et al. [1991].

An interesting consequence of multimode transitional flow is that multiple regions of laminar and turbulent flow can simultaneously occur on the same surface. That is, an instantaneous snapshot of the flow over the surface may show a laminar boundary layer near the leading edge, a wake-induced turbulent strip farther downstream, followed by a second region of laminar flow, and another transition to turbulent flow by any one of the usual modes, be it natural, bypass, or separated-flow transition. Furthermore, since becalmed zones also follow turbulent strips (Pfeil et al. [1983], Orth [1991, 1992], Gostelow et al. [1997]), intervals of becalmed flow appear as well. This, by the way, explains the complete suppression of wake-affected normal transition found in some cases by Dong and Cumpsty.

A simplified picture of the flow just described is shown in the xt-diagram of Fig. 8.9. Cuts parallel to the x axis describe the instantaneous state of flow on the surface, while vertical cuts describe the unsteady behavior of the flow at a fixed position on the surface. The dark gray areas in the figure represent the wake-induced turbulent strips generated at x_{tw}. Consequently, this diagram is drawn for the case when the streamwise distance is less than $1/\sigma_w$, otherwise, as explained in §8.1.2, the dark areas associated with the turbulent strips would interfere with one another. Their period of course corresponds to the wake-passing period τ.

The white areas in the figure represent the laminar portions of the flow on the surface. They have been separated into two regions to distinguish the laminar becalmed flow from the undisturbed flow. The line separating these regions represents the trailing edge of the *wake-induced becalmed* flow following the strips. For the becalmed zone behind a forced two-dimensional band of turbulence traveling downstream, Schubauer and Klebanoff [1955] measured a trailing edge velocity of $0.29u_\infty$, which in their case was equal to the Tollmien-Schlichting wave speed. From Orth's data, Mayle [1993] estimated that the trailing-edge velocity of the becalmed flow behind a turbulent strip is about 0.30 to $0.38u_\infty$, the higher of which nearly corresponds to the Tollmien-Schlichting wave speed for the critical Reynolds number of instability in his experiment. In their work, Hodson et al. [1992] estimated the velocity to be about 0.25 to $0.30u_\infty$. From these values, it is easy to calculate that the residence time of the becalmed flow over a point on the surface is about the same as the residence time of the turbulent strip itself.

---

[1] The interested reader may find more articles about "wake-disturbed" separation bubbles referenced in the Bibliography.

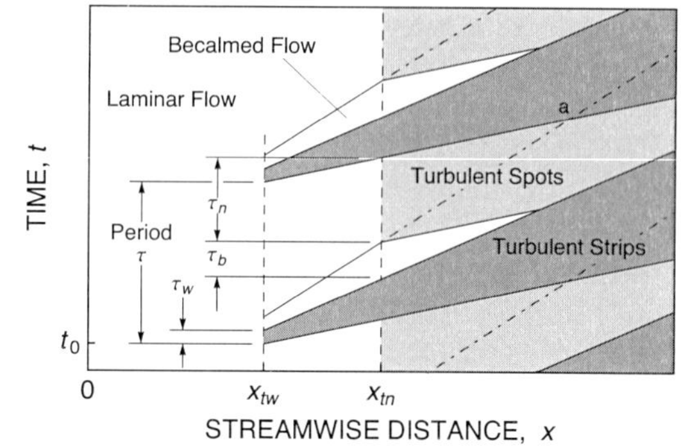

Figure 8.9.
Multimode transition in the xt-plane.

The light gray areas in Fig. 8.9 represent the ensemble averaged area of the turbulent spots and their attending becalmed zones generated between the turbulent strips and their accompanying becalmed zones. The spots may be caused by either natural, bypass, or separated-flow transition. Since they are only produced at x_{tn} and only after the wake-induced becalmed zone has passed, this transition can only occur within the skewed quadrilateral area shown. Obviously, multimode transition is a complex unsteady flow that not only involves the physics of turbulent strips and spots, but also the state of the flow and interactions following them. As it turns out, even after two decades of research, very little is known about the actual physics. Consequently, we are presently relegated to using some ad hoc model to calculate the phenomenon.

Considering the simplified picture shown in Fig. 8.9, we can quickly estimate when *multimode* transition occurs. The interval of time, τ_n, during which *spots* can be produced is easily determined from geometry to be

$$\tau_n = \tau - \tau_w - \left(x_{tn} - x_{tw}\right)\left(u_{te,b}^{-1} - u_{le,w}^{-1}\right); \quad (x_{tn} \geq x_{tw}) \tag{8.16}$$

where τ_w is the interval of time during which turbulent strips are produced by the wake at x_{tw}, $u_{te,b}$ is the trailing edge velocity of its becalmed region, and $u_{le,w}$ is the leading-edge velocity of the turbulent strips. Using $u_{te,b} \approx 0.34u_\infty$ and $u_{le,w} = 0.88u_\infty$, this expression simplifies to

$$\tau_n = \tau - \tau_w - 1.80\left(x_{tn} - x_{tw}\right)/u_\infty$$

By setting τ_n equal to zero, the maximum downstream distance at which transition by any other mode occurs can be obtained. After introducing the wake-passing Strouhal number based on the characteristic length of the surface L, this condition may be written as

$$\frac{\Delta x_{tn,\max}}{L} = \frac{x_{tn,\max} - x_{tw}}{L} = \frac{0.55}{S}\left(1 - \frac{\tau_w}{\tau}\right) \tag{8.17}$$

This is plotted in Fig. 8.10 for $\tau_w/\tau = 0$ and 0.4. When the distance from the onset of wake-induced transition to the onset of transition for normal transition is larger than $\Delta x_{tn\,max}$, transition cannot occur between the strips, that is, only wake-induced transition occurs. On the other hand, when $\Delta x_{tn} < \Delta x_{tn\,max}$, multimode transition can occur. The difficulty with this analysis is its inherent assumption that the onset to normal transition is unaffected by the passing wakes, which more than likely is incorrect. That is, x_{tn} in Eq. (8.16) probably depends on the characteristics of the wakes and their Strouhal number. Estimated differences between the onset of normal and wake-induced transition for the data shown in Fig. 8.8 are plotted as the filled circles in the figure using $L = c$. Since $\tau_w/\tau \leq 0.40$ for these tests,[1] all the points lie within the multimode transition region as they should.

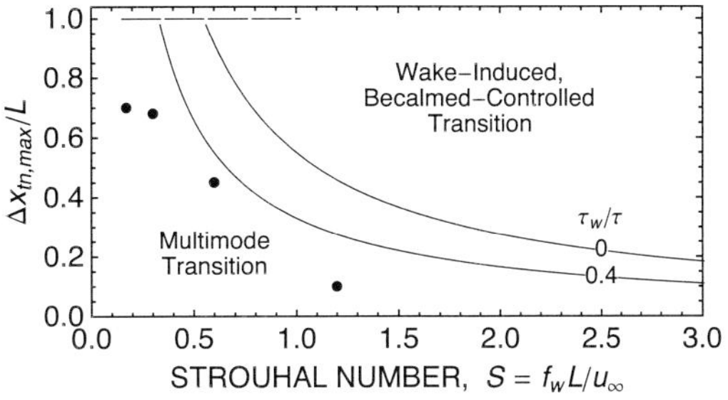

Figure 8.10.
Condition for multimode transition when $\tau_{tw} = 0$. Data from Dullenkopf et al. [1991].

Becalmed-controlled transition. An interesting feature about the maximum downstream distance for multimode transition, is that $\Delta x_{tn} = \Delta x_{tn\,max}$ also corresponds to the distance where the becalmed zone begins interfering with the leading edge of the following turbulent strip; that is, point "a" in Fig. 8.9. Farther downstream, the two regions overlap as indicated by the triangular area to the right of "a." Therefore, the wake-induced transition region in Fig. 8.10 is highly dependent on the interaction taking place between the turbulent flow within the strips and the laminar flow within the becalmed zone. This flow might best be referred to as "wake-induced, becalmed-controlled transitional" flow. Clearly, when the turbulence within the strips is unaffected, the time-averaged intermittency continues to increase and is given by either Eq. (8.8), its approximation Eq. (8.11), or Eq. (8.14) for a distributed source. However, when the turbulence within the strips is dissipated by the becalmed flow, the intermittency depends on how far the strips penetrate into it. As seen by subtracting the overlapping triangular areas to the right of "a" in Fig. 8.9 from the dark areas, the time-averaged intermittency could even decrease!

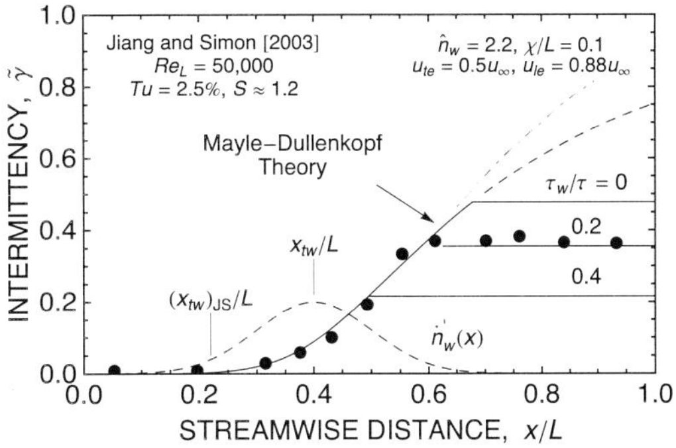

Figure 8.11.
Comparison of the Mayle-Dullenkopf wake-induced theory for becalmed-controlled transition with experiments on the suction side of a gas-turbine airfoil. Data from Jiang and Simon [2003]

An example of becalmed-controlled transition is nicely provided by Jiang and Simon [2003], whose time-averaged intermittency data is plotted in Fig. 8.11. The abscissa in this figure is the distance along the suction side of an airfoil divided by its surface length. Although the data was obtained on an airfoil which displayed attached laminar and turbulent flow, separated flow, and transitional flow at various times within each period, transition was found to be prevented within the becalmed zone over

[1] See Dullenkopf [1992].

the last forty percent of the surface.[1]

The theory presented in the previous section is also plotted in the figure. These results were obtained by using a theory that accounts for both the distribution of the wake-induced production rate and the effects of the becalmed zones. Without going into the details, the curve through the data on the forward portion of the surface was calculated using Eq. (8.14) for a Gaussian distribution with an integrated production rate $\hat{n}_w = 2.2$ and a standard deviation of $\chi/L = 0.10$. The value of $\hat{n}_w$ is that previously determined from measurements. The value for χ/L was chosen after noting that the rise in intermittency, as observed by Jiang and Simon, and shown in the figure, first occurs near $x/L = 0.25$. The best fit over the first half of the surface was obtained using $x_{tw}/L = 0.4$.

According to Eq. (8.17) and Fig. 8.10, the calculations for wake-induced transition are valid as long as

$$\frac{x - x_{tw}}{L} \leq \frac{0.55}{S}\left(1 - \frac{\tau_w}{\tau}\right) \tag{8.18}$$

For distances greater than this, the interaction between the flow within the turbulent strips and that within the laminar becalmed zones must be known. Schubauer and Klebanoff [1955] indicate that turbulent spots propagate only so far into the becalmed zone before they decay. Gostelow et al. [1997] show that the becalmed zones behind turbulent strips are remarkably stable and persistent. If we assume that the width of the turbulent strips interacting with the becalmed zones doesn't increase once the upper limit of Eq. (8.18) is surpassed, then the time-averaged intermittency beyond the upper limit, which depends on the Strouhal number and wake duration, becomes independent of distance. Its dependence on the wake duration ratio in this case, calculated by using the experimental Strouhal number of $S = 1.2$, is shown by the horizontal lines in Fig. 8.11. The dashed line extension above the highest horizontal line corresponds to the intermittency distribution without becalmed zones using the Mayle-Dullenkopf theory. The overall agreement of this wake-induced, becalmed-controlled transition theory with data is either truly amazing or fortuitous, and suggests that the "effective" wake duration in Jiang and Simon's case is about twenty percent of the wake passing period, that is, $\tau_w/\tau = 0.2$.

The gray dot-dashed line shown in the figure is the wake-induced transition result using Ramesh and Hodson's [1999] model as calculated by Jiang and Simon. As explained in §5.4.3, it is essentially a turbulent *spot* model that purports to account for the unsteady formation of turbulent spots and their becalmed zones by constructing a spot production distribution that fits their data. While the calculated result over the forward portion of the surface using their model is nearly the same as that given by Mayle-Dullenkopf's, and hence not shown, the result aft is clearly in error despite their claim of "incorporating the calming effect." This is because they neglect the time-averaged interference between successive wakes and their becalmed zones.

8.2.2 Multimode Transition Theory.

The original theory for multimode transition was presented by Mayle and Dullenkopf [1989, 1990], and later expanded upon by Mayle [1991, 1993], Dullenkopf [1992], and Dullenkopf and Mayle [1994] to include the effects of spot-source distributions, becalmed zones, separated-flow transition, and free-stream turbulence. About the same time, Hodson et al. [1992] presented a model based on the linear analysis described in §8.1.2, which is valid only when the streamwise distance is less than $1/\sigma_w$. Now a research topic much in vogue, the most recent numerical analyses solve the unsteady Reynolds-averaged equations including the effects of wakes. To the author's knowledge, the first of these models was presented by Cho et al. [1993]. Some of the later variations were already referenced

[1] See the original papers for details.

in §7.3.2. In this section, we will present the Mayle-Dullenkopf theory which combines their wake-induced turbulent strip theory with Emmons' turbulent spot theory.

Consider Emmons' general result for the intermittency given by Eq. (5.9), namely,

$$\gamma(P) = 1 - \exp\left\{-\int_R g(P')dV'\right\} \tag{8.19}$$

where g is the production function and R is the volume of the dependence cone associated with the point P (see Fig. 5.17). Assume that the production function for multimode transition is the sum of those for both wake-induced and "normal-mode" transition. We will suppose the latter, which could occur by either natural or bypass transition, begins downstream from that induced by the wakes. Furthermore, we will assume that the production rates of the spots and strips are statistically two-dimensional.

Then, after substituting Eq. (8.6) into Eq. (8.4), the production function for wake-induced transition is

$$g_w(x',t') = \dot{n}_w \delta(x' - x_{tw}) \sum_{n=-\infty}^{\infty} \left[\mathrm{U}(t' - n\tau) - \mathrm{U}(t' - \tau_w - n\tau)\right]$$

where $\dot{n}_w$ is the number of strips produced per unit time. Likewise, since spots can be produced only in the intervals between the turbulent strips and their becalmed zones (see Fig. 8.9), the production function associated with the normal mode of transition is

$$g_n(x',t') = \dot{n}_n' \delta(x' - x_{tn}) \sum_{n=-\infty}^{\infty} \left\{\mathrm{U}\left(t' + \tau_n(x') - n\tau\right) - \mathrm{U}\left(t' - n\tau\right)\right\}$$

where $\dot{n}_n'$ is the number of turbulent spots per unit time per unit span within the intervals, and τ_n is the interval of time associated with that portion of the flow where turbulent spots can be produced. Substituting these expressions into Eq. (8.19) and integrating, the time-averaged intermittency for multimode transition is given by

$$\tilde{\gamma}(x; x_{tn}, x_{tw}) = 1 - \frac{1}{\tau}\int_0^\tau \left[\exp\left\{-\dot{n}_w \sum_{n=-\infty}^{\infty} \int_{R_w} \left[\mathrm{U}(t' - n\tau) - \mathrm{U}(t' - \tau_w - n\tau)\right] dt'\right\}\right.$$

$$\left. \times \exp\left\{-\dot{n}_n' \int_{R_n} \delta(x' - x_{tn}) \sum_{n=-\infty}^{\infty} \left\{\mathrm{U}\left(t' + \tau_n(x') - n\tau\right) - \mathrm{U}\left(t' - n\tau\right)\right\} dx'\,dz'\,dt'\right\}\right] dt \tag{8.20}$$

This expression is difficult to evaluate, especially the integral in the second exponential function, which to the author's knowledge has never been evaluated. But by comparing the above expression with Eqs. (8.7) and (8.8), the integral in the first exponential is $G_n(x, t; x_{tw})$. Hence, the first exponential equals $1-\gamma_w(x, t; x_{tw})$, and by inference the second must equal $1-\gamma_n(x, t; x_{tn}, x_{tw})$. Consequently, the time-averaged intermittency for multimode transition is given by

$$\tilde{\gamma}(x; x_{tn}, x_{tw}) = 1 - \tilde{\gamma}_w - \tilde{\gamma}_n + \frac{1}{\tau}\int_0^\tau \gamma_w \gamma_n \, dt$$

where $\tilde{\gamma}_w$ and $\tilde{\gamma}_n$ are the time-averaged intermittency functions for wake-induced and normal transition.

To evaluate Eq. (8.20), assume that the primary influence of the sum within the integral over R_n is to reduce the production rate g_n by the fraction of time that normal transition between the wake-induced transition and becalmed zones can occur. Since $\tilde{\gamma}_w(x_{tn})$ is the fraction of time the flow is turbulent at the onset of normal transition, and since $\tilde{f}_{bw}(x_{tn})$ is the fraction of time the laminar flow at onset is becalmed (using the same notation as in §5.4.2), the fraction of time turbulent spots can be produced at onset is given by the product of one minus these quantities. Thus, the second exponential function in Eq. (8.20) may be approximated by $1-\overline{\gamma}_n(x;x_{tn},x_{tw})$ where

$$\overline{\gamma}_n(x;x_{tn},x_{tw})=1-\exp\left\{-\dot{n}'_n\left[1-\tilde{\gamma}_w(x_{tn})\right]\left[1-\tilde{f}_{bw}(x_{tn})\right]\int_{A'_{s,zt}}dz'\,dt'\right\} \tag{8.21}$$

is the average of the *wake-disturbed* intermittency, and $A'_{s,zt}$ is the cross-sectional area of the spot's dependence cone in the x' equal-constant plane (Fig. 5.17). Since this intermittency factor is independent of time, Eq. (8.20) simplifies to

$$\tilde{\gamma}(x;x_{tn},x_{tw})\approx 1-\left[1-\tilde{\gamma}_w(x;x_{tw})\right]\left[1-\overline{\gamma}_n(x;x_{tn},x_{tw})\right] \tag{8.22}$$

Furthermore, if we substitute Eq. (8.8) for the wake-induced intermittency and introduce Emmons' propagation parameter into Eq. (8.21) using Eq. (5.15), it then follows that

$$\tilde{\gamma}(x;x_{tn},x_{tw})\approx 1-\left\{\frac{1}{\tau}\int_0^{\tau}\exp\left[-\dot{n}_w\sum_{n=-\infty}^{\infty}G_n(x,t;x_{tw})\right]dt\right\}$$
$$\times\exp\left\{-\frac{\dot{n}'_n\sigma}{u_\infty}\left[1-\tilde{\gamma}_w(x_{tn})\right]\left[1-\tilde{f}_{bw}(x_{tn})\right](x-x_{tn})^2\,\mathrm{U}(x-x_{tn})\right\} \tag{8.23}$$

where $G_n(x,\,t;\,x_{tw})$ is given by Eq. (8.9). This is Mayle and Dullenkopf's multimode intermittency solution.[1] Using the approximate solution given by Eq. (8.11) for the wake-induced component, we obtain

$$\tilde{\gamma}(x;x_{tn},x_{tw})\approx 1-\exp\left[-\hat{n}_w\sigma_w\frac{(x-x_{tw})}{u_\infty\tau}\mathrm{U}(x-x_{tw})\right]$$
$$\times\exp\left\{-\frac{\dot{n}'_n\sigma}{u_\infty}\left[1-\tilde{\gamma}_w(x_{tn})\right]\left[1-\tilde{f}_{bw}(x_{tn})\right](x-x_{tn})^2\,\mathrm{U}(x-x_{tn})\right\} \tag{8.24}$$

The becalmed zones are considered in this analysis only in their role of reducing the production of turbulent spots at the onset of normal transition and only to the first approximation. However, the propagation of the turbulent spots into them from behind, as shown in Fig. 8.9, is permitted by properly accounting for their role in reducing turbulence. When the becalmed zones behind turbulent strips force the turbulent spots to laminarize, the regions covered by turbulent spots decrease and the overall time-averaged intermittency reduces accordingly. The dot-dashed line in the light gray areas of Fig. 8.9 show the maximum reduction of this area. Although the actual effect of becalmed zones can be taken into account as we did in §5.4 or by guessing as we did in the previous section, more information regarding the laminarization of turbulent spots in becalmed zones behind spots and wakes is required to do it correctly than that provided by Schubauer and Klebanoff [1955].

8.2.3 Some Examples and Comparisons with Data.

There are numerous multimode transition situations to consider. Specifically, they include different wake-induced and wake-disturbed transition arrangements, production distributions, propagation parameters, wake-durations, becalmed zones, besides considering the different modes of natural, bypass, and separated-flow transition. In this section, we will only examine the effect of the wake-induced production rate and make a cursory comparison between theory and experiment.

---

[1] A similar analysis has been presented by Schulte and Hodson [1998], but they double bookkeep the mutual exclusivity of overlapping turbulent spots and turbulent strips.

Some examples. For the sake of simplicity, consider only multimode transition occurring without the calming effect on turbulent spots entering the wake-induced becalmed zones. The effect of them entering the becalmed zone will be shown in the following subsection.

The results of a calculation using Eq. (8.23) is shown in Fig. 8.12. The calculated streamwise variation of the near-wall intermittency is shown in part (a) of the figure for a case where bypass transition occurs immediately before the first possible interference between the wake-induced turbulent strips. In this case, we have arbitrarily chosen the onset of bypass transition to occur at a distance Reynolds number of 10^5 downstream from the onset of wake-induced transition, that is, Re_{xtn}–$Re_{xtw} = 10^5$. The wake-induced and normal-mode transition contributions to the intermittency downstream of this location are shown by the dashed lines. These contributions were obtained from Eqs. (8.8) and (8.21). When the calming effect on turbulent spots entering the wake-induced becalmed zones are taken into account, both the multimode and normal intermittency distributions downstream of the onset of normal transition are reduced accordingly. The parameters used to calculate the wake-induced result are $\hat{n}_w = 0.5$, $\tau_w/\tau = 0.2$, $u_{te}/u_\infty = 0.5$, and $u_{le}/u_\infty = 0.9$. The spot production parameter used to calculate the normal transition result is $\hat{n}_n\sigma = 7.5(10)^{-11}$, which corresponds to the value for bypass transition with a free-stream turbulence of about two-and-a-half percent.

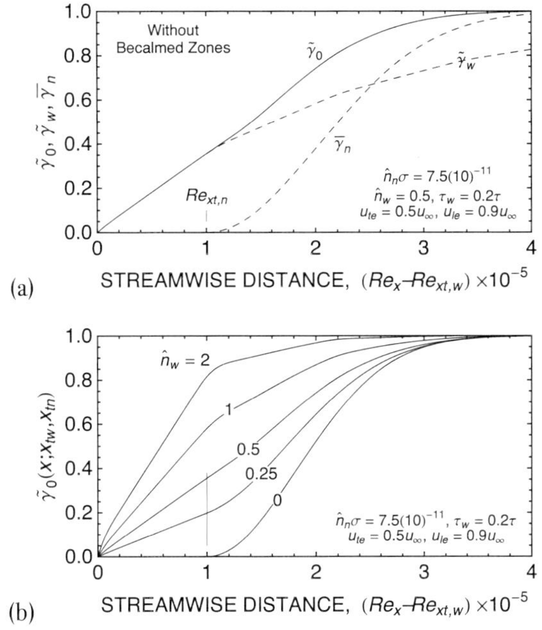

Figure 8.12a,b.
Calculated examples of multimode transition.

The effect of increasing the wake-induced production rate $\hat{n}_w$ is shown in part (b) of the figure. Here, the wake-induced production rate parameter $\hat{n}_w$ varies from zero (no wakes) to 2, which is only slightly less the value of 2.2 used to correlate wake-induced transition data. The variation shown here is easy to understand. Namely, when the wake-induced production rate is zero, only bypass transition occurs and when it is high, only wake-induced transition occurs. For intermediate production rates, the time-averaged multimode intermittency distribution morphs from one intermittency distribution into the other. As previously noted, numerous other examples may be constructed.

Comparisons with data. The above theory is compared with the data of Dullenkopf et al. in Fig.

8.13.[1] As with all comparisons, the onset and length of transition were adjusted to provide the best fit to the data. The calculated result for $S = 0$, corresponding to a steady free-stream flow and bypass transition, was obtained from Eq. (5.18) using $x_{tn}/c = 0.92$ and $\hat{n}\sigma = 2.26(10)^{-10}$. Those for the periodic unsteady flows were obtained from Eq. (8.24) using the standard wake-induced production and propagation parameters of $\hat{n}_w = 2.2$ and $\sigma_w = 0.86$. The fraction of time the laminar flow is becalmed was determined using the becalmed and turbulent strip velocities of $u_{te,b} = 0.34u_\infty$, $u_{te,w} = 0.5u_\infty$, and $u_{le,w} = 0.88u_\infty$. A calculation for $S = 1.2$ not including the effect of the becalmed zone is shown by the dashed line in the figure. The difference between these two results decreases for the smaller Strouhal numbers because the becalmed zones occupy a smaller fraction of the laminar flow.

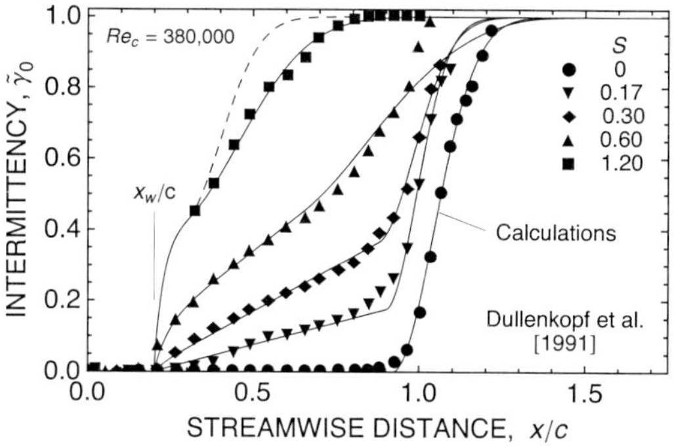

Figure 8.13.
A comparison of Mayle-Dullenkopf's multimode transition theory with data from the suction side of a gas turbine airfoil. Data from Dullenkopf et al. [1991].

In this case, the best fit to the data is given by $x_{tw}/c = 0.2$ and $x_{tn}/c = 0.30, 0.65, 0.88, 0.90$ for decreasing Strouhal numbers. That is, the onset of wake-induced transition is fixed for this series of experiments, while the onset of normal-mode transition decreases with increasing Strouhal number or wake-duration.[2] Although a connection between the onset for normal-mode transition and Strouhal number certainly exists, the author has yet to find it. But, as it turns out, the onset of wake-induced transition for these experiments occurs near the position of minimum pressure on the suction side of the airfoil. Nevertheless, the upshot of the comparisons shown in Fig. 8.13 is that the time-averaged intermittency data can be calculated quite well using Mayle-Dullenkopf's multimode theory providing the onset of wake-induced and wake-disturbed transition are known a priori.

A comparison between the measured and calculated Nusselt number (heat transfer) distributions for Dullenkopf's et al. experiments is shown in Fig. 8.14. The calculations were obtained by using the calculated intermittency distributions provided in the previous figure and substituting $Q = Nu_c$ in Eq. (8.1), where Nu_c is the Nusselt number based on the airfoil's chord. Since the "measured" intermittency distributions were obtained using the measured Nusselt number distributions, and since the intermittency distributions are calculated reasonably well, the excellent agreement shown in this figure is to be expected. Even so, it provides a good account of the basic ingredients required to predict transitional flow.

---

[1] This comparison is slightly different than that presented in the original papers by Mayle and Dullenkopf [1989] and Mayle [1993]. In both, as herein, the approximate solution given by Eq. (8.11) was used. In the former, the becalmed zones were not taken into account and an estimated wake-duration parameter $\tau_w/\tau = 0.14$ was used. In the latter, the becalmed zones were taken into account, but for the sake of simplicity τ_w/τ was set equal to zero. As a result, the values given in those papers for the onset of wake-induced and bypass transition differ from the values presented herein.

[2] The ratios of the wake-duration, determined by the wake's half width, to wake period in these experiments are 0.06, 0.1, 0.2, and 0.4 for the increasing Strouhal numbers (see Dullenkopf [1992]).

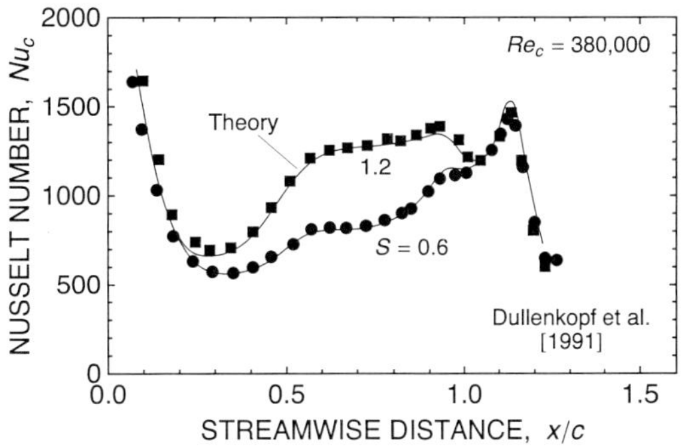

Figure 8.14.
A comparison of measured and calculated Nusselt number distributions on a gas turbine airfoil using Mayle-Dullenkopf's multimode transition theory. Data from Dullenkopf et al. [1991].

8.3 Separated-Flow Transition.

Some aspects of a separation bubble. Onset of transition. Short-bubble transition correlations.

Laminar boundary layers in decelerating flows most often separate and when they do, they can separate either before or after transition begins. When the flow separates first, transition occurs in the unstable laminar free-shear layer between the relatively stagnant flow near the wall and the free stream. In this case, the turbulent spots generated within the free-shear layer grow as they propagate downstream eventually coalescing and reattaching as a turbulent boundary layer to form an enclosed separation "bubble." When transition occurs first, separation occurs in the laminar portions of the flow *between* the turbulent spots. In this case, three-dimensional laminar separation bubbles develop and collapse unsteadily as the turbulent spots and their attending becalmed zones randomly pass over the time-averaged locus of separation.

As shown by Hatman and Wang [1998], the progression of separation preceding transition to where it follows transition occurs continuously as either the adverse pressure gradient is decreased or the free-stream turbulence level is increased. A guideline to whether transition precedes separation or not was first offered by Mayle [1991]. The guideline was developed for moderate to high free-stream turbulence levels where the onset of transition is presumably given by Eq. (6.7). Using this equation together with Thwaites criterion for separation, namely, $\lambda_{\delta 2,s} = -0.082$, where $\lambda_{\delta 2} = Re_{\delta 2}^2 K$, it is easy to determine that transition will always occur before laminar separation when

$$K_{xt} > K_{crit} \approx -1.62(10)^{-4}\, Tu^{5/4} \; ; \quad (x_t < x_s, Tu > 0.005) \tag{8.25}$$

The expression for K_{crit} is plotted in Fig. 8.15 together with the data of Gostelow et al. [1994], Hatman and Wang [1998], and Roberts and Yaras [2003] where transition precedes separation, that is, when $x_t < x_s$. Undeniably, all their data for turbulence levels above two percent lie above the K_{crit} curve. For lower turbulence levels, Hatman and Wang found that transition precedes separation when $K_{xt} > K_{crit} \approx -0.3(10)^{-6}$. This is shown by the dashed line in the figure. Since separation reduces the local free-stream deceleration (relieves the adverse pressure gradient), we cannot expect data where separation precedes transition to lie below the K_{crit} curve. In fact, the data of Hatman and Wang, Roberts and Yaras, and Yaras [2001, 2002] indicate that the average acceleration parameter at separation for *all* free-stream turbulence levels and velocity distributions is roughly $K_s \approx -0.8(10)^{-6}$. Consequently, the criterion plotted in Fig. 8.15 can only be used to estimate when separation in attached flows with elevated free-stream turbulence is imminent.

Although simple, the criterion given by Eq. (8.25) has some far reaching implications, particularly for the aircraft and gas turbine industries. First, for those conducting either small or large scale experiments where transitional flow with adverse pressure gradients is being investigated, both the turbulence level and spectrum (length scales) must be appropriately modeled, otherwise, unlike the flow on the prototype, that on the model may separate before transition or vice versa yielding completely misleading results.[1] Second, for flows periodically disturbed by wakes where large rapid changes in turbulence occur (or shock waves where large rapid changes in pressure occur), a separation bubble that normally appears can periodically disappear and reappear as the disturbance passes over them.[2] Third, for situations where the flow under normal operation extends over a large Reynolds number range that includes transition, the flow in any region with an adverse pressure gradient will most likely change from transitional to separated-transitional as it passes through the range.[3] This is easy to understand once it is realized that the acceleration parameter varies inversely proportional to the Reynolds number of the flow about any object, that is, $K \propto 1/Re_L$. Consequently, designs that operate satisfactorily at one end of the Reynolds number range may completely fail at the other.

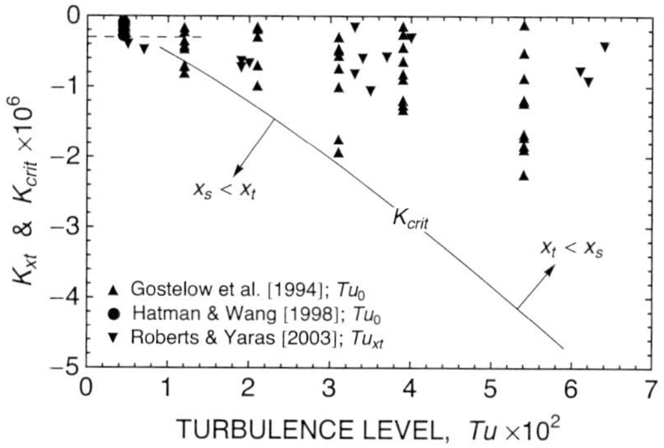

Figure 8.15.
A criterion for whether transition precedes separation or not and its comparison with data.

Nevertheless, whether transition or separation happens first, it is now customary to refer to transition occurring with laminar separation as *separated-flow* transition. This path of transition from laminar-to-turbulent flow is one of the three primary transition modes; the other two being *natural* and *bypass* transition. As with transition by these other two modes, separated-flow transition can also be affected by periodic disturbances in the free-stream. Consequently, multimode transition in flows with adverse pressure gradients often involve separated-flow transition.

Somewhat ignored until about the 1990s, the number of investigations, and hence publications, related to separated-flow transition have since increased dramatically.[4] Since separated-flow transition is also one of the most difficult problems in fluid mechanics and beyond the scope of this book, the presentation in the following sections will be introductory in nature. In particular, we will first discuss some general aspects of separation bubbles and then provide some correlations for the onset of transition in separation bubbles. Since little is known about the case where laminar separation occurs after the onset of transition, we will only consider the situation where separation precedes transition. Furthermore, since the flow in "long" bubbles strongly interacts with the free-stream flow, we will primarily be concerned with separated flow transition in "short" bubbles.

[1] This problem plagued early turbomachinery designers using low-turbulence cascade test results.

[2] See Dong and Cumpsty [1990], Schröder [1989], and Johnson et al.[1990] for early examples of this phenomenon.

[3] This problem continues to plague designers of small gas turbines and the low-pressure turbine in large engines where airfoil Reynolds numbers are transitional and change over the operating range of the machine.

[4] A brief description of all the pre-1996 work, both experimental and analytical, may be found in Hatman [1997].

8.3.1 Some aspects of a separation bubble.

Long and short bubbles. Velocity profiles. Boundary-layer thicknesses.

The basic features of a laminar-separating/turbulent-reattaching flow and its effect on the free-stream velocity distribution are sketched in Fig. 8.16. Since the actual flow associated with separation bubbles is highly three dimensional and unsteady, this sketch is only a time- and spanwise-averaged view of the flow. Hence, the distances x_s, x_t, and x_r correspond to the average temporal and spatial positions of separation, transition, and reattachment, while x_m corresponds to the position of the average maximum displacement of the external flow by the bubble.

Because the laminar free-shear layer from separation to the onset of transition grows slowly, the fluid entrained into it is small. Consequently, the flow between the shear layer and wall within this portion of the bubble remains rather stagnant, which further implies that the pressure in this region remains constant. Moreover, since the relatively thin shear layer cannot support a transverse pressure gradient, the pressure and consequently the velocity in the nearby free stream must be constant in this region. And furthermore, since the transverse pressure gradient is negligible, the streamlines within and near the laminar shear layer between x_s and x_t must be relatively straight, as depicted in the figure by the streamline dividing the recirculating flow within the bubble from the external flow. Once transition ensues, however, a marked increase in the free-shear layer thickness and entrainment occurs, which in turn forces the flow within the bubble to reverse. The velocities associated with the reverse flow typically reach about twenty percent of that in the free stream. The skin-friction coefficient in this region is of course negative. With the shear layer now being turbulent, or nearly so, it can reattach under a stronger adverse pressure gradient than when it separated, thus enclosing the reverse flow in the bubble to form its recirculating zone.

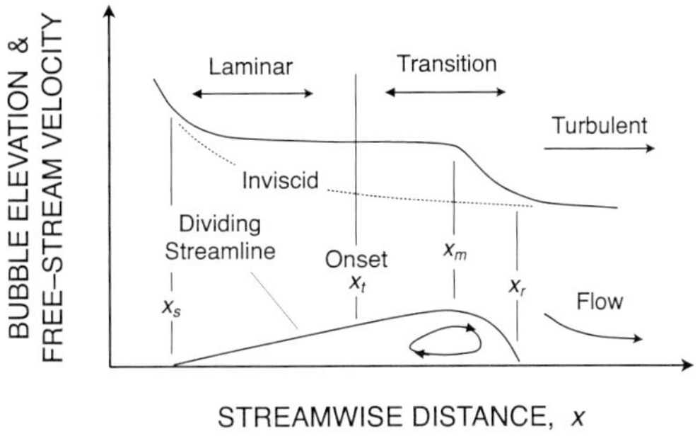

Figure 8.16.
The flow and free-stream velocity distribution associated with a typical separation bubble.

In general, there are two types of separation bubbles – long and short – with the distinction between the two being their effect on the free-stream velocity distribution. Short bubbles have only a local displacement effect on the external flow, so that the free-stream velocity distribution before and after is nearly that predicted without separation. The sketch in Fig. 8.16 is for a short bubble. On the other hand, long bubbles interact with the exterior flow to such an extent that the free-stream velocity distribution over an extended portion of the surface is appreciably different from that predicted without them. Since very small changes in either Reynolds number, free-stream turbulence level, or pressure gradient[1] can cause a bubble to catastrophically change from short to long in a process called "burst-

[1] For airfoils, think "angle of attack" here.

ing," it becomes of utmost importance to understand the physics associated with laminar-to-turbulent transition before successful predictions of separated flow can be realized. In particular, we must understand the physics of free-shear layer instability, the formation, growth and propagation of turbulent spots within a near-wall free-shear layer, and the viscid-inviscid interaction between the bubble and free-stream flow.[1]

The early experimental investigations[2] on separation bubbles concentrated more on the effect of the bubbles on lift, drag and the mean flow downstream of the bubble than on transition, whereas the later investigations[3] concentrated on separated-flow transition and the effects of moving wakes. In general, these experiments were conducted on either airfoils, thick plates with circular leading edges, or flat plates in a variable area duct. As a result, data is available for numerous conditions at separation, a variety of bubbles, various types of bubble/free-stream interactions, and therefore a variety of transition conditions.

Early attempts at correlating the data were made by Crabtree [1959], Gaster [1969], Roberts [1980], Schmidt and Mueller [1989], and Mayle [1991], while the attempts by Hatman and Wang [1999], and Yaras [2001, 2002] are more recent. Methods to predict separated-flow transition range from the early ad hoc shear-layer models to the latest direct numerical simulations using the full Navier-Stokes equations. From their models, both Doenhoff [1938] and Horton [1969] demonstrated that reattachment strongly depends on where transition begins in the laminar free shear layer. Kwon and Pletcher [1979] and Wherle [1986] developed viscid-inviscid boundary-layer analyses to determine the point of laminar separation. The more recent analyses typically use a Reynolds-averaged Navier-Stokes approach, large-eddy simulations, and direct numerical simulations.[4]

A series of velocity profiles within a typical "leading-edge" separation bubble are shown in Fig. 8.17. Measured by Bellows and Mayle [1986], the bubble was produced on a thick, parallel-sided body with a semicircular leading edge. Although the leading edge joined the body smoothly at $x/R = \pi/2$, the surface has a step change in curvature at the joint that produces a local adverse pressure gradient and a short laminar separation bubble. Since these profiles were obtained using a single hot-wire where the direction of the flow cannot be sensed, the reverse velocity in the separated-recirculating region also appears to be positive. Furthermore, even where the mean velocity should vanish, the hot-wire senses a small but nonzero mean caused by large velocity fluctuations. This is particularly true near reattachment where the velocity fluctuations are the greatest. Nevertheless, the profiles shown in the figure are for a Reynolds number based on the incident velocity and leading edge radius of 50,000. In addition, the data are plotted against the distance away from the wall normalized by the local momentum thickness with each consecutive profile shifted to the right by an amount equal to 0.2.

While the first profile corresponds to that for laminar attached flow, the last three correspond to turbulent reattached flow. Between these are the profiles showing almost stagnant flow near the wall (below the dashed line) that is characteristic of separated recirculating flow. From these measurements, separation appears to occur near $x/R = 1.5$, while reattachment seems to occur near $x/R = 1.9$.

---

[1] For example, as pointed out by Hourmouziadis [1989] and Mayle [1991], large gains in gas-turbine performance can be obtained by not only understanding separated-flow transition, but also by using short separation bubbles to force transition at a desirable position.

[2] See for example Doenhoff [1938], Gault [1955], Gaster [1969], Roberts [1980], Bellows [1985], and Bellows and Mayle [1986].

[3] See for example Malkiel [1994], Malkiel and Mayle [1996], Hazarika and Hirsch [1997], Qiu and Simon [1997], Hatman [1997], Hatman and Wang [1998, 1999], Lou and Hourmouziadis [2000], Funazaki and Kato [2002], Volino [2002], Roberts and Yaras [2003], Lang et al. [2004], McAuliffe and Yaras [2005], Funazaki and Koyabu [2006], and Yaras [2001, 2002, 2008].

[4] See Vicedo et al. [2003] for an example of the RANS approach, which was briefly described in §7.3.2, Yang and Voke [2001], and Roberts and Yaras [2006] for examples of the LES approach, and Alam and Sandham [2000], Wissink and Rodi [2002], and McAuliffe and Yaras [2006] for examples of the DNS approach.

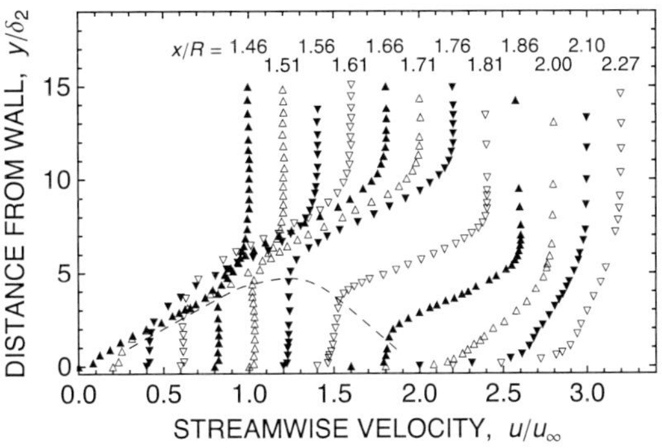

Figure 8.17.

Velocity profiles through a short leading-edge separation bubble. Data from Bellows and Mayle [1986]; $Re_R = 50{,}000$, $R = 76.2$mm.

Although the last three profiles may be classified as turbulent, Bellows and Mayle report that none of them possess a recognizable logarithmic region. Slightly downstream, however, beginning at $x/R = 2.7$, the profiles did have a logarithmic region but were highly non-equilibrium. They found that this was the case even at their last measurement position of $x/R = 6.5$, which is about 200 boundary-layer thicknesses or twelve bubble lengths downstream from reattachment. This behavior may be attributed to the slow relaxation of the turbulence structure in the outer layer of the reattached boundary layer; a phenomenon found by Coles [1964] for a large variety of "tripped" boundary-layer flows. In fact, the wake strength for the profile at their last measurement station is reported to be about $\Pi = 0.2$, which is about half of that expected for an equilibrium turbulent boundary layer with the same momentum thickness Reynolds number.[1]

If we fit a straight line to the free shear layer portion of these profiles, its y-intercepts at $u/u_\infty = 0$ and 1 may be considered the inner and outer limits of the shear layer. Furthermore, the average of the intercepts may be considered to be the position of maximum shear (vorticity) within the layer. The loci of these positions for all the profiles in the previous figure are plotted in Fig. 8.18a. Furthermore, the separation and reattachment locations, as well as the location of maximum displacement are shown. None of these lines are streamlines, but the time-averaged streamline dividing the recirculating flow within the bubble from that external to it lies between the inner limit of the shear layer and the locus of maximum shear. Considering this information alone, the height of the bubble is estimated to be about $0.04R$ or about ten percent its length. Hence, a rather straightforward viscid-inviscid *boundary-layer* analysis could be used to calculate this flow. Near reattachment, the lines of maximum shear and average position of the free-shear layer separate from one another (not shown) with the line of maximum shear turning downward to meet the wall at reattachment. The line representing the average position of the shear layer continues downstream as the average position of Coles' wake in the reattached turbulent boundary layer. After reattachment, the shear is maximum at the wall where the sublayers of the reattached turbulent flow develop (dark-shaded area in the figure).

The difference between the y-intercepts, which is roughly equivalent to the height of the shear layer, is plotted in part (b) of the figure. In this case, the change in the growth rate of the shear layer caused by transition begins near the position of maximum displacement and continues to a position somewhat downstream of reattachment. Hatman and Wang [1998] and Yaras [2001, 2002] found that the onset of transition in a short bubble nearly coincides with the position of maximum displacement and that it ends shortly thereafter.

[1] See §3.2.2 about the law of the wake and wake strength.

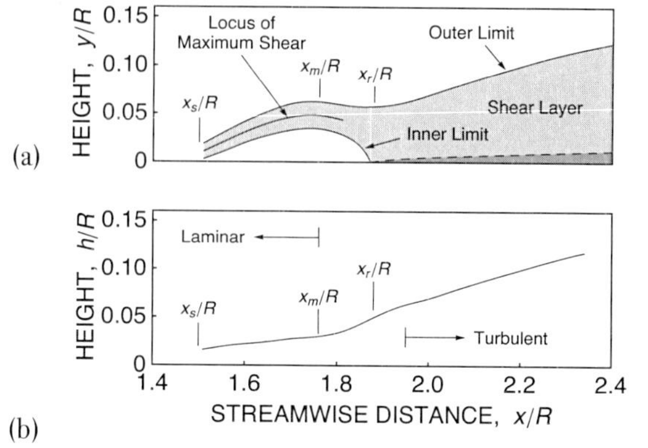

Figure 8.18a,b.
Loci of the extremities and maximum shear position in the free shear-layer for a short leading-edge separation bubble. Data from Bellows and Mayle [1986]; $Re_R = 50{,}000$, $R = 76.2$mm.

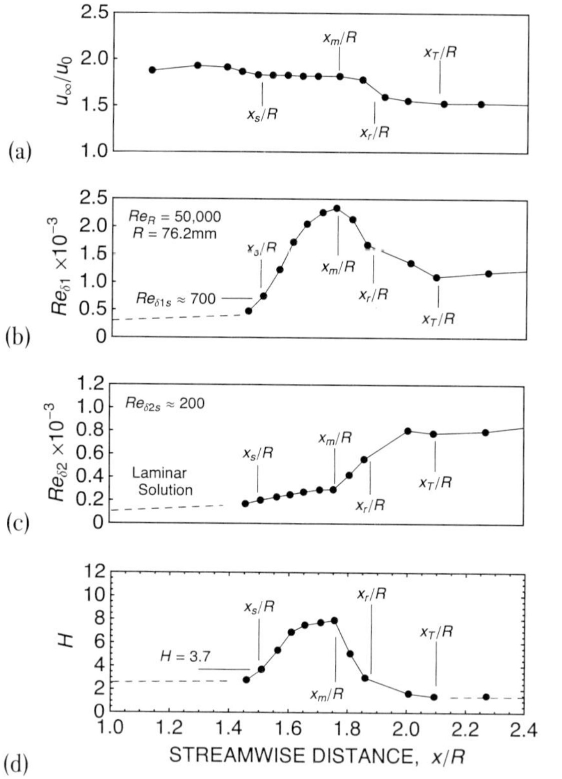

Figure 8.19a,b,c,d.
Streamwise variation of (a) the free-stream velocity, (b) the displacement thickness Reynolds number, (c) the momentum thickness Reynolds number, and (d) the shape factor through a short leading-edge bubble. Data from Bellows and Mayle [1986].

The streamwise variation of the free-stream velocity and integral boundary-layer parameters in the vicinity of the bubble are presented in Fig. 8.19. The plateau of constant velocity up to transition in part (a) of the figure reflects the region of relatively stagnant constant-pressure flow within the bubble. Here, u_0 is the velocity incident to the test body. Upstream of separation, the free-stream velocity equals that calculated using inviscid flow theory (not shown). Downstream, after the rapid deceleration and reattachment, the velocity quickly returns to the inviscid calculation (also not shown) which is characteristic of a short separation bubble.

The streamwise variation in displacement- and momentum-thickness Reynolds numbers are plotted in parts (b) and (c) of the figure. Their values at separation are about 700 and 200 respectively. The shape factor is plotted in part (d). Its value at separation is about the same as that given by Thwaites for laminar separation, that is, $H_s = 3.7$. The rapid growth of the displacement thickness downstream separation reflects the displacement effect of the relatively slow moving fluid in the bubble on the external flow. Once transition occurs, the displacement thickness decreases as the flow reattaches. The maximum displacement of the flow in this case occurs near $x_{xm} = 1.75R$. Initially, the effect of separation on the momentum thickness is hardly detectable. Since the free-stream velocity is nearly constant and the wall shear stress τ_0 is essentially zero in this region, we can expect little change in the momentum thickness through the first portion of the bubble.[1] However, once transition occurs, the decrease in the external velocity (adverse pressure gradient) causes a rapid increase in the momentum thickness in the last part of the bubble. Downstream of reattachment, the shape factor quickly attains a value close to its equilibrium value for a turbulent boundary layer on a flat plate. However, if H is considered to be the only relevant profile shape parameter, one would incorrectly conclude that the boundary layer is completely relaxed at the position $x_T = 2.1R$, or about half the length of the bubble downstream from reattachment, since, as noted earlier, Coles' wake strength is far from its equilibrium value at this position.

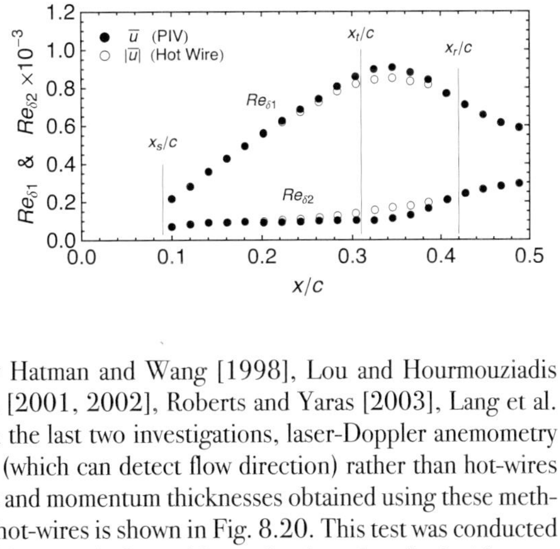

Figure 8.20.
A comparison of results for a short bubble using particle-image velocimetry with that which would be obtained using hot-wires. Data from McAuliffe and Yaras [2005]; Test case R40α8, $Re_c = 40{,}000$.

Similar results have since been obtained by Hatman and Wang [1998], Lou and Hourmouziadis [2000], Volino and Hultgren [2001], Yaras [2001, 2002], Roberts and Yaras [2003], Lang et al. [2004], and McAuliffe and Yaras [2005]. In the last two investigations, laser-Doppler anemometry (LDA) and particle-image velocimetry (PIV) (which can detect flow direction) rather than hot-wires were used. A comparison of the displacement and momentum thicknesses obtained using these methods with that which *would be* obtained using hot-wires is shown in Fig. 8.20. This test was conducted in a tow tank on an airfoil with a chord $c = 203$mm and a Reynolds number based on the incident velocity and chord of $Re_c = 40{,}000$. Clearly, the discrepancies are greatest near the position of maximum displacement where the recirculating flow is the strongest. As may be anticipated either from basic physics or their definitions in Eq. (1.75), the negative velocity within the bubble increases the displacement thickness and decreases the momentum thickness. Here the momentum thickness through the first portion of the bubble, compared with the hot-wire results shown in Fig. 8.19c, is

[1] See Eq. (1.76).

indeed nearly constant.

8.3.2 Onset of Transition.

Tollmien-Schlichting and Kelvin-Helmholtz instabilities. Intermittency profiles.

Transition in separated flow occurs first within the free shear layer where the vorticity is a maximum and the flow is least stable. If the upstream flow is stable, it quickly becomes unstable in the adverse pressure gradient before separation and afterwards when it leaves the surface as a free shear layer. Before separation, where the flow is attached and the effect of viscosity is important, a boundary layer with low free-stream turbulence is susceptible to the Tollmien-Schlichting type of instabilities (§4.2.1). After separation, where the detached flow is hardly affected by viscosity and the velocity profiles contain an inflection point, the shear layer is susceptible to the inviscid flow instabilities known as *Kelvin-Helmholtz waves* (Helmholtz [1882] and Kelvin [1910]). However, since velocity profiles in an adverse pressure gradient always contain an inflection point, Kelvin-Helmholtz instabilities may also develop before separation and Tollmien-Schlichting waves may persist downstream of separation. Indeed, both types of instabilities have been detected in separation bubbles.[1] Hence, separated-flow transition for low free-stream turbulence depends on a complicated interplay of stability mechanisms. For high free-stream turbulence, the scenario changes according to the turbulence level and length scale.[2]

Despite the numerous spectral analyses of time-dependent hot-wire measurements in many of the recent investigations, the role of the two instability mechanisms is not particularly clear. Although part of the problem appears to be associated with the fact that Tollmien-Schlichting waves have been observed to initiate Kelvin-Helmholtz instabilities, some of the confusion can be attributed to the fact that the characteristics of the most amplified wave for both mechanisms *near separation* are roughly the same through a good portion of the Reynolds number range. This can be understood better if we consider the Reynolds number effect on the most amplified instabilities near separation.

At laminar separation, the wave number of the most amplified Tollmien-Schlichting instability is $k_{TS} = 5.6(10)^{-4}u_\infty/\nu$ and its corresponding growth rate is $\omega_{TS} = 6.7(10)^{-5}\, u_\infty{}^2/\nu$, using the results from a temporally-amplified solution (see Appendix D, Table D.3 for $m = -0.91$). Whereas for a free shear layer of height h, the wave number and growth rate of the most amplified Kelvin-Helmholtz instability are $k_{KH} = 0.8/h$ and $\omega_{KH} = 0.2\, u_\infty/h$ (see Appendix E).[3] Thus,

$$k_{KH}/k_{TS} \approx 1400/Re_h \quad \text{and} \quad \omega_{KH}/\omega_{TS} \approx 3000/Re_h$$

where Re_h is the Reynolds number based on the free-stream velocity outside the shear layer. Since the velocity profiles in the shear layer and at separation are about the same, the celerity of both instabilities u_c will be about the same, and hence their ratio of frequencies $u_c k$ will be approximately equal to their wave number ratio. In addition, since the height of the shear layer near separation is about 1.2 times the displacement thickness and the shape factor at separation is about four, the shear-layer Reynolds number will be roughly $Re_h \approx 4.8 Re_{\delta 2_s}$, or larger allowing for the growth in the shear layer downstream of separation.[4] Hence, the above relations become

---

[1] See the experiments of Malkiel [1994], Malkiel and Mayle [1996], Watmuff [1999], and the numerical simulations of Yang and Voke [2001], and Abdalla and Yang [2004] about the Kelvin-Helmholtz instability mechanism, and the experiments of Volino and Hultgren [2001], Volino [2002], Volino and Bohl [2004], Roberts and Yaras [2004], and McAuliffe and Yaras [2005, 2006], and the numerical simulations of Roberts and Yaras [2006] and McAuliffe and Yaras [2007] about the Tollmien-Schlichting instability mechanism in separated-flow transition.

[2] See Roberts [1980], Bellows [1985], Volino [2002], Yaras [2002], Roberts and Yaras [2003], Volino and Bohl [2004] about the effects of elevated free-stream turbulence levels on separated-flow transition.

[3] Here h is defined as the difference between the y-intercepts with the $u/u_\infty = 0$ and 1 axes of the straight-line fit to the velocity profile through the central portion of the shear layer.

[4] According to the result shown in Fig. 8.18b, we can expect $4.8 < h/\delta_{2,s} < 8$.

$$\left(k_{KH}/k_{TS}\right)_{\max} \approx 290/Re_{\delta 2_s} \quad \text{and} \quad \left(\omega_{KH}/\omega_{TS}\right)_{\max} \approx 625/Re_{\delta 2_s} \tag{8.26}$$

where the subscript "max" has been added to emphasize the fact that these values correspond to the conditions for the *most* amplified instability. Clearly, considering the growth rates, the Kelvin-Helmholtz mechanism dominates when the separation Reynolds number is low and the Tollmien-Schlichting mechanism dominates when it is high. The difficulty is that the separation Reynolds numbers for most investigations cover this range, which in turn leads to the confusion about which mechanism is responsible for transition. Values of the above ratios across the range of relevant Reynolds numbers are provided in the first two columns of Table 8.1.

Table 8.1
Ratios of the most amplified Kelvin-Helmholtz and Tollmien-Schlichting instabilities in separation bubbles.

$Re_{\delta 2_s}$	$(k_{KH}/k_{TS})_{max}$	$(\omega_{KH}/\omega_{TS})_{max}$	$(S_{\delta 2_s})_{KH}$	$(S_{\delta 2_s})_{TS}$
100	2.9	6.25	0.013†	0.004
290	1	2.15	↓	0.013
625	0.46	1	↓	0.028
1000	0.29	0.62	0.013	0.045

† Allowing for the growth in the shear layer and using $Re_h \approx 8Re_{\delta 2_s}$ as the upper limit, this number reduces to 0.008.

According to these results, the ratios change most rapidly for Reynolds numbers $Re_{\delta 2_s}$ less than about three hundred. In this range, since the growth of Kelvin-Helmholtz waves is significantly greater than Tollmien-Schlichting waves, any pre-separation Tollmien-Schlichting instability that induces a Kelvin-Helmholtz wave will be quickly overwhelmed by it in the subsequent free shear layer. This behavior has been observed by Roberts and Yaras [2006] in their "large-eddy" numerical simulation of transition in a separation bubble. On the other hand, for separation Reynolds numbers greater than about three hundred, the growth rate of the two are about the same as previously noted.

Near $Re_{\delta 2_s}$ = 300, the frequencies (wavelengths) of the most amplified Kelvin-Helmholtz and Tollmien-Schlichting instabilities are about the same. For smaller Reynolds numbers, the frequency of the Kelvin-Helmholtz instability becomes two to three higher than the Tollmien-Schlichting instability, while for larger Reynolds numbers, the Tollmien-Schlichting frequency becomes two to three times higher. As it turns out, frequencies both above and below the primary Kelvin-Helmholtz frequency have been detected in separation bubbles and noticed in numerical simulations. In particular, the first subharmonic found is usually that caused by a rolling over and pairing of the primary Kelvin-Helmholtz vortices. Common in laminar free shear layers,[1] this pairing is also observed in separation bubbles,[2] but usually only when the separation Reynolds numbers $Re_{\delta 2_s}$ are about two hundred and less, that is, when the Kelvin-Helmholtz instability dominates.

The Strouhal numbers corresponding to the most amplified Kelvin-Helmholtz and Tollmien-Schlichting instabilities are listed in last two columns of Table 8.1. Based on the conditions at separation, they are defined by

[1] See for example Huang and Ho [1990], and Estevadeordal and Kleis [1999].

[2] See for example Gleyzes et al. [1985], Malkiel and Mayle [1996], and McAuliffe and Yaras [2005, Fig. 9].

$$S_{\delta 2,s} = f_{\max}\delta_{2,s}/u_{\infty,s} = \left(u_c k/2\pi\right)_{\max}\left(\delta_2/u_\infty\right)_s$$

where f is the circular frequency of the instability. Considering that the instabilities travel at the average speed of the shear-layer, these Strouhal numbers are given by

$$(S_{\delta 2,s})_{kh} \approx 0.064\delta_{2,s}/h \quad \text{and} \quad (S_{\delta 2,s})_{ts} \approx 4.5(10)^{-5}\,Re_{\delta 2,s} \tag{8.27}$$

Malkiel [1994] measured $S_{\delta 2,s} \approx 0.016$ when transition occurred via the Kelvin-Helmholtz mechanism, while Watmuff [1999] measured 0.0085, and Talan and Hourmouziadis [2002], and McAuliffe and Yaras [2005] measured $0.010 \leq S_{\delta 2,s} \leq 0.014$. In their numerical simulations, Ripley and Pauley [1993], and Muti Lin and Pauley [1996] obtained $0.005 \leq S_{\delta 2,s} \leq 0.008$, McAuliffe and Yaras [2007] obtained $S_{\delta 2,s} = 0.011$, while Yang and Voke [2001] found $0.005 \leq S_{\delta 2,s} \leq 0.011$ in their large-eddy simulations. In these investigations, even though Tollmien-Schlichting instabilities were detected, the dominant transition mechanism was perceived to be the inviscid Kelvin-Helmholtz mechanism. When both the measured and numerical results are compared with the results in Table 8.1, it appears that the dominant transition mechanism across the whole range of separation Reynolds number is also the inviscid Kelvin-Helmholtz mechanism.

When the Kelvin-Helmholtz mechanism dominates, the transition process is similar to that for a plane free-shear layer. First, the primary most amplified Kelvin-Helmholtz two-dimensional wave, often containing spanwise non-uniformities stemming from a secondary spanwise instability,[1] rolls up along the line of maximum shear into a series of two-dimensional vortices. Accompanying the rollup, whether the vortices pair or not, small three-dimensional fluctuations are generated within the shear layer between the vortices which subsequently produce the small scale turbulence associated with the onset of transition. For high separation Reynolds numbers, this system becomes fully turbulent within one wavelength of the primary instability from transition, or soon thereafter, and reattaches. For low Reynolds numbers, the system remains almost in tact as it converts downstream producing an extended periodic train of multiple separation and reattachment regions until the flow becomes mostly turbulent. In this case, the unsteady reattachment "zone" may be several wavelengths long. Furthermore, the large scale vortices associated with the primary Kelvin-Helmholtz wave may still be detected ten or more wavelengths downstream from the time-averaged point of reattachment.

When the Tollmien-Schlichting mechanism dominates, a similar scenario occurs except that harmonics, instead of subharmonics, seem to prevail. In either case, that is, whether the Kelvin-Helmholtz or Tollmien-Schlichting mechanism dominates, the small three-dimensional laminar fluctuations generated between the vortices in the shear layer are those that quickly become turbulent (similar to spot formation in attached flows). From here on, these nascent spots of turbulence grow, coalesce, and reattach randomly to eventually form a non-stationary undeveloped turbulent boundary layer. As in attached boundary layers, this process can be traced by measuring the intermittency.

Intermittency profiles. The first intermittency measurements in separation bubbles were obtained by Malkiel [1994]. Using the same test configuration as Bellows [1985], whose data were presented in Figs. 8.17–8.19, he obtained data for both short and long bubbles by varying the angle of attack of the test body. His data for a long bubble with low free-stream turbulence are presented in Figs. 8.21 and 8.22. In this case, the test was conducted for a Reynolds number based on the incident velocity and leading edge radius of $Re_R = 29{,}000$, an incident turbulence level of $Tu = 0.3$ percent, and an angle of attack of $\alpha = 8$ degrees. Similar data has since been recorded by Hatman and Wang [1998], Volino [2002], Roberts and Yaras [2003], Jiang and Simon [2003], Volino and Bohl [2004], and Yaras [2008] for a wide range of turbulence levels and bubbles.

---

[1] See Pierrehumbert and Widnall [1982].

In Fig. 8.21, the intermittency profiles using both a temporal and spatial discrimination scheme are plotted across the bubble at three streamwise positions through transition. The maximum intermittency at each streamwise location, measured equally as well by either discrimination scheme, occurs where the vorticity in the free-shear layer is a maximum. This is where the turbulence is first produced and continually generated. Hence, it is also where the time-averaged turbulence intensity is a maximum. The difference between the temporal and spatial measurements near the wall is a result of the stagnant flow within the bubble which renders the temporal method of detecting turbulent spots unreliable. The reader is referred to the work by Malkiel [1994] for further information about intermittency measurements in the recirculating zone.

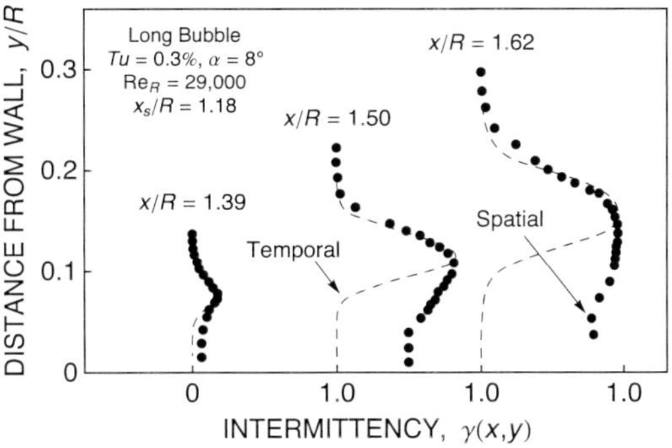

Figure 8.21.
Intermittency distribution in the free-shear layer of a long separation bubble. Data from Malkiel [1994]; also Malkiel and Mayle [1996].

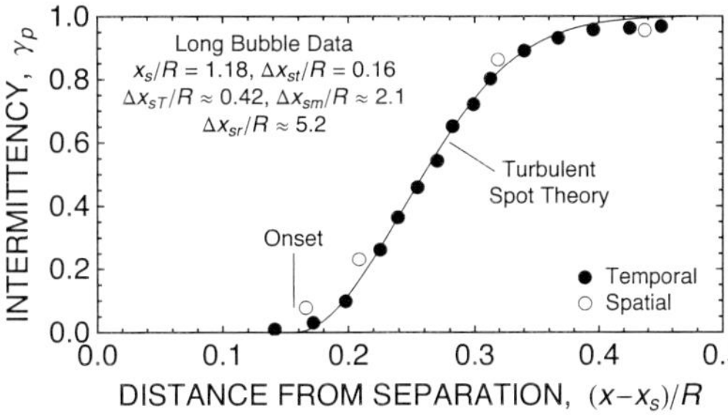

Figure 8.22.
Streamwise distribution of the maximum intermittency in the free-shear layer of a long separation bubble. Data from Malkiel [1994]; also Malkiel and Mayle [1996].

The streamwise variation of the "peak" shear-layer intermittency $\gamma_p(x)$ is plotted in Fig. 8.22, where all distances are measured from separation. As noted by Mayle [1991], Emmons turbulent spot theory should also apply to separated-flow transition since the presence of a wall is irrelevant to his theory. A curve fit using Narasimha's solution for a line source of turbulent spots, Eq. (5.18), is also plotted in the figure. If we introduce $Re_{xt,s}$ as the transition Reynolds number based on the velocity at separation and $(n\sigma)_s$ as the spot-rate parameter for transition in a free-shear layer, Eq. (5.18) takes the form

$$\gamma(x;x_t) = 1 - \exp\left[-(\hat{n}\sigma)_s \left(Re_x - Re_{xt,s}\right)^2\right] \tag{8.28}$$

where for the curve fit shown in the figure $Re_{xt,s}$ and $(n\sigma)_s$ are approximately 9200 and $1.9(10)^{-8}$. Comparing this spot-rate parameter with that in Fig. 6.15 for the pressure-gradient parameter of $\lambda_{\delta 2,s} = -0.082$ corresponding to separated flow, indicates that turbulent spots in a shear layer grow roughly an order of magnitude faster than those in a flow continually on the verge of separating. Consequently, transition in this separation bubble is rather short. The same is also found for short bubbles (see for example Hatman and Wang [1998]).

The physics associated with transition and reattachment in long bubbles is beyond the scope of this book, and frankly, still relatively unknown. Nevertheless, it should be noted from the data listed in Fig. 8.22 that transition begins shortly after separation, $\Delta x_{st}/\Delta x_{sr} = 0.03$, and ends long before the time-averaged flow reattaches, $\Delta x_{sT}/\Delta x_{sr} = 0.08$. This behavior has also been reported by Hatman and Wang [1998], who furthermore observed that the turbulent portions of the flow in long bubbles sometimes reattach and separate again before finally reattaching farther downstream. On the other hand, as measured by Gaster [1969], Yaras [2001, Cases 1, 2 and 3], and Jiang and Simon [2003, Part I], the distance between separation and transition in long bubbles can be substantial. In any case, the reader is referred to these articles for further details.

8.3.3 Bubble Transition Correlations.

Short bubbles. Bursting and long bubbles.

Calculating separated-flow transition suffers from the same problems we found in calculating transition in attached flows – namely, predicting the onset and rate of transition. In the past, calculating separated flow and the bubble's displacement effect on the mainstream flow was also a problem. However, with the appearance of modern fluid-dynamic computation methods, this is no longer the case except for long separation bubbles which are not only highly unsteady and three-dimensional, but also affect large portions of the flow. Nevertheless, for flows with low free-stream turbulence levels, the difficulties in determining the onset of transition are associated with calculating the nonlinear growth of the instabilities and the subsequent generation of the small-scale turbulence. For elevated turbulence levels, the difficulties are associated with knowing the "effective" frequency of turbulence associated with the free-shear layer and the subsequent generation of the small-scale turbulence caused by the interaction between these eddies and the flow. As a consequence, most practical methods for calculating separated-flow transition use correlations for the onset and rate of transition.

The earliest correlations for separated-flow transition assumed that transition from laminar to turbulent flow took place instantaneously within the shear layer of the bubble, and that this occurred at the end of the constant pressure or velocity plateau. Furthermore, experiments showed that the distance between laminar separation and this "point" of transition was characterized by a constant Reynolds number

$$Re_{st} = u_{\infty,s}\left(x_t - x_s\right)/\nu = \text{constant} \qquad (8.29)$$

where $u_{\infty,s}$ is the free-stream velocity at separation. However, later compilations by Tani [1964] and Horton [1969] showed that Re_{st} is not a constant but varies between 30,000 and 60,000. Roberts [1980] removed some of the scatter in the data by scaling Re_{st} with a turbulence factor that not only depends on turbulence level, but also on its length scale. In an attempt to further reduce the scatter, Schmidt and Mueller [1989] correlated Re_{st} against the momentum thickness Reynolds number at separation $Re_{\delta 2,s}$.[1] However, they also reported a large scatter which they attributed to the variety of techniques researchers used to evaluate the location of transition. Since $Re_{\delta 2,s}$ is not well predicted by boundary layer theory near separation, a significant improvement in the correlations was found by Mayle [1991] when only *measured* values were used. It seems incredible that such a complex, interacting flow can be described by a single parameter (as long as we don't consider free-stream turbulence and other external disturbances), but apparently it both reflects the feedback effect of the bub-

[1] Presently, since the thickness of the separated shear layer h is nearly equal to the displacement thickness, it appears that the displacement-thickness Reynolds number should be used.

ble on the external flow and provides an appropriate scale for the instabilities and turbulence generated in the separated free shear layer.

Short bubbles. When using only the measured values, Mayle determined that many of the distances from separation seem to scale according to the seven-tenths power of the momentum-thickness Reynolds number at separation. His plot of the distance between separation and maximum displacement (end of the velocity plateau) in short bubbles with low free-stream turbulence is reproduced in Fig. 8.23a. The line plotted in the figure is given by

$$Re_{sm} = 780 Re_{\delta 2s}^{0.7} \qquad \text{(short bubble)}$$

where

$$Re_{sm} = u_{\infty,s}\left(x_m - x_s\right)/\nu$$

The coefficient in Mayle's original correlation is 700.[1] Roberts and Yaras [2005] suggest using 835. The coefficient in the above expression, which best correlates the data, is roughly an average of the two.

(a)

(b)

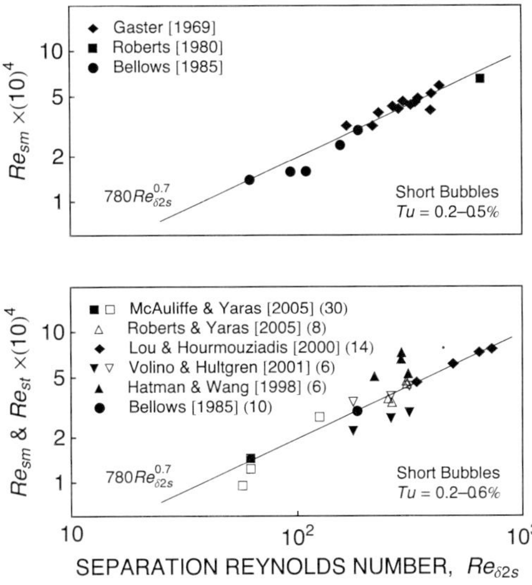

Figure 8.23a,b.
Reynolds number based on the constant pressure portion of a short bubble as a function of the momentum thickness Reynolds number at separation. Part (a) from Mayle [1991].

A comparison with more recent data is shown in part (b) of the figure, where Bellows' data point in part (a) corresponding to the data plotted in Figs. 8.17–8.19 is also shown. In this figure, the data for the distance from separation to the positions of maximum displacement Re_{sm} are plotted using the filled symbols, while that for the transition Reynolds number Re_{st} are plotted (if available) using the empty symbols. The location of maximum displacement is relatively easy to determine providing that enough velocity profiles are taken within the bubble. Therefore, the average number of profiles, including those directly before and after the bubble, is provided for each investigator in the legend.[2]

[1] At the time, transition was considered to be complete at the end of the pressure plateau rather than just beginning. Therefore, the coefficient here must be compared to that in Mayle's original correlation for Re_{sT}.

[2] It can be shown that the error in Re_{sm} will equal roughly one or one-half times $(N_{sm}-1)/(N_{sr}-1)$ depending on whether N_{sr} is odd or even, where N is the number of profiles.

The data for Re_{st} were determined by a variety of means. Hatman and Wang used the position where either the intermittency first exceeded a zero value, the structure of the shear layer first deviated from the laminar behavior, or the dissipation spectrum showed an increase in magnitude at selective frequencies. For short bubbles, they observed that Re_{st} coincides with Re_{sm}. On the other hand, Roberts and Yaras identified the onset of transition by the first streamwise location where turbulent spots were observed in the velocity traces, while McAuliffe and Yaras, considering linear stability theory, identified it as the position where the logarithmic growth of the turbulent shear stress became nonlinear. However, despite these different methods of determining transition, it appears 1) that the onset of transition is approximately where the displacement effect of the bubble is a maximum; and 2) that the above correlation may be used to determine it. Hence, for short bubbles with low free-stream turbulence, the onset of separated-flow transition is given by

$$Re_{st} \approx 780\, Re_{\delta 2,s}^{0.7}\ ; \qquad (\text{short bubble}, 0.002 < Tu < 0.006) \qquad (8.30)$$

This is different from Mayle's original correlation, which was obtained using the measured streamwise variation of turbulence intensity rather than intermittency to determine the onset of transition. It is also different from that proposed by Walker [1989] based on stability considerations and a minimum turbulent spot merging distance.

For elevated levels of turbulence, we would expect the distance between separation and transition to decrease. Presumably, the effect will not only depend on both the level and length scale of turbulence, but also on the distortion of the large-scale and decay of the small-scale turbulent eddies. In this regard, and as noted before, distortion typically leads to large-scale vortices stretched into the streamwise direction that will significantly affect the results if separation is preceded by an accelerating flow. However, descriptions regarding the flow leading to the creation and development of turbulent spots for elevated free-stream turbulence levels are scarce. Here, the concerned reader is referred to the work by Watmuff [1999] and D'Ovidio et al. [2001], who examine the growth of turbulent spots induced by a traveling wave packet.

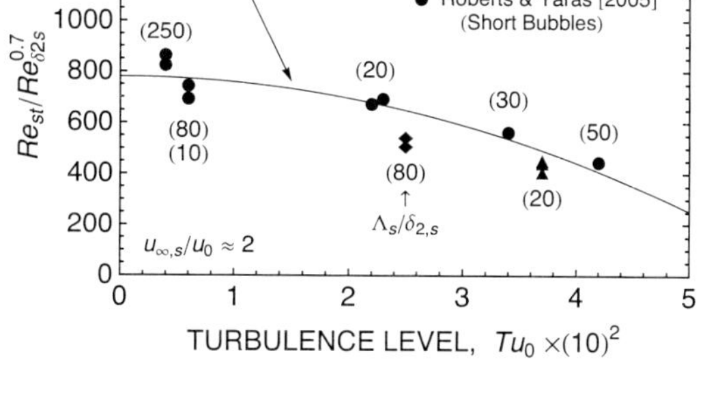

Figure 8.24.
Effect of free-stream turbulence on separated flow transition in short bubbles.

Despite these concerns, the measured effect of free-stream turbulence on the onset of separated-flow transition in short bubbles is shown in Fig. 8.24. The abscissa of the plot is the *incident* turbulence level, whereas the ordinate is the transition Reynolds number scaled by the momentum-thickness Reynolds number at separation raised to the seven-tenths power. Consequently, the intercept for $Tu_0 = 0$ should be 780. The data for this figure was obtained for an accelerating/decelerating flow around the circular leading edge of either a thick flat plate or airfoil where the velocity at separation was about twice the incident velocity. A correlation of these data, which reduces to Eq. (8.30) when $Tu_0 = 0$, is given by

$$Re_{st} = 780\left[1 - 0.027\left(100Tu_0\right)^2\right]Re_{\delta 2,s}^{0.7} \quad \text{(short bubble, } Tu_0 \leq 0.05\text{)} \tag{8.31}$$

Here it should be noted that while the four data points for the low turbulence levels in this figure seem to be problematic, the farthest still lies only twelve percent from the correlation. A much different perspective of these data points is found in Fig. 8.23b (the upward pointing empty triangles). The above correlation differs from that given by Roberts and Yaras, who attempted to include the effect of turbulent length scale in their correlation by using the ratio of the integral length scale of turbulence Λ_s divided by the length of their plate L. Since the length of their test plate should be irrelevant, a more appropriate parameter might be Λ_s divided by the momentum thickness at separation. The values of this ratio are given in the parentheses of Fig. 8.24, and since they don't indicate any particular trend when plotted in this format, it may be concluded that the effect of turbulent length scale on the onset of separated-flow transition is surprisingly small.

Finally, the correlations created by Hatman and Wang [1999], and Yaras [2002] should be addressed. Reminiscent of the early correlations given by Eq. (8.29), they are

$$Re_{xt} = 1.08Re_{xs} + 26{,}800 \quad \text{(Hatman and Wang)}$$
$$Re_{xt} = 1.04Re_{xs} + 35{,}700 \quad \text{(Yaras)}$$

for low free-stream turbulence. In both experiments, however, the range of separation Reynolds numbers is not only about the same, roughly $220 < Re_{\delta 2,s} < 380$, but small. Consequently, their correlations represent a linear approximation to the data within this range.

Another type of correlation using the distance from the peak free-stream velocity to transition has been proposed by Volino and Bohl [2004], who investigated separated-flow transition disturbed by a trip attached to the wall upstream of the velocity peak. Their correlation, which also includes the effect of free-stream turbulence, is based on the measured growth of small disturbances in the pre-transitional boundary layer. Here, the reader is referred to the original paper for the details.

Bursting and long bubbles. Below $Re_{\delta 2,s} \approx 300$, Gaster [1969] reports that a short bubble *may* suddenly "burst" into a long one – which in most cases would be an aerodynamic disaster.[1] Malkiel [1994] observed that just a two degree increase in the angle of attack caused this to happen on his test body. Hatman and Wang [1998] observed that either a long or short bubble may occur when the separation Reynolds number was less than about three hundred depending on the pressure gradient. In particular they found short bubbles when $240 < Re_{\delta 2,s} < 320$ and $-1.3(10)^{-6} < K_s < -0.3(10)^{-6}$, that is, for mildly adverse pressure gradient, and long bubbles when $Re_{\delta 2,s} < 240$ and $K_s < -1.3(10)^{-6}$, that is, for strongly adverse gradients. Furthermore, they found that the bubbles could change from long to short, or vise versa, by simply changing the *downstream* pressure distribution, sometimes referred to as "back" pressure. In other words, these are flows with a strong "feedback loop."

Transition and maximum displacement Reynolds numbers for *long* bubbles are plotted in Fig. 8.25. In addition, the correlation given by Eq. (8.30) for short bubbles and the value below which either long or short bubbles may occur are also shown. Furthermore, the original transition Reynolds number correlation given by Mayle [1991] for long bubbles, namely,

$$Re_{st} \approx 1300 Re_{\delta 2,s}^{0.7} \quad \text{(long bubble, } 0.002 < Tu < 0.006\text{)} \tag{8.32}$$

is shown by the dashed line. Again, as in Fig. 8.23b, the data for Re_{sm} are plotted using the filled symbols while those for Re_{st} are plotted using the empty symbols. Since the two coincided in Hatman and Wang's test, only one of them is shown. However, Malkiel observed a large difference between these

[1] Note that his criterion corresponds to the region where Kelvin-Helmholtz waves are the dominant transition mechanism.

two positions. Furthermore, he observed a large difference between the end of transition Re_{sT}, shown by the circled-cross data point in the figure, and the position of maximum displacement.

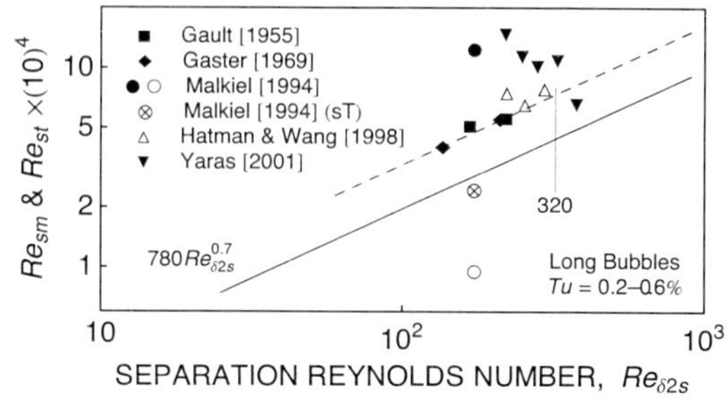

Figure 8.25.
Transition Reynolds number for a long bubble. Filled symbols denote the Re_{sm} data, while the empty symbols denote the data for Re_{st}.

Clearly, the separation Reynolds number alone is not sufficient to correlate transition in long bubbles. Furthermore, considering that Malkiel's transition Reynolds number is much less than that for a short bubble, the physics associated with the feedback loop in long and short bubbles must be substantially different. For long bubbles, this has barely been investigated.

8.4 Post- and Reverse Transition.

Early post-transitional flow. Reverse transition.

In this section, the flow immediately downstream transition, known as *early post-transitional flow*, and transition from turbulent to laminar flow, known as *reverse transition* will be discussed. Since little is known about the former, and since the latter is easily calculated, both discussions are brief.

8.4.1 Post-Transitional Flow.

Despite evidence to the contrary, it is generally accepted that a "normal" two-dimensional turbulent boundary layer follows once the last remnants of laminar flow within the boundary layer can be detected, that is, once transition as indicated by the near-wall intermittency ends. In reality, however, once transition "ends," remnants of the turbulent spots remain. Therefore, this phase of post-transition may be considered to be a *spot-dependent* turbulent flow. As shown by Zilberman et al. [1977], turbulent spots can be detected far downstream within a turbulent boundary layer. As might be expected, this is particularly noticeable in the outer portions of the newly formed turbulent boundary layer before the caps of individual spots finally merge and become undetectable within the wake of the turbulent boundary layer. The whole process is random, unsteady, and results in a turbulent boundary layer with a wake parameter Π substantially less than that for a fully-developed turbulent boundary layer.[1] Since the latter isn't attained until $Re_{\delta 2} \approx 4000$, a quick calculation using Eq. (3.41) and a comparison with the results shown in Fig. 5.18 indicates that the region of early post-transition for flow over a flat plate with low free-stream turbulence is about as long as transition itself. In spite of this, the region has been thoroughly neglected; probably because the flow in the wake of a turbulent boundary layer barely affects that near the wall and this can be calculated quite well using modern computational schemes.

Nevertheless, if we consider the region of early post-transitional flow to be part of the whole transition process, that is, included with the pre-transitional and transitional regions of flow, then the true extent of transition from laminar to fully turbulent flow becomes substantial.

---

[1] Here the reader may wish to consult §3.2.2 regarding Coles' law of the wake, and §8.3.1 regarding post-transitional flow downstream separation bubbles.

8.4.2 Reverse Transition.

Sometimes known as *re-laminarization*, reverse transition was first noticed in flows though nozzles with very strong acceleration. It was later studied by Julien et al. [1969] and Jones and Launder [1972]. Since accelerations on the pressure side of most gas turbine airfoils near the trailing edge, in the exit ducts of combustion chambers, and on the suction side of turbine airfoils near the leading edge are generally high, this phenomenon is particularly important to the gas turbine designer.

In any case, it is not difficult to imagine how the turbulent eddies within an accelerating boundary layer stretch and dissipate to a point where the dissipation of turbulent kinetic energy becomes greater than its production within the boundary layer with a result that the flow eventually reverts to one that is viscously dominated. Since these processes are easily taken into account by all modern computational programs, reverse transition is no longer an issue for these designers.

As a rule of thumb, however, we note that reverse transition can be expected to occur at low turbulence levels when the acceleration parameter exceeds $3(10)^{-6}$, that is, when

$$K = \frac{\nu}{u_\infty^2}\frac{du_\infty}{dx} \geq 3(10)^{-6} \tag{8.33}$$

This criterion has been previously plotted in terms of the pressure gradient parameter $\lambda_{\delta 2}$ in Figs. 6.9, 6.11 and 6.12. In this case, the criterion is given by

$$\lambda_{\delta 2} \geq 3(10)^{-6} Re_{\delta 2}^2 \qquad \text{(reverse transition)} \tag{8.34}$$

For high free-stream turbulence levels, this value is expected to be about the same since without any mechanism for production in the free stream, the turbulence there will dissipate before that in the boundary layer.

A corollary of the above result is that *forward* transition cannot occur when K exceeds $3(10)^{-6}$. This was initially investigated by Schraub and Kline [1965], who observed that turbulent streaks in the sublayers of a turbulent boundary layer were completely suppressed at this level of acceleration. In addition, it has been the author's experience that forward transition begins almost immediately once K falls below this value. Hence, designers should always pay attention to the variation of the acceleration parameter when calculating transitional flow in highly accelerated flows. In this regard, the interested reader may want to consult Mayle [1991, Part III].

References

Abdalla, I.E., and Yang, Z., 2004. "Numerical Study of the Instability Mechanism in Transitional Separating- Reattaching Flow," Intl. Jour. Heat and Fluid Flow, 25, pp. 593–605.

Abu-Ghannam, B.J., and Shaw, R., 1980, "Natural Transition of Boundary Layers–The Effects of Turbulence, Pressure Gradient and Flow History," J. Mech. Engng. Sci., 22, pp. 213–228.

Addison, J.S., and Hodson, H.P., 1990, "Unsteady Transition in an Axial Flow Turbine: Part 1–Measurements on the Turbine Rotor, Part 2–Cascade Measurements and Modeling," J. Turbomachinery, 112, pp. 206–214, 215–221.

Alam, M., and Sandham, N.D., 2000, "Direct Numerical Simulation of 'Short' Laminar Separation Bubbles with Turbulent Reattachment," J. Fluid Mech., 403, pp. 223–250.

Bellows, W.J., 1985, "An Experimental Study in Leading Edge Separating-Reattaching Boundary Layer Flows," Ph.D. Dissertation, Rensselaer Polytechnic Institute, Troy.

Bellows, W.J. and Mayle, R.E., 1986, "Heat Transfer Downstream of a Leading Edge Separation Bubble," J. Turbomachinery, 108, pp.131–136.

Chakka, P., and Schobeiri, M.T., 1999, "Modeling of Unsteady Boundary Layer Transition on a

Curved Plate under Periodic Unsteady Flow Condition: Aerodynamic and Heat Transfer Investigations," J. Turbomachinery, 121, pp. 88–97.

Cho, N.H., Liu, X., Rodi, W., and Schönung, B., 1993, "Calculation of Wake-Induced Unsteady Flow in a Turbine Cascade," J. Turbomachinery, 115, pp. 675–686.

Coles, D.E., 1964, "The Turbulent Boundary Layer in a Compressible Fluid," Phys. Fluids, 7, pp. 1403–1423; also 1962, RAND Corporation Report B-403-PR.

Crabtree, L.F., 1959, "The Formation of Regions of Separated Flow on Wing Surfaces," ARC R&M No. 3122, pp. 1–28.

Doenhoff, A.E. von, 1938, "A Method of Rapidly Estimating the Position of the Laminar Separation Point," NACA TH 671.

Dong, Y., and Cumpsty, N.A., 1990, "Compressor Blade Boundary Layers: Part 1–Test Facility and Measurements with No Incident Wakes, and Part 2–Measurements with Incident Wakes," J. Turbomachinery, 112, pp. 222–240.

Doorly, D.J., 1988, "Modelling the Unsteady Flow in a Turbine Rotor Passage," J. Turbomachinery, 110, pp. 27–37.

Doorly, D.J., and Oldfield, M.L.G., 1985, "Simulation of the Effects of Shock Wave Passing on a Turbine Rotor Blade," J. Engng. for Gas Turbines and Power, 107, pp. 998–1006.

D'Ovidio, A., Harkins, J.A., and Gostelow, J.P., 2001, "Turbulent Spots in Strong Adverse Pressure Gradients: Part 1–Spot Behavior," ASME Paper 2001-GT-0194.

Dring, R.P., Blair, M.F., Joslyn, H.D., Power, G.D., and Verdon, J.M., 1986, "The Effects of Inlet Turbulence and Rotor/Stator Interactions on the Aerodynamics and Heat Transfer of a Large-Scale Rotating Turbine Model," NASA CR 4079.

Dullenkopf, K., 1992, "Untersuchungen zum Einfluß periodisch instationärer Nachlaufströmungen auf den Wärme-übergang konvektiv gekühlter Gasturbinenschaufeln," Dr.-Ing. Dissertation, Universität Karlsruhe, Germany.

Dullenkopf, K., Schulz, A., and Wittig, S., 1991, "The Effect of Incident Wake Conditions on the Mean Heat Transfer on an Airfoil," J. Turbomachinery, 113, pp. 412–418.

Dullenkopf, K., and Mayle, R.E., 1994, "The Effects of Incident Turbulence and Moving Wakes on Laminar Heat Transfer in Gas Turbines," J. Turbomachinery, 116, pp. 23–28.

Emmons, H.W., 1951, "The Laminar-Turbulent Transition in a Boundary Layer, Part I," J. Aero. Sci., 18, pp. 490–498.

Estevadeordal, J., and Kleis, S.J., 1999, "High-Resolution Measurements of Two-Dimensional Instabilities and Turbulence Transition in Plane Mixing Layers," Exp. Fluids, 27, pp. 378–390.

Funazaki, K., 1996, "Unsteady Boundary Layers on a Flat Plate Disturbed by Periodic Wakes: Part I–Measurement of Wake-Affected Heat Transfer and Wake-Induced Transition Model, and Part II–Measurement of Unsteady Boundary Layers and Discussion," J. Turbomachinery, 118, pp. 327–336, 337–346.

Funazaki, K., and Aoyama, Y., 2000, "Studies on Turbulence Structure of Boundary Layers Disturbed by MovingWakes," ASME Paper 2000-GT-0272.

Funazaki, K., and Kato, Y., 2002, "Studies on a Blade Leading Edge Separation Bubble Affected by Periodic Wakes: Its Transitional Behavior and Boundary Layer Loss Reduction," ASME Paper GT-2002-30221.

Funazaki, K., and Koyabu, E., 2006, "Studies on Wake-Induced Bypass Transition of Flat-Plate Boundary Layers under Pressure Gradients and Free-Stream Turbulence," ASME Paper GT2006-

91103.

Funazaki, K., Kitazawa, T., Koizumi, K., and Tanuma, T., 1997, "Studies on Wake-Disturbed Boundary Layers under the Influences of Favorable Pressure Gradient and Free-Stream Turbulence; Part II–Effect of Free-Stream Turbulence," ASME Paper 97-GT-453.

Gaster, M., 1969, "The Structure and Behavior of Laminar Separation Bubbles," ARC R&M 3595.

Gault, D.E., 1955, "An Experimental Investigation of Regions of Separated Laminar Flow," NACA TN 3505.

Gleyzes, C., Cousteix, J., and Bonnet, J.L., 1985, "Theoretical and Experimental Study of Low Reynolds Number Transitional Separation Bubbles," Proceedings of the Conference on Low Reynolds Number Airfoil Aerodynamics, pp. 137–152.

Gostelow, J.P., Blunden, A.R., and Walker, G.J., 1994, "Effects of Free-Stream Turbulence and Adverse Pressure Gradients on Boundary Layer Transition," J. Turbomachinery, 116, pp. 392–404.

Gostelow, J.P., Walker, G.J., Solomon, W.J., Hong, G., and Melwani, N., 1997, "Investigation of the Calmed Region Behind a Turbulent Spot," J. Turbomachinery, 119, pp. 802–809.

Halstead, D.E., Wisler, D.C., Okiishi, T.H, Walker, G.J., Hodson, H.P., and Shin, H.-W., 1997, "Boundary Layer Development in Axial Compressors and Turbines; Part 1–Composite Picture," J. Turbomachinery, 119, pp. 114–127.

Hatman, A., 1997, "Laminar-Turbulent Transition in Separated Boundary Layers," Ph.D. Thesis, Clemson University, Clemson, SC.

Hatman, A., and Wang, T., 1998, "Separated Flow Transition," Report No. GTL-9801, Gas Turbine Laboratory, Dept. Mech. Engng., Clemson University, Clemson, SC; also "Separated-Flow Transition: Part I–Experimental Methodology and Mode Classification, and Part II–Experimental Results," ASME Papers 98-GT-461 and 98-GT-462.

Hatman, A., and Wang, T., 1999, "A Prediction Model for Separated-Flow Transition," J. Turbomachinery, 121, pp. 594–602.

Hazarika, B.K., and Hirsch, C., 1997, "Transition Over C4 Leading Edge and Measurement of Intermittency Factor Using PDF of Hot-Wire Signal," J. Turbomachinery, 119, pp. 412–425.

Helmholtz, H., 1882, "Über discontinuirliche Flüssigkeitsbewegungen, Wissenschaftliche Abhandlungen, J.A. Barth, Leipzig, pp. 146–157.

Herbst, R., 1980, "Entwicklung von Grenzschichten bei instationärer Zuströmung," Ph.D Thesis, D17, Technische Hochschule Darmstadt, Germany.

Hodson, H.P., 1990, "Modelling Unsteady Transition and Its Effects on Profile Loss," AGARD-CP-468, Paper 18.

Hodson, H.P., Addison, J.S., and Shepherdson, C.A., 1992, "Models for Unsteady Wake-induced Transition in Axial Turbomachines," J. Physique III of France, 2, pp. 545–574.

Horton, H.P., 1969, "A Semi-Empirical Theory for the Growth and Bursting of Laminar Separation Bubbles," ARC CP No. 1073.

Hourmouziadis, J., 1989, "Aerodynamic Design of Low Pressure Turbines," AGARD Lecture Series, 167.

Huang, L.S., and Ho, C.M., 1990, "Small-Scale Transition in a Plane Mixing Layer," J. Fluid Mech., 210, pp. 475–500.

Jiang, N., and Simon, T.W., 2003, "Modeling Laminar-to-Turbulent Transition in a Low-Pressure Turbine Flow which is Unsteady due to Passing Wakes: Part I–Transition Onset, Part II–Transition Path," ASME Papers No. GT2003-38787 and GT2003-38963.

Johnson, A.B., Oldfield, M.L.G., Rigby, M.J., and Giles, M.B., 1990, "Nozzle Guide Vane Shock Wave Propagation and Bifurcation in a Transonic Turbine Rotor," ASME Paper 90-GT-310.

Jones, W.P., and Launder, B.E., 1972, "The Prediction of Laminarization with a Two-Equation Model of Turbulence," Int. J. Heat Mass Transfer, 15, pp. 301–314.

Julien, H.L., Kays, W.M., and Moffat, R.J., 1969, "The Turbulent Boundary Layer on a Porous Plate: Experimental Study of the Effects of a Favorable Pressure Gradient," Thermo. Sci. Div. Rep. HMT-4, Stanford University, Palo Alto, CA.

Kelvin, Lord, 1910, "Hydrodynamics and General Dynamics," Mathematical and Physical Papers, IV, Cambridge, UK.

Kerrebrock, J.L., and Mikolajczak, A.A., 1970, "Intra-Stator Transport of Rotor Wakes and Its Effect on Compressor Performance," J. Engng. for Power, 92, pp. 359–369.

Kwon, C.L., and Pletcher, R.H., 1979, "Prediction of Incompressible Separated Boundary Layers Including Viscous-Inviscid Interaction," J. Fluids Engng., 101, pp. 466–472.

Lang, M., Rist, U., and Wagner, S., 2004. "Investigations on Controlled Transition Development in a Laminar Separation Bubble by Means of LDA and PIV," Experiments in Fluids, 36, pp. 43–52.

Liu, X., and Rodi, W., 1991, "Experiments on Transitional Boundary Layers with Wake-Induced Unsteadiness," J. Fluid Mech., 231, pp. 229–256.

Lou, W., and Hourmouziadis, J., 2000, "Separation Bubbles under Steady and Periodic-Unsteady Main Flow Conditions," J. Turbomachinery, 122, pp. 634–643.

Malkiel, E., 1994, "An Experimental Investigation of a Separation Bubble," Ph.D. Dissertation, Rensselaer Polytechnic Institute, Troy, NY.

Malkiel, E., and Mayle, R.E., 1996, "Transition in a Separation Bubble," J. Turbomachinery, 118, pp. 752–759.

Mayle, R.E., 1991, "The Role of Laminar-Turbulent Transition in Gas Turbine Engines," J. Turbomachinery, 113, pp. 509–537.

Mayle, R.E., 1993, "Unsteady, Multimode Transition in Gas Turbine Engines," Heat Transfer and Cooling in Gas Turbines, AGARD CP 527.

Mayle, R.E. and Dullenkopf, K., 1989, "A Theory for Wake-Induced Transition," J. Turbomachinery, 112, pp. 188–195.

Mayle, R.E. and Dullenkopf, K., 1990, "More on the Turbulent-Strip Theory for Wake-Induced Transition," J. Turbomachinery, 113, pp. 428–432.

McAuliffe, B.R., and Yaras, M.I., 2005, "Separation-Bubble-Transition Measurements on a Low RE Airfoil Using Particle Image Velocimetry," ASME Paper GT2005-68663.

McAuliffe, B.R., and Yaras, M.I., 2006, "Numerical Study of Instability Mechanisms Leading to Transition in Separation Bubbles," ASME Paper GT2006-91018.

McAuliffe, B.R., and Yaras, M.I., 2007, "Transition Mechanisms in Separation Bubbles Under Low and Elevated Freestream Turbulence," ASME Paper GT2007-27605.

Michelassi, V., Martelli, F., Dénos, R., Arts, T., and Severing, C.H., 1999, "Unsteady Heat Transfer in Stator-Rotor Interaction by Two-Equation Turbulence Model," J. Turbomachinery, 121, pp. 436–447.

Muti Lin, J.C., and Pauley, L.L., 1996, "Low-Reynolds-Number Separation on an Airfoil," AIAA Journal, 34, pp. 1570–1577.

Narasimha, R., 1985, "The Laminar-Turbulent Transition Zone in the Boundary Layer," Prog. Aerospace Sci., 22, pp. 29–80.

Orth, U., 1991, "Untersuchung des Umschlagvorganges von Platten- und Zylindergrenzschichten bei ungestörter und stationär oder periodisch gestörter Zuströmung," Dr.-Ing. Dissertation, Technische Hochschule Darmstadt, Germany.

Orth, U., 1992, "Unsteady Boundary Layer Transition in Flow Periodically Disturbed by Wakes," ASME Paper 92-GT-283.

Pierrehumbert, R.T., and Widnall, S.E., 1982, "The Two- And Three-Dimensional Instabilities of a Spatially Periodic Shear Layer," J. Fluid Mech., 114, pp. 59–82.

Pfeil, H., and Herbst, R., 1979, "Transition Procedure of Instationary Boundary Layers," ASME Paper 79-GT-128.

Pfeil, H., Herbst, R., and Schröder, T., 1983, "Investigation of the Laminar-Turbulent Transition of Boundary Layers Disturbed by Wakes," J. Engng. Power, 105, pp. 130–137.

Qiu, S., and Simon, T.W., 1997, "An Experimental Investigation of Transition as Applied to Low Pressure Turbine Suction Surface Flows," ASME Paper 97-GT-455.

Ramesh, O.N., and Hodson, H.P., 1999, "A New Intermittency Model Incorporating the Calming Effect," 3rd European Conference on Turbomachinery, IMechE, London, UK.

Ripley, M.D., and Pauley, L.L., 1993, "The Unsteady Structure of Two-Dimensional Steady Laminar Separation," Phys. Fluids A, 5, pp. 3099–3106.

Roberts, W.B., 1980, "Calculation of Laminar Separation Bubbles and Their Effect on Airfoil Performance," AIAA Journal, 18, pp. 25–31.

Roberts, S.K., and Yaras, M.I., 2003, "Measurements and Prediction of Free-Stream Turbulence Effects on Attached-Flow Boundary-Layer Transition," ASME Paper GT2003-38261.

Roberts, S.K., and Yaras, M.I., 2004, "Boundary-Layer Transition in Separation Bubbles over Rough Surfaces," ASME Paper GT2004-53667.

Roberts, S.K., and Yaras, M.I., 2005, "Modeling Transition in Separated and Attached Boundary Layers," J. Turbomachinery, 127, pp. 402–411.

Roberts, S.K., and Yaras, M.I., 2006, "Large-Eddy Simulation of Transition in a Separation Bubble," J. Fluids Engng., 128, pp. 232–238.

Schmidt, G.S., and Mueller, T.J., 1989, "Analysis of Low Reynolds Number Separation Bubbles Using Semiempirical Methods," AIAA Journal, 27, pp. 993–1001.

Schobeiri, M.T., 2005, "Intermittency Based Unsteady Boundary Layer Transition Modeling, Implementation into Navier-Stokes Equations," ASME Paper GT2005-68375.

Schobeiri, M.T., and Radke, R., 1994, "Effects of Periodic Unsteady Wake Flow and Pressure Gradient on Boundary Layer Transition Along The Concave Surface of a Curved Plate," ASME Paper 94-GT-327.

Schraub, F.A., and Kline, S.J., 1965, Mech. Engng. Dept. Rep. MD-12, Stanford University, Palo Alto, CA.

Schröder, T., 1985, "Entwicklung des instationären Nachlaufs hinter quer zur Strömungsrichtung bewegten Zylindern und dessen Einfluß auf das Umschlagverhalten von ebenen Grenzschichten stromabwärts angeordneter Versuchskörper," Dr.-Ing. Dissertation, Technische Hochschule Darmstadt, Germany.

Schröder, T., 1989, "Measurements with Hot-Film Probes and Surface-Mounted Hot-Film Gauges in a Multistage Low-Pressure Turbine," 1989 European Propulsion Forum, Bath, UK, 15, pp.1–27.

Schubauer, G.B., and Klebanoff, P.S., 1955, "Contribution to the Mechanism of Boundary-Layer Transition," NACA TN 3489.

Schulte, V., and Hodson, H.P., 1998, "Prediction of the Becalmed Region for LP Turbine Profile Design," J. Turbomachinery, 120, pp. 839–846.

Solomon, W.J., Walker, G.J., and Gostelow, J.P., 1996, "Transition Length Prediction for Flows with Rapidly Changing Pressure Gradients," J. Turbomachinery, 118, pp. 744–751.

Steelant, J., and Dick, E., 2001, "Modeling of Laminar-Turbulent Transition for High Freestream Turbulence," J. Fluids Engng., 123, pp. 22–30.

Talan, M., and Hourmouziadis, J., 2002, "Characteristic Regimes of Transitional Separation Bubbles in Unsteady Flow," Flow, Turbulence and Combustion, 69, pp. 207–227.

Tani, I., 1964, "Low-Speed Flows Involving Bubble Separations," Progress in Aeronautical Science, 5, pp. 70–103.

Vicedo, J., Vilman, S., Dawes, W.N., and Savill, A.M., 2003, "Intermittency Transport Modeling of Separated Flow Transition," ASME Paper GT2003-38719.

Volino, R.J., 2002, "Separated Flow Transition under Simulated Low-Pressure Turbine Airfoil Conditions: Part 1–Mean Flow and Turbulence Statistics, Part 2–Turbulence Spectra," J. Turbomachinery, 124, pp. 645–655, 656–664.

Volino, R.J., and Bohl, D.G., 2004, "Separated Flow Transition Mechanism and Prediction with High and Low Freestream Turbulence Under Low Pressure Turbine Conditions," ASME Paper GT2004-53360.

Volino, R.J., and Hultgren, L.S., 2001, "Measurements in Separated and Transitional Boundary Layers Under Low-Pressure Turbine Airfoil Conditions," J. Turbomachinery, 123, pp. 189–197.

Walker, G.J., 1974, "The Unsteady Nature of Boundary Layer Transition on an Axial Compressor Blade," ASME Paper 74-GT-135.

Walker, G.J., 1989, "Modeling of Transitional Flow in Laminar Separation Bubbles," 9th Int. Sym. Air Breathing Engines, pp. 539–548.

Watmuff, J.H., 1999. "Evolution of a Wave Packet into Vortex Loops in a Laminar Separation Bubble," J. Fluid Mech., 397, pp. 119–169.

Wherle, V.A., 1986, "Determination of the Separation Point in Laminar Boundary Layer Flows," AIAA Journal, 25, pp. 1636–1642.

Wissink, J.G., and Rodi, W., 2002, "DNS of Transition in a Laminar Separation Bubble," Advances in Turbulence IX, Proceedings of the Ninth European Turbulence Conference (eds: I. P. Castro and P. E. Hancock), Southampton, UK.

Wittig, S., Schulz, A., Dullenkopf, K., and Fairbank, J., 1988, "Effects of Free-Stream Turbulence and Wake Characteristics on the Heat Transfer Along a Cooled Gas Turbine Blade," ASME Paper 88-GT-179.

Yang, Z., and Voke, P.R., 2001, "Large-Eddy Simulation of Boundary-Layer Separation and Transition at a Change of Surface Curvature," J. Fluid Mech., 439, pp. 305–333.

Yaras, M.I., 2001, "Measurements of the Effects of Pressure-Gradient History on Separation-Bubble Transition," ASME Paper GT-2001-0193.

Yaras, M.I., 2002, "Measurements of the Effects of Freestream Turbulence on Separation-Bubble Transition," ASME Paper GT-2002-30232.

Yaras, M.I., 2008, "Instability and Separation in a Separation Bubble under a Three-Dimensional Free-stream Pressure Gradient," ASME Paper GT2008-51256.

Zilberman, M., Wygnanski, I., and Kaplan, R.E., 1977, "Transitional Boundary Layer Spot in a Fully Turbulent Environment," Phys. Fluids Supplement, 20, pp. 258–271.

Appendices,
Supplement on Cartesian Tensors,
Bibliography & Indices

Appendix A – Blasius' Function

Solution to Blasius' equation $ff''+2f''' = 0$ for laminar flow on a flat plate with a zero pressure gradient where $\psi(x, y) = f(\eta)\sqrt{(\nu\bar{u}_\infty x)}$ and $\eta = y/\sqrt{(\nu x/\bar{u}_\infty)}$.

Table A.1
Solution to Blasius' equation.

η	f	f'	f''	η	f	f'	f''
0	0	0	0.33206	4.6	2.88826	0.98269	0.02948
0.2	0.00664	0.06641	0.33199	4.8	3.08534	0.98779	0.02187
0.4	0.02656	0.13276	0.33147	5	3.28329	0.99155	0.01591
0.6	0.05974	0.19894	0.33008	5.2	3.48188	0.99425	0.01134
0.8	0.10611	0.26471	0.32739	5.4	3.68094	0.99616	0.00793
1	0.16557	0.32978	0.32301	5.6	3.88031	0.99748	0.00543
1.2	0.23795	0.39378	0.31659	5.8	4.0799	0.99838	0.00365
1.4	0.32298	0.45626	0.30787	6	4.27964	0.99898	0.0024
1.6	0.42032	0.51676	0.29666	6.2	4.47948	0.99937	0.00155
1.8	0.52952	0.57476	0.28293	6.4	4.67938	0.99962	0.00098
2	0.65003	0.62977	0.26675	6.6	4.87932	0.99977	0.00061
2.2	0.7812	0.68131	0.24835	6.8	5.07928	0.99987	0.00037
2.4	0.9223	0.72899	0.22809	7	5.27926	0.99993	0.00022
2.6	1.07251	0.77246	0.20646	7.2	5.47925	0.99996	0.00013
2.8	1.23098	0.81151	0.18401	7.4	5.67924	0.99998	0.00007
3	1.39682	0.84605	0.16136	7.6	5.87924	0.99999	0.00004
3.2	1.5691	0.87609	0.13913	7.8	6.07924	1.00000	0.00002
3.4	1.74696	0.90177	0.11788	8	6.27924	1.00000	0.00001
3.6	1.92954	0.92333	0.09809	8.2	6.47923	1.00000	0.00001
3.8	2.11604	0.94112	0.08013	8.4	6.679236	1.00000	0.00000
4	2.30576	0.95552	0.06423	8.6	6.87923	1.00000	0.00000
4.2	2.49805	0.96696	0.05052	8.8	7.07923	1.00000	0.00000
4.4	2.69237	0.97587	0.03897	9.0	7.27923	1.00000	0.00000

Appendix B – Turbulence Spectral Functions

With the three-dimensional energy spectral density for isotropic turbulence given by[1]

$$E(k)=\alpha\left(\varepsilon v^5\right)^{1/4}\left(k_d/k_e\right)^{5/3}\left(k/k_e\right)^4\exp\left[-\tfrac{3}{2}r\alpha\left(k/k_d\right)^{4/3}\right]\left[1+\left(k/k_e\right)^2\right]^{-17/6}$$

Eqs. (3.12), (3.13), (3.14) and (3.15) may be expressed respectively as

$$\frac{\overline{u'^2}}{(\varepsilon v^5)^{1/4}k_e}=\tfrac{1}{3}\alpha\left(k_d/k_e\right)^{5/3}I_u=\tfrac{1}{3}\left(k_d/k_e\right)^3 I_u/I_\varepsilon$$

$$\alpha=\left(k_d/k_e\right)^{4/3}I_\varepsilon^{-1}\,,\qquad \frac{\overline{u'^2}\Lambda}{(\varepsilon v^5)^{1/4}}=\tfrac{1}{4}\pi\alpha\left(k_d/k_e\right)^{5/3}I_\Lambda=\tfrac{1}{4}\pi\left(k_d/k_e\right)^3 I_\Lambda/I_\varepsilon$$

and

$$\frac{E_1(k_1)}{(\varepsilon v^5)^{1/4}}=\tfrac{1}{2}\alpha\left(k_d/k_e\right)^{5/3}I_E=\tfrac{1}{2}\left(k_d/k_e\right)^3 I_E/I_\varepsilon$$

where

$$I_u=\int_0^\infty\frac{z^{3/2}\exp(-bz^{2/3})}{(1+z)^{17/6}}dz\,,\quad I_\varepsilon=\int_0^\infty\frac{z^{5/2}\exp(-bz^{2/3})}{(1+z)^{17/6}}dz\,,\quad I_\Lambda=\int_0^\infty\frac{z\exp(-bz^{2/3})}{(1+z)^{17/6}}dz$$

$$I_E=\left(k_1/k_e\right)^4\int_0^\infty\frac{(z-1)\exp\left[-b(k_1/k_e)^{4/3}z^{2/3}\right]}{(1+z)^{17/6}}dz$$

and

$$b=\tfrac{3}{2}r\alpha(k_e/k_d)^{4/3}=\tfrac{3}{2}rI_\varepsilon^{-1}$$

The numerical values of the integrals I_ε, I_u, and I_Λ are provided in Table B.1 for several values of b. Since the frequency parameter $f\Lambda/U=(k_1/k_e)(\Lambda k_e/2\pi)$ is often used in spectral plots, the quantity $\Lambda k_e/2\pi=3I_\Lambda/8I_u$ is also provided. Note for $b\to 0$, corresponding to large Reynolds numbers, $\Lambda k_e\to 0.75$ as generally accepted. It should also be noted that the results presented in Table B.1 are independent of Kolmogorov's constant α.

Once α is chosen, the turbulence parameters $k_d/k_e=(\alpha I_\varepsilon)^{3/4}$, $A=(3/\alpha I_u)^{3/2}$, and the turbulent Reynolds number $Re_\lambda=(15^{1/2}/3)\alpha^{3/2}I_\varepsilon^{1/2}I_u$ can be determined. These quantities are provided in Table B.2 for $\alpha=1.7$. The quantity $(A\Lambda k_e)^{-1}$, which is the ratio of the dissipation and integral length scales, is also tabulated. Experimental values for this quantity typically lie between 1.1 and 1.5. Batchelor [1953] quotes values between 0.8 and 1.3.

The integral $I_E(k_1)$ was evaluated for $b=0.02$, 0.0034, and 0.0008, which for $\alpha=1.7$ correspond to $Re_\lambda=34$, 106, and 237, respectively. The results of these calculations are presented in Table B.3. Since the one-dimensional frequency spectrum $E_1(f)$ is often used, the quantity $UE_1(f)/\Lambda=2\pi E_1(k_1)/\Lambda=4I_E/I_\Lambda$ is also given in the table. Note that for $k_1/k_e\to 0$, corresponding to low frequencies, $2\pi E_1/u'^2\Lambda\to 4$ as it should. Except for Re_λ, the values given in this table are independent of the value assigned to α. For those interested in performing other calculations, replacing the values of Re_λ with their corresponding values of b from Table B.2 makes the whole table independent of α.

[1] See Eq. (3.10) in text.

Table B.1
Turbulence spectrum integrals.

b	I_ε	I_u/I_ε	I_Λ/I_ε	r	$\Lambda k_e/2\pi$
$3.0\ 10^{-2}$	$4.45\ 10^{1}$	$2.93\ 10^{-2}$	$1.30\ 10^{-2}$	0.890	0.1671
$2.0\ 10^{-2}$	$6.91\ 10^{1}$	$2.05\ 10^{-2}$	$8.70\ 10^{-3}$	0.921	0.1587
$7.5\ 10^{-3}$	$1.94\ 10^{2}$	$8.48\ 10^{-3}$	$3.25\ 10^{-3}$	0.970	0.1438
$3.4\ 10^{-3}$	$4.34\ 10^{2}$	$4.09\ 10^{-3}$	$1.48\ 10^{-3}$	0.984	0.1359
$2.0\ 10^{-3}$	$7.43\ 10^{2}$	$2.47\ 10^{-3}$	$8.71\ 10^{-4}$	0.991	0.1320
$8.0\ 10^{-4}$	$1.87\ 10^{3}$	$1.03\ 10^{-3}$	$3.49\ 10^{-4}$	0.996	0.1272
$5.0\ 10^{-4}$	$2.99\ 10^{3}$	$6.52\ 10^{-4}$	$2.18\ 10^{-4}$	0.998	0.1256
$1.0\ 10^{-4}$	$1.50\ 10^{4}$	$1.34\ 10^{-4}$	$4.36\ 10^{-5}$	0.999	0.1219
$3.0\ 10^{-5}$	$5.00\ 10^{4}$	$4.07\ 10^{-5}$	$1.31\ 10^{-5}$	1.000	0.1205

Table B.2
Turbulence parameters ($\alpha = 1.7$).

b	k_d/k_e	A	Re_λ	$(A\Lambda k_e)^{-1}$
$3.0\ 10^{-2}$	26	1.58	25	0.60
$2.0\ 10^{-2}$	36	1.39	34	0.72
$7.5\ 10^{-3}$	77	1.11	66	1.00
$3.4\ 10^{-3}$	142	0.992	106	1.18
$2.0\ 10^{-3}$	212	0.941	143	1.28
$8.0\ 10^{-4}$	423	0.882	237	1.42
$5.0\ 10^{-4}$	602	0.861	305	1.47
$1.0\ 10^{-4}$	2020	0.821	705	1.59
$3.0\ 10^{-5}$	4980	0.807	1300	1.64

Table B.3
One-dimensional energy spectrum.

	$Re_\lambda = 34$		$Re_\lambda = 106$		$Re_\lambda = 237$	
k_1/k_d	I_E/I_ε	$4I_E/I_\Lambda$	I_E/I_ε	$4I_E/I_\Lambda$	I_E/I_ε	$4I_E/I_\Lambda$
0.0251	$8.69\ 10^{-3}$	3.99	$1.48\ 10^{-3}$	4.00	$3.49\ 10^{-4}$	4.00
0.0398	$8.68\ 10^{-3}$	3.99	$1.48\ 10^{-3}$	3.99	$3.48\ 10^{-4}$	4.00
0.0631	$8.66\ 10^{-3}$	3.98	$1.48\ 10^{-3}$	3.99	$3.48\ 10^{-4}$	3.99
0.100	$8.62\ 10^{-3}$	3.96	$1.47\ 10^{-3}$	3.97	$3.46\ 10^{-4}$	3.97
0.158	$8.5\ 10^{-3}$	3.91	$1.45\ 10^{-3}$	3.92	$3.42\ 10^{-4}$	3.92
0.251	$8.23\ 10^{-3}$	3.78	$1.41\ 10^{-3}$	3.80	$3.31\ 10^{-4}$	3.80
0.398	$7.62\ 10^{-3}$	3.50	$1.31\ 10^{-3}$	3.53	$3.08\ 10^{-4}$	3.54
0.631	$6.43\ 10^{-3}$	2.96	$1.12\ 10^{-3}$	3.01	$2.63\ 10^{-4}$	3.02
1.00	$4.65\ 10^{-3}$	2.14	$8.23\ 10^{-4}$	2.22	$1.95\ 10^{-4}$	2.24
1.58	$2.75\ 10^{-3}$	1.27	$5.08\ 10^{-4}$	1.37	$1.22\ 10^{-4}$	1.40
2.51	$1.35\ 10^{-3}$	$6.23\ 10^{-1}$	$2.70\ 10^{-4}$	$7.28\ 10^{-1}$	$6.56\ 10^{-5}$	$7.53\ 10^{-1}$
3.98	$5.70\ 10^{-4}$	$2.62\ 10^{-1}$	$1.29\ 10^{-4}$	$3.48\ 10^{-1}$	$3.23\ 10^{-5}$	$3.71\ 10^{-1}$
6.31	$2.07\ 10^{-4}$	$9.52\ 10^{-2}$	$5.76\ 10^{-5}$	$1.56\ 10^{-1}$	$1.51\ 10^{-5}$	$1.74\ 10^{-1}$
10.0	$6.21\ 10^{-5}$	$2.86\ 10^{-2}$	$2.42\ 10^{-5}$	$6.53\ 10^{-2}$	$6.88\ 10^{-6}$	$7.90\ 10^{-2}$
15.8	$1.42\ 10^{-5}$	$6.52\ 10^{-3}$	$9.38\ 10^{-6}$	$2.53\ 10^{-2}$	$3.03\ 10^{-6}$	$3.48\ 10^{-2}$
25.1	$2.10\ 10^{-6}$	$9.68\ 10^{-4}$	$3.24\ 10^{-6}$	$8.75\ 10^{-3}$	$1.28\ 10^{-6}$	$1.47\ 10^{-2}$
39.8	$1.53\ 10^{-7}$	$7.01\ 10^{-5}$	$9.38\ 10^{-7}$	$2.53\ 10^{-3}$	$5.08\ 10^{-7}$	$5.83\ 10^{-3}$
63.1	$3.36\ 10^{-9}$	$1.54\ 10^{-6}$	$2.04\ 10^{-7}$	$5.51\ 10^{-4}$	$1.84\ 10^{-7}$	$2.11\ 10^{-3}$
100	$9.12\ 10^{-12}$	$4.19\ 10^{-9}$	$2.83\ 10^{-8}$	$7.65\ 10^{-5}$	$5.71\ 10^{-8}$	$6.55\ 10^{-4}$
158	–	–	$1.85\ 10^{-9}$	$5.00\ 10^{-6}$	$1.40\ 10^{-8}$	$1.60\ 10^{-4}$
251	–	–	–	–	$2.33\ 10^{-9}$	$2.68\ 10^{-5}$

References

Batchelor, G.K., 1953, The Theory of Homogeneous Turbulence, Cambridge University Press, Cambridge, UK.

Appendix C – Another Reynolds-Stress Closure

Another method of closing the Reynolds-stress equation[1] is to define the dissipation tensor by

$$\hat{\varepsilon}_{ij} = f_{\varepsilon 1}\left[\hat{\varepsilon}\left(\frac{\overline{u_i'u_j'}}{k}-\frac{2}{3}\delta_{ij}\right)+\hat{\varepsilon}_{ij,w}\right]+\frac{2}{3}\delta_{ij}\hat{\varepsilon} \tag{C.1}$$

instead of using Eq. (3.60), and develop an equation for $\hat{\varepsilon}$. Again the function $f_{\varepsilon 1}$ in this definition equals unity at the wall and zero in the free stream. In this case, however, it also modifies the wall component of the dissipation which is defined specifically to yield the behavior given in Eq. (1.67) for the various components of the dissipation. Through the years this wall component eventually evolved into its most recent form, namely,

$$\hat{\varepsilon}_{ij,w} = f_{\varepsilon,w}\frac{\hat{\varepsilon}}{k}\left\{\frac{\overline{u_i'u_k'}n_j n_k+\overline{u_j'u_k'}n_i n_k+\overline{u_k'u_m'}n_k n_m n_i n_j-\frac{3}{2}\overline{u_i'u_j'}\left(\overline{u_p'u_q'}/k\right)n_p n_q}{1+\frac{3}{2}\left(\overline{u_p'u_q'}/k\right)n_p n_q f_{s,w}}\right\} \tag{C.2}$$

where x_n is the direction normal to the wall and $f_{\varepsilon,w}$ is a function that equals one for small Reynolds numbers and zero for large. The upshot of these definitions is that the trace of the dissipation tensor, and therefore $\hat{\varepsilon}$, which now represents the actual dissipation of turbulent energy, is not zero at the wall.

The natural choice for the dissipation equation is Eq. (3.65), but with $\tilde{\varepsilon}$ replaced by $\hat{\varepsilon}$ and the viscous wall dissipation term eliminated. This provides

$$\overline{u}_j\frac{\partial\hat{\varepsilon}}{\partial x_j}=c_{\varepsilon 1}\frac{\hat{\varepsilon}}{k}P+\frac{\partial}{\partial x_i}\left[\nu\frac{\partial\hat{\varepsilon}}{\partial x_i}+c_{\varepsilon}\overline{u_i'u_j'}\frac{k}{\hat{\varepsilon}}\frac{\partial\hat{\varepsilon}}{\partial x_j}\right]-c_{\varepsilon 2}f_{\varepsilon 2}\frac{\hat{\varepsilon}^2}{k}+c_{\varepsilon 3}\nu\frac{k\overline{u_l'u_j'}}{\hat{\varepsilon}}\left(\frac{\partial^2\overline{u}_i}{\partial x_l\partial x_k}\right)\left(\frac{\partial^2\overline{u}_i}{\partial x_j\partial x_k}\right)$$

However, since $\hat{\varepsilon}$ is finite at the wall whereas k equals zero, the third term on the right-hand side now presents a problem. This is readily solved by replacing one of the $\hat{\varepsilon}$'s in this term by

$$\tilde{\varepsilon}=\hat{\varepsilon}-2\nu\left(\partial k^{1/2}/\partial x_n\right)^2 \tag{C.3}$$

which according to Eq. (1.65) and the definition of $\hat{\varepsilon}$ given in Eq. (1.66) approaches zero at the same rate as k at the wall. As a consequence, the third term now becomes finite there.

If all the other terms in the equations are modeled as in §3.3 of the text, Eqs. (3.66)–(3.71) based on this method of closure transform to

$$\begin{aligned}\overline{u}_k\frac{\partial\overline{u_i'u_j'}}{\partial x_k}&=P_{ij}+\frac{\partial}{\partial x_k}\left[\nu\frac{\partial\overline{u_i'u_j'}}{\partial x_k}+c_s\frac{k}{\hat{\varepsilon}}\left(\overline{u_i'u_l'}\frac{\partial\overline{u_j'u_k'}}{\partial x_l}+\overline{u_j'u_l'}\frac{\partial\overline{u_k'u_i'}}{\partial x_l}+\overline{u_k'u_l'}\frac{\partial\overline{u_i'u_j'}}{\partial x_l}\right)\right]\\&\quad-\left(c_1 f_{\phi 1}A+f_{\varepsilon 1}\right)\frac{\hat{\varepsilon}}{k}\left[\overline{u_i'u_j'}-\frac{2}{3}\delta_{ij}k\right]-c_2\sqrt{A}\left[P_{ij}-\frac{2}{3}\delta_{ij}P\right]-\frac{2}{3}\delta_{ij}\hat{\varepsilon}+\phi_{ij,w}-f_{\varepsilon 1}\hat{\varepsilon}_{ij,w}\end{aligned} \tag{C.4}$$

$$\overline{u}_j\frac{\partial k}{\partial x_j}=P+\frac{\partial}{\partial x_j}\left[\nu\frac{\partial k}{\partial x_j}+c_s\frac{k}{\hat{\varepsilon}}\left(\overline{u_k'u_l'}\frac{\partial\overline{u_j'u_k'}}{\partial x_l}+\overline{u_j'u_l'}\frac{\partial k}{\partial x_l}\right)\right]-\hat{\varepsilon} \tag{C.5}$$

[1] See Eq. (3.46) in text.

$$\bar{u}_j \frac{\partial \hat{\varepsilon}}{\partial x_j} = c_{\varepsilon 1} \frac{\hat{\varepsilon}}{k} P + \frac{\partial}{\partial x_i}\left[\nu \frac{\partial \hat{\varepsilon}}{\partial x_i} + c_\varepsilon \overline{u_i' u_j'} \frac{k}{\hat{\varepsilon}} \frac{\partial \hat{\varepsilon}}{\partial x_j} \right] - c_{\varepsilon 2} f_{\varepsilon 2} \frac{\tilde{\varepsilon}\hat{\varepsilon}}{k} + c_{\varepsilon 3} \nu \frac{k \overline{u_l' u_j'}}{\hat{\varepsilon}} \left(\frac{\partial^2 \bar{u}_i}{\partial x_l \partial x_k} \right) \left(\frac{\partial^2 \bar{u}_i}{\partial x_j \partial x_k} \right) \tag{C.6}$$

where

$$P_{ij} = -\overline{u_i' u_k'} \frac{\partial \bar{u}_j}{\partial x_k} - \overline{u_j' u_k'} \frac{\partial \bar{u}_i}{\partial x_k}, \qquad P = \frac{P_{kk}}{2} \tag{C.7}$$

$$\begin{aligned} \phi_{ij,w} = f_{\phi,w} \Bigg[& c_{1w} \frac{\hat{\varepsilon}}{k} \left(\overline{u_k' u_l'} n_k n_l \delta_{ij} - \frac{3}{2} \overline{u_k' u_i'} n_k n_j - \frac{3}{2} \overline{u_k' u_j'} n_k n_i \right) \\ & + c_{2w} \left(\phi_{kl,2} n_k n_l \delta_{ij} - \frac{3}{2} \phi_{ki,2} n_k n_j - \frac{3}{2} \phi_{kj,2} n_k n_i \right) \Bigg] \end{aligned} \tag{C.8}$$

$$\hat{\varepsilon}_{ij,w} = f_{\varepsilon,w} \frac{\hat{\varepsilon}}{k} \left\{ \frac{\overline{u_i' u_k'} n_j n_k + \overline{u_j' u_k'} n_i n_k + \overline{u_k' u_m'} n_k n_m n_i n_j - \frac{3}{2} \overline{u_i' u_j'} \left(\overline{u_p' u_q'} / k \right) n_p n_q}{1 + \frac{3}{2} \left(\overline{u_p' u_q'} / k \right) n_p n_q f_{\varepsilon,w}} \right\} \tag{C.9}$$

with

$$\tilde{\varepsilon} = \hat{\varepsilon} - 2\nu \left(\partial k^{1/2} / \partial x_n \right)^2$$

Naturally, some of the constants c and functions f will change, but generally only those attached to the wall correction terms. To use Daly and Harlow's turbulent diffusion model, disregard all but the last diffusion term in Eqs. (C.4) and (C.5).

Appendix D – Stability Results

The results of some stability calculations for flows with and without a pressure gradient are provided in this appendix.

D.1 Zero Pressure Gradient Results.

The values of $\beta = \omega\delta_1/\bar{u}_\infty$ and $\alpha_r = k_r\delta_1$ on the neutral stability curve as calculated by Jordinson [1970] for a Blasius' profile are given in Table D.1.

Two solutions to the Orr-Sommerfeld equation, Eq. (4.10), for $\phi(\eta)$ and $\phi'(\eta) = d\phi/d\eta$ as calculated by Jordinson are given in Table D.2. These correspond directly with his Case 1 for a stable perturbation and Case 3 for an unstable one. The parameters for these solutions are:

Case 1: $Re_{\delta 1} = 336$, $\beta = 0.1297$, $\alpha_r = 0.3084$, and $\alpha_i = 0.0079$.

Case 3: $Re_{\delta 1} = 998$, $\beta = 0.1122$, $\alpha_r = 0.3086$, and $\alpha_i = -0.0057$.

Since the solutions for Re{$\phi(\eta)$} are nearly the same, only the average result of the two is given in this table.

Table D.1

Parameters on the neutral stability curve $\alpha_i = 0$.

Upper Branch			Lower Branch		
$Re_{\delta 1}$	β	α_r	$Re_{\delta 1}$	β	α_r
$1\ 10^6$	0.0153	0.102	$1\ 10^6$	0.0010	0.017
$5\ 10^5$	0.0178	0.111	$5\ 10^5$	0.0015	0.021
$2\ 10^5$	0.0224	0.126	$2\ 10^5$	0.0024	0.027
$1\ 10^5$	0.0257	0.136	$1\ 10^5$	0.0035	0.032
$5\ 10^4$	0.0267	0.140	$5\ 10^4$	0.0050	0.040
$2\ 10^4$	0.0393	0.176	$2\ 10^4$	0.0083	0.053
$1\ 10^4$	0.0534	0.212	$1\ 10^4$	0.0122	0.066
5000	0.0724	0.254	5000	0.0184	0.085
4000	0.0975	0.269	4000	0.0211	0.092
3000	0.0894	0.2878	2964	0.0255	0.1034
2600	0.0947	0.2973	2564	0.0280	0.1097
2200	0.1010	0.3084	2164	0.0313	0.1178
1800	0.1088	0.3212	1764	0.0360	0.1289
1400	0.1185	0.3359	1364	0.0433	0.1455
1000	0.1306	0.3512	1164	0.0489	0.1576
800	0.1368	0.3559	964	0.0568	0.1746
600	0.1380	0.3466	764	0.0697	0.2011
531	0.1293	0.3233	671	0.0793	0.2205
520	0.1193	0.3012	604	0.0893	0.2405
			559	0.0993	0.2604
			531	0.1093	0.2806

Table D.2

Values of $\phi(\eta)$ and $\phi'(\eta)$ for a stable (Case 1) and unstable (Case 3) perturbation.

$\eta = y/\delta$	Re{ϕ}	Im_1{ϕ}	Im_3{ϕ}	Re_1{ϕ'}	Re_3{ϕ'}	Im_1{ϕ'}	Im_3{ϕ'}
0	0	0	0	0	0	0	0
0.1	0.045	–0.148	0.227	–0.105	–0.289	0.453	–0.225
0.2	0.091	–0.266	0.435	–0.12	–0.46	0.684	–0.215
0.3	0.150	–0.328	0.589	–0.084	–0.52	0.84	–0.087
0.4	0.230	–0.325	0.686	–0.023	–0.523	0.886	0.026
0.5	0.312	–0.279	0.753	0.026	–0.482	0.891	0.069
0.6	0.389	–0.189	0.809	0.090	–0.392	0.871	0.077
0.7	0.474	–0.081	0.824	0.136	–0.304	0.824	0.074
0.8	0.557	0.007	0.811	0.123	–0.237	0.755	0.061
0.9	0.629	0.105	0.780	0.082	–0.193	0.673	0.046
1.0	0.698	0.146	0.731	0.046	–0.163	0.590	0.035
1.1	0.758	0.169	0.667	0.015	–0.134	0.500	0.028
1.2	0.804	0.174	0.592	–0.006	–0.111	0.428	0.025
1.3	0.850	0.164	0.518	–0.016	–0.093	0.346	0.020
1.4	0.894	0.141	0.436	–0.024	–0.075	0.264	0.017
1.5	0.923	0.113	0.353	–0.027	–0.057	0.194	0.014
1.6	0.951	0.085	0.266	–0.029	–0.044	0.135	0.012
1.7	0.969	0.059	0.173	–0.032	–0.031	0.076	0.009
1.8	0.982	0.036	0.101	–0.032	–0.02	0.032	0.009
1.9	0.987	0.011	0.045	–0.03	–0.01	–0.014	0.009
2.0	0.990	–0.017	–0.004	–0.027	0.001	–0.066	0.006
2.1	0.988	–0.030	–0.053	–0.025	0.006	–0.097	0.006
2.2	0.983	–0.040	–0.089	–0.022	0.011	–0.128	0.006
2.3	0.975	–0.048	–0.125	–0.023	0.016	–0.159	0.006
2.4	0.965	–0.053	–0.154	–0.02	0.022	–0.184	0.003
2.5	0.947	–0.058	–0.177	–0.018	0.024	–0.195	0.003
2.6	0.924	–0.060	–0.2	–0.015	0.027	–0.208	0
2.7	0.899	–0.060	–0.218	–0.013	0.030	–0.221	0
2.8	0.878	–0.060	–0.226	–0.01	0.032	–0.229	0
2.9	0.853	–0.063	–0.231	–0.008	0.038	–0.231	0
3.0	0.830	–0.062	–0.234	–0.008	0.040	–0.234	0
3.2	0.786	–0.062	–0.232	–0.006	0.046	–0.232	–0.001
3.4	0.740	–0.065	–0.222	–0.003	0.049	–0.222	–0.001
3.6	0.694	–0.064	–0.211	–0.001	0.054	–0.211	–0.001
3.8	0.653	–0.064	–0.201	–0.001	0.057	–0.201	–0.001
4.0	0.620	–0.064	–0.189	0	0.060	–0.189	0
4.4	0.551	–0.063	–0.169	0	0.063	–0.169	0
4.8	0.480	–0.060	–0.149	0	0.066	–0.149	0
5.2	0.421	–0.057	–0.131	0	0.069	–0.131	0

D.2 Results for the Falkner-Skan Profiles.

Some of the pertinent parameters for the Falkner-Skan profiles are given in Table D.3. The profiles are characterized by the exponent m in the expression $u_\infty = ax^m$ for the free-stream velocity when plotted (as they are in Fig. 4.10) against the similarity variable η defined in Eq. (4.20). The quantity $\beta = 2m/(1+m)$, called the "Falkner-Skan parameter," is the included wedge angle divided by π. The quantities $\eta_{\delta 2}$ and $\lambda_{\delta 2}$ are the values of η where y equals the momentum thickness δ_2 and the pressure-gradient parameter $(\delta_2^2/\nu)du_\infty/dx$, both of which depend on m only. These quantities are easily determined by solving the boundary-layer equations assuming a similar solution for the above free-stream velocity distribution. For details, see Schlichting [1979] or any number of books on laminar boundary layers.

The neutral stability results in Table D.3 were calculated by Wazzan et al. [1968] and include the critical Reynolds numbers based on both the displacement and momentum thickness, the maximum temporal growth rate, and its corresponding wave number.

Table D.3

Profile parameters and stability results for the Falkner-Skan Profiles with $u_\infty = ax^m$.

Profile parameters					Stability results			
m	β	δ_1/δ_2	$\eta_{\delta 2}$	$\lambda_{\delta 2}$	$Re_{\delta 1,crit}$	$Re_{\delta 2,crit}$	$(\omega_i/u_\infty k)_{max}$	$(k\nu/u_\infty)_{max}$ $\times(10)^7$
−0.091	−0.1988	3.94	0.579	−0.067	67	17	0.120	5,600
−0.065	−0.14	2.94	0.538	−0.041	138	47	0.0525	963
−0.048	−0.10	2.80	0.515	−0.027	199	71	0.0388	450
−0.024	−0.05	2.67	0.490	−0.012	318	119	0.0275	186
0	0	2.59	0.470	0	520	201	0.0196	74
0.026	0.05	2.53	0.451	0.010	865	342	0.0154	32
0.053	0.1	2.48	0.435	0.019	1,380	556	0.0129	15.7
0.111	0.2	2.41	0.408	0.033	2,830	1,174	0.0104	6.0
0.176	0.3	2.36	0.386	0.045	4,550	1,927	0.0095	3.45
0.25	0.4	2.33	0.367	0.054	6,230	2,679	0.0085	2.42
0.33	0.5	2.30	0.350	0.061	7,680	3,344	0.0080	1.92
0.43	0.6	2.27	0.335	0.068	8,890	3,909	0.0075	1.67
0.67	0.8	2.25	0.309	0.077	10,920	4,874	0.0070	1.35
1.0	1.0	2.22	0.291	0.085	12,490	5,636	0.0065	1.14

References

Jordinson, R., 1970, "The Flat Plate Boundary Layer. Part I: Numerical Integration of the Orr-Sommerfeld Equation," J. Fluid Mech., 43, pp. 801–811.

Schlichting, H., 1979, Boundary-Layer Theory, McGraw-Hill, New York.

Wazzan, A.R., Okamura, T.T, and Smith, A.M.O.,1968, "Spatial and Temporal Stability Charts for the Falkner-Skan Boundary-Layer Profiles," Douglas Aircraft Report No. DAC-67086.

Appendix E – Free-Shear-Layer Instability

Consider a free shear layer of height h and constant vorticity U/h sandwiched between a quiescent fluid below and fluid moving at the velocity U above.[1] Without going into the details, a solution to the Kelvin-Helmholtz equation, that is, Eq. (4.10) when $Re_{\delta 1}^{-1}=0$, only exists (see Chandrasekhar [1961], §102) when

$$\alpha^2 v^4-\left(2\alpha^2-2\alpha+1-e^{-2\alpha}\right)v^2+\left[\left(1-e^{-\alpha}\right)-\alpha\right]\left[\left(1+e^{-\alpha}\right)-\alpha\right]=0$$

with

$$v=2\beta/\alpha \quad \text{with} \quad \alpha=kh \quad \text{and} \quad \beta=\omega h/U$$

where $\alpha=kh$ and $\beta=\omega h/U$ are the dimensionless wave number and complex growth rate corresponding to a disturbance proportional to $\exp[i(kx-\omega t)]$. The solution to this equation for v has two real and two complex roots. The real roots are $v=\pm 1$ while the complex are

$$v=\pm\frac{1}{\alpha}\sqrt{1-e^{-2\alpha}-2\alpha+\alpha^2}$$

It is relatively easy to show that $v=\pm i$ when $\alpha=0$ and zero when $\alpha=1.27846$, between which v is pure imaginary. Since the dimensionless growth rate is given by

$$i\omega h/U=i v\alpha/2$$

the solutions for neutral stability are $\alpha=0$ and 1.2784. A plot of the *unstable* growth rate is given in Fig. E.1. It has a maximum of 0.2012 when $\alpha=0.7968$. Consequently, the most amplified Kelvin-Helmholtz instability has the wave number and growth rate of

$$k_{max}=0.7968/h \quad \text{and} \quad \omega_{max}=0.2012U/h$$

while its wavelength is

$$\lambda_{max}=2\pi/k_{max}=7.89h$$

The subscript of the latter refers to the wavelength of the *most* amplified instability and not a maximum wavelength.

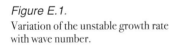

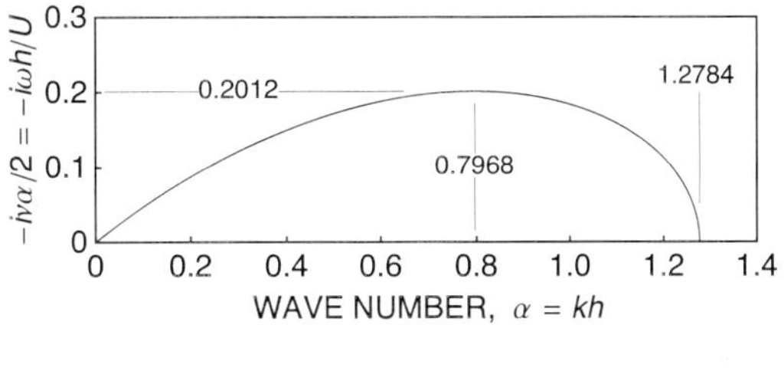

Figure E.1.
Variation of the unstable growth rate with wave number.

References

Chandrasekhar, S., 1961, Hydrodynamic and Hydromagnetic Stability, Oxford University Press; also 1981, Dover Publications.

Dovgal, A.V., Kozlov, V.V., and Michalke, A., 1994, "Laminar Boundary Layer Separation: Instabil-

[1] The analysis using a profile given by $u=\frac{1}{2}U\tanh(2y/h)$ is provided by Michalke [1991] and Dovgal et al. [1994].

ity and Associated Phenomena," Progress Aerospace Sciences, 30, pp. 61–94.

Michalke, A., 1991, "On the Instability of Wall Boundary Layers Close to Separation," IUTAM Symposium on Separated Flows and Jets, Springer-Verlag, Berlin, pp. 557–564.

Appendix F – Transition with Anisotropic Turbulence

The analysis presented in §4.3.2 was specifically tailored for a laminar boundary layer with isotropic free-stream turbulence, or at least turbulence that is initially isotropic. As noted, a few adjustments were made for "nearly" isotropic turbulence, but the point here is that all the production terms for the inner layer were scaled by the square root of the free-stream dissipation which was introduced using the isotropic relations given in Eq. (4.27). Since this rate can be obtained directly from turbulent-kinetic-energy equation, the production terms for isotropic free-stream turbulence become relatively easy to determine. However, for some tests with high free-stream turbulence, particularly those with turbulence grids placed upstream of a nozzle preceding the test section, the turbulence is highly anisotropic and its repartitioning dominates the flow through most of the test section.[1] In this case, the relations given in Eq. (4.29) do not apply and another tack becomes necessary.

F.1 Equations for Non-Isotropic Free-Stream Turbulence.

The other tack is obvious. Instead of introducing the isotropic relations, we simply write the production-rate terms using the effective wave number component and corresponding energy density. That is, from Eqs. (4.19)–(4.21), the production of the streamwise normal Reynolds stress becomes

$$\overline{u'\frac{\partial u'_\infty}{\partial t}} = R_{\phi 11}\bar{u}_\infty k_{1,eff}\sqrt{k_{1,eff}E_1(k_{1,eff},x)}\sqrt{\overline{u'u'}}$$

where the coefficient $R_{\phi 11}$ differs from that in Eq. (4.29), and $k_{1,eff}$ is a function of x only. Accordingly, the set of inner equations, Eqs. (4.31)–(4.34), become

$$\bar{u}\frac{\partial\overline{u'u'}}{\partial x}+\bar{v}\frac{\partial\overline{u'u'}}{\partial y} = R_{\phi 11}\bar{u}_\infty k_{1,eff}\sqrt{k_{1,eff}E_1(k_{1,eff},x)}\sqrt{\overline{u'u'}}+\nu\frac{\partial^2\overline{u'u'}}{\partial y^2}-\tilde{\varepsilon}_{11,w}$$

$$\bar{u}\frac{\partial\overline{v'v'}}{\partial x}+\bar{v}\frac{\partial\overline{v'v'}}{\partial y} = R_{\phi 22}\bar{u}_\infty k_{2,eff}\sqrt{k_{2,eff}E_2(k_{2,eff},x)}\sqrt{\overline{v'v'}}+\nu\frac{\partial^2\overline{v'v'}}{\partial y^2}-\tilde{\varepsilon}_{22,w}$$

$$\bar{u}\frac{\partial\overline{w'w'}}{\partial x}+\bar{v}\frac{\partial\overline{w'w'}}{\partial y} = R_{\phi 33}\bar{u}_\infty k_{3,eff}\sqrt{k_{3,eff}E_3(k_{3,eff},x)}\sqrt{\overline{w'w'}}+\nu\frac{\partial^2\overline{w'w'}}{\partial y^2}-\tilde{\varepsilon}_{33,w}$$

$$\bar{u}\frac{\partial\overline{u'v'}}{\partial x}+\bar{v}\frac{\partial\overline{u'v'}}{\partial y} = -R_{\phi 12}\bar{u}_\infty k_{2,eff}\sqrt{k_{2,eff}E_2(k_{2,eff},x)}\sqrt{\overline{v'v'}}+\nu\frac{\partial^2\overline{u'v'}}{\partial y^2}-\tilde{\varepsilon}_{12,w}$$

where, in general, the components of the effective wave number depends on direction.

The complexity of the situation (both theoretical and experimental) is now significantly increased. Not only do we have to calculate the variation of the three one-dimensional spectral functions and their corresponding effective wave numbers, but they must also be measured. This is an extensive undertaking, which, to the author's knowledge has yet to be accomplished even for an isotropic turbulent free stream.

F.2 Pre-Transition Spectral Results.

Nevertheless, some free-stream spectral data for pre-transitional flow is available. Usually this amounts to the one-dimensional spectrum $E_1(f)$ of the incident turbulence.[2] However, for one of his

---

[1] The investigations of Blair [1982, 1983] and Zhou and Wang [1995], where the transverse components of turbulence are significantly higher than the streamwise component, are typical here.

[2] See Eq. (3.21) for its definition.

grids, Blair [1983] presents this distribution at several streamwise positions.[1] But the most extensive information is provided by Dyban and Epik [1985][2]. For this case, measurements of the three principal spectral distributions throughout the boundary layer are presented at the streamwise position of $x = 0.225$m for a turbulence level of about 2.7 percent with $u_\infty = 5$m/s. The distance Reynolds number at this location is about 76,000 and the boundary-layer thickness is about 4mm. Some of these spectra are replotted (using wave number instead of frequency) in Fig. F.1, where η is Blasius' similarity variable.

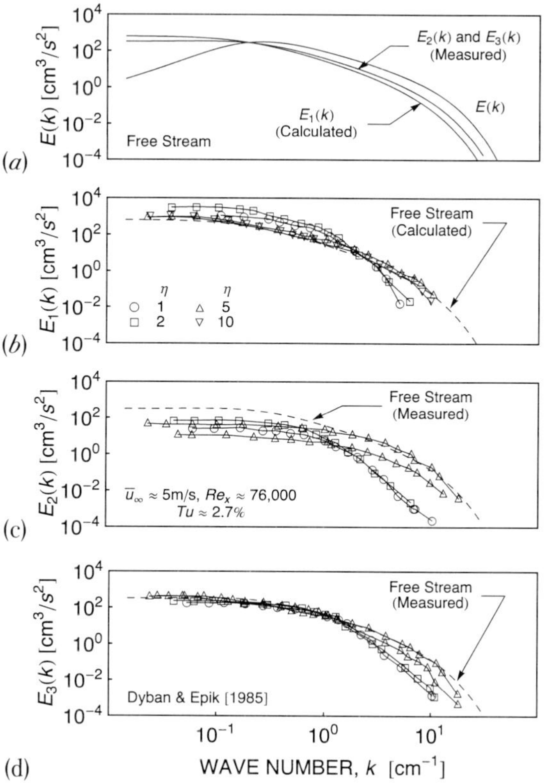

Figure F.1.
The principal components of the energy spectrum in a pre-transitional boundary layer. Data from Dyban and Epik [1985] (Fig. 54).

The free-stream spectra for the transverse fluctuations E_2 and E_3 were measured at $\eta \approx 135$, which is about twenty-seven *viscous* boundary-layer thicknesses away from the wall and at the edge of the *kinematic* boundary layer. The spectra for both of these components are virtually the same and are shown by a single line in part (a) of the figure. Assuming isotropic turbulence, the corresponding spectral distributions E_1 and E calculated from Eqs. (3.19) and (3.20) are also shown. From these and other relations given in §3.1, the turbulent-kinetic-energy, dissipation, and various wave numbers and Reynolds numbers can be calculated. The results of these calculations are presented in Table F.1. The last entry in this table is the ratio of the characteristic size of the large eddies to the boundary

[1] Although presented in a paper about the effect of free-stream turbulence on turbulent boundary layers, the same grid is used in his experiments on pre-transitional flow (Blair [1982]).

[2] See their Fig. 54.

layer thickness, which is roughly equal to the ratio of the kinematic-to-viscous boundary-layer thicknesses.

Table F.1
Turbulence quantities calculated from the free-stream E_2 and E_3 spectrums shown in Fig. F.1a.

$k\,[m^2/s^2]$	$\varepsilon\,[m^2/s^3]$	$k_e\,[cm^{-1}]$	$k_{eff}\,[cm^{-1}]$	$k_d\,[cm^{-1}]$	Re_T	Re_λ	L_e/δ
0.028	0.042	0.16	5.6	19	1260	92	27

The spectra measured within the boundary layer are shown in parts (b)–(d) of the figure. These are shown at the distances of roughly $\delta/5$, $2\delta/5$, δ, and 2δ; all well within the kinematic boundary layer. As should be expected, the energy contained at the low wave numbers (large eddies) is transferred from the fluctuations v' normal to the wall to the fluctuations u' parallel to the flow as the wall is approached, while the energy of the transverse fluctuations w' parallel to the wall remain roughly unchanged. In addition, the energy contained at the high wave numbers (small eddies) is dissipated near the wall.

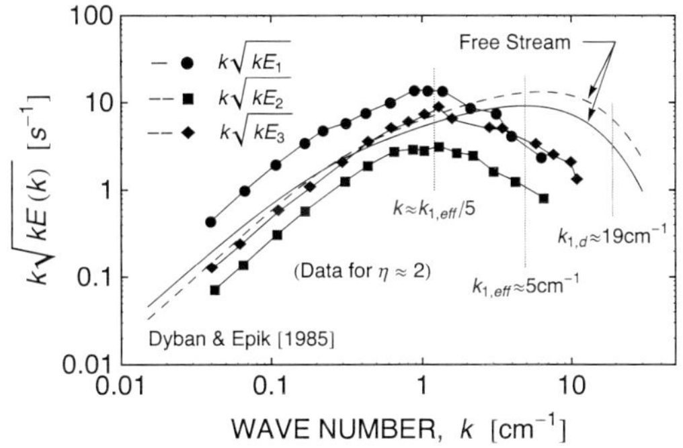

Figure F.2.
Effective wave numbers in the free stream and boundary layer. Data from Dyban and Epik [1985] (Fig. 54).

With the shift of energy from the high to low wave numbers as the wall is approached, a similar shift should also occur in the *effective* wave number. That is, following our reasoning in §4.3.2, the maximum in the quantity $k\sqrt{kE(k)}$ should shift toward the lower wave numbers as the wall is approached. A plot of this quantity in the free stream and within the viscous layer is provided in Fig. F.2 for each principal direction. It follows from the spectra plotted in part (a) of the previous figure, that the maximum of the quantity $k\sqrt{kE(k)}$ for the fluctuations in the transverse directions will be at a higher wave number than that for the fluctuations in the streamwise direction. In this case, the ratio of the wave numbers is about 1.3. In the boundary layer the maximum occurs at a wave number one-fifth that in the free stream and its value is about the same for all components. Although shown for the $\eta = 2$ data only, the value is about the same for $1 \le \eta \le 3$, which is the lower portion of the boundary layer where the highly amplified fluctuations are found (see Fig. 4.25). Whether this value remains the same for all components at all streamwise locations throughout pre-transition or not remains to be seen. If it does, then using the effective wave number of the free stream is sufficient to characterize all production rates in Eqs. (4.31)–(4.34) for the inner layer. Otherwise, the production terms would have to be suitably modified. The factor $(L_e/\delta_1)^{1/2}$ in the production terms of Eqs. (4.41) and (4.42) seems to indicate that the value does not remain constant.

References

Blair, M.F., 1982, "Influence of Free-Stream Turbulence On Boundary Layer Transition in Favorable Pressure Gradients," J. Engng. for Power, 104, pp. 743–750.

Blair, M.F., 1983, "Influence of Free-Stream Turbulence On Turbulent Boundary Layer Heat Transfer and Mean Profile Development: Part 1–Experimental Data, and Part 2–Analysis of Results," J. Heat Transfer, 105, pp. 33–40, 41–47.

Dyban, E., and Epik, E., 1985, Тепломссобмен и Гидродинамика Тубулизированных Потоков, Киев Наукоа Думка (in Russian).

Zhou, D., and Wang, T., 1995, "Effects of Elevated Free-Stream Turbulence on Flow and Thermal Structures in Transitional Boundary Layers," J. Turbomachinery, 117, pp. 407–417.

Appendix G – A Line of Point Sources

Using Eqs. (5.31) and (5.33), the near-wall intermittency for any number of point sources of equal strength located at $(x_{s,k}, z_{s,k})$ is given by

$$\gamma_0(x,z) = 1 - \exp\left[-\frac{\dot{n}\sigma'}{\bar{u}_\infty}\sum_{k=1}^{N} F_0(z;z_{s,k})\left(x - x_{s,k}\right)\right] \tag{G.1}$$

where $\dot{n}$ is the number of turbulent spots produced per unit time at each source, and the transit function F_0 is given by Eq. (5.32), namely,

$$F_0(\zeta_k) = \left(1 - 0.85\zeta_k^2 - 0.15\zeta_k^8\right)\left[\mathrm{U}(\zeta_k + 1) - \mathrm{U}(\zeta_k - 1)\right]$$

where $\zeta_k = 2(z - z_{s,k})/b$ and $b = 2(x - x_s)\tan\alpha$.[1]

When all the sources are distributed along the line $x = x_s$, Eq. (G.1) yields

$$\gamma_0(x,z) = 1 - \exp\left[-\frac{\dot{n}\sigma'}{\bar{u}_\infty}\left(x - x_s\right)\sum_{k=1}^{N} F_0(z;\zeta_k)\right]$$

Moreover, when the sources are spaced along the line a distance s apart from one another, their position is given by $z_{s,k} = ks$ where k now assumes the values 0, ±1, ±2, ..., ±N. Then the above equation becomes

$$\gamma_0(x,z) = 1 - \exp\left[-\frac{\dot{n}\sigma'}{\bar{u}_\infty}\left(x - x_s\right)\sum_{k=-N}^{N} F_0(z;\zeta_k)\right] \tag{G.2}$$

where

$$\zeta_k = \frac{(z - ks)}{(x - x_s)\tan\alpha}$$

Intermittency contours downstream a line of point sources as calculated from Eq. (G.2) are plotted in Fig. G.1. The results are shown for a dimensionless source strength $\dot{n}\nu\sigma'/\bar{u}_\infty^2 = 4(10)^{-7}$ and angle $\alpha = 10°$, which are the same values used in Fig. 5.24 of the text. In this example, the sources are placed a distance equal to $Re_s = 750{,}000$ apart.

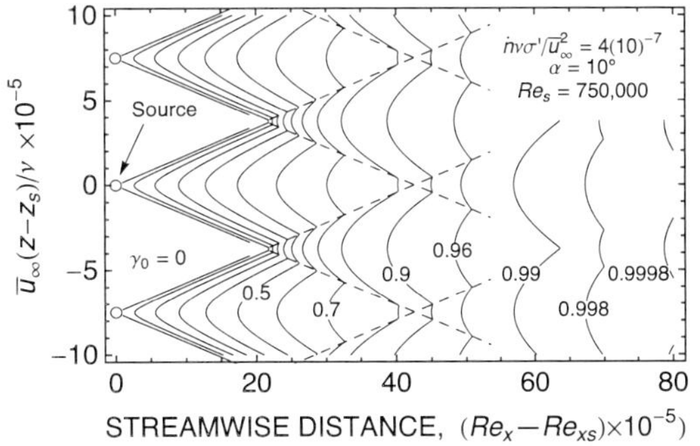

Figure G.1.
Contour plot of the intermittency downstream a line of discrete point sources.

[1] The expressions in this appendix apply to spots expanding linearly across the surface as they propagate downstream.

The straight lines from the sources, drawn solid before they intersect and dashed thereafter, show the areas of influence for each source. They are also the loci of minimums in the near-wall intermittency distribution. The first set of intersections is where the nearest neighboring spots (first neighbors) begin to interact, the second set of intersections is where every second of the spot's neighbors begin to interact, etc. Hence, the intermittency contours within the first diamond shaped areas downstream of the sources are identical to those shown in Fig. 5.24 for a single spot source.

An interesting point to note here is that if transition occurs at $Re_{xs} \approx 2.6(10)^6$, which corresponds to Schubauer and Klebanoff's [1955] value for $Tu = 0.03$ percent (see Fig. 5.18 of the text), the boundary-layer thickness Reynolds number would be about 8,000. Then the spacing shown in Fig. G.1 would correspond to a spacing-to-boundary-layer-thickness ratio of about 96. Reducing this ratio to five, which corresponds to the average spacing ratio for the "natural" spanwise irregularities and breakdowns observed by Schubauer [1957] and Klebanoff and Tidstrom [1959] in their experiments, reduces the source spacing in Fig. G.1 by a factor of sixteen. Without even calculating and plotting the result, it immediately becomes clear that the intermittency distribution in this case will be identical to that for a line source of turbulent spots as given by Eq. (5.18).

The streamwise variation of the spanwise-averaged intermittency for several source spacings are plotted in Fig. G.2. Because of symmetry, these results were obtained by evaluating the integral of Eq. (G.2) over half the spacing. That is, using

$$\bar{\gamma}_0(x) = \frac{2}{s}\int_0^{s/2}\left\{1-\exp\left[-\frac{\hat{n}\sigma'}{\bar{u}_\infty}\left(x-x_s\right)\sum_{k=-N}^{N} F_0(z;\zeta_k)\right]\right\}dz$$

As expected, the spanwise-averaged intermittency increases as the number of spot sources per unit span increase. If transition occurred at $Re_{xs} \approx 2.6(10)^6$, the curves correspond to the spacing-to-boundary-layer-thickness ratios of about 96, 48, and 24. For comparison, the intermittency distribution for a line source of spots, given by Eq. (5.18) in the text, was fit to the line-point-source result for $Re_s = 20{,}000$ and plotted as the dashed lines in the figure for the larger spacings using $\hat{n}\sigma Re_s$ = constant. Clearly, the results are barely discernible from the point source result even when they are spaced far apart. For the largest spacing, the value of $\hat{n}\sigma$ is $1.33(10)^{-13}$.

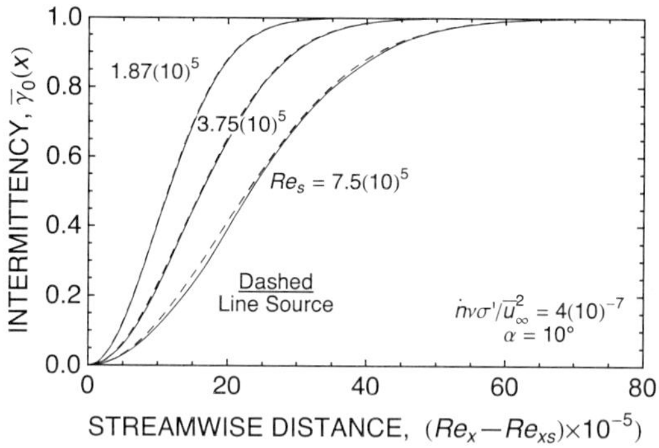

Figure G.2.
Contour plot of the intermittency downstream a line of discrete point sources.

References

Klebanoff, P.S., and Tidstrom, K.D., 1959, "Evolution of Amplified Waves Leading to Transition in a Boundary Layer with Zero Pressure Gradient," NASA TN D-195.

Schubauer, G.B., 1957, "Mechanism of Transition at Subsonic Speeds," Grenzschlichtforschung Symposium Freiburg im Brieslau (ed., H. Görtler, 1958), Springer-Verlag, Berlin, pp. 85–109.

Schubauer, G.B., and Klebanoff, P.S., 1955, "Contribution to the Mechanism of Boundary-Layer Transition," NACA TN 3489.

Appendix H – Intermittency Measurements

The measured intermittency profiles of Owen [1970], Acharya [1985], Kim, Simon and Kestoras [1989], and Blair [1992], all display shapes similar to those shown in Fig. 5.25 of the text. In particular, it is found that the intermittency remains nearly constant as the wall is approached. Using wall mounted shear-stress gages, both Owen and Acharya measured an intermittency at the wall that fit within the scatter of their near-wall values. The conclusion from these investigations is that the derivative of the intermittency with respect to distance away from the wall is nearly zero at the wall. All of these investigators discriminated between the laminar and turbulent portions of the flow by setting a threshold to the fluctuations above which the flow was deemed turbulent. Some changed the threshold at each measurement station. Blair used one threshold throughout.

Contrary to these results, Kuan and Wang [1990], Gostelow and Walker [1990], and most notably Keller and Wang [1995], Zhou and Wang [1995], and Wang and Zhou [1998], report measurements showing a peak in the intermittency at a height of the local laminar boundary layer thickness, where the spot's leading-edge bulge occurs, and a decrease as the wall is approached. The profiles obtained by Zhou and Wang for transition with an elevated free-stream turbulence level are presented in Fig. H.1.[1] Wang and his coworkers developed a "dual-slope" method to determine a discrimination threshold that not only varies from one measurement station to another, but also across the profile. Presumably, this provides for a more sensitive discrimination scheme than those that use a constant value. After a thorough investigation of various discrimination schemes, Keller and Wang concluded that the maximum value of the intermittency is independent of the discrimination scheme and therefore it should be used as the representative near-wall intermittency.[2]

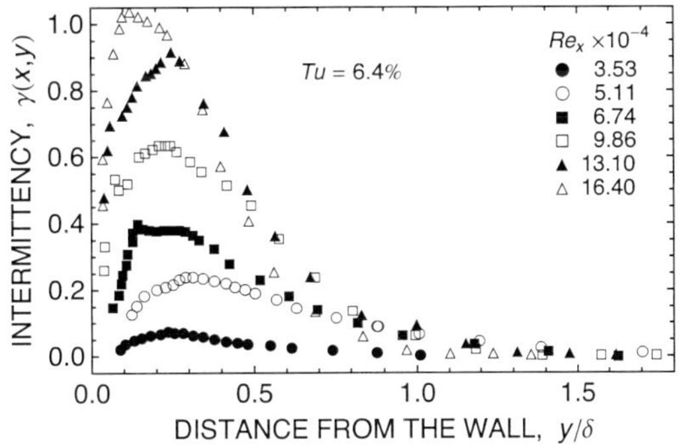

Figure H.1.

Wang and Zhou's [1998] intermittency profiles through transition for $dp/dx = 0$ with a free-stream turbulence level of 6.4 percent.

Since wall-shear gages and hot-wire anemometers yield the same intermittency, for those investigators that measured a maximum intermittency at the wall, the disparity between near-wall behavior measured by the two groups of investigators must be caused, as pointed out by the author [1991], by the discrimination schemes. All investigators claim good comparisons between the intermittency function as determined by "eye" and their scheme, but the dual-slope scheme is apparently "more

---

[1] These data display profile half-widths $y_{1/2}$ that are less than any previously measured. For comparison, the half-width of the profile at the end of transition ($Re_x = 164{,}000$ in the figure) is about $y_{1/2}/\delta \approx 0.45$, which is about fifty-six percent of Corrsin and Kistler's [1955] result for a fully-developed turbulent boundary layer and seventy-five percent of Acharya's result shown and discussed in Appendix I. Whether this is a result of the discrimination scheme or normalizing by a boundary-layer thickness that is too large remains to be investigated.

[2] Accordingly, Wang and Zhou's intermittency data plotted in Fig. 7.19b of the text correspond to the maximum profile values shown here.

sensitive" than others. An interesting corollary to these observations is that even though the intermittency may decrease near the wall, the *near-wall* intermittency can still be obtained by using "less sensitive" discrimination schemes. A further discussion regarding these schemes is presented by Hatman and Wang [1998].

Until now we presumed that the intermittency function is determined by discriminating between the behavior of the streamwise velocity fluctuation in a laminar and turbulent flow, but investigators have used fluctuations in vorticity, Reynolds shear stress, and temperature as well. Using their dual-slope discrimination scheme, Keller and Wang found a difference between the intermittency measured in transitional boundary layers using the Reynolds normal-stress and shear-stress fluctuations, while Blair found none.

References

Acharya, M., 1985, "Pressure-Gradient and Free-Stream Turbulence Effects on Boundary-Layer Transition," Forschungsbericht KLR 85-127 C, Brown Boveri Forschungszentrum, Baden, Switzerland; also Feiereisen, W.J., and Acharya, M., 1986, "Modeling of Transition and Surface Roughness Effects in Boundary-Layer Flows," AIAA Journal, 24, pp. 1642–1649.

Blair, M.F., 1992, "Boundary Layer Transition in Accelerating Flows with Intense Free-Stream Turbulence: Part 1–Disturbances Upstream of Transition Onset, and Part 2–The Zone of Intermittent Turbulence," J. Fluids Engng., 114, pp. 313–332.

Corrsin, S., and Kistler, A.L., 1955, "Free-Stream Boundaries of Turbulent Flows," NACA TR 1244; also 1954, NACA TN 3133.

Gostelow, J.P., and Walker, G.J., 1990, "Similarity Behavior in Transitional Boundary Layers over a Range of Adverse Pressure Gradients and Turbulence Levels," ASME Paper 90-GT-130.

Hatman, A., and Wang, T., 1998, "Separated Flow Transition," Report No. GTL-9801, Gas Turbine Laboratory, Dept. Mech. Engng., Clemson University, Clemson, SC.

Keller, F.J., and Wang, T., 1995, "Effects of Criterion Functions on Intermittency in Heated Transitional Boundary Layers With and Without Streamwise Acceleration," J. Turbomachinery, 117, pp. 154–165.

Kim, J., Simon, T.W., and Kestoras, M., 1989, "Fluid Mechanics and Heat Transfer Measurements in Transitional Boundary Layers Conditionally Sampled on Intermittency," Heat Transfer in Convective Flows, 1989 Nat'l Heat Transfer Conf., ASME Publication HTD-107, pp. 69–81.

Kuan, C., and Wang, T., 1990, "Some Intermittent Behavior of Transitional Boundary Layers," Exp. Thermal and Fluid Sci., 3, pp. 157–170.

Mayle, R.E., 1991, "The Role of Laminar-Turbulent Transition in Gas Turbine Engines," J. Turbomachinery, 113, pp. 509–537.

Owen, F.K., 1970, "Transition Experiments On a Flat Plate at Subsonic and Supersonic Speeds," AIAA Journal, 8, pp. 518–523.

Wang, T., and Zhou, D., 1998, " Conditionally Sample Flow andThermal Behavior of a Transitional at Elevated Free-Stream Turbulence," Int. J. Heat and Fluid Flow, 19, pp. 348–357.

Zhou, D., and Wang, T., 1995, "Effects of Elevated Free-Stream Turbulence on Flow and Thermal Structures in Transitional Boundary Layers," J. Turbomachinery, 117, pp. 407–417.

Appendix I – Intermittency Profile Corrections

Unfortunately, it is usual to plot boundary-layer profiles of any dependent quantity using the distance from the wall normalized by the boundary-layer thickness. Since δ is an ill-defined quantity, comparisons of measured profiles reported by different investigators, and of measured and theoretical profiles, using this format are always suspect. The remedy to this problem is simple – for flows with $dp/dx = 0$, normalize by either the displacement or momentum thickness because they are well defined, and for flows with $dp/dx \neq 0$, normalize by either one and report the shape factor. In this appendix, the corrections used for Acharya's y/δ-profile data in Fig. 5.25 are determined.

The variation of intermittency-profile half widths against streamwise distance are plotted in Fig. I.1. The abscissa is Narasimha's universal coordinate ξ defined by Eq. (5.20) in the text, while the ordinate is half-width $y_{1/2}$, defined as the value of y where $\gamma_0 = 0.5$, divided by the boundary-layer thickness. The half-widths were determined by using a fourth-order polynomial to fit the central portion of the y/δ-profile data reported by Owen [1970], Acharya [1985], and Kim et al. [1989] and solving it for y/δ with $\gamma_0 = 0.5$. The abscissa ξ was obtained by fitting the near-wall intermittency distribution using Eq. (5.21) to determine the quantities x_{xt} and $(x_{0.75}–x_{0.25})$ as described in §6.1.2. As noted in §5.2.2, transition is virtually complete when $\xi = 3.34$. Also included in the figure is Corrsin and Kistler's [1955] result for a fully-developed turbulent boundary layer ($y_{1/2}/\delta = 0.8$), and curves representing the averages of Owen and Kim's et al. (solid) and Acharya's (dashed) results.

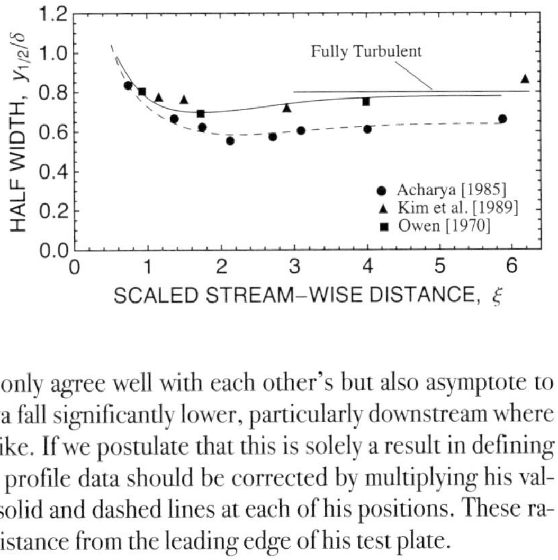

Figure I.1.
Variation of the intermittency profile half-widths with streamwise distance.

Clearly, the data of Kim et al. and Owen not only agree well with each other's but also asymptote to the fully-turbulent result, while that of Acharya fall significantly lower, particularly downstream where the boundary layer becomes more turbulent-like. If we postulate that this is solely a result in defining the boundary-layer thickness, then Acharya's profile data should be corrected by multiplying his values of y/δ by the ratio of values given by the solid and dashed lines at each of his positions. These ratios are provided in Table I.1, where x is the distance from the leading edge of his test plate.

Table I.1
Correction factors for Acharya's [1985] test TRA intermittency profile data.

x [mm]	1036	1200	1300	1400	1500	1600	1836
ξ	0.74	1.37	1.74	2.13	2.72	3.09	4.01
Ratio	1.04	1.07	1.11	1.26	1.27	1.24	1.24

References

Acharya, M., 1985, "Pressure-Gradient and Free-Stream Turbulence Effects on Boundary-Layer Transition," Forschungsbericht KLR 85-127 C, Brown Boveri Forschungszentrum, Baden, Switzerland; also Feiereisen, W.J., and Acharya, M., 1986, "Modeling of Transition and Surface Roughness Effects in Boundary-Layer Flows," AIAA Journal, 24, pp. 1642–1649.

Corrsin, S., and Kistler, A.L., 1955, "Free-Stream Boundaries of Turbulent Flows," NACA TR 1244; also 1954, NACA TN 3133.

Kim, J., Simon, T.W., and Kestoras, M., 1989, "Fluid Mechanics and Heat Transfer Measurements in Transitional Boundary Layers Conditionally Sampled on Intermittency," Heat Transfer in Convective Flows, 1989 Nat'l Heat Transfer Conf., ASME Publication HTD-107, pp. 69–81.

Owen, F.K., 1970, "Transition Experiments On a Flat Plate at Subsonic and Supersonic Speeds," AIAA Journal, 8, pp. 518–523.

Appendix J – Translating Between Spot Areas

The relation between the spot areas in the x and t planes is the same whether we consider either the spot's propagation or dependence cone (see Figs. 5.16 and 5.17). In the following, we will develop a relation between the two areas using the propagation cone. An exploded version of Fig. 5.16 showing these areas in more detail along with some of the required geometry is provided in Fig. J.1. As in Fig. 5.16, all the lines connecting related points in the x and t planes of the cone extrapolate back to the origin $P_s(x_s, t_s, z_s)$ of the spot. For the footprint of a spot expanding in a linear fashion across the surface, these lines are straight.

Using the notation developed in §5.3, the cross-sectional area $A_{s,zt}$ of the cone in the t plane at $x = x_1$ is

$$A_{s,zt}(x_1) = \int_{-w(t;x_1)/2}^{w(t;x_1)/2} dz(x_1) \int_{t_{te1}}^{t_{le1}} dt(x_1) = \int_{t_{te1}}^{t_{le1}} w(t;x_1)\, dt(x_1)$$

where $w(t; x_1)dt(x_1)$ is the element of the area in the x_1 plane, and the subscripts "$te1$" and "$le1$" refer to the trailing and leading edge of the spot at the position x_1. Our task is now to determine a relation between this area and the area of the spot in the physical plane $A_{s,xz}$ at the time say t_2, that is, the area $A_{s,xz}(t_2)$.

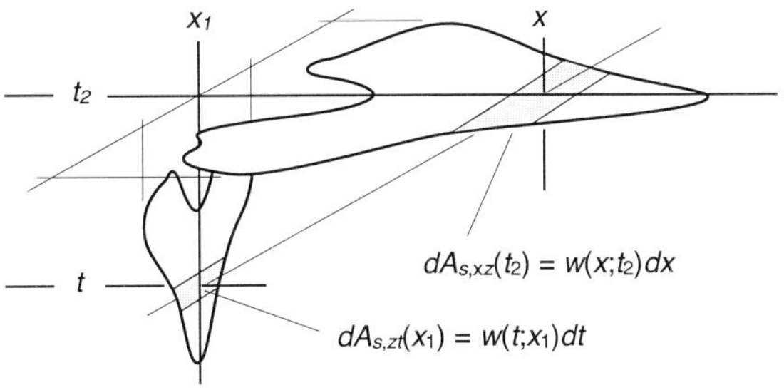

Figure J.1.
Spot footprints in a zt-plane and an xz-plane showing the elemental areas.

Assume the spot's footprint expands linearly as it propagates downstream and consider the relation between the elements of area at x_1 and x_2. The width changes from $w(t; x_1)$ to $w(t; x_2) = (x_2/x_1)w(t; x_1)$, while the differential dt changes to $dt(x_2) = (x_2/x_1)dt(x_1)$. Consequently, the area at the position x_1 is also given by

$$A_{s,zt}(x_1) = \int_{t_{te2}}^{t_{le2}} \frac{x_1^2 w(t;x_2)}{x_2^2(t)} dt(x_2)$$

In addition, since the velocity of any point within the footprint in the t_2 plane is given by $u_s(x_2; t_2) = dx(t_2)/dt(x_2)$, the integral can be changed from one in the t plane to one in the x plane. Realizing that x_2 can be any position x within the physical footprint in any t equal constant plane, the subscripts and the parametric notation may be dropped, yielding

$$A_{s,zt}(x_1) = x_1^2 \int_{x_{te}}^{x_{le}} \frac{w(x)}{x^2 u_s(x)} dx \tag{J.1}$$

where "x_{te}" and "x_{le}" refer to the trailing and leading edge positions of the physical footprint at any time.

Now define[1] the average distance to the spot and its velocity using

$$\bar{x}_s^2 \bar{u}_s = \int_{x_{te}}^{x_{le}} w(x)dx \Big/ \int_{x_{te}}^{x_{le}} \frac{w(x)dx}{x^2 u_s(x)} \tag{J.2}$$

where for a linearly expanding spot

$$\bar{x}_s = x_{te}\left(\bar{u}_s / u_{te}\right) = x_{le}\left(\bar{u}_s / u_{le}\right)$$

Then, noting that the numerator of Eq. (J.2) equals the physical area of the footprint at any time, we obtain the sought after relation,

$$A_{s,zt}(x) = \frac{x^2}{\bar{x}_s^2 \bar{u}_s} A_{s,xz}(t) \tag{J.3}$$

where, because it was defined arbitrarily, x_1 has been replaced by x. The task is now to determine the average distance and velocity which, of course, depends on the shape of the spot.

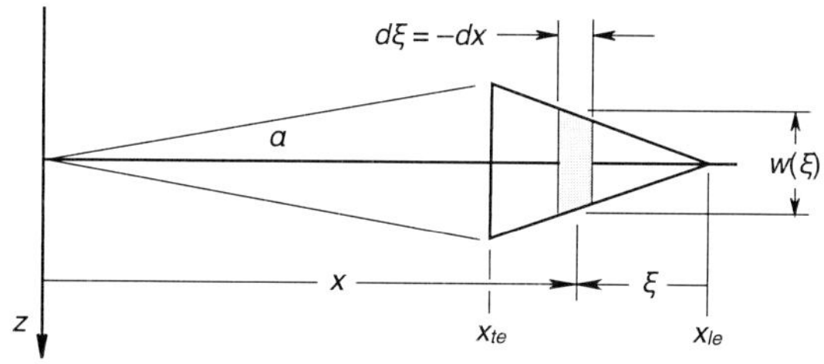

Figure J.2.
A triangular shaped footprint in an xz-plane.

An elementary example. Let's consider a triangular shaped footprint of length l and width b. Then $A_{s,xz} = \frac{1}{2}bl$. Introducing the coordinate $\xi = x_{le} - x$ measured from the leading edge of the spot, as shown in Fig. J.2, we obtain

$$w = b\frac{\xi}{l} \quad \text{and} \quad x = x_{le}\left[1 - \left(1 - x_{te}/x_{le}\right)\frac{\xi}{l}\right]$$

Furthermore, assuming a linear variation in the spot-velocity from its leading to trailing edge, we obtain

$$u_s(\xi) = u_{le}\left[1 - \left(1 - u_{te}/u_{le}\right)\frac{\xi}{l}\right]$$

since $u_{te}/u_{le} = x_{te}/x_{le}$. Define

$$a = 1 - u_{te}/u_{le} = 1 - x_{te}/x_{le}$$

and introduce the variable $\xi'' = \xi/l$. Then

$$w = b\xi'', \quad x = x_{le}\left(1 - a\xi''\right), \quad u_s = u_{le}\left(1 - a\xi''\right)$$

[1] For readers familiar with Emmons' paper, this definition is different from his. Emmons defines $\bar{x}_s$ and $\bar{u}_s$ as the distance from the apex of the cone to the center of area and the propagation velocity of the center of area of the spot. In general, the quantities defined herein are not the values at the center of area of the spot.

and Eq. (J.2) becomes

$$\bar{x}_s^2 \bar{u}_s = x_{le}^2 u_{le} \left[2 \int_0^1 \frac{\xi'' d\xi''}{(1-a\xi'')} \right]^{-1}$$

Evaluating the integral yields

$$\bar{x}_s^2 \bar{u}_s = x_{le}^2 u_{le} \left(u_{te} / u_{le} \right)^2 = x_{te}^2 u_{le} \quad \text{(J.4)}$$

which in turn provides

$$\bar{x}_s^2 = x_{le} \left(u_{te} / u_{le} \right)^{2/3} = x_{te} \left(u_{le} / u_{te} \right)^{1/3} \quad \text{and} \quad \bar{u}_s^3 = u_{le} u_{te}^2$$

For $u_{le} = 0.89\bar{u}_\infty$ and $u_{te} = 0.58\bar{u}_\infty$, we obtain $\bar{x}_s = 1.15x_{te} = 0.75x_{le}$ and $\bar{u}_s = 0.67\bar{u}_\infty$. These should be contrasted with the values at the center of the spot, which are $\bar{x}_{s,center} = 1.27x_{te}$ and $\bar{u}_{s,center} = 0.75\bar{u}_\infty$.

Eliminating the mean values between Eq. (J.3) and Eq. (J.4), we obtain

$$A_{s,zt}(x) = \frac{x^2}{x_{te}^2 u_{le}} A_{s,xz}(t) \quad \text{(triangular footprint)}$$

For $x = x_{te}$ and x_{le}, it quickly follows that

$$A_{s,zt}(x_{te}) = \frac{1}{u_{le}} A_{s,xz}(t) \quad \text{and} \quad A_{s,zt}(x_{le}) = \frac{1}{u_{te}} \frac{u_{le}}{u_{te}} A_{s,xz}(t)$$

Here, the reader may wish to consult Fig. 5.30 and the relations given in Eq. (5.62) of the text to show that the above relations for this elementary example can be obtained directly from geometry, providing due consideration is given to the breadth at the different locations.

Finally, from Eq. (5.101), the spot propagation parameter using the present notation is

$$\sigma(x) = \frac{\bar{u}_\infty A_{s,zt}(x)}{x^2}$$

Substituting $A_{s,xz} = ½bl$ and using

$$b = 2x_{te} \tan\alpha \quad \text{and} \quad l = x_{le} - x_{te}$$

we obtain

$$\sigma = \frac{\bar{u}_\infty}{u_{le}} \frac{u_{le} - u_{te}}{u_{te}} \tan\alpha \quad \text{(triangular footprint)}$$

as given by Eq. (5.106) with $f_s = 1$.

Starting with Eq. (J.1), the propagation parameter for any shape of footprint may be determined as long as the footprint expands linearly with distance. For example, the propagation parameter for a rectangular shaped footprint is given by

$$\sigma = \frac{\bar{u}_\infty}{u_{le}} \frac{u_{le}^2 - u_{te}^2}{u_{le} u_{te}} \tan\alpha \quad \text{(rectangular footprint)}$$

which differs from that for a triangular footprint by the factor $1+(u_{te}/u_{le})$. Of course, the relation between the areas is also different (as it will be for any other shape).

Appendix K – Transition Conversion Tables

Approximate conversions from the wall-shear- to intermittency-based definitions of the onset and length of transition are given in this appendix using

$$f_1(Re_{xt},\hat{n}\sigma)=\left(Re_{cf,\min}-Re_{xt}\right)/\left(Re_{cf,\max}-Re_{cf,\min}\right)$$
$$f_2(Re_{xt},\hat{n}\sigma)=Re_{LT}/\left(Re_{cf,\max}-Re_{cf,\min}\right) \tag{K.1}$$

where f_1 and f_2 are the conversion functions to be determined, Re_{xt} and Re_{LT} are the transition and transition-length Reynolds number based on intermittency measurements, and $Re_{cf,min}$ and $Re_{cf,max}$ are the distance Reynolds numbers at the positions of the minimum and maximum in the measured skin-friction coefficient distribution. According to Eq. (6.2) in the text, the relation between the length and rate of transition is given by

$$Re_{LT}\approx 2.15/\sqrt{\hat{n}\sigma}$$

The conversion uses the assumption described in §7.1 that the local time-averaged skin-friction coefficient $\bar{c}_f(x)$ can be obtained from its laminar and turbulent components by using

$$\bar{c}_f=\left(1-\gamma_0\right)\bar{c}_{f,L}+\gamma_0\bar{c}_{f,T}$$

and the intermittency solution for a line-source given by Eq. (5.18), namely,

$$\gamma_0(x)=1-\exp\left[-\hat{n}\sigma\left(Re_x-Re_{xt}\right)^2\right];\quad (\sigma \text{ constant}, Re_x\geq Re_{xt})$$

For these calculations the skin-friction coefficients were determined from the laminar and turbulent boundary-layer solutions given by Eqs. (2.10) and (3.40), namely,

$$c_{f,L}=0.664Re_x^{-1/2}\quad \text{and} \quad c_{f,T}=0.0574\left(Re_x-Re_{x0}\right)^{-1/5}$$

where Re_{x0} is the virtual origin of the turbulent boundary layer. As discussed in §5.2.5, and given by Eq. (5.45), it is approximately equal to

$$Re_{x0}=Re_{xt}\left(1-25.9r^{5/4}Re_{xt}^{-3/8}\right)$$

where r is the ratio of the initial spot height to the laminar boundary layer thickness at the onset of transition.

The results of these calculations are given in Table K.1 for r equal to zero and one, values suggested respectively by Dhawan and Narasimha [1958] and Schubauer and Klebanoff [1955]. The values in the cells shaded gray correspond to the expected values, from top to bottom, for transition with free-stream turbulence levels of roughly 0.2, 1, 3, and 10 percent.

When the correlation given by Eq. (6.18) in the text is used together with Blasius' solution to eliminate $\hat{n}\sigma$ and $Re_{\delta 2,t}$ in favor of Re_{xt}, then the functions f_1 and f_2 in Eq. (K.1), when weighted toward the higher turbulence levels, are approximately given by

$$f_1(Re_{xt})=15Re_{xt}^{-0.5},\quad f_2(Re_{xt})=2Re_{xt}^{-0.04}\ ;\quad (r=0) \tag{K.2}$$

where for $r=1$, only the coefficient in the first relation changes to twenty.

These relations have sometimes been used in the text to convert from one definition or method of determining transition to the other.

(Table on following page.)

Table K.1
Conversion functions f_1 and f_2 in Eq. (K.1) for $r = 0$ and $r = 1$.

$\hat{n}\sigma$ $Re_{LT}\times(10)^{-6}$	$Re_{xt}\times(10)^{-6}$	$r = 0$		$r = 1$	
		f_1	f_2	f_1	f_2
$(10)^{-12}$ 2.2	0.03	0.19	1.40	0.19	1.39
	0.1	0.16	1.35	0.16	1.35
	0.3	0.087	1.27	0.090	1.26
	1.0	0.014	1.18	0.020	1.17
	3.0	0.0018	1.15	0.0040	1.13
$(10)^{-11}$ 0.68	0.03	0.22	1.45	0.22	1.44
	0.1	0.12	1.33	0.13	1.32
	0.3	0.026	1.21	0.035	1.19
	1.0	0.0025	1.16	0.0059	1.12
	3.0	0.00035	1.15	0.0012	1.08
$(10)^{-10}$ 0.22	0.03	0.20	1.46	0.21	1.44
	0.1	0.041	1.25	0.056	1.22
	0.3	0.0045	1.18	0.011	1.12
	1.0	0.00048	1.16	0.0018	1.06
	3.0	0.00010	1.14	0.00035	1.00
$(10)^{-9}$ 0.068	0.03	0.085	1.35	0.12	1.32
	0.1	0.0068	1.21	0.017	1.13
	0.3	0.00082	1.17	0.0031	1.04
	1.0	0.00013	1.15	0.00051	0.97
	3.0	0.00003	1.14	0.00010	0.92

References

Dhawan, S., and Narasimha, R., 1958, "Some Properties of Boundary Layer Flow During Transition from Laminar to Turbulent Motion," J. Fluid Mech., 3, pp. 418–436.

Schubauer, G.B., and Klebanoff, P.S., 1955, "Contribution to the Mechanism of Boundary-Layer Transition," NACA TN 3489.

Appendix L – A Conditioned Reynolds-Stress Equation

Following the procedure described in §7.2.3, the conditioned Reynolds-stress equation for the *turbulent portion* of the flow is given by

$$\rho A_{ij} = -\rho B_{ij} - \rho C_{ij} + \rho D_{ij} - E_{ij} + F_{ij} \tag{L.1}$$

where

$$A_{ij} = \bar{u}_k \overline{I u_j' \frac{\partial u_i'}{\partial x_k}} + \bar{u}_k \overline{I u_i' \frac{\partial u_j'}{\partial x_k}}, \quad B_{ij} = \overline{I u_j' u_k'}\frac{\partial \bar{u}_i}{\partial x_k} + \overline{I u_i' u_k'}\frac{\partial \bar{u}_j}{\partial x_k}, \quad C_{ij} = \overline{I u_j' u_k' \frac{\partial u_i'}{\partial x_k}} + \overline{I u_i' u_k' \frac{\partial u_j'}{\partial x_k}}$$

$$D_{ij} = \overline{\overline{I u_j' u_k'}\frac{\partial u_i'}{\partial x_k}} + \overline{\overline{I u_i' u_k'}\frac{\partial u_j'}{\partial x_k}}, \quad E_{ij} = \overline{I u_j' \frac{\partial p'\delta_{ik}}{\partial x_k}} + \overline{I u_i' \frac{\partial p'\delta_{jk}}{\partial x_k}}, \quad \text{and} \quad F_{ij} = \overline{I u_j' \frac{\partial \tau_{ik}'}{\partial x_k}} + \overline{I u_i' \frac{\partial \tau_{jk}'}{\partial x_k}} \tag{L.2}$$

Initially it appears that D_{ij} equals zero because it is an average of a fluctuating and time-averaged quantity, but this would be a mistake. Since the *jump component* in u_j' is time *independent*, it turns out that D_{ij} is not zero. Consequently, evaluating the quantities in Eq. (L.2) using the various definitions and rules in §7.2.1 yields

$$A_{ij} = \gamma \bar{u}_k \frac{\partial \overline{(u_i' u_j')}_T}{\partial x_k} + \gamma(1-\gamma)^2 \bar{u}_k \frac{\partial \Delta\bar{u}_i \Delta\bar{u}_j}{\partial x_k} + \left[2(1-\gamma)\left(\tfrac{1}{2}-\gamma\right)\bar{u}_k \Delta\bar{u}_i \Delta\bar{u}_j + \tfrac{1}{2}\bar{u}_k\left(\overline{u_{i,T}'\Delta u_j'} + \overline{u_{j,T}'\Delta u_i'}\right)\right]\frac{\partial \gamma}{\partial x_k}$$

$$B_{ij} = \gamma \overline{(u_i' u_k')}_T \frac{\partial \bar{u}_j}{\partial x_k} + \gamma \overline{(u_j' u_k')}_T \frac{\partial \bar{u}_i}{\partial x_k} + \gamma(1-\gamma)^2 \Delta\bar{u}_k \left(\Delta\bar{u}_i \frac{\partial \bar{u}_j}{\partial x_k} + \Delta\bar{u}_j \frac{\partial \bar{u}_i}{\partial x_k}\right)$$

$$\begin{aligned}
C_{ij} = {} & \gamma \overline{u_{i,T}' u_{k,T}' \frac{\partial u_{j,T}'}{\partial x_k}} + \gamma \overline{u_{j,T}' u_{k,T}' \frac{\partial u_{i,T}'}{\partial x_k}} + \gamma(1-\gamma)\left[\overline{(u_i' u_k')}_T \frac{\partial \Delta\bar{u}_j}{\partial x_k} + \overline{(u_j' u_k')}_T \frac{\partial \Delta\bar{u}_i}{\partial x_k}\right] \\
& + \gamma(1-\gamma)\Delta\bar{u}_k \frac{\partial \overline{(u_i' u_j')}_T}{\partial x_k} + \gamma(1-\gamma)\left[\Delta\bar{u}_i \overline{u_{k,T}' \frac{\partial u_{j,T}'}{\partial x_k}} + \Delta\bar{u}_j \overline{u_{k,T}' \frac{\partial u_{i,T}'}{\partial x_k}}\right] + \gamma(1-\gamma)\Delta\bar{u}_k \frac{\partial \Delta\bar{u}_i \Delta\bar{u}_j}{\partial x_k} \\
& + \tfrac{1}{2}\left[\overline{(u_i' u_k')_T \Delta u_j'} + \overline{(u_j' u_k')_T \Delta u_i'}\right]\frac{\partial \gamma}{\partial x_k} + \left(\tfrac{1}{2}-\gamma\right)\left[\overline{(u_i' u_k')}_T \Delta\bar{u}_j + \overline{(u_j' u_k')}_T \Delta\bar{u}_i\right]\frac{\partial \gamma}{\partial x_k} \\
& + \tfrac{1}{2}(1-\gamma)\Delta\bar{u}_k \left[\overline{u_{i,T}'\Delta u_j'} + \overline{u_{j,T}'\Delta u_i'}\right]\frac{\partial \gamma}{\partial x_k} + 2\left(\tfrac{1}{2}-\gamma\right)(1-\gamma)^2 \Delta\bar{u}_i \Delta\bar{u}_j \Delta\bar{u}_k \frac{\partial \gamma}{\partial x_k}
\end{aligned}$$

$$\begin{aligned}
D_{ij} = {} & \gamma(1-\gamma)\Delta\bar{u}_i \frac{\partial}{\partial x_k}\left[(1-\gamma)\overline{(u_j' u_k')}_L + \gamma\overline{(u_j' u_k')}_T + \gamma(1-\gamma)\Delta\bar{u}_j \Delta\bar{u}_k\right] \\
& + \gamma(1-\gamma)\Delta\bar{u}_j \frac{\partial}{\partial x_k}\left[(1-\gamma)\overline{(u_i' u_k')}_L + \gamma\overline{(u_i' u_k')}_T + \gamma(1-\gamma)\Delta\bar{u}_i \Delta\bar{u}_k\right]
\end{aligned} \tag{L.3}$$

$$E_{ij} = \gamma \overline{u_{i,T}' \frac{\partial p_T' \delta_{jk}}{\partial x_k}} + \gamma \overline{u_{j,T}' \frac{\partial p_T' \delta_{ik}}{\partial x_k}} + \gamma(1-\gamma)^2 \left[\Delta\bar{u}_i \frac{\partial \Delta\bar{p}\delta_{jk}}{\partial x_k} + \Delta\bar{u}_j \frac{\partial \Delta\bar{p}\delta_{ik}}{\partial x_k}\right]$$

$$+(1-\gamma)\left(\tfrac{1}{2}-\gamma\right)\Delta\bar{p}\left(\Delta\bar{u}_i\delta_{jk}+\Delta\bar{u}_j\delta_{ik}\right)\frac{\partial\gamma}{\partial x_k}+\tfrac{1}{2}\left[\overline{u_i'\Delta p'\delta_{jk}}+\overline{u_j'\Delta p'\delta_{ik}}\right]\frac{\partial\gamma}{\partial x_k}$$

$$F_{ij}=\gamma\overline{u'_{i,T}\frac{\partial\tau'_{jk,T}}{\partial x_k}}+\gamma\overline{u'_{j,T}\frac{\partial\tau'_{ik,T}}{\partial x_k}}+\gamma(1-\gamma)^2\left[\Delta\bar{u}_i\frac{\partial\Delta\bar{\tau}_{jk}}{\partial x_k}+\Delta\bar{u}_j\frac{\partial\Delta\bar{\tau}_{ik}}{\partial x_k}\right]$$

$$+(1-\gamma)\left(\tfrac{1}{2}-\gamma\right)\left(\Delta\bar{u}_i\Delta\bar{\tau}_{jk}+\Delta\bar{u}_j\Delta\bar{\tau}_{ik}\right)\frac{\partial\gamma}{\partial x_k}+\tfrac{1}{2}\left[\overline{u_i'\Delta\tau'_{jk}}+\overline{u_j'\Delta\tau_{ik}}\right]\frac{\partial\gamma}{\partial x_k}$$

where

$$\Delta\bar{u}_i=\bar{u}_{i,T}-\bar{u}_{i,L}\ ,\quad \Delta u_i'=u'_{i,T}-u'_{i,L}\ ,\quad \text{etc.}$$

Although it is definitely not obvious, writing this for the laminar component of the equation, by exchanging the subscripts "T" and "L" and the quantities γ and $(1-\gamma)$, and adding to the above yields after much effort the global Reynolds-stress equation given by Eq. (7.57) in the text. That this is true is most easily seen by considering term B_{ij} of the equation.

For those seeking other forms of this equation, the following identities should be noted:

$$\gamma(1-\gamma)\frac{\partial\Delta\bar{u}_i}{\partial x_i}=-\left(\tfrac{1}{2}-\gamma\right)\Delta\bar{u}_i\frac{\partial\gamma}{\partial x_i}$$

$$\overline{u'_{j,T}\frac{\partial u'_{i,T}}{\partial x_i}}=-\tfrac{1}{2}\overline{u'_{j,T}\left(u'_{i,T}-u'_{i,L}\right)}\frac{\partial\gamma}{\partial x_i} \qquad \text{(L.4)}$$

The first of which is a consequence of the conditioned continuity equations given by Eq. (7.45) in the text. The second is given by Eq. (7.46).

Supplement on Cartesian Tensors

This supplement is provided for those readers who want to refresh their memory on Cartesian tensor notation, algebra, and calculus. For those who wish to learn more about tensors and their invariance under coordinate transformations, the books by Jeffreys [1952] and Temple [1960], or the Appendix in Hinze [1975] are recommended.

Notation. A tensor is a mathematical object that remains invariant under transformations from one frame of reference to another. The *elements* or *components* of a tensor are generally arranged in a multi-dimensional array *indexed* according to their position within the array. In particular, a tensor denoted by the symbol $A_{ijk...pq}$ corresponds to a tensor of *rank* or *order* given by the included number of indices from i to q. The *dimensions* of a tensor are the largest integers within the range of integers that each index can assume. Thus, A_{ij} with $i = 1, 2, 3$ and $j = 1, 2, 3, 4$ is a three-by-four second-order tensor. It contains or is described by 12 *elements* or *components*, which equals the product of its dimensions. As might be imagined, the complexity of a tensor increases with increasing order and dimensions. The tensors most familiar to those interested in fluid mechanics have three dimensions and are of zero, first, and second order. Higher order tensors are usually a result of mathematically manipulating these.

A tensor of *zero* order is a scalar object such as a constant, or a variable of state like the density ρ, pressure p, or temperature T. Defined solely by its *magnitude*, it is determined by only one component which is represented in tensor notation by a symbol without any index.

A tensor of *first* order is a vector object defined by its *magnitude* and *direction*. In vector notation, it is usually represented by a bold typeface symbol such as $\boldsymbol{A}$. In tensor notation, it is represented by the indexed symbol A_i. In a three-dimensional space, A_i is determined by its components $\{A_1, A_2, A_3\}$, that is, A_i with $i = 1, 2, 3$, which in turn correspond to the three mutually-perpendicular components of the vector $\boldsymbol{A}$. In physics, the largest integer of the index equals the number of dimensions needed to describe the space. The most familiar of these tensors in fluid mechanics are the position, velocity, and force vectors, x_i, u_i, and f_i, respectively. Mathematically, a first-order tensor of three dimensions is treated as a column matrix with three elements. That is,

$$A_i = \begin{bmatrix} A_1 \\ A_2 \\ A_3 \end{bmatrix} \tag{1}$$

A tensor of *second* order is an object defined by its *magnitude* and *two directions*. In a three-dimensional space, it is determined by an *ordered set* of three groups of three for a total of nine components, namely, $\{\{A_{11}, A_{12}, A_{13}\}, \{A_{21}, A_{22}, A_{23}\}, \{A_{31}, A_{32}, A_{33}\}\}$. These components are represented more concisely in tensor notation by the symbol A_{ij} where both i and j take the values of 1, 2, and 3. The most familiar of these tensors in fluid mechanics are the strain-rate and stress tensors where, to be specific, the components of the stress tensor σ_{ij} define its magnitude, the direction of the stress (index j), and the direction of the normal to the surface that the stress acts upon (index i). Mathematically, a second-order tensor of three dimensions is treated as a three-by-three matrix where the component A_{ij} resides in the ith row of the jth column. That is,

$$A_{ij} = \begin{bmatrix} A_{11} & A_{12} & A_{13} \\ A_{21} & A_{22} & A_{23} \\ A_{31} & A_{32} & A_{33} \end{bmatrix} \tag{2}$$

Conventions and definitions. Let's now consider some related conventions and definitions. The foremost of these is Einstein's *summation convention* which asserts that any quantity or combination of quantities with *two repeated* indices is to be interpreted as the *sum* of all the values that quantity can take as the index assumes all of its possible values. For example,

$$A_{ii} = \sum_{i=1}^{3} A_{ii} = A_{11} + A_{22} + A_{33}$$

Applying this convention to a product of the two quantities A_i and B_i we obtain

$$A_i B_i = \sum_{i=1}^{3} A_i B_i = A_1 B_1 + A_2 B_2 + A_3 B_3 \qquad (3)$$

For the slightly more complex example $A_{ij}B_{ji}C_k$, the summation convention implies that

$$A_{ij} B_{ji} C_k = \sum_{i,j=1}^{3} A_{ij} B_{ji} C_k = \sum_{j=1}^{3} \left(A_{1j} B_{j1} C_k + A_{2j} B_{j2} C_k + A_{3j} B_{j3} C_k \right)$$

$$= \left(A_{11} B_{11} C_k + A_{12} B_{21} C_k + A_{13} B_{31} C_k \right) + \left(A_{21} B_{12} C_k + A_{22} B_{22} C_k + A_{23} B_{32} C_k \right) \qquad (4)$$

$$+ \left(A_{31} B_{13} C_k + A_{32} B_{23} C_k + A_{33} B_{33} C_k \right)$$

which contains a total of twenty-seven terms when expanded for $k = 1, 2, 3$. Moreover, since $A_{ij}B_{ji}C_k = A_{ji}B_{ij}C_k$, any letter other than k may be used for the *repeated* indices in this example. That is,

$$A_{ij} B_{ji} C_k = \sum_{i,j=1}^{3} A_{ij} B_{ji} C_k = \sum_{l,m=1}^{3} A_{lm} B_{ml} C_k$$

A tensor quantity with more than two indices the same is undefined. Thus, the quantities A_{kkk} and $A_{jk}B_{kk}$ have no meaning even though components such as A_{222} and $A_{j2}B_{22}$ with $j = 1,2,3$ do.

The *trace* of a second-order tensor is the sum of the elements having the same index numbers; in other words, as seen in Eq. (2), the sum of the elements on its diagonal. Thus, the trace of A_{ij} is the scalar $A_{11} + A_{22} + A_{33}$. For higher order tensors, the trace is still the sum of its diagonal elements unless the trace is defined relative to a pair of its indices. Thus, the trace of A_{ijk} is the scalar $A_{111} + A_{222} + A_{333}$ unless it is defined relative to the indices i and j, in which case it becomes the first-order tensor $A_{iik} = A_{11k} + A_{22k} + A_{33k}$.

A second-order tensor is *symmetric* when the tensor remains unchanged upon exchanging its indices, that is, when $A_{ij} = A_{ji}$. It is *anti-symmetric* when $A_{ij} = -A_{ji}$. In the latter case, the elements on the diagonal are equal to zero, since zero is the only number that equals its negative. For higher order tensors, symmetry and anti-symmetry are defined relative to pairs of indices. Thus, the tensor A_{ijk} is symmetric or anti-symmetric relative to the indices j and k when it satisfies $A_{ijk} = A_{ikj}$, or $A_{ijk} = -A_{ikj}$.

Position in space. Let's define the position of a point P in space by using a Cartesian coordinate system of mutually perpendicular axes X_1, X_2, and X_3 with its origin at the point O. As shown in Fig. 1a, let $\boldsymbol{x}$ be the *position* or *radius vector* of P relative to the origin and denote its magnitude by the quantity r. Also let $\boldsymbol{x}_1$, $\boldsymbol{x}_2$ and $\boldsymbol{x}_3$ be the vector components of $\boldsymbol{x}$ along each axis. Then the distance from O to P according to Pythagoras' theorem is given by

$$r = |\boldsymbol{x}| = \sqrt{\boldsymbol{x} \cdot \boldsymbol{x}} = \sqrt{\boldsymbol{x}_1 \cdot \boldsymbol{x}_1 + \boldsymbol{x}_2 \cdot \boldsymbol{x}_2 + \boldsymbol{x}_3 \cdot \boldsymbol{x}_3} \qquad (5)$$

where $\boldsymbol{x} \cdot \boldsymbol{x}$ represents the dot product of the vector $\boldsymbol{x}$ with itself, and thus is equal to the magnitude of $\boldsymbol{x}$ squared.

The same applies when using tensor notation where x_i and $\{x_1, x_2, x_3\}$ represent the position vector and its components in the $\{X_1, X_2, X_3\}$ directions. Hence, by inspection, the distance from O to P according to Eq. (5) is given by

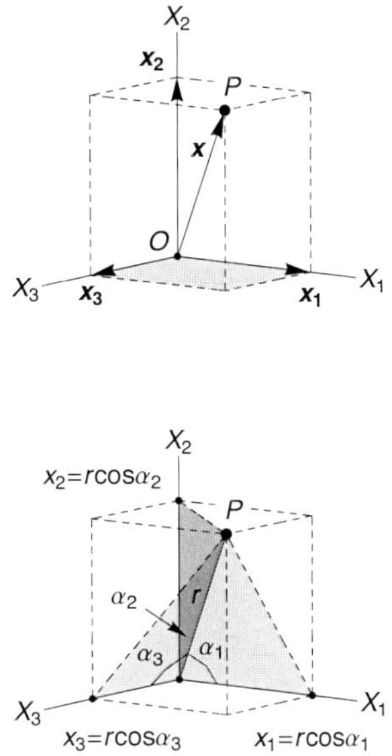

(a)

(b)

Figure 1a,b.
The coordinates of a point $P(x_i)$ as described using a Cartesian coordinate system and the angles between the line OP and the individual axes of the coordinate system $\{X_1, X_2, X_3\}$.

$$r = \sqrt{x_i x_i} = \sqrt{x_1^2 + x_2^2 + x_3^2} \tag{6}$$

If the angles between the position vector $\boldsymbol{x}$ and the individual axes are $\{\alpha_1, \alpha_2, \alpha_3\}$ as shown in Fig. 1b, then $\{x_1, x_2, x_3\} = \{r\cos\alpha_1, r\cos\alpha_2, r\cos\alpha_3\}$. This equation of ordered sets may be expressed more compactly in tensor notation by the single equation

$$x_i = r\cos\alpha_i \; ; \qquad (i = 1, 2, 3)$$

Substituting the above expressions for $x_1, x_2,$ and x_3 into Eq. (6) yields

$$\cos^2\alpha_1 + \cos^2\alpha_2 + \cos^2\alpha_3 = 1$$

Furthermore, if we define the *direction cosines* of the position vector by

$$l_i = \cos\alpha_i \; ; \qquad (i = 1, 2, 3) \tag{7}$$

then the previous expression may be written as

$$l_1 l_1 + l_2 l_2 + l_3 l_3 = 1$$

That is, the direction cosines l_1, l_2, and l_3 are the components of a *unit vector* lying on the line OP. Using Einstein's summation convention, this relation becomes

$$l_i l_i = 1 \tag{8}$$

Known as the *orthogonality relation* between direction cosines, it implies that only two direction cosines of the position vector, or the line OP, can be independently chosen.

Coordinate transformation. Let's now consider the change in the components of a vector under transforming from one frame of reference to another, and establish the conditions under which the vector remains invariant. Limiting ourselves to a Cartesian coordinate system with its three mutually perpendicular axes $\{X_1, X_2, X_3\}$, let $\boldsymbol{f}(P)$ be a vector quantity, such as a force or velocity, measured at the point P. Furthermore, let $\boldsymbol{f}_1$, $\boldsymbol{f}_2$ and $\boldsymbol{f}_3$ be the vector components of $\boldsymbol{f}$ in the direction of these axes as shown in Fig. 2a. Clearly, *translating* the origin of the coordinate system in any direction changes only the coordinates of P and not the magnitudes nor directions of $\boldsymbol{f}_1$, $\boldsymbol{f}_2$ and $\boldsymbol{f}_3$. In other words, even though the coordinates of the position where $\boldsymbol{f}$ is measured will be different, the components of $\boldsymbol{f}$ remain invariant with regard to coordinate translation. However, as shown in Fig. 2b, the same cannot be said about the components of $\boldsymbol{f}(P)$ in the directions $\{X_1', X_2', X_3'\}$ under a *rotation* of the coordinate system about its origin.

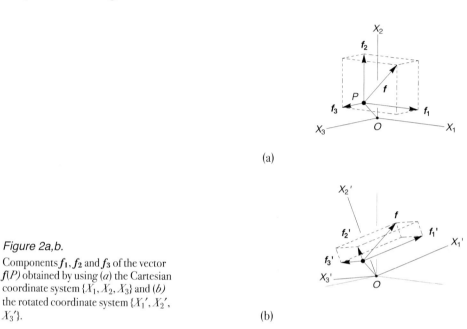

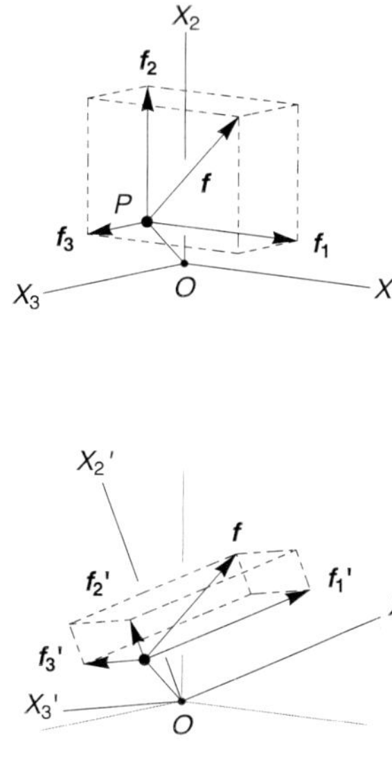

(a)

(b)

Figure 2a,b.
Components $\boldsymbol{f}_1$, $\boldsymbol{f}_2$ and $\boldsymbol{f}_3$ of the vector $\boldsymbol{f}(P)$ obtained by using (*a*) the Cartesian coordinate system $\{X_1, X_2, X_3\}$ and (*b*) the rotated coordinate system $\{X_1', X_2', X_3'\}$.

Since the vector f_i and its components (using tensor notation) are invariant with respect to coordinate translation, translate both the $\{X_i\}$- and $\{X_i'\}$-coordinate systems from point O to P for convenience. To determine the new components of f_i when the coordinate system is rotated, let the second-order tensor α_{ij} with $i, j = 1, 2, 3$ be the angle between the positive X_i-axis of the old system and the positive X_j'-axis of the new (rotated) system as illustrated in Fig. 3a. Note, as shown by the angles α_{12} and α_{21} in the figure, that $\alpha_{ij} \neq \alpha_{ji}$ unless i and j are numerically the same.

Furthermore, let $\{f_1, f_2, f_3\}$ and $\{f_1', f_2', f_3'\}$ be the components of the vector f_i in the old and new coordinate systems. Then a component of f_i along one of the axes in the new system, say X_1', is obtained by summing the projections of all the old components $\{f_1, f_2, f_3\}$ onto the X_1'-axis. These components together with their projections and sum f_1' are shown in Fig. 3b. It then follows that

$$f_1' = f_1 \cos\alpha_{11} + f_2 \cos\alpha_{21} + f_3 \cos\alpha_{31}$$
$$f_2' = f_1 \cos\alpha_{12} + f_2 \cos\alpha_{22} + f_3 \cos\alpha_{32}$$
$$f_3' = f_1 \cos\alpha_{13} + f_2 \cos\alpha_{23} + f_3 \cos\alpha_{33}$$

which in tensor notation is written simply as

$$f_i' = f_j \cos\alpha_{ji}$$

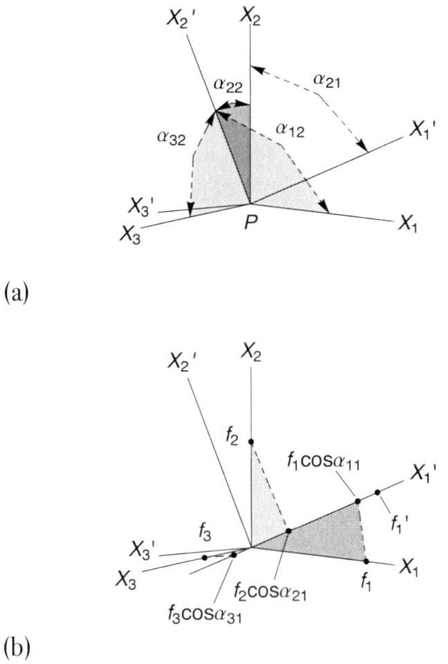

Figure 3a,b.
(a) Angles between the axes of the old coordinate system $\{X_1, X_2, X_3\}$ and the X_2'-axis of the new rotated system, and (b) the components of f_i in the old coordinate system projected onto the X_1'-axis of the new.

If we define the *direction cosines* between the axes of the old and new coordinate systems by

$$m_{ij} = \cos\alpha_{ij} \tag{9}$$

then the components of f_i in the new (rotated) coordinate system in terms of the old are given by

$$f_i' = f_j m_{ji}$$

Similarly, an expression for the old components in terms of the new can be obtained. Realizing that the definition of m_{ij} regarding the order of its indices for the old and new coordinate system must be retained, this transformation is

$$f_j = f_i' m_{ji}$$

Thus, any quantity whose coordinates transforms according to these relations remains invariant with regard to coordinate rotation. As it turns out, this is also true with regard to coordinate translation, since in this case m_{ij} equals one when $i = j$ and zero when $i \neq j$. Moreover, since the coordinate system may consist of any number of orthogonal axes, these relations also hold for a space having any number of dimensions.

Hence, the components of any first-order tensor, say A_i, in n-dimensional space will transform according to

$$A_i' = A_j m_{ji} \quad \text{and} \quad A_j = A_i' m_{ji} \tag{10}$$

providing the axes of the coordinate system remain orthogonal under its transformation.

Orthogonality relations. As we found in Eq. (8), the direction cosines l_i between a line and the axes of an orthogonal three-dimensional coordinate system must satisfy an orthogonality relation and that, in this case, only two of the three direction cosines could be chosen arbitrarily. A similar situation exists for the direction cosines m_{ij}.

To determine the relations between them, again consider the three-dimensional coordinate systems shown in Fig. 3a. In this case m_{ij} contains nine direction cosines. Since two of them are required to specify the direction of one new axis relative to the old coordinate system and another is required to specify the attitude of the new system relative to the old, only three direction cosines can be chosen arbitrarily when the coordinate systems are orthogonal. Thus, six equations for the six unknown components of m_{ij} must be found.

These relations may be obtained directly from Eq. (10) by requiring that the magnitude of A_i remains invariant with a change in coordinate systems. Since this is equivalent to requiring that its square A_iA_i in one coordinate system equals that in another, we obtain

$$A_iA_i = A_i'A_i' = \left(A_jm_{ji}\right)\left(A_km_{ki}\right) = A_jA_km_{ji}m_{ki}$$

The most left- and right-hand side expressions in this series contain three and twenty-seven terms, respectively, which when expanded yields

$$\begin{aligned} A_1^2 + A_2^2 + A_3^2 = {} & A_1^2\left(m_{11}m_{11} + m_{12}m_{12} + m_{13}m_{13}\right) + A_2^2\left(m_{21}m_{21} + m_{22}m_{22} + m_{23}m_{23}\right) \\ & + A_3^2\left(m_{31}m_{31} + m_{32}m_{32} + m_{33}m_{33}\right) + 2A_1A_2\left(m_{11}m_{21} + m_{12}m_{22} + m_{13}m_{23}\right) \\ & + 2A_1A_3\left(m_{11}m_{31} + m_{12}m_{32} + m_{13}m_{33}\right) + 2A_2A_3\left(m_{21}m_{31} + m_{22}m_{32} + m_{23}m_{33}\right) \end{aligned}$$

Since this identity must hold for any values of A_1, A_2, and A_3, the coefficients of the first three terms on its right-hand side must equal one, while the coefficients of the last three must equal zero. Hence, the six required orthogonality relations are

$$\begin{aligned} m_{11}m_{11} + m_{12}m_{12} + m_{13}m_{13} &= 1 \\ m_{21}m_{21} + m_{22}m_{22} + m_{23}m_{23} &= 1 \\ m_{31}m_{31} + m_{32}m_{32} + m_{33}m_{33} &= 1 \\ m_{11}m_{21} + m_{12}m_{22} + m_{13}m_{23} &= 0 \\ m_{11}m_{31} + m_{12}m_{32} + m_{13}m_{33} &= 0 \\ m_{21}m_{31} + m_{22}m_{32} + m_{23}m_{33} &= 0 \end{aligned}$$

In other words, given three, the remaining six components of m_{ij} can be determined from these equations. Using the summation convention, they may be written as

$$\begin{aligned} m_{ji}m_{ki} &= 1; \qquad \text{for } j = k \\ m_{ji}m_{ki} &= 0; \qquad \text{for } j \neq k \end{aligned}$$

Furthermore, introducing the *Kronecker delta* δ_{jk}, defined by

$$\begin{aligned} \delta_{jk} &= 1; \qquad \text{for } j = k \\ \delta_{jk} &= 0; \qquad \text{for } j \neq k \end{aligned} \tag{11}$$

the above set of orthogonality relations may be expressed by a single equation, namely,

$$m_{ji}m_{ki} = \delta_{jk} \tag{12}$$

In a similar fashion, starting with the condition

$$A_i'A_i' = A_iA_i = \left(A_j'm_{ij}\right)\left(A_k'm_{ik}\right) = A_j'A_k'm_{ij}m_{ik}$$

we obtain another form of the relations, namely,

$$m_{ij}m_{ik} = \delta_{jk} \tag{13}$$

It is now possible to show that one of the transformations in Eq. (10) follows from the other as long as the above orthogonality relations apply. For example, multiplying both sides of the first transformation in Eq. (10) by m_{ji} and using Eqs. (11) and (12) yields

$$A_i'm_{ji} = A_k m_{ki} m_{ji} = A_k \delta_{kj} = A_j$$

or by reversing its order $A_j = A_i' m_{ji}$, which is the second transformation.

Tensor algebra. It is now easy to show that the sum of two first-order tensors transform as another first-order tensor. For example, transforming the sum $A_i + B_i$ from one coordinate system to another yields

$$A_i + B_i = A_j' m_{ij} + B_j' m_{ij} = (A_j' + B_j') m_{ij}$$

Since the sum transforms according to Eq. (10), it is another first-order tensor, say C_i.

In similar fashion we can determine that the product of a scalar and a first-order tensor, which is equivalent to multiplying each component of the tensor by the scalar, transforms as another first-order tensor that differs only in magnitude, not in direction. For example, transforming the product cA_i from one coordinate system to another yields

$$cA_i = c(A_j' m_{ij}) = (cA_j') m_{ij}$$

Since the product transforms according to Eq. (10), it is also first-order tensor, say B_i, with the same direction as A_i, but with the magnitude

$$|B_i| = \sqrt{B_i B_i} = \sqrt{(cA_i)(cA_i)} = c\sqrt{A_i A_i} = c|A_i|$$

By definition the product of two first-order tensors is obtained by multiplying each component of one tensor by each component of the other. For example, if we consider the two tensors B_i and C_j with $i, j = 1, 2, 3$, then their product is the quantity

$$B_i C_j = \begin{bmatrix} B_1C_1 & B_1C_2 & B_1C_3 \\ B_2C_1 & B_2C_2 & B_2C_3 \\ B_3C_1 & B_3C_2 & B_3C_3 \end{bmatrix}$$

which is a second-order tensor, say A_{ij}. In this case, transforming $A_{ij} = B_i C_j$ from one coordinate system to another yields

$$B_i C_j = (B_k' m_{ik})(C_l' m_{jl}) = (B_k' C_l') m_{ik} m_{jl} = A_{kl}' m_{ik} m_{jl}$$

while transforming the product $B_k' C_l' = A_{kl}'$ back into the original system yields

$$B_k' C_l' = (B_i m_{ik})(C_j m_{jl}) = (B_i C_j) m_{ik} m_{jl} = A_{ij} m_{ik} m_{jl}$$

In other words, any object A_{ij} determined by the nine components $\{\{A_{11}, A_{12}, A_{13}\}, \{A_{21}, A_{22}, A_{23}\}, \{A_{31}, A_{32}, A_{33}\}\}$ that transforms according to

$$A_{kl}' = A_{ij} m_{ik} m_{jl} \quad \text{and} \quad A_{ij} = A_{kl}' m_{ik} m_{jl} \tag{14}$$

will remain invariant with regard to coordinate translation and rotation. This object is *defined* as a tensor of second order.

Definition of an *n*-order tensor. Following the above process for multiplying tensors, it is not too difficult to deduce that the product of two tensors of order n and m is a tensor of order $(n + m)$. Furthermore, the above results are easily extended to raising a tensor to an integer power. In addition, it can be deduced that a tensor of order n in three-dimensional space contains 3^n components, and that it transforms according to

$$A'_{\underbrace{ijkl\ldots}_{n\text{ indices}}} = A_{\underbrace{pqrs\ldots}_{n\text{ indices}}} m_{pi} m_{qj} m_{rk} m_{sl} \ldots$$

By extending the range of integers the indices can take, this transformation also holds for a space having any number of dimensions.

Special tensors. The Kronecker delta as defined in Eq. (11) is one of many special tensors frequently occurring in tensor algebra. As a second-order tensor, it possesses the property of transforming into itself under any coordinate transformation. In other words, it is an *isotropic* tensor. This is easily shown by writing the first expression in Eq. (14) for the Kronecker delta and realizing that δ_{ij} is nonzero only when $i=j$. Accordingly, we obtain

$$\delta'_{kl} = \delta_{ij} m_{ik} m_{jl} = m_{ik} m_{il} = \delta_{kl}$$

where the last equality arises from Eq. (13).

As it turns out, the Kronecker delta and numerical multiples of it are the only isotropic tensors of second order. Mathematically, δ_{ij} is treated as the three-by-three matrix

$$\delta_{ij} = \begin{bmatrix} 1 & 0 & 0 \\ 0 & 1 & 0 \\ 0 & 0 & 1 \end{bmatrix}$$

From this, it follows that

$$\delta_{ii} = \delta_{11} + \delta_{22} + \delta_{33} = 3$$
$$\delta_{ij}\delta_{jk} = \delta_{ik}$$

Since any numerical multiple of δ_{ij} is an isotropic tensor, and since $A_{kk}/3$ is the average of A_{ij}'s trace, it follows that $\delta_{ij}A_{kk}/3$ must be the *isotropic part* of A_{ij}. Consequently, the *anisotropic part* of A_{ij} is given by

$$a_{ij} = A_{ij} - \delta_{ij} A_{kk}/3 \qquad (15)$$

which is easily shown to have a trace equal to zero, that is, $a_{ii} = 0$.

Because δ_{ij} equals one only when $i=j=1$, 2, or 3, the product of δ_{ij} and a tensor with one of its indices equal to either i or j yields

$$\delta_{ij} A_j = A_i$$
$$\delta_{ij} A_{jk} = A_{ik}$$

This process simply *substitutes* the index i for j yielding a tensor of the same order as the original. On the other hand, the product of δ_{ij} and a tensor with two of its indices equal to i and j yields

$$\delta_{ij} A_{ij} = A_{jj} = A_{11} + A_{22} + A_{33}$$
$$\delta_{ij} A_{ik} B_j = A_{jk} B_j = A_{1k} B_1 + A_{2k} B_2 + A_{3k} B_3$$

which reduces or "contracts" the order of the original tensor by two. Referred to as "contracting," this process is usually carried out by simply substituting one of the contracted indices in the tensor quantity with the other (i with j in these examples) without actually multiplying by the Kronecker delta.

A particular contraction to note is that of the product A_iB_j. Recalling vector algebra, it is equivalent to taking the *dot* product of the two vectors $\boldsymbol{A}$ and $\boldsymbol{B}$ as carried out nonchalantly in Eq. (5). That is,

$$\delta_{ij}A_iB_j = A_iB_i = A_1B_1 + A_2B_2 + A_3B_3 = \boldsymbol{A}\cdot\boldsymbol{B}$$

Another tensor that transforms into itself under any coordinate transformation is the *anti-symmetric* or *alternating* third-order tensor ε_{ijk}. Also know as the "Levi-Civita" tensor of third order, it equals zero when any *two* of the indices are the same, and either +1 or –1 depending on whether the indices (i, j, k) are respectively either an even or odd permutation of (1, 2, 3). As it turns out, the alternating tensor and numerical multiples of it are the only isotropic tensors of third order. Mathematically, ε_{ijk} is treated as the 3×3×3 matrix

$$\varepsilon_{ijk} = \begin{bmatrix} \varepsilon_{1jk} \\ \varepsilon_{2jk} \\ \varepsilon_{3jk} \end{bmatrix} = \begin{bmatrix} \varepsilon_{11k} & \varepsilon_{12k} & \varepsilon_{13k} \\ \varepsilon_{21k} & \varepsilon_{22k} & \varepsilon_{23k} \\ \varepsilon_{31k} & \varepsilon_{32k} & \varepsilon_{33k} \end{bmatrix} = \begin{bmatrix} \begin{bmatrix} 0 \\ 0 \\ 0 \end{bmatrix} & \begin{bmatrix} 0 \\ 0 \\ 1 \end{bmatrix} & \begin{bmatrix} 0 \\ -1 \\ 0 \end{bmatrix} \\ \begin{bmatrix} 0 \\ 0 \\ -1 \end{bmatrix} & \begin{bmatrix} 0 \\ 0 \\ 0 \end{bmatrix} & \begin{bmatrix} 1 \\ 0 \\ 0 \end{bmatrix} \\ \begin{bmatrix} 0 \\ 1 \\ 0 \end{bmatrix} & \begin{bmatrix} -1 \\ 0 \\ 0 \end{bmatrix} & \begin{bmatrix} 0 \\ 0 \\ 0 \end{bmatrix} \end{bmatrix} \tag{16}$$

It is much easier to remember the elements of ε_{ijk} as described in words than in matrix form, at least for the author. From the above definitions, it can be shown that

$$\delta_{ij}\varepsilon_{ijk} = 0$$
$$\varepsilon_{ijk}\varepsilon_{ijk} = 6$$
$$\varepsilon_{ijm}\varepsilon_{klm} = \delta_{ik}\delta_{jl} - \delta_{il}\delta_{jk}$$

The product of ε_{ijk} and a tensor of first order is a tensor of fourth order. For example,

$$\varepsilon_{ijk}A_l = B_{ijkl}$$

If this is contracted with respect to the k and l indices, we obtain a tensor of second order, namely,

$$\delta_{kl}\varepsilon_{ijk}A_l = \varepsilon_{ijk}A_k = \varepsilon_{ij1}A_1 + \varepsilon_{ij2}A_2 + \varepsilon_{ij3}A_3 = B_{ijkk} = C_{ij}$$

In matrix form, C_{ij} is given by

$$C_{ij} = \begin{bmatrix} 0 & A_3 & -A_2 \\ -A_3 & 0 & A_1 \\ A_2 & -A_1 & 0 \end{bmatrix}$$

which is an anti-symmetric second-order tensor. That is $C_{ij} = -C_{ji}$.

In similar fashion, the product of ε_{ijk} and two first-order tensors A_p and B_q is a tensor of fifth order, which when contracted with respect to the i and p indices and the j and q indices, yields a first-order tensor with the index k. In matrix form, it is given by

$$C_k = \varepsilon_{ijk}A_iB_j = \begin{bmatrix} A_2B_3 - A_3B_2 \\ A_3B_1 - A_1B_3 \\ A_1B_2 - A_2B_1 \end{bmatrix}$$

This is equivalent in vector notation to taking the *cross product* of the two vectors $\boldsymbol{A}$ and $\boldsymbol{B}$, which yields another vector $\boldsymbol{C}$ perpendicular to $\boldsymbol{A}$ and $\boldsymbol{B}$ having a direction following the right-hand rule.

That is, the product $\varepsilon_{ijk} A_i B_j$ is equivalent to $\boldsymbol{C} = \boldsymbol{A} \times \boldsymbol{B}$ with $\boldsymbol{C}$ pointing in the direction indicated by the right-hand rule as $\boldsymbol{A}$ turns toward $\boldsymbol{B}$.

Tensor calculus. Since translating or rotating a coordinate system doesn't affect the distance between two position vectors, the differential change of a position vector, that is, dx_i, remains invariant regarding coordinate transformation and obeys the rules of a first-order tensor. This follows directly by writing Eq. (10) for x_i and taking its derivative which yields

$$dx'_j = (dx_i) m_{ij}$$

In general, differentiation with respect to a coordinate leads to a tensor of order one higher. For example, the derivative of a scalar A with respect to x_i is given by

$$\frac{\partial A}{\partial x_i} = \frac{\partial A}{\partial x'_j}\frac{dx'_j}{dx_i} = \frac{\partial A}{\partial x'_j} m_{ij} \tag{17}$$

which follows the transformation rule for a first-order tensor. Known as the *gradient* of A, it is expressed in vector notation as ∇A where $\nabla = \boldsymbol{i}\partial/\partial x_1 + \boldsymbol{j}\partial/\partial x_2 + \boldsymbol{k}\partial/\partial x_3$ and $\boldsymbol{i}, \boldsymbol{j}$ and $\boldsymbol{k}$ are the unit vectors in the x_1, x_2, and x_3, directions, respectively.

Likewise, the second derivative of a scalar with respect to the coordinates x_i and x_j is a tensor of second order. That is,

$$\frac{\partial^2 A}{\partial x_i \partial x_j} = \frac{\partial}{\partial x'_k}\left(\frac{\partial A}{\partial x'_l}\frac{dx'_l}{dx_i}\right)\frac{dx'_k}{dx_j} = \frac{\partial^2 A}{\partial x'_k \partial x'_l} m_{il} m_{kj}$$

Contracting this yields a scalar, namely,

$$\delta_{ij}\frac{\partial^2 A}{\partial x_i \partial x_j} = \frac{\partial^2 A}{\partial x_i \partial x_i} = \frac{\partial^2 A}{\partial x_1 \partial x_1} + \frac{\partial^2 A}{\partial x_2 \partial x_2} + \frac{\partial^2 A}{\partial x_3 \partial x_3} \tag{18}$$

which is known as the *Laplacian* of A, and written in vector notation as $\nabla^2 A$.

The derivative of a first-order tensor A_i with respect to x_j, that is, the gradient of a vector, is a second-order tensor. For example,

$$\frac{\partial A_i}{\partial x_j} = \frac{\partial}{\partial x'_l}\left(A'_k m_{ik}\right) m_{jl} = \frac{\partial A'_k}{\partial x'_l} m_{ik} m_{jl}$$

which follows the transformation rule for a second-order tensor. Contracting this expression provides

$$\delta_{ij}\frac{\partial A_i}{\partial x_j} = \frac{\partial A_i}{\partial x_i} = \frac{\partial A_1}{\partial x_1} + \frac{\partial A_2}{\partial x_2} + \frac{\partial A_3}{\partial x_3} \tag{19}$$

which is known as the *divergence* of A_i, and expressed in vector notation as $\nabla \cdot \boldsymbol{A}$.

Finally, the product of ε_{ijk} and the derivative of a first-order tensor is a tensor of fifth order. For example,

$$\varepsilon_{ijk}\frac{\partial A_l}{\partial x_m} = B_{ijklm}$$

When contracted with respect to (j, m) and (k, l) this yields the first-order tensor

$$C_i = \delta_{jm}\delta_{kl} B_{ijklm} = \delta_{jm}\delta_{kl}\varepsilon_{ijk}\frac{\partial A_l}{\partial x_m} = \varepsilon_{ijk}\frac{\partial A_k}{\partial x_j} \tag{20}$$

Known as the *curl* or *rot* of A_k, it is mathematically treated as the column matrix

$$C_i = \begin{bmatrix} \dfrac{\partial A_3}{\partial x_2} - \dfrac{\partial A_2}{\partial x_3} \\ \dfrac{\partial A_1}{\partial x_3} - \dfrac{\partial A_3}{\partial x_1} \\ \dfrac{\partial A_2}{\partial x_1} - \dfrac{\partial A_1}{\partial x_2} \end{bmatrix}$$

In vector notation, it is written as $\boldsymbol{C} = \text{curl}\,\boldsymbol{A} = \nabla \times \boldsymbol{A}$.

Stokes' and Gauss' theorems. Let A be a surface bounded by a closed curve C within a vector field $\boldsymbol{f}$ that is defined everywhere as a function of the position vector $\boldsymbol{x}$. Furthermore, let $\boldsymbol{n}$ denote the *unit normal* vector at a point $P(\boldsymbol{x})$ on the surface pointing in the direction defined by the right-hand rule as C is navigated counterclockwise around the surface. That is, the components of both $\boldsymbol{f}(\boldsymbol{x})$ and $\boldsymbol{n}(\boldsymbol{x})$ transform according to Eq. (10), and the magnitude of $\boldsymbol{n}$ equals one.

If the *magnitude* of the elemental area of the surface at the point P is given by dA, the elemental area vector at P is given by $\boldsymbol{dA} = \boldsymbol{n}dA$. Then according to *Stokes' theorem*, the projected component of $\boldsymbol{f}$ onto the curve, that is $\boldsymbol{f}\cdot\boldsymbol{dx}$, integrated around the curve C equals the normal component of the curl of f, which is $\boldsymbol{n}\cdot(\nabla \times f)$, integrated over the bounded surface A. In tensor notation, Stokes' theorem is given by

$$\int_C f_i\,dx_i = \int_A n_i \varepsilon_{ijk} \frac{\partial f_k}{\partial x_j} dA \tag{22}$$

Finally, let V be any volume within the vector field $\boldsymbol{f}$. Then according to *Gauss' theorem*, the normal component of $\boldsymbol{f}$ at any point P on the surface of the volume, that is $\boldsymbol{f}\cdot\boldsymbol{n}$, integrated over the the enclosing surface A is equal to the divergence of $\boldsymbol{f}$, which is $\nabla\cdot\boldsymbol{f}$, integrated over the volume. In tensor notation, Gauss' theorem is given by

$$\int_A f_i n_i\,dA = \int_V \frac{\partial f_i}{\partial x_i} dV \tag{23}$$

As it turns out, Gauss' theorem can be applied to any tensor field by substituting it for the tensor f_i.

References

Hinze, J.O., 1975, Turbulence, McGraw-Hill, New York.

Jeffreys, H., 1952, Cartesian Tensors, Cambridge University Press, Cambridge.

Temple, G., 1960, Cartesian Tensors: An Introduction, John Wiley & Sons, New York; also, 2004, Dover Publications, New York.

Bibliography

Chapter 1

Batchelor, G.K., 1967, An Introduction to Fluid Mechanics, Cambridge University Press, Cambridge.

Bradshaw, P., 1971, An Introduction to Turbulence and its Measurement, Pergamon Press, Braunschweig.

Goldstein, S., 1938, Modern Developments in Fluid Dynamics, Oxford University Press, Oxford.

Prandtl, L., 1914, "Der Luftwiderstand von Kugeln," Nachr. Ges. Wiss. Göttingen, Math. Phys. Klasse, pp. 177–190 (also Collected Works II, pp. 597–608).

Reynolds, O., 1883, "An Experimental Investigation of the Circumstances which Determine whether the Motion of Water Shall Be Direct or Sinuous, and of the Law of Resistance in Parallel Channels," Phil. Trans. Roy. Soc., 174, pp. 935–982.

Chapter 2

Moore, F.K., 1957, "On the Separation of the Unsteady Laminar Boundary Layer,"IUTAM Symposium on Boundary Layers (ed., H. Görtler), Freiburg.

Nikuradse, J., 1942, "Laminare Reibungsschichten an der Längsangeströmten Platte," Monograph. Zentrale f. wiss. Berichtwesen, Berlin.

Chapter 3

Baines,W.D., and Peterson, E.G., 1951, "An Investigation of Flow Through Screens," ASME Trans., 73, pp. 467-480.

Batchelor, G.K., 1946, "The Theory of Axisymmetric Turbulence," Proc. Roy. Soc. London, A 186, pp. 480–502.

Borello, D., Rispoli, F., and Hanjalić, K., 2004, "Prediction of Turbulence and Transition in Turbomachinery Flows Using an Innovative Second Moment Closure Modeling," ASME Paper GT2004-53706.

Boussinesq, J., 1896, "Théorie de l'écoulement tourbillonnant et tumultueux des liquides dans les lits rectilignes á grande section (tuyaux de conduite et canaux découverts), quand cet écoulement s'est régularisé en un régime uniforme, c'est-á-dire, moyennement pareil a travers toutes les sections normales du lit," Comptes Rendus de l'Académie des Sciences CXXII, pp. 1290–1295.

Bradshaw, P., Ferriss, D.H., and Atwell, N.P., 1967. "Calculation of Boundary-Layer Development Using the Turbulent Energy Equation," J. Fluid Mech., 28, pp. 593–616.

Deissler, R.G., 1954, NACA Aero. Tech. Notes No. 3154.

Driest, E.R. van, 1956, "On Turbulent Flow Near a Wall," J. Aerosp. Sci., 23, pp. 1007–1012.

Durbin, P.A., 1996, "On the k-ε Stagnation Point Anomaly," Int. J. Heat Fluid Flow, 17, pp. 89–90.

Glushko, G.S., 1965, "Turbulent Boundary Layer on a Flat Plate in an Incompressible Fluid," Izv. Ak. Nauk. SSSR, Ser. Mekh. No. 4, pp. 13–23 (in Russian); also NASA TTF 10, 080 {in English).

Jones, W.P., and Launder, B.E., 1973, "The Calculation of Low-Reynolds-Number Phenomena With a Two-Equation Model of Turbulence," Int. J. Heat and Mass Transfer, 16, pp. 1119–1130.

Kármán, Th. von, and Howarth, L., 1938, "On the Statistical Theory of Isotropic Turbulence," Proc. Roy. Soc. London A, 164, pp. 192–215.

Laufer, J., 1951, "Investigation of Turbulent Flow in a Two-Dimensional Channel," NACA Rep.

1053.

Laufer, J., 1954, "The Structure of Turbulence in Fully Developed Pipe Flow," NACA Rep. 1174.

Naot, D., Shavit, A., and Wolfshtein, M., 1970, "Interaction between Components of the Turbulent Velocity Correlation Tensor," Israel J. Tech., 8, pp. 259–269.

Prandtl, L., 1942, "Bemerkungen zur Theorie der freien Turbulenz," ZAMM 22, pp. 241–243.

Prandtl, L., 1945, "Über ein neues Formelsystem für die ausgebildete Turbulenz," Nachr. Akad. Wiss. Göttingen, Math.-Phys. Klasse, pp. 6–19.

Reichardt, H., 1938, "Messungen turbulenter Schwankungen," Naturwissenschaften, 404; also 1933, ZAMM, 13, pp. 177–180, and 1939, ZAMM, 18, pp. 358–361.

Roach, P.E., 1987, "The Generation of Nearly Isotropic Turbulence by Means of Grids," Int. J. Heat Fluid Flow, 8, pp. 82–92.

Rodi, W., and Spalding, D.B., 1970, "A Two-Parameter Model of Turbulence, and Its Application to Free Jets," Wärme- und Stoffübertragung, 3, pp. 85–95.

Sjögren, T., and Johansson, A.V., 2000, "Development and Calibration of Algebraic Nonlinear Models for Terms in the Reynolds Stress Transport Equations," Phys. Fluids, 12, pp. 1554–1572.

Taylor, G.I., 1936, "Correlation Measurements in Turbulent Flow Through a Pipe," Proc. Roy. Soc. London A, 157, pp. 537–546.

Chapter 4

Ackerberg, R.C., and Phillips, J.H., 1972, "The Unsteady Laminar Boundary Layer on a Semi-Infinite Flat Plate Due to Small Fluctuations in The Magnitude of the Free Stream," J. Fluid Mech., 51, pp. 137–157.

Betchov, R., 1960, "On the Mechanism of Turbulent Transition," Phys. Fluids, 3, pp. 1026–1027.

Betchov, R., and Criminale, W.O., 1967, Stability of Parallel Flow, Academic Press, New York.

Boiko, A.V., Westin, K.J.A., Klingmann, B.G.B., Kozlov, V.V., and Alfredsson, P.H., 1994, "Experiments in a boundary layer subjected to free stream turbulence; Part 2. The role of TS-waves in the transition process," J. Fluid Mech., 281, pp. 219–245.

Davis, S., 1976, "The Stability of Periodic Flows," Annual Review Fluid Mech., 8, pp. 57–74.

Gaster, M., 1962, "A Note on the Relation between Temporally-Increasing and Spatially-Increasing Disturbances in Hydrodynamic Stability," J. Fluid Mech., 7, pp. 222–224.

Gaster, M., 1965, "The Role of Spatially Growing Waves in the Theorv of Hydrodynamic Stability," Progress in Aero. Sciences (ed., D. Kiichemann), 6, pp. 251–270.

Gaster, M., 1974,"On the Effect of Boundary Layer Growth on Flow Stability," J. Fluid Mech., 66, pp. 465–480.

Gaster, H., and Jordinson, R., 1975, "On the Eigen-Values of the Orr-Sommerfeld Equation," J. Fluid Mech., 72, pp. 121–133.

Habermann, R., 1976. "Nonlinear Perturbations of the Orr-Sommerfeld Equation Asymptotic Expansion of the Logarithmic Phase Shift across the Critical Layer," SIAM J. Math. Analysis, 7, pp. 70–81.

Hall, D.J., and Gibbings, J.G., 1972, "Influence of Stream Turbulence and Pressure Gradient upon Boundary Layer Transition," J. Mech. Engng. Sci., 14, pp. 134–146.

Heisenberg, W., 1924. "Über Stabilitiit und Turbulenz von Flüssigkeitsströmen," Ann. d. Phys., 71, pp. 577–627.

Kármán, Th. von, 1924, "Über die Stabilität der Laminarströmung und die Theorie der Turbulenz,"

Proc. 1st Int. Congr. Appl. Mech., Delft, pp. 97–112.

Landahl, M.T., 1980, "A Note on an Algebraic Instability on Inviscid Parallel Shear Flows," J. Fluid Mech., 98, pp. 243–251.

Mack, L.M., 1976, "A Numerical Study of the Temporal Eigenvalue Spectrum of the Blasius Boundary Layer," J. Fluid Mech., 73, pp. 497–520.

Mack, L.M., 1977, "Transition Prediction and Linear Stability Theory," AGARD CP 224, pp. 1–1 to 1–22.

Miller, J.A., and Fejer, A.A., 1964, "Transition Phenomena in Oscillating Boundary Layer Flows," J. Fluid Mech.,18, pp. 438–449.

Orszag, S.A., 1983, "Secondary Instability of Wall-Bounded Shear Flows," J. Fluid Mech., 128, pp. 347–385.

Prandtl, L., 1935. Aerodynamic Theory (ed., W.F. Durand), Springer Verlag, Berlin, 3, pp. 178–90.

Savill, A.M., 1991, "Synthesis of T3 Test Case Computations," Proc. 1st ERCOFTAC Workshop on Numerical Simulation of Unsteady Flows, Transition to Turbulence and Combustion, Lausanne.

Sieger, K., Schulz, A., Wittig, S., and Crawford, M.E., 1993, "An Evaluation of Low-Reynolds Number k-ε Turbulence Models for Predicting Transition Under the Influence of Free-Stream Turbulence and Pressure Gradient," Proc. 2nd Int. Sym. Engng. Turb. Modeling & Measurements, Florence, Italy, pp. 593–602.

Smith, A.M.O., 1956, "Transition, Pressure Gradient and Stability Theory," IX Int. Congr. Appl. Mech., Brussels.

Sohn, K.H., and Reshotko, E., 1991, "Experimental Study of Boundary Layer Transition with Elevated Freestream Turbulence on a Heated Plate," NASA CR 187068.

Stuart, J.T., 1956, "On the Effects of the Reynolds Stress on Hydrodynamic Stability," ZAMM-Sonderheft, pp. 32–38.

Stuart, J.T., 1960, "Non-Linear Effects in Hydrodynamic Stability," Proc. Xth Intern. Congress of Appl. Mech., Stresa, pp. 63–97.

Stuart, J.T., 1963, "Hydrodynamic Stability," Laminar Boundary Layers (ed., L. Rosenhead), Clarendon Press., Oxford, pp. 482–579.

Volino, R.J., 2002, "An Investigation of the Scales in Transitional Boundary Layers under High Free-Stream Turbulence Conditions," ASME Paper GT-2002-30233.

Chapter 5

Benjamin, T.B., 1961,"The Development of Three-Dimensional Disturbances in an Unstable Liquid Flowing Down an Inclined Plane," J. Fluid Mech., 10, 401–419.

Breuer, K.S., and Landahl, M.T., 1990, "The Evolution of a Localized Disturbance in a Laminar Boundary Layer; Part 2. Strong disturbances," J. Fluid Mech., 220, pp. 595–621.

Criminale, W.O., and Kovasznay, L.S.G., 1962, "The Growth of Localized Disturbances in a Laminar Boundary Layer," J. Fluid Mech., 14, pp. 59–80.

Dryden, H.L., 1946, "Some Recent Contributions to the Study of Transition and Turbulent Boundary Layers," Sixth Intern. Congress Appl. Mech., Paris; also 1947, NACA TN 1168.

Dryden, H.L., 1948, "Recent Advances in the Mechanics of Boundary Layer Flow," Adv. Appl. Mech., 1, pp. 2–40.

Dryden, H.L., 1956, "Recent Investigation of the Problem of Transition," ZFW, 4, pp. 89–95.

Dryden, H.L., 1959, "Transition from Laminar to Turbulent Flow," High Speed Aerodynamics and Jet Propulsion, 5, Princeton and Oxford Press, pp. 3–74.

Falco, R.E., and Gendrich, C.P., 1990, "The Turbulence Burst Detection Algorithm of Z. Zarić," Nearwall Turbulence 1988 Z. Zarić Memorial Conference (eds., S.J. Kline and N.H. Afgan), Hemisphere, pp. 911–931.

Gaster, M., 1982, "The Development of a Two-Dimensional Wavepacket in a Growing Boundary Layer," Proc. Roy. Soc. London A, 384, pp. 317–332.

Hughes, J.D., and Walker, G.J., 2000, "Natural Transition Phenomenon on an Axial Compressor Blade," ASME Paper 2000-GT-0264.

Kachanov, Y.S., and Levchenko, V.Y., 1984, "The Resonant Interaction of Disturbances at Laminar Turbulent Transition in a Boundary Layer," J. Fluid Mech., 138, pp. 209–247.

Keller, F.J., and Wang, T., 1995, "Effects of Criterion Functions on Intermittency in Heated Transitional Boundary Layers With and Without Streamwise Acceleration," J. Turbomachinery, 117, pp. 154–165.

Landahl, M.T., 1975, "Wave Breakdown and Turbulence," SIAM J. Appl. Math., 28, pp. 735–756.

Landahl, M.T., 1980, "A Note on an Algebraic Instability of Inviscid Parallel Shear Flows," J. Fluid Mech., 98, pp. 243–251.

Rotta, J.C., 1956, "Experimenteller Beitrag zur Entstehung turbulenter Strömung im Rohr," Ing.-Arch., 24, pp. 258–281.

Sreenivasan, K.R., Prahbu, A., and Narasimha, R., 1983, "Zero-Crossings in Turbulent Signals," J. Fluid Mech., 137, pp. 251–272.

Tani, I., 1962, "Some Aspects of Boundary Layer Transition at Subsonic Speeds," Advances in Aeronautical Sciences (ed., T. von Kármán), 3, Pergamon Press, New York and London, pp. 143–160.

Tani, I., 1969, "Boundary Layer Transition," Annual Rev. Fluid Mech., 7, pp. 169–196.

Tso, J., Chang, S.I., and Blackwelder, R.F., 1989, "On the Breakdown of a Localized Disturbance in a Laminar Boundary Layer," IUTAM Symposium on Laminar-Turbulent Transition, (eds: D. Arnal and R. Michel), Springer-Verlag, Berlin, pp. 199–214.

Watson, J., 1960, "Three-Dimensional Disturbances in Flow Between Parallel Planes," Proc. Roy. Soc. London A, 254, pp. 562–569.

Zhong, S., Chong, T.P., and Hodson, H.P., 2002, "On the Spreading Angle of Turbulent Spots in Non-Isothermal Boundary Layers with Favorable Pressure Gradients," ASME Paper GT-2002-30222.

Chapter 6

Bayley, F.J., and Priddy, W.J., 1981, "Effects of Free-Stream Turbulence Intensity and Frequency on Heat Transfer to Turbine Blading," J. Engng. for Power, 103, pp. 60–64.

Fashifar, A., and Johnson, M.W., 1992, "An Improved Boundary Layer Transition Correlation," ASME Paper ASME-92-GT -245.

Prandtl, L., 1921, "Bemerkungen über die Entstehung der Turbulenz, ZAMM, 1, pp. 431–436; also 1922, Phys. Z., 23, pp. 19–25.

Turbulence and Pressure Gradients.

Kim, J., 1990, "Freestream Turbulence and Concave Curvature Effects on Heated, Transitional Boundary Layers," Ph. D. Thesis, University of Minnesota, St Paul.

Martin, B.W., Brown, A., and Garrett, S.E., 1978, "Heat Transfer to a PVD Rotor Blade at High-Subsonic Passage Throat Mach Number," Proc. Inst. Mech Engrs., 192, pp. 225–235.

Priddy, W.J., and Bayley, F.J., 1987, "Turbulence Measurements in Turbine Blade Passages and Implications for Heat Transfer," ASME Paper 87-GT-195.

Taulbee, D.B., Tran, L., and Dunn, M.G., 1988, "Stagnation Point and Surface Heat Transfer for a Turbine Stage: Prediction and Comparison with Data," ASME Paper 88-GT-30.

Walker, G.J., 1993, "The Role of Laminar-Turbulent Transition in Gas Turbine Engines: A Discussion," J. Turbomachinery, 115, pp. 207–217.

Curvature.

Schobeiri, M.T., John, J., and Pappu, K., 1996, "Development ofTwo-Dimensional Curved Channels: Theoretical Framework and Experimental Investigation," J. Turbomachinery, 118, pp. 506–518.

Schultz, M.P., and Volino, R.J., 2001, "Effects of Concave Curvature on Boundary Layer Transition under High Free-Stream Turbulence Conditions," ASME Paper 2001-GT-0191.

Volino, R.J., and Simon, T.W., 1995, "Bypass Transition in Boundary Layers Including Curvature and Favorable Pressure Gradient Effects," J. Turbomachinery, 117, pp. 166–174.

Wang, T., 1984, "An Experimental Investigation of Curvature and Freestream Turbulence Effects on Heat Transfer and Fluid Mechanics in Transitional Boundary Layer Flows," Ph. D. Thesis, University of Minnesota, St Paul.

Roughness.

Abuaf, N., Bunker, R.S., and Lee, C.P., 1997, "Effects of Surface Roughness on Heat Transfer and Aerodynamic Performance of Turbine Airfoils," ASME Paper 97-GT-10.

Bammert, K., and Sandstede, H., 1980, "Measurements of the Boundary Layer Development along a Turbine Blade with Rough Surfaces," J. Engng. for Power, 102, pp. 978–983.

Bogard, D.G., Schmidt, D.L., and Tabbita, M., 1996, "Characterization and Laboratory Simulation of Turbine Airfoil Surface Roughness and Associated Heat Transfer," ASME Paper 96-GT-386.

Bons, J.P., McICain, S.T., Taylor, R.P., and Rivir, R.B., 2001, "The Many Faces of Turbine Surface Roughness," ASME Paper 2001-GT-0163.

Braslow, A.L., 1960, "Review of the Effect of Distributed Surface Roughness on Boundary-Layer Transition," AGARD Report 254.

Bunker, R.S., 1997, "Separate and Combined Effects of Surface Roughness and Turbulence Intensity on Vane Heat Transfer," ASME Paper 97-GT-135.

Guo, S.M., Jones, T.V., Lock, G.D., and Dancer, S.N., 1998, "Computational Prediction of Heat Transfer to Gas Turbine Nozzle Guide Vanes with Roughened Surfaces," J. Turbomachinery, 120, pp. 343–350.

Hoffs, A., Drost, U., and Bolcs, A., 1996, "Heat Transfer Measurements on a Turbine Airfoil at Various Reynolds Numbers and Turbulence Intensities Including Effects of Surface Roughness," ASME Paper 96-GT-169.

Pinson, M., 1992, "The Effects of Initial Conditions on Heat Transfer in Transitional Boundary Layers," M.S. Thesis, Clemson University, Clemson.

Rice, M.C, 1999, "Effect of Elevated Free-Stream Turbulence on Heat Transfer over Two-Scaled Rough Surfaces," Master Thesis, Clemson University, Clemson.

Schlichting, H., 1936, "Experimental Untersuchungen zum Rauhigkeitsproblem," Ingenieu Archiv, Archive of Applied Mechanics, 7, pp. 1–34; also "Experimental Investigation of the Problem of Surface Roughness," NACA TM 823.

Stripf, M., Schulz, A., and Wittig, S., 2005, "Surface Roughness Effects on External Heat Transfer of a HP Turbine Blade," J. Turbomachinery, 127, pp. 200–208.

Chapter 7

Dey, J., and Narasimha, R., 1990, "Integral Method for the Calculation of Incompressible Two-Dimensional Transitional Boundary Layers," J. Aircraft, 27, pp. 859–865.

Dopazo C., 1977, "On Conditioned Averages for Intermittent Turbulent Flows," J. Fluid Mech., Vol. 81, pp. 433–438.

Drela, M., 1995, "MISES Implementation of Modified Abu-Ghanam/Shaw Transition Criterion," MIT Aero-Astro.

Hazarika, B.K., and Hirsch, C., 1997, "Transition over C4 Leading Edge and Measurement of Intermittency Factor using PDF of Hot-Wire Signal," J. Turbomachinery, 119, pp. 412–425.

Johnson, M.W. and Ercan, A.H., 1999, "A Physical Model of Bypass Transition," Intl. J. of Heat and Fluid Flow, 20, pp. 95–104.

Johnson, M.W., 2002, "Prediction Transition Without Empiricism or DNS," ASME Paper GT-2002-30238.

Keller, F.J., and Wang, T., 1995, "Effects of Criterion Functions on Intermittency in Heated Transitional Boundary Layers With and Without Streamwise Acceleration," J. Turbomachinery, 117, pp. 154–165.

Lodefier, K., and Dick, E., 2005, "Modelling of Unsteady Transition in Low-Pressure Turbine Blade Flows with Two Dynamic Intermittency Equations," Flow, Turbulence and Combustion, 76, pp. 103–132.

Medic, G., and Durban, P.A., 2002, "Toward Improved Prediction of Heat Transfer on Turbine Blades," J. Turbomachinery, 124, pp. 187–192.

Palma, P.D., 2002, "Numerical Analysis of Turbomachinery Flows with Transitional Boundary Layers," ASME Paper GT-2002-30223.

Papanicolaou E.L. and Rodi W., 1999, "Computation of Separated-Flow Transition Using a Two-Layer Model of Turbulence," J. Turbomachinery, 121, pp. 78–87.

Pecnik, R., Sanz, W., and Pieringer, P., 2004, "Numerical Investigation of Unsteady Boundary Layer Transition Induced by Periodically Passing Wakes with an Intermittency Transport Equation," ASME Paper GT–2004–53204.

Roux, J.M., Mahe, P., Sauthier, B., and Duboue, J. M., 2001, "Aerothermal Predictions with Transition Models for High-Pressure Turbine Blades," Proc. Inst. Mech. Engrs., 215, pp. 735–742.

Roux, J.M., Lefebvre, M., and Liamis, N., 2002, "Unsteady and Calming Effects Investigation on a Very High Lift LP Turbine Blade; Part 2–Numerical Analysis," ASME Paper GT-2002-30228.

Savill, A.M., 1993, "Some Recent Progress in the Turbulence Modeling of By-pass Transition," Near-Wall Turbulent Flows (eds., R.M.C. So, C.G. Speziale and B.E. Launder), Elsevier, pp. 829–848.

Savill, A.M., 2001, "By-Pass Transition Using Conventional Closures and New Strategies in Modelling By-Pass Transition," Closure Strategies for Turbulent and Transitional Flows (eds., B.E. Launder and O. Sharma), Cambridge University Pres, pp. 464–519.

Savill, A.M., 2002, "New Strategies in Modelling By-Pass Transition," Closure Strategies for Turbulent Transitional Flows (eds., B.E. Launder and N. Sandham), Cambridge University Press, pp. 493–

521.

Sharma, O., 1987, "Momentum and Thermal Boundary Layer Development on Turbine Airfoil Suction Surfaces," AIAA Paper No. 87-1918.

Volino, R.J., Schultz, M.P. and Pratt, C.M., 2001, "Conditional Sampling in a Transitional Boundary Layer under High Free-Stream Turbulence Conditions," ASME Paper 2001- GT-0192.

Zhou, D., and Wang, T., 1995, "Effects of Elevated Free-Stream Turbulence on Flow and Thermal Structures in Transitional Boundary Layers," J. Turbomachinery, 117, pp. 407–417.

Chapter 8

Hourmouziadis, J., Buckl, F., Bergmann, P., 1986, "The Development of the Profile Boundary Layer in a Turbine Environment," ASME Paper 86-GT-244.

Van Dresar, N., and Mayle, R.E., 1989, "A Quasi-Steady Approach to Leading Edge Transfer Rates," J. Turbomachinery, 111, pp. 483-490.

Separated-flow transition.

Castro, I.P., and Epik, E., 1998, "Boundary Layer Development after a Separated Region," J. Fluid Mech., 374, pp. 91–116.

Choi, D.H., and Kang, D.J., 1991, "Calculation of Separation Bubbles using a Partially Parabolized Navier-Stokes Procedure," AIAA Journal, 29, pp. 1266–1272.

Choi, D.H., and Lee, E.H., 1995, "Prediction of Separation Bubbles Using Improved Transition Criterion with Two-Equation Turbulence Model," AIAA Journal, 33, pp. 1512–1514.

Crimi, P., and Reeves, B.L., 1976, "Analysis of Leading-Edge Separation Bubbles on Airfoils," AIAA Journal, 14, pp. 1548–1555.

Davis, R.L., Carter, J.E., and Reshotko, E., 1985, "Analysis of Transitional Separation Bubbles on Infinite Swept Wings," AIAA Paper AIAA-85-1685.

Dovgal, A.V., Kozlov, V.V., and Michalke, A., 1996, "Laminar Boundary Layer Separation: Instability and Associated Phenomena," Progress Aerospace Sci., 30, pp. 61–94.

Driver, D.M., 1991, "Reynolds Stress Measurements in a Separated Boundary Layer Flow," AIAA Paper AIAA-91-787.

Papanicolaou, E.L., and Rodi, W., 1999, "Computation of Separated-Flow Transition Using a Two-Layer Model of Turbulence," J. Turbomachinery, 121, pp. 78–87.

Sanz, W. and Platzer, M.F., 1997, "On the Calculation of Laminar Separation Bubbles Using Different Transition Models," ASME Paper GT-97-GT-453.

Sanz, W. and Platzer, M.F., 1998, "On the Navier-Stokes Calculation of Separation Bubbles with a New Model," J. Turbomachinery, 120, pp. 36–42.

Sanz, W. and Platzer, M.F., 2002, "On the Numerical Difficulties in Calculating Laminar Separation Bubbles," ASME Paper GT-2002-30235.

Simon, T.W., Qiu, S., and Yuan, K., 2000, "Measurements in a Transitional Boundary Layer Under Low-Pressure Turbine Airfoil Conditions," NASA CR 2000-209957.

Walraevens, R.E., and Cumpsty, N.A., 1995, "Leading Edge Separation Bubbles on Turbomachine Blades," J. Turbomachinery, 117, pp. 115–125.

Wake-induced transition.

Ashworth, D.A., LaGraff, J.E., Schultz, D.L., and Grindrod, K.J., 1985, "Unsteady Aerodynamic and Heat Transfer Processes in a Transonic Turbine Stage," J. Engng. for Gas Turbines and Power, 107, pp. 1022– 1030.

Ashworth, D.A., LaGraff, J.E., and Schultz, D.L., 1987, "Unsteady Interaction Effects on a Transitional Turbine Blade Boundary Layer," Proc. 1987 ASME-JSME Thermal Engng. Conf. (eds., P.J. Marto and I. Tanasawa), 2, pp. 603–610.

Cardamone, P., Stadtmüller, P. and Fottner, L., 2002, "Numerical Investigation of the Wake-Boundary Layer Interaction on a Highly Loaded LP Turbine Cascade Blade," ASME Paper GT-2002-30367.

Chernobrovkin, A., and Lakshminarayana, B., 2000, "Unsteady Viscous Flow Causing Rotor-Stator Interaction in Turbines, Part 2: Simulation, Integrated Flowfield, and Interpretation," J. Propulsion and Power, 16, pp. 751–759.

Coton, T., Arts, T., Lefebvre, M., and Liamis, N., 2002, "Unsteady and Calming Effects Investigation on a Very High Lift LP Turbine Blade; Part 1–Experimental Analysis," ASME Paper GT-2002-30227.

Doorly, D.J., and Oldfield, M.L.G., 1985, "Simulation of Wake Passing in a Stationary Turbine Rotor Cascade," J. Propulsion and Power, 1, pp. 316–318.

Doorly, D.J., Oldfield, M.L.G., and Scrivener, C.T.J., 1985, "Wake-Passing in a Turbine Rotor Cascade," Heat Transfer and Cooling in Gas Turbines, AGARD-CP-390, pp. 7–1 to 7–18.

Dring, R.P., Joslyn, H.D., Hardin, L.W., and Wagner, J.H., 1982, "Turbine Rotor-Stator Interaction," J. Engng. for Power, 104, pp. 729–742.

Dunn, M.G., 1986a, "Heat-Flux Measurements for the Rotor of a Full-Stage Turbine; Part 1–Time-Averaged Results," J. Turbomachinery, 108, pp. 90–97.

Dunn, M.G., George, W.K., Rae, W.J., Woodward, S.H., Moller, J.C., and Seymour, P.J., 1986b, "Heat-Flux Measurements for the Rotor of a Full-Stage Turbine; Part 2–Description of Analysis Technique and Typical Time-Resolved Measurements," J. Turbomachinery, 108, pp. 98–107.

Dunn, M.G., Martin, H.L., and Stanek,M.J., 1986, "Heat Flux and Pressure Measurements and Comparison with Prediction for a Low-Aspect-Ratio Turbine Stage," J. Turbomachinery, 108, pp. 108–115.

Dunn, M.G., Seymour, P.J., Woodward, S.H., George, W.K., and Chupp, R.E., 1988, "Phase-Resolved Heat-Flux Measurements on the Blade of a Full-Scale Rotating Turbine," ASME Paper 88-GT-173.

Elsner W., Vilmin S., Drobniak S., and Piotrowski W., 2004, "Experimental Analysis and Prediction of Wake-Induced Transition in Turbomachinery," ASME Paper GT2004-53757.

Eulitz, F., and Engel, K., 1998, "Numerical Investigation of Wake Interaction in a Low Pressure Turbine," ASME Paper 98-GT-563.

Funazaki, K., Kitazawa, T., Koizumi, K., and Tanuma, T., 1997, "Studies on Wake-Disturbed Boundary Layers under the Influences of Favorable Pressure Gradient and Free-Stream Turbulence; Part I–Experimental Setup and Discussions on Transition Model," ASME Paper 97-GT-452.

Funazaki, K., and Koyabu, E., 1999, "Effects of Periodic Wake Passing Upon Flat-Plate Boundary Layers Experiencing Favorable and Adverse Pressure Gradients," J. Turbomachinery, 121, pp. 333–340.

Gostelow, J.P., and Thomas, R.L., 2006, "Interactions between Propagating Wakes and Flow Instabilities in the Presence of a Laminar Separation Bubble," ASME Paper GT2006-91193.

Guenette, G.R., Epstein, A.H., Norton, R.J.G., and Yuzhang, C., 1985, "Time Resolved Measurements of a Turbine Rotor Stationary Tip Casing Pressure and Heat Transfer Field," AIAA Paper No. 85-1220.

Halstead, D.E., Wisler, D.C., Okiishi, T.H., Walker, G.J., Hodson, H.P., and Shin, H-W, 1997,

"Boundary Layer Development in Axial Compressors and Turbines: Part 2–Compressors, Part 3–LP Turbines, and Part 4–Computations and Analyses," J. Turbomachinery, 119, pp. 426–444, pp. 234–246, pp. 128–139.

Hodson, H.P., 1984, "Boundary Layer and Loss Measurements on the Rotor of an Axial-Flow Turbine," J. Engng. Gas Turbines and Power, 106, pp. 391–399.

Hodson, H.P., 1991, "Aspect of Unsteady Blade-Surface Boundary Layers and Transition in Axial Turbomachines," VKI Lecture Series, von Kármán Institute, Rhode St. Genese, Belgium.

Hodson, H.P., 1998, "Blade Row Interference Effects in Axial Turbomachinery Stages," Von Karman Institute for Fluid Dynamics Lecture Series, Rhode St. Genese, Belgium.

Höhn, W., and Heinig, K., 2001, "Numerical and Experimental Investigation of Unsteady Flow Interaction in Low Pressure Multistage Turbine," ASME Paper 2001-GT-437.

Höhn, W., Gombert, R., and Kraus, A., 2001 "Unsteady Aerodynamical Blade Row Interaction in a New Multistage Research Turbine; Part 2–Numerical Investigation," ASME Paper 2001-GT-0307.

Kaszeta, R.W., Simon, T.W., and Ashpis, D.E., 2001, "Experimental Investigation of Transition to Turbulence as Affected by Passing Wakes," ASME Paper 2001-GT-0195.

Kaszeta, R.W., and Simon, T.W., 2002, "Experimental Investigation of Transition to Turbulence as Affected by Passing Wakes," NASA-CR-2002-212104.

Kim, K., and Crawford, M.E., 1998, "Prediction of Unsteady Wake-passing Effects on Boundary Layer Development," Proc. ASME Heat Trans. Div., ASME Publication HTD-361-3.

Kim, K., and Crawford, M.E., 2000, "Prediction of Transitional Heat Transfer Characteristics of Wake-Affected Boundary Layers," J. Turbomachinery, 122, pp. 78–87.

Kittichaikarn, C., Ireland, P.T., Zhong, S. and Hodson, H.P., 1999, "An Investigation on the Onset of Wake-Induced Transition and Turbulent Spot Production Rate using Thermochromic Liquid Crystals," ASME Paper 99-GT-126.

LaGraff, J.E., Ashworth, D.A., and Schultz, D.L., 1989, "Measurement and Modeling of the Gas Turbine Blade Transition Process as Disturbed by Wakes," J. Turbomachinery, 111, pp. 315–322.

Lakshminarayana, B., Chernobrovkin, A., and Ristic, D., 2000, "Unsteady Viscous Flow Causing Rotor-Stator Interaction in Turbines; Part 1–Data, Code Pressure," J. Propulsion and Power, 16, pp. 744–750.

Lardeau S., and Leschziner M., 2005, "Unsteady RANS Modelling of Wake–Blade Interaction: Computational Requirements and Limitations," Computers and Fluids, 34, pp. 3–21.

Lardeau S., and Leschziner M., 2006, "Modeling of Wake-Induced Transition in Linear Low-Pressure Turbine Cascades," AIAA Journal, 44, pp. 1854–1865

Liu, X., and Rodi, W., 1989, "Measurements of Unsteady Flow Over and Heat Transfer from a Flat Plate," ASME-Paper 89-GT-2.

Lodefier, K., Dick, E., Piotrowski, W., and Elsner, W., 2005, "Modelling of Wake Induced Transition with Dynamic Description of Intermittency," Proc. 6th European Turbomachinery Conference, pp. 730–739.

Lodefier, K., and Dick, E., 2006, "Modelling of Unsteady Transition in Low-Pressure Turbine Blade Flows with Two Dynamic Intermittency Equations," Flow, Turbulence and Combustion, 76, pp. 103–132.

Lodefier, K., Dick, E., Piotrowski, W., and Elsner, W., 2007, "Validation of a Dynamic Intermittency Model for the Prediction of Wake-Induced Transition on Turbine Blades," Proc. of 7th European Turbomachinery Conference, pp. 1379–1388.

Lou, W., and Hourmouziadis, J., 1999, "Experimental Investigation of Periodic-Unsteady Flat Plate Boundary Layers With Pressure Gradients," The 3rd ASME/JSME Joint Fluids Engineering Division Summer Meeting, Paper No. FEDSM99-7190.

Michelassi, V., Wissink, J., and Rodi, W., 2002, "Analysis of DNS and LES of Flow in a Low Pressure Turbine Cascade with Incoming Wakes and Comparison with Experiments," Flow, Turbulence and Combustion, 69, pp. 295–330.

Paxson, D.E., and Mayle, R.E., 1990, "Laminar Boundary Layer Interaction with an Unsteady Passing Wake," ASME Paper 90-GT-120.

Roberts, S.K., and Yaras, M.I., 2003, "Effects of Periodic-Unsteadiness, Free-Stream Turbulence, and Flow Reynolds Number on Separation-Bubble Transition," ASME Paper GT2003–38262.

Schobeiri, M.T., Read, K., and Lewalle, J., 1995, "Effect of Unsteady Wake Passing Frequency on Boundary Layer Transition," ASME Paper 95-GT-437.

Schobeiri, M.T., Chakka, P., and Pappu, K., 1998, "Unsteady Wake Effects on Boundary Layer Transition and Heat Transfer Characteristics of a Turbine Blade," ASME Paper 98-GT-291.

Schobeiri, M.T., and Chakka P., 2002, "Prediction of Turbine Blade Heat Transfer and Aerodynamics Using a New Unsteady Boundary Layer Transition Model," Int. J. Heat and Mass Transfer, 45, pp. 815–829.

Stadtmuller, P., Fottner, L., and Fiala, A., 2000, "Experimental and Numerical Investigation of Wake-induced Transition on a Highly Loaded LP Turbine at Low Reynolds Numbers," ASME Paper 2000-GT-0269.

Thurso, J., and Stoffel, B., 2001, "Numerical Simulation of Wake Velocity and Wake Turbulence Effects on Unsteady Boundary Layer Transition," J. Power and Energy, 215, pp. 753–762.

Tiedemann, M., and Kost, F., 1999, "Unsteady Boundary Layer Transition on a High Pressure Turbine Rotor Blade," ASME Paper 99-GT-194.

Victor, X.S., and Houdeville, R., 2000, "Influence of Periodic Wakes on the Development of a Boundary Layer," Aerospace Sci. Technology, 4, pp. 371–381.

Vilmin, S., Savill, M.A., Hodson, H.P., and Dawes, W.N., 2003, "Predicting Wake–Passing Transition In Turbomachinery Using a Intermittency-Conditioned Modelling Approach," 33rd AIAA Fluid Dynamics Conference and Exhibit, Orlando.

Wissink, J.G., and Rodi, W., 2003, "DNS of a Laminar Separation Bubble in the Presence of Oscillating External Flow," Flow, Turbulence and Combustion, 71, pp. 311–331.

Wolff, S., Brunner, S. and Fottner, L., 2000, "The Use of Hot-Wire Anemometry to Investigate Unsteady Wake-Induced Boundary-Layer Development on a High Lift LP Turbine Cascade," ASME Paper 2000-GT-49.

Wu, X., and Durbin, P.A., 2000, "Boundary Layer Transition Induced by Periodic Wakes," J. Turbomachinery, 122, pp. 442–449.

Wu, X., and Durbin, P.A., 2000, "Numerical Simulation of Heat Transfer in a Transitional Boundary Layer with Passing Wakes," J. Heat Transfer, 122, pp. 248–257.

Zarzycki, R., and Elsner, W., 2005, "The Effect of Wake Parameters on the Transitional Boundary Layer on Turbine Blade," J. Power and Energy, 219, pp. 471–480.

Index

About the Author

Professor Mayle presently lives in Roque Bluffs, Maine. A Fellow of the American Society of Mechanical Engineers, an Alexander Humboldt Senior Scientist Awardee, and the 1st International Gas Turbine Scholar, he received his Ph.D. degree in Mechanics at Harvard University in 1971. From 1970 to 1977, he held a Research Engineer position in the Turbine Research Group at Pratt & Whitney Aircraft, East Hartford, Connecticut, after which he became a Professor of Mechanical Engineering and Director of the Gas Turbine Laboratory at Rensselaer Polytechnic Institute, Troy, New York. Retiring from Rensselaer in 1994, he divided his time between the Institut für Thermische Strömungsmaschinen, Universität Karlsruhe, Karlsruhe, Germany, as a Visiting Professor, Asea Brown-Boveri, Baden, Switzerland, as an Engineering Consultant, and sailing the waters of southeastern United States and The Bahamas as Captain of his sixteen ton yellow cutter "Magier." In 2004, Professor Mayle retired to Roque Bluffs with his wife, artist Penni Carlton.